Pure

Baking Soda

Over 100 Helpful Household Hints

Christine Halvorson

 Publications International, Ltd.

Writer

Christine Halvorson is the author of *Solve It with Salt & Vinegar*, *100s of Helpful Hints: Practical Uses for Arm & Hammer Baking Soda*, and *The Home Hints Calendar 2000*, and she is the coauthor of *Amazing Uses for Brand-Name Products* and *Clean & Simple: A Back-to-Basics Approach to Cleaning Your Home*. She is a frequent contributor to *The Old Farmer's Almanac* publications, including *Home Owner's Companion, Gardener's Companion*, and *Good Cook's Companion*. She works as a freelance writer from her home in Hancock, New Hampshire.

Illustrations by Bot Roda.

ISBN-13: 978-1-4127-1547-8
ISBN-10: 1-4127-1547-4

Manufactured in China.

8 7 6 5 4 3 2 1

Contents

Introduction

In 1846, John Dwight and his brother-in-law, Dr. Austin Church, founded Church & Dwight Co., Inc., the manufacturer of ARM & HAMMER® Baking Soda. They took trona, or soda ash, out of the ground and turned it into sodium bicarbonate for use both outside and inside the house.

Baking soda is a naturally occurring, very versatile substance that's environmentally safe and inexpensive. Not only is baking soda nontoxic, it's actually a food so—unlike many commercial household products—it is safe to use around children and pets.

For this book, we've gathered more than 100 ways average people can—and do—use baking soda around the house. Whether in the kitchen, the bathroom, the laundry room, or even the garage, you'll be amazed at what baking soda can do.

BAKING SODA GUIDELINES

The hundreds of tips included in this book use baking soda in three different ways: directly, in a solution, and in a paste. Follow these guidelines for each use when we don't otherwise supply specific measurements.

DIRECTLY

Baking soda is sprinkled directly onto something or onto a sponge. It is not diluted.

SOLUTION

Use 4 tablespoons baking soda for each quart of warm water. Sometimes solutions may contain a liquid other than water. Follow directions carefully.

PASTE

To make a baking-soda paste, add just enough water to give it a consistency that will not run if applied to a vertical surface. Sometimes we may suggest pastes made of baking soda and a liquid other than water, such as lemon juice.

Clean and Fresh Kitchens

*B*aking soda can safely tackle kitchen jobs above and beyond the legendary box at the back of the refrigerator. Once you discover the versatility of baking soda, you'll do away with all those cleaners under your sink, and you'll never use oven spray  again. Of course, you can still use baking soda for baking, but we'll leave those tips for "Cooking," beginning on page 64.

RUBBER, PLASTIC, AND WOOD

A baking-soda paste removes stains from plastic and rubber utensils. Apply with a scouring pad or sponge.

Scrub stained plastic storage containers with a paste of lemon juice and baking soda.

Renew old sponges, nylon scrubbers, and scrub brushes by soaking them overnight in a solution of 4 tablespoons baking soda to 1 quart water.

Deodorize and remove stains from wooden bowls or utensils with a baking-soda solution.

POTS, PANS, AND COOKWARE

Clean encrusted grease and food on roasting pans by dampening with hot water and sprinkling with baking soda. Let sit for an hour, and sponge clean.

To loosen baked- or dried-on food in pans, gently boil water and baking soda in the pans. When food is loosened, cool and wipe clean.

Clean stains off aluminum cookware with baking-soda paste and scouring pad. Rinse and wipe clean.

For enamel cookware: Apply a baking-soda paste and let sit for an hour, then clean with a synthetic scrubber, and rinse.

Remove stains from a nonstick pan by boiling 1 cup water, 2 tablespoons baking soda, and ½ cup liquid bleach in the pan for several minutes. Wash as usual, and use cooking oil to reseason.

To polish stainless-steel cookware, sprinkle baking soda directly onto a wet pan, and scrub with a synthetic scrubber.

Cover burned-on stains on cookie sheets with baking soda, then with hot water, and let soak for 10 minutes. Next, scour with baking soda and a scrubber.

SHINY SURFACES

Use General Purpose Cleanser (page 9) to clean mildew stains on the inside of a refrigerator.

Stainless-steel sinks and other surfaces can be cleaned with a baking-soda paste or by sprinkling baking soda directly onto a sponge or clean cloth and scrubbing the surface. Rinse and buff dry.

Clean the exterior of your refrigerator and most other surfaces in your kitchen by using the General Purpose Cleanser. You'll find the recipe below.

To clean sticky refrigerator door gaskets, mix 4 tablespoons baking soda with 1 quart water, and apply with a toothbrush. Wipe clean.

GENERAL PURPOSE CLEANSER

This homemade concoction can replace most of the commercial cleaners you probably have on your shelf.

1 tsp borax
1 tsp baking soda
2 tsp vinegar or lemon juice
¼ tsp liquid dish soap
2 cups hot water

Be sure to wear rubber gloves when working with this mixture. Mix and store in a squirt or spray bottle.

COUNTERTOPS

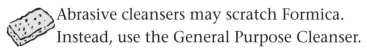

Abrasive cleansers may scratch Formica. Instead, use the General Purpose Cleanser.

Remove stains on laminated countertops with a baking-soda paste. Apply, let dry, then rub off and rinse.

Clean a countertop made of tile and grout with a mixture of ½ cup vinegar, 1 cup clear ammonia, ¼ cup baking soda, and 1 gallon warm water. Apply with a sponge. Protect your hands with rubber gloves.

White appliances or enamel sinks that are yellowing can be brightened with ¼ cup baking soda, ½ cup bleach, and 1 quart warm water, applied with a sponge. Wear rubber gloves. Let sit 10 minutes, then rinse.

FLOORS

Clean tile floors with ½ cup baking soda in a bucket of warm water. Mop and rinse clean.

Remove black heel marks on linoleum or vinyl floors with a damp sponge or scrubber dipped in baking soda.

OVENS AND STOVES

Clean induction and glass stovetops with a baking-soda solution, using a toothbrush to get into tight corners. Wipe clean.

For a thorough oven cleaning, leave 1 cup of ammonia in a cold, closed oven overnight to loosen dirt. In the morning, wipe away ammonia, then wipe surfaces with baking soda.

MORE TIPS AND TRICKS AROUND THE KITCHEN

COFFEE AND TEA

Dip a damp sponge in baking soda and rub coffee cup and teacup stains away. Stubborn stains may also require a little salt.

GREASE CUTTER CLEANUP

Use this homemade solution to cut greasy buildup on stoves, backsplashes, or glossy enamel surfaces.

¼ cup baking soda
½ cup white vinegar
1 cup ammonia
1 gallon hot water

Wear rubber gloves and use in a well-ventilated area.

To clean teapots and stovetop percolators, fill with water, add 2 or 3 tablespoons of baking soda, and boil for 10 to 15 minutes. After cooling, scrub and rinse thoroughly.

To remove rust stains and mineral deposits from teapots, fill the pot with water, and add 2 tablespoons baking soda and the juice of half a lemon. Boil gently for 15 minutes, and rinse.

SMALL APPLIANCES

Use a toothbrush dipped in baking soda to scrub the cutting wheel of a handheld or electric can opener.

DISHWASHERS AND DRAINERS

Clean a dish drainer in a sink filled with hot water. Sprinkle baking soda over all surfaces, let sit, then scrub with a vegetable brush or toothbrush. Rinse.

CLOG PREVENTION AND ELIMINATION

To prevent clogs, periodically pour ½ cup baking soda down your kitchen sink, followed by hot water. You can use the old box from your refrigerator when you replace it.

Baking soda and vinegar will foam your drain clean and help prevent clogs. Use ½ cup baking soda, followed by 1 cup vinegar. When foam subsides, rinse with hot water.

OTHER HANDY TRICKS

Sprinkle baking soda on a damp sponge, and scrub your fruits and vegetables to remove dirt, wax, or pesticide residue. Rinse well.

Clean the oil out of a salad dressing cruet by shaking baking soda inside, then rinsing it clean with warm water.

To clean scratches in stoneware, apply a baking-soda paste to the cracks, let stand a few minutes, then wash as usual.

You can hide cuts in countertops by keeping them clean with a baking-soda paste.

GET THE SMELL OUT

REFRIGERATORS AND FREEZERS

An open box of baking soda in the refrigerator absorbs odors for up to three months. The same is true of freezers.

To remove any unpleasant taste in ice cubes from an automatic ice-cube maker, clean removable parts of the unit with baking soda and water.

CUTTING BOARDS

 Rub a wooden cutting board with a baking-soda paste to remove odors.

GARBAGE CANS

Reduce garbage-can smells by sprinkling baking soda in each time you add garbage.

Periodically wash out and deodorize garbage cans with a solution of 1 cup baking soda per 1 gallon water.

MICROWAVES

The smell from microwave accidents (like burnt popcorn) can be removed with liberal applications of baking soda and water.

DISHWASHERS

If you're waiting for a full load before running your dishwasher, avoid odor buildup by sprinkling baking soda directly onto the dishes or in the bottom of the machine.

PUTTING OUT FIRES

 Keep a box of baking soda within reach of the stove (but far enough away to be out of range of a fire). Pour baking soda directly on the flames to extinguish the fire.

 Do not use baking soda to extinguish a fire in a deep-fat fryer; the fat may splatter.

 Do not use baking soda on any fire involving combustibles, such as wood or paper.

 Do not hesitate to call 911 if you think the fire is out of hand.

When the fire is extinguished, allow pots and their contents to cool before removing and cleaning them.

You may want to keep an Emergency Fire Pail in your kitchen. Request a label with instructions on how to make one by sending a self-addressed, stamped envelope to ARM & HAMMER® Fire Pail Brochures, P.O. Box 7468, Princeton, NJ 08543.

Baking Soda Throughout the House

*E*ven the most meticulous housekeeper will find unwanted scuffs, smudges, and smells around every corner, in every room of the house. Whether it's cleaning windows or stained piano keys, removing water marks from a wood floor, or shining up the family silver and china, baking soda can be used in dozens of everyday household chores.

CLEANING WALLS AND OTHER SURFACES

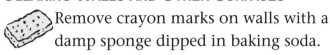 Remove crayon marks on walls with a damp sponge dipped in baking soda.

Wax from dripped candles can be removed from most hard surfaces with a baking-soda paste. Scrub with a nylon scrubber.

A baking-soda paste is best for cleaning chrome surfaces. Apply, then buff dry.

Sprinkle baking soda directly onto stainless-steel surfaces, and clean with a damp sponge.

BASIC WALL CLEANER

Keep this mixture on hand for cleaning walls or other painted surfaces.

1 cup ammonia
1 cup baking soda
1 gallon water

Mix thoroughly, then apply with a sponge. Wear gloves to protect your hands from the ammonia. Scrub marks gently. Rinsing is unnecessary.

Scour soot and ash from fireplace bricks with a baking-soda solution. Rub into bricks with a stiff brush.

TABLETOPS AND WOOD

Remove alcohol stains from a wood table with a paste of baking soda and mineral, linseed, or lemon oil. Rub in the direction of the grain, then wipe with linseed oil.

Rub out white rings on wood tables with a paste made of equal amounts toothpaste and baking soda.

CARPETS AND FLOORS

Remove water spots on wood floors with a sponge dampened in a baking-soda solution.

Shampoo your carpet with 1 cup baking soda mixed with 1 gallon warm water. Scrub by hand with a brush, or use this mixture in a carpet-cleaning machine.

Pretreat carpet stains with a small amount of baking soda sprinkled directly onto the dampened area. Let stand. Be sure to test carpet for colorfastness first.

Remove wine stains from carpeting by sprinkling with baking soda. Dab with a damp sponge. Repeat until stain is gone.

Clean up pet accidents or vomit with baking soda. Scrub gently into rug with warm water and a brush. Vacuum when dry.

CARPET FRESHENER

1 cup crushed, dried herbs (rosemary, southernwood, lavender, etc.)
1 teaspoon ground cloves
1 teaspoon cinnamon
1 teaspoon baking soda

Combine ingredients, and sprinkle over carpet. Allow to sit for a few minutes, then vacuum.

CARPET FRESHENER VARIATIONS

Mix 1 small box baking soda with your favorite potpourri oil, using just a few drops, and sprinkle on as a carpet freshener. Leave on carpet 10 to 20 minutes, then vacuum.

Use 1 cup baking soda, 1 cup cornstarch, and 15 drops essential oil fragrance. Leave on carpet 10 to 20 minutes, then vacuum. Store mixture in a glass jar or airtight container.

Control odor from pet accidents by leaving a thin layer of baking soda after cleaning. Vacuum when dry.

 Add baking soda to a vacuum bag to fight smells that can accumulate there.

UPHOLSTERY

 A fresh stain of oily or greasy food on a cloth chair can be absorbed with equal parts baking soda and salt. Sprinkle, rub lightly, leave on for a few hours, then vacuum.

 Clean old stains on upholstery with 3 parts baking soda and 1 part water. Rub paste into the stain, let dry, then brush away.

 Clean vinyl upholstery, such as a recliner or kitchen chair, with a baking-soda paste rubbed on, dried, then wiped off.

CONTROL HOUSEHOLD ODORS

Freshen blankets that have been in storage by sprinkling with baking soda and rolling them up for a couple of hours. Then fluff in the dryer without heat.

In a closet with bare floors, freshen by washing floor with 1 cup baking soda and ½ cup vinegar in 1 gallon warm water.

Mix equal amounts baking soda and borax to fill a shoe box halfway. Tape the shoe box shut, and cut holes in the lid. Store in closet to control odors.

Before storing luggage or travel trunks, place an open box of baking soda inside, and close the luggage overnight. Repeat this when removing luggage from long-term storage.

If your waterbed mattress develops a musty odor, rinse it inside and out with a baking-soda solution, then refill as usual. Use a sponge to gently scrub the outside of the mattress with the solution.

Eliminate residue and smells from mops or rags by soaking them in a mixture of 4 tablespoons baking soda and 1 gallon water.

Fill the toes of old panty hose with baking soda, cut off the foot, and tie to secure. Hang the sachet to absorb musty odors.

Sprinkle baking soda in a thin layer at the bottom of any household trash can to control odors.

JEWELRY

Clean gold and silver jewelry with 3 parts baking soda to 1 part water. Rub on gently, then rinse. Buff with a soft cloth.

Shine platinum jewelry with dry baking soda buffed on with a soft cloth.

To clean tin, dampen the tin object, and sprinkle a small amount of baking soda on it. Rub with a plastic scouring pad, and rinse.

GLASSWARE AND KNICKKNACKS

Clean stained china with a paste of baking soda and water.

SILVER TARNISH REMOVER

Boil water and ½ teaspoon salt with 1 to 2 teaspoons of baking soda. Place tarnished silverware in a pan with the boiled mixture and a piece of aluminum foil. Simmer for 2 to 3 minutes. Rinse the silverware well, then use a soft cloth to buff dry.

To clean glass vases or other containers, fill three-quarters full with hot water, add a teaspoon of baking soda, and shake. Let sit, then rinse.

Stubborn spots on porcelain surfaces such as lamps, vases, and candlesticks can be cleaned by dipping a damp cloth in baking soda and rubbing.

WINDOWS AND BLINDS

Wash windows with a sponge dipped in baking soda. To avoid dry haze on the windows, rinse them with a clean sponge and plenty of water, and dry.

Put dirty venetian blinds in a tub of warm water and ½ cup baking soda, soak for half an hour, then scrub and rinse.

When cleaning blinds, rubbing the cords with baking soda will whiten them.

TIPS AND TRICKS

Once a month, sprinkle carpets with baking soda, let sit overnight, then vacuum.

Trick Stained piano keys can be cleaned with a
damp sponge dipped in baking soda.
Wipe off, and buff.

Tip Use baking soda to simulate snow on your
Christmas tree.

Trick A permanent filler for nail holes on
white walls is a mixture of baking soda
and white glue formed into paste.

Tip Keep a box of baking soda near a fireplace
or wood-burning stove in case of fire.

Trick Deodorize musty books that are not
damp by sprinkling baking soda into the
pages and leaving them closed for several days.
Brush out baking soda, and repeat if necessary.

Tip Canvas tote bags can be cleaned with dry
baking soda sprinkled on and rubbed with
a brush.

Trick Freshen ashtrays with ½ inch baking
soda at the bottom. This also aids in
extinguishing the butts.

Beautifying Your Bathroom

*F*or everyday cleaning in the bathroom, a sprinkling of baking soda on a damp sponge can often do the trick. Baking soda mixed with the right ingredients can clean mildew stains that plague tubs and showers and make your powder room smell clean and look shiny at all times without much effort.

THE BASICS

Any time you use a baking-soda paste, add your favorite essential oil to customize the smell.

Apply a baking-soda paste to chrome fixtures, and buff dry to clean.

TOILET

A ½ cup of baking soda in the toilet bowl will work for light-duty cleaning. Let sit for 30 minutes, then brush.

BASIC BATHROOM CLEANER

3 tablespoons baking soda
½ cup ammonia
2 cups warm water

Mix and use for everyday cleaning. Be sure to wear rubber gloves, and use in a well-ventilated area.

VARIATION: THIS VERSION LEAVES OUT THE AMMONIA.

1 box (16 ounces) baking soda
4 tablespoons dishwashing liquid
1 cup warm water

Mix well, and store in a spray or squeeze container.

Soap Scum Cleaner

This mixture is great for removing soap scum buildup around your tub and sink.

¼ cup baking soda
½ cup vinegar
1 cup ammonia
1 gallon warm water

Mix well, and apply liberally. Wear rubber gloves. Be sure the area is well ventilated. Rinse well.

Remove stubborn toilet stains by scrubbing with fine steel wool dipped in baking soda.

Tubs and Showers

Cleaning fiberglass tubs and showers requires a bit of caution since fiberglass scratches easily. Make a paste of baking soda and dishwashing liquid, and wipe on with a sponge.

Adding baking soda to bathwater will reduce the ring around the tub and soften your skin. Use 2 tablespoons for a tubful of water.

For tough grout or tile stains, use a paste of 1 part bleach to 3 parts baking soda.

Stains on nonskid strips or appliqués in the tub can be removed by dampening the area, then sprinkling with baking soda. Let sit for 20 minutes, then scrub and rinse.

Use a baking-soda paste to remove mildew stains on grout. Apply, scrub with an old toothbrush, and rinse.

Clean mildew stains and do light cleaning of a shower curtain by sprinkling baking soda on a sponge and scrubbing. Rinse well.

SHOWERHEAD TREATMENT

Remove mineral buildup and improve performance of your showerhead with this remedy.

½ cup baking soda
1 cup vinegar

Mix in a sturdy plastic bag, then secure the bag around the showerhead with a rubber band so that the showerhead is submerged in solution. Keep on, and soak for one hour. Remove, and run very hot water through the showerhead for several minutes.

Brush a stained porcelain tub with a paste made with baking soda and hydrogen peroxide, then cover with damp paper towels. The towels keep the paste from sliding down into the tub. Let sit for 30 minutes, then scrub.

FLOORS

A tile or no-wax bathroom floor can be cleaned with ½ cup baking soda in a bucket of warm water. Mop and rinse.

To remove odors from floors, sprinkle the area with baking soda, let it sit, then vacuum or sweep up. The stronger the odor, the more baking soda you should use and the longer you should leave it on.

DRAIN MAINTENANCE

For routine cleaning of sink and tub drains, pour in ½ cup baking soda followed by 1 cup vinegar. Let sit for 10 to 20 minutes, then flush with very hot water.

ODOR CONTROL

Musty-smelling bath towels should be deodorized by machine washing. Add ½ cup baking soda to the rinse cycle.

Add a perpetual air freshener to the toilet area by keeping baking soda in a pretty dish on the back of the tank. Add your favorite scented bath salts to the mix as well. Change every three months.

Add baking soda to the toe of old panty hose or nylon knee-highs, tie, and cut off the excess stocking. Hang around pipes under the sink for ongoing odor control.

Sprinkle baking soda in the bathroom trash can after each emptying.

COUNTERS AND VANITIES

Clean marble surfaces with a paste made of baking soda and white vinegar. Wipe clean, and buff.

A simple baking-soda paste will attack hard water or rust stains on ceramic tile. Use a nylon scrubber, then rinse.

Care for Clothing

Baking soda makes a great laundry product because of its mild alkali qualities. Dirt and grease are easily dissolved, while

clothes are softened. It is especially helpful in homes with hard water because it not only cleans clothes  better, it also prevents the stain buildup that can come with hard water.

SMELLY HAMPERS

Freshen laundry hampers by sprinkling baking soda over dirty clothes as they await washing.

IN THE WASH

Add ½ cup baking soda with your detergent to freshen your laundry and help liquid detergents work harder.

Use baking soda instead of fabric softener. Add ½ cup at the rinse cycle.

BOOST YOUR BLEACH

Add ½ cup baking soda (only ¼ cup for front-loading machines) with the usual amount of bleach to increase whitening power.

PERSPIRATION STAINS

For perspiration stains, scrub in a paste of baking soda and water, let sit for 1 hour, then launder.

Treat stubborn perspiration stains around the collar with a paste of 4 tablespoons baking soda and ¼ cup water. Rub in, then add a little vinegar to the collar, and wash.

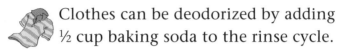 Using baking soda in the wash softens water, which will keep perspiration stains from forming.

REMOVING ODORS

Clothes can be deodorized by adding ½ cup baking soda to the rinse cycle.

Add ½ cup baking soda to rinse water to remove mothball odors.

Remove cigarette smoke odors in clothes by soaking them in a baking-soda solution before washing.

Deodorize knit caps that may have accumulated perspiration odors by soaking them in baking soda and water.

Before laundering, remove gas and oil odors in clothes and rags by sprinkling them with baking soda and securing them in a plastic trash bag for several days.

DELICATES

Prevent nylon items from yellowing by adding baking soda to both the wash and the rinse water.

Eliminate the stale smell in stored-away handwashables by soaking in 4 table-spoons baking soda and a quart of water. Rinse well, squeeze, then air dry.

Yellowed linens can be brightened by adding 4 tablespoons baking soda to the wash water.

LAUNDRY ACCIDENTS

Clean the drum of your washer or dryer—for example, to remove ink stains—with a plastic scrubber and baking soda.

PRETREATING STAIN REMOVER

½ cup ammonia
½ cup white vinegar
¼ cup baking soda
2 tablespoons liquid soap
2 quarts water

Mix all ingredients in a spray bottle. Spray the liquid onto the stain, and let it sit for a few minutes. Launder as usual.

A VARIATION: Use soap flakes and baking soda when washing stained natural fabrics. Add 1 tablespoon of vinegar to the rinse to keep colors bright.

If you've washed a crayon with a load of clothes, rewash the load with the hottest possible water, adding ½ to 1 full box of baking soda. Repeat if necessary.

If you've stained your white clothes by washing them with colored ones, undo the damage by soaking them in warm water to which you have added baking soda, salt, and detergent.

SPECIAL STAINS AND SPECIAL CLOTHING

Remove tar from clothing by rubbing in a baking-soda paste. Then wash with baking soda rather than detergent.

Eliminate alcohol stains caused by perfume with a paste of baking soda and ammonia. Test for colorfastness first. Wear rubber gloves, and use this mixture in a well-ventilated area. Dry the fabric in the sun, then wash as usual.

To remove blood stains, dampen the area, then rub with baking soda. Follow by dabbing with hydrogen peroxide until the stain is gone. Test for colorfastness first.

Rinse pool chlorine out of bathing suits in a sink full of water with 1 tablespoon baking soda added.

Remove crayon marks on clothing by rubbing gently with baking soda sprinkled on a damp cloth.

Soak up wine or fresh fruit spills on table linens immediately by sprinkling with baking soda. Then stretch the fabric over a large bowl, and pour boiling water through the stain and baking soda.

The chemical finish in new clothes can bother sensitive skin. Soak them in water and ½ to 2 cups vinegar. Rinse. Then add ½ cup baking soda to the wash load.

DRY-CLEANABLES

Some dry-clean-only items can be cleaned with a solution of 4 tablespoons baking soda in cold water. Test for colorfastness first.

CLEANING WITHOUT WATER

Rub dry baking soda into polyester fabrics to remove a grease spot. Brush off, and the stain should be gone.

Clean suede with baking soda applied with a soft brush. Let it sit, and then brush it off.

LEATHER

An ink stain on leather can be removed by laying the item flat and sprinkling baking soda on the stain. Leave on until ink is absorbed, brush off, and repeat if necessary.

SHOES AND SOCKS

Remove black scuff marks on shoes with a baking-soda paste, rubbed on and wiped off, before applying polish.

Clean the rubber on athletic shoes with baking soda sprinkled on a sponge or washcloth.

Sprinkle baking soda into clean socks before wearing to control odor and moisture.

Keep smelly feet at bay by sprinkling baking soda into athletic shoes and street shoes to control odor and moisture.

Health and Beauty with Baking Soda

*O*nce upon a time, baking soda was one of the few products on the market for cleaning your teeth or settling an upset stomach. While we have many more choices today, baking soda still does the trick for these and dozens of other health and beauty tasks. Try it for shaving and shampooing, for minor burns and cuts, or for relaxation in the bath.

HAIR

SQUEAKY CLEAN HAIR

Add a teaspoon of baking soda to your usual shampoo bottle to help remove buildup from conditioners, mousses, and sprays, and to improve manageability.

In emergencies, use baking soda as a dry shampoo on oily hair. Sprinkle on and comb through, then fluff with a blow dryer.

CHLORINE REMOVER

Rinse hair with ½ teaspoon baking soda in 1 pint water to remove the dullness or discoloration caused by chlorinated pools.

Processed hair in danger of turning green from chlorine can be rinsed with ½ cup baking soda dissolved in 2 quarts lemon juice. Wet hair, and pour solution over head promptly.

COMBS AND BRUSHES

Hair spray and oil buildup on combs and brushes can be removed by soaking them in a sink of warm water and adding 3 tablespoons baking soda and 3 tablespoons bleach.

SKIN

HANDS AND ELBOWS

 Remove fish, onion, or garlic odor from hands with a solution of 3 parts baking soda to 1 part water or liquid soap. Rub, and rinse.

 Prevent dry, chapped hands by sprinkling baking soda into dishwater.

The dirt that gets imbedded in the elbows and knees of children can be scrubbed away with baking soda sprinkled on a damp washcloth.

Rub a baking-soda paste onto elbows to smooth away rough skin.

FEET

 Soak tired feet in a basin of warm water with 3 tablespoons baking soda.

Add 4 tablespoons baking soda to 1 quart warm water, and soak feet for 10 minutes to relieve foot itch.

Smooth rough and hardened calluses and heels by massaging with a paste of 3 parts baking soda per 1 part water.

RELAXING BATHS

Baking soda added to bathwater has a softening effect on the skin. Add ½ cup to a full bath, and soak.

Make bubbling bath salts with 2½ cups baking soda, 2 cups cream of tartar, and ½ cup cornstarch. Mix, and store in a covered container. Use ¼ cup per bath.

A DAY SPA AT HOME

MUSCLE SOOTHER BATH

Mix 2 cups baking soda, 1 cup Epsom salts, and ½ cup salt. Add ½ cup of mixture to bathwater to sooth sore and tired muscles. Store in covered container.

STRESS BALL

Make your own stress ball by taking a durable balloon and adding baking soda. Tie tightly, and squeeze.

AFTER BATH POWDER

1 cup baking soda
1 cup cornstarch
10 drops of essential oil

Mix the baking soda and cornstarch. Add the essential oil and mix. Sprinkle over body and gently rub into skin.

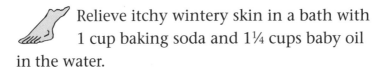 Relieve itchy wintery skin in a bath with 1 cup baking soda and 1¼ cups baby oil in the water.

NAIL CARE

Clean fingernails and toenails by scrubbing with a nailbrush dipped in baking soda. This also softens cuticles.

SPONGE BATH

Freshen up with a washcloth dipped in a solution of 4 tablespoons baking soda to 1 quart water.

MAKEUP

Soak cosmetic sponges or makeup applicators in a baking-soda solution overnight to clean.

DEODORANT

 Apply cornstarch to underarms with a powder puff first, then apply baking soda.

 For a simple daily deodorant, dust baking soda under arms using a powder puff.

HERBAL DEODORANT

 Mix baking soda with an equal amount of cornstarch. Add a fragrant dried herb or a few drops of peppermint, anise, or cinnamon essential oil.

TOOTH AND MOUTH CARE

TOOTHPASTE

Sprinkle baking soda into your palm, dip a damp toothbrush into it, and brush.

CINNAMON TOOTH POWDER

2 tablespoons baking soda
2 tablespoons powdered cinnamon
2 tablespoons oil of cinnamon

Mix and apply with damp toothbrush.

OLD-FASHIONED TOOTH POWDER

2 tablespoons dried lemon or orange rind
¼ cup baking soda
2 teaspoons salt

Grind rinds in food processor until peel becomes a fine powder. Add baking soda and salt, and process again. Store in an airtight container. Dip moistened toothbrush into mixture, and brush as usual.

Mix 3 parts baking soda with 1 part salt. Add 3 teaspoons of glycerin. Add 10 to 20 drops of flavoring (peppermint, wintergreen, anise, or cinnamon) and enough water to make a paste. Spoon into a small, refillable squeeze bottle.

REFRESHING MOUTHWASH

To freshen breath, use 1 teaspoon baking soda in ½ glass of water, swish the solution through your teeth, and rinse.

DENTURES AND OTHER DENTAL APPLIANCES

Soak dentures in a solution of 2 teaspoons baking soda dissolved in warm water.

Use baking soda to soak athletic mouth-guards, retainers, or other oral appliances.

Scrub dentures, mouthguards, and retainers with a toothbrush dipped in baking soda.

Soak toothbrushes in a baking-soda solution overnight.

Brushing braces with baking soda alone will make them shine.

FACE

 Make a paste of 3 parts baking soda to 1 part water, and use as a gentle, exfoliating facial scrub after washing with soap and water. Rinse clean.

 Mix baking soda with oatmeal in your blender; it makes a great facial scrub.

ACNE

Treat pimples with a paste of baking soda and peroxide applied at bedtime.

BLACKHEADS

To loosen blackheads, mix equal parts baking soda and water, rub gently 2 to 3 minutes, then rinse with warm water.

SHAVING

 Use a dab of baking soda on a shaving cut to stem bleeding.

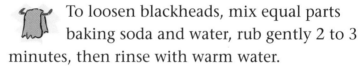 Men with sensitive skin may find that a solution of 1 tablespoon baking soda in 1 cup water makes a great preshave treatment or a soothing aftershave rinse.

For instant relief of razor burn, dab on a baking-soda solution.

SOOTHE MINOR MISHAPS

 For sunburn pain, saturate a washcloth with a solution of 4 tablespoons baking soda in 1 quart water. Apply to affected area.

Ease windburn or poison ivy irritation with a paste of 3 parts baking soda and 1 part water. Do not use on broken skin.

CASTS

Blow baking soda into a leg or arm cast to relieve itching and eliminate odor.

ANTACID

Take ½ teaspoon baking soda in ½ glass of water to relieve acid indigestion or heartburn. Read antacid use information on the baking soda package before using. **WARNING:** *People who must restrict salt intake should not use baking soda as an antacid. Do not take for nausea, stomachache, gas, cramps, or distention from overeating.*

Dogs, Cats, and Other Critters

*B*aking soda is nontoxic and, therefore, safe for use around your dogs, cats, and other pets. It makes it easy to control odors and clean up accidents that are a natural part of having pets in the family. You can also use baking soda as a dry wash, to tend bee stings, and to help maintain healthy teeth.

WASHING AND CLEANUP

BATHING

 Give your dog a dry bath by sprinkling it with baking soda. Rub in, then brush out.

For a wet wash, combine 3 tablespoons baking soda with 1 teaspoon dishwashing liquid and 1 teaspoon baby oil in a spray bottle. Spritz your pet, then wipe dry.

KITTY LITTER

 To eliminate odors from kitty litter boxes, sprinkle in ½ cup baking soda.

Make your own natural litter by mixing a small box of baking soda with 2 to 3 inches of dry, sandy clay.

Clean the kitty litter pan by removing litter and pouring in ½ inch vinegar. Let stand, then pour out, and dry. Sprinkle baking soda over the bottom before adding fresh kitty litter.

PET ACCIDENTS

To clean up after a pet's "accident," scrub the area with club soda, and let dry. Then sprinkle with baking soda, and let stand. Vacuum.

HEALTHY TEETH AND GUMS

You can brush your pet's teeth by dipping a damp, soft brush in baking soda and brushing gently.

Maintain your pet's dental hygiene by rinsing its mouth regularly with a solution of ½ teaspoon salt and ½ teaspoon baking soda in 1 cup warm water.

MORE TIPS AND TRICKS

SKUNKS

If your pet has a run-in with a skunk, wash the pet in a bath containing 1 quart 3-percent hydrogen peroxide, ¼ cup baking soda, and 1 teaspoon liquid soap. Rinse well, and dry. Discard unused cleaner.

BEE STING RELIEF

After making sure the stinger is removed, cover a bee sting on your pet with a baking-soda paste.

DEODORIZE BEDDING

Sprinkle the bedding area with baking soda, let stand 15 minutes, and vacuum.

TOENAIL TRIMMING TIP

If you trim your pet's toenails yourself, you may accidentally draw blood by cutting too close to the quick. Dip the affected nail in baking soda, and apply pressure to stop bleeding.

OTHER CRITTERS

BIRDCAGE

Clean the bottom of your bird's cage by sprinkling baking soda on a damp sponge and scrubbing. Wipe clean to dry.

FERRET CAGE DEODORIZER

To reduce odors in a ferret cage, sprinkle a layer of baking soda over the bottom of the cage after cleaning. Cover with appropriate bedding.

SALTWATER AQUARIUM MAINTENANCE

To maintain the proper pH level in your saltwater aquarium, mix 1 tablespoon baking soda in 1 cup dechlorinated water. Add this to the tank slowly, over a couple hours. The pH balance should be around 8.2.

Baby Care

Babies may be little, but the cleaning, deodorizing, and minor health challenges they present certainly aren't. Baking soda can help you deal with this bundle of chal- lenges. As an effective and inexpensive  product, it's safe for you and your baby. From deodorizing baby bottles to cleaning up the smell and mess of diapers and spit- up, baking soda can help.

BEFORE THE BABY COMES

Pregnant women suffering from itchy dry skin can find relief in a bath of baking soda and water.

CLEANING

Sprinkle baking soda on a damp sponge to wipe cribs, changing tables, baby mattresses, and playpens. Rinse thoroughly, and allow to dry.

Clean and deodorize baby spills or accidents on carpeting by soaking up as much as possible. When dry, sprinkle with baking soda, and let sit 15 minutes before vacuuming.

Use baking soda directly on metal, plastic, or vinyl strollers, car seats, and high chairs, using a damp sponge. Rinse, and wipe.

Remove odors from cloth strollers or car seats by sprinkling baking soda on the fabric. Wait 15 minutes (longer for strong odors) and vacuum.

Whiten changing pads with solution of 1 tablespoon baking soda per cup of water. Wipe on, rinse, and dry.

When urine accidents occur on mattresses, sprinkle with baking soda, let dry thoroughly, then vacuum.

If your baby spits up on his shirt or yours, moisten a cloth, dip it in baking soda, and dab at the spot. The odor will be controlled until you can change clothes.

Take a small spray bottle of a baking-soda solution with you on outings with baby for quick cleanups.

BABY COMBS

Rinse baby combs and brushes by swishing them in a small basin of water with 1 teaspoon baking soda. Rinse, and allow to dry.

TOY CARE

Clean and deodorize baby toys using a baking-soda solution. Wash toys with a damp sponge or cloth, rinse, and dry.

Cloth toys can get grungy. To clean without water, sprinkle on baking soda as a dry shampoo, let sit 15 minutes, then brush off.

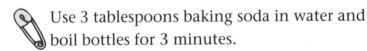 Wipe vinyl toys clean with 1 tablespoon baking soda per cup of water. Remove stains with baking soda sprinkled on a damp sponge.

Scour ink or crayon marks on plastic toys with a paste of 3 parts baking soda to 1 part dishwashing liquid or liquid soap.

DEODORIZE

BABY BOTTLES

Fill bottles with warm water, and add 1 teaspoon baking soda. Shake, rinse, and then clean as usual.

Freshen bottle nipples and bottle brushes overnight by soaking in a mixture of 4 tablespoons baking soda per 1 quart hot water. Drain, rinse, and clean as normal in the morning.

Use 3 tablespoons baking soda in water and boil bottles for 3 minutes.

DIAPER PAILS

 Keep diaper pails smelling fresh by sprinkling baking soda over dirty cloth diapers.

 Line the bottom of a diaper pail with baking soda to control odors after you empty it.

BABY LAUNDRY

Add ½ cup baking soda to powder or liquid laundry detergent to freshen clothes and help improve the detergent's performance on baby-food stains. With powder laundry detergent, add baking soda in the rinse cycle only.

Remove chemical residues from new baby clothes by washing them first in mild soap and ½ cup baking soda.

TODDLERS

Use baking soda as toothpaste—or add a little to your child's regular toothpaste—to help take care of plaque buildup.

GRANDPARENTS' GIFT OR FAMILY MEMENTO

Capture your baby's handprints and footprints in Play Clay (see recipe, page 58), and decorate with his or her name, birth date, and vital statistics at birth. You need only rinse off the feet or hands after making the impression in the damp clay. Make a set of prints for yourself and one for each set of grandparents. The clay plaque can be hung, set on a stand, or made into a refrigerator magnet.

Baking Soda for Kids Big and Small

Mixing a liquid with a solid can form a gas. That may sound boring but it can be a lot of fun to watch. The various chemical properties of baking soda make for some interesting rainy-day projects for kids and may just pique an interest in science. Build a volcano, make a picture frame, or create a piece of jewelry. Use these tips to demonstrate some basic principles or just to have fun.

JUST FOR FUN

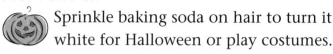 Sprinkle baking soda on hair to turn it white for Halloween or play costumes.

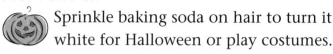 Baking soda mixed with superglue can be used to fill large gaps when assembling plastic models.

EASY-CLEANUP STENCILING

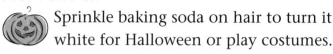 On a dark-colored floor or carpet, use stencils to sprinkle baking soda into fun shapes for holiday celebrations or parties. Make footprints or hearts, or spell out a message.

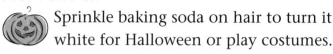 Stencil a "Welcome" sign onto your front porch or front hallway mat to greet special guests.

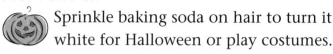 Make a paste of baking soda and water, add food coloring, and decorate windows with holiday messages or designs. The designs will clean up easily.

MAGIC BEANS

Fill a vase with water, add food coloring and ¼ cup vinegar, then add 3 teaspoons baking soda. Drop in buttons, rice, or pasta, and watch them rise and fall like magic.

PLAY CLAY

Make this clay, and then go wild with all the things you can create. The clay hardens as it dries to form a lasting keepsake.

2 cups baking soda
1 cup cornstarch
1¼ cups cold water
    Food coloring (optional)

Mix baking soda and cornstarch in saucepan. Add water, stir to mix, then cook over medium heat, stirring constantly, 10 to 15 minutes. Add food coloring to the water to make colored clay. Don't overcook. Clay should have the consistency of mashed potatoes. Remove to a plate. Cover with a damp cloth to cool.

Make it ahead of time, and store it for up to one week. Keep it refrigerated in a plastic container, but bring it to room temperature before using.

THREE WAYS TO DRY PLAY CLAY ART

Air: Set on a wire rack overnight.

Oven: Preheat to 350°F, turn off, then place finished objects on a cooking sheet. Leave in until oven is cold.

Microwave: Place objects on a paper towel, bake at medium power for 30 seconds, turn over, bake for another 30 seconds. Repeat until dry.

JUMPING SEEDS

Dissolve ⅔ teaspoon baking soda in ½ cup water in a large glass. Add apple seeds from one apple and 1 tablespoon lemon juice, then stir the mixture. The bubbles will carry the seeds up and down.

PLAY CLAY PROJECTS

FRAMEUPS

Cut a square or rectangle from Play Clay, then cut a frame opening the size of a favorite photograph. Leave a ½-inch border. Use another piece of clay for a stand to attach to the back. Decorate the frame when dry.

Make a mirror frame by following the same instructions above and adding a mirror with a clay back.

Make Play Clay into refrigerator magnets. Mold the clay into the shapes you want, dry, and glue magnets onto the back after decorating or painting.

Create a name plaque for a child's room by cutting out the shapes of letters and attaching them to a rectangular piece of Play Clay as the background. Paint and finish when dry.

Capture a child's handprint in Play Clay (see recipe on page 58) by pressing into damp clay. When dry, paint and add the child's name and date on the back, then attach a picture hanger.

One square of Play Clay makes a great pad or trivet for hot dishes. Cut out, and paint or decorate when dry. Finish with clear acrylic spray. Glue felt to the bottom to prevent scratching the table surface.

YOUR OWN VOLCANO

Shape a piece of cardboard into a cone. Insert a 4-ounce cup in the top of the cone to make the crater of the volcano. Stand the cone on a baking sheet. Cover the cone with plaster of paris. Don't get any in the cup. Let the cone dry completely. Paint or decorate the cone to look like a volcano.

MAKING THE ERUPTION

Mix ¼ cup vinegar with 1 teaspoon dishwashing liquid and a little red food coloring. Put 1 teaspoon baking soda into the crater cup. Pour the vinegar mixture into the crater.

VOLCANO VARIATION

Make a mini-volcano in a sandbox. Set a juice glass with vinegar in a mountain made of sand. Add 1 tablespoon baking soda to the glass.

JEWELRY

Shape beads for a necklace by rolling Play Clay into oval or round shapes. Press a toothpick through to make holes for stringing.

String Play Clay beads on thread, shoe-laces, yarn, kite string, or fishing line. Tie knots between beads to hold them in place.

Cut a heart, four-leaf clover, or other pattern. Use a skewer or toothpick to make a large hole at the top. String leather or thick cord through hole to use as a necklace.

To make an earring or brooch, create small shapes with a flat backside, and glue to earring or pin backings.

Shape a tube of clay around a glass to form a bracelet. Decorate or add stones.

String clay beads together on a short string to make a bracelet.

Glue a barrette clip onto the back of a clay shape for a customized hair accessory.

PLAY CLAY HOLIDAY IDEAS

Make fancy napkin rings by rolling out a long, narrow rectangle of clay, then piecing the ends together into a ring.

Use cookie cutters to make tree ornaments. While the ornament is still damp, make a hole near the top for hanging. Add an ornament hook or ribbon to hang the ornament.

Make one white and one red batch of Play Clay. Roll some of each color into equal-length "snakes." Twist red and white together into a candy-cane ornament.

Create placeholders for the Thanksgiving table using turkey- or leaf-shaped cookie cutters. While clay is still damp, write each guest's name into the shape.

Attach clay shapes to Easter baskets. Make egg, bunny, duck, and chick shapes, and glue on jelly beans or hard candy.

Create Valentine's Day pins or earrings using a variety of heart shapes.

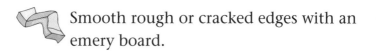 Make a small flowerpot for a Mother's Day gift. Use shellac or other finish on the inside of the clay pot when it is dry to waterproof it.

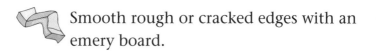 For Father's Day, create a set of Play Clay bookends decorated with themes from Dad's favorite hobby.

PLAY CLAY FINISHING TOUCHES

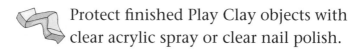 Paint dry pieces with watercolor, poster, or acrylic paints. Draw with a felt-tip pen or waterproof marker. Apply glitter to wet paint.

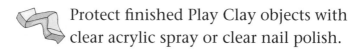 Smooth rough or cracked edges with an emery board.

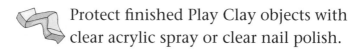 Protect finished Play Clay objects with clear acrylic spray or clear nail polish.

Cooking

*I*f you've done nothing else with baking soda, you've probably baked with it. Baking soda is a great leavening agent when heated.

It can be used to replace yeast in many recipes, and it can also  help sweeten tart fruits, fix some cooking mishaps, and perform tricks to make a good cook look like a better cook.

THE BASICS

 Test baking soda freshness by pouring a small amount of vinegar or lemon juice over ½ teaspoon of baking soda. If it doesn't actively bubble, it's too old to use.

 A batter using baking soda should be mixed and put in the oven quickly to retain the best leavening action.

BAKING POWDER SUBSTITUTIONS

 If you need 1 teaspoon baking *powder,* use one of the following substitutions:

¼ teaspoon baking soda plus ⅝ teaspoon (½ teaspoon plus ⅛ teaspoon) cream of tartar

¼ teaspoon baking soda plus ½ cup sour milk or buttermilk or yogurt (decrease liquid called for in recipe by ½ cup)

¼ teaspoon baking soda plus ½ tablespoon vinegar or lemon juice used with enough milk to make ½ cup (decrease liquid called for in recipe by ½ cup)

¼ teaspoon baking soda plus ¼ to ½ cup molasses (decrease liquid in recipe by 1 to 2 tablespoons)

OTHER SUBSTITUTIONS

Make self-rising flour with 3½ cups flour, 1¾ teaspoons baking powder, 1¾ teaspoons baking soda, and 1¾ teaspoons salt.

Yeast can be replaced in a recipe with equal parts baking soda and powdered vitamin C. The dough will rise during baking.

In a recipe calling for sour milk or buttermilk, substitute fresh milk, and add ¾ teaspoon baking soda to each cup needed.

To substitute honey for sugar in cookies, quick breads, or cakes, use ⅔ cup honey for each cup sugar. Add ½ teaspoon baking soda for every cup of honey. Reduce liquid by ¼ cup. Bake at 25 degrees less than the recipe recommends.

CAKES

Adding 1 teaspoon baking soda to the other dry ingredients in a chocolate cake will give the cake a darker color.

In a recipe for sour-cream cake, combining the baking soda and sour cream before mixing with other ingredients will activate the soda more quickly.

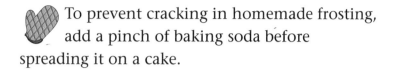 To prevent cracking in homemade frosting, add a pinch of baking soda before spreading it on a cake.

Substitute 1 teaspoon baking soda and 2 teaspoons vinegar for 2 eggs in any fruitcake or ginger cake recipe.

When making fruitcake, add a teaspoon of baking soda to darken the cake and soften the fruit a bit.

KITCHEN TRICKS

Add a pinch of baking soda to water when soaking dried beans. It helps make them more digestible.

Add 1 teaspoon baking soda to the cooking water for rice to improve fluffiness.

If you add too much vinegar to a recipe, add a pinch of baking soda to counteract.

Avoid lumps in batter by mixing baking soda with 1 teaspoon vinegar.

Omelets get fluffier if you add ½ teaspoon baking soda for every 3 eggs.

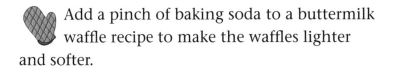 Add a pinch of baking soda to a buttermilk waffle recipe to make the waffles lighter and softer.

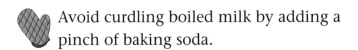 Avoid curdling boiled milk by adding a pinch of baking soda.

When gravy separates, a pinch of baking soda may get oils and fats to stick back together.

When making a boiled syrup, add a pinch of baking soda to prevent crystallizing.

MEAT, POULTRY, AND FISH

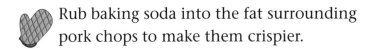 Tenderize tough meat by rubbing it with baking soda. Let it stand for several hours, then rinse and cook.

Rub baking soda into the fat surrounding pork chops to make them crispier.

Reduce fishy taste by soaking raw fish at least half an hour in 2 tablespoons baking soda and 1 quart water. Rinse, and cook.

Tenderize fowl by rubbing the cavity with baking soda, then refrigerating overnight.

When scalding a whole, fresh chicken, add 1 teaspoon baking soda to the boiling water. Feathers will come off more easily, and the flesh will be clean and white.

Before cooking poultry, rinse it in cold water and sprinkle baking soda inside and out. Rinse well.

Soften the pungent taste of wild game by soaking it in baking soda and water overnight. Rinse and dry before cooking.

VEGETABLES

Add a pinch of baking soda to water when boiling cabbage to tenderize and avoid overcooking.

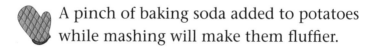

A pinch of baking soda added to potatoes while mashing will make them fluffier.

Test the acidic level of canned tomatoes. Dip a moist teaspoon in baking soda, then stir the tomatoes. Bubbling means their acid level is high.

Cut the acidic level of tomato sauce or chili by adding a pinch of baking soda.

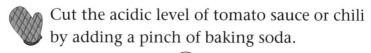

Eliminate the gaseous side effects of baked beans by adding a dash of baking soda while cooking.

Keep cauliflower white and its odor under control when boiling or steaming by adding 1 teaspoon baking soda.

FRUITS

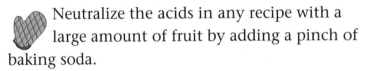 Neutralize the acids in any recipe with a large amount of fruit by adding a pinch of baking soda.

When making fresh cranberry sauce, cover cranberries with water, and boil. Add 1 tablespoon baking soda, stir, drain, and return to heat. You'll need less sugar than usual to complete the sauce.

Tart blackberries can be sweetened by adding ½ teaspoon baking soda before adding any sugar when making pies or cobblers.

Soak rhubarb in cold water and a pinch of baking soda prior to making sauce. The water will turn black. Drain. You'll need less sugar in the sauce.

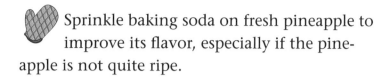 Sprinkle baking soda on fresh pineapple to improve its flavor, especially if the pineapple is not quite ripe.

LIQUIDS

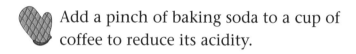 Any time you might have to boil water before drinking it, soften it with 1 tablespoon baking soda per gallon of water.

Add a pinch of baking soda to a cup of coffee to reduce its acidity.

Add ¼ teaspoon baking soda to 8 ounces of orange juice, grapefruit juice, or lemonade, and stir. This will add a fizz to the drink and reduce its acidic level.

CANNING

Home-canned tomato juice may become too acidic. Add a bit of baking soda before using it in cooking to cut the acid level.

Clean mineral deposits and neutralize any acids in old canning jars by shaking a baking-soda solution inside. Rinse thoroughly, then sterilize as usual.

Outdoors and the Open Road

*B*aking soda's versatility makes it one of the best pieces of equipment to put in a backpack, boat, trunk, or garage. Closer to home it can clean your screens and aluminum siding and keep your garden green and thriving.

ON THE LAWN AND IN THE GARDEN
GREENER GARDENS
Brighten fading green outdoor plants and bushes by mixing 1 teaspoon baking soda, 1 teaspoon Epsom salts, ½ teaspoon clear ammonia, and 1 gallon water. Apply 1 quart per average rosebush-size shrub.

ACIDITY TEST FOR SOIL
To test the acidity level of your garden soil, add a pinch of baking soda to 1 tablespoon of soil. If it fizzes, the soil's pH level is probably less than 5.0.

Flower species that prefer alkaline soil such as geranium, begonia, and hydrangea should be watered occasionally with a weak baking-soda solution.

Sprinkle baking soda lightly around tomato plants. This will sweeten the tomatoes by lowering acidity.

RAISE ALKALINITY IN POTTED PLANT SOIL
Carnations, mums, and petunias prefer neutral soil. To raise potting soil alkalinity, apply a baking-soda solution. Use sparingly.

FLOWERS AND PLANTING POTS

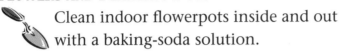 Clean indoor flowerpots inside and out with a baking-soda solution.

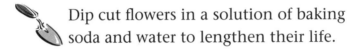 Coat clay pots with a thin layer of baking soda when transplanting plants, before adding the soil. This helps keep dirt fresh.

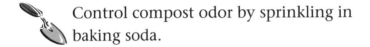 Dip cut flowers in a solution of baking soda and water to lengthen their life.

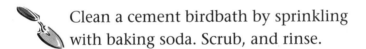 Control compost odor by sprinkling in baking soda.

Clean a cement birdbath by sprinkling with baking soda. Scrub, and rinse.

Clean white landscaping stones by scrubbing with a baking-soda solution.

ON THE DECK

Give a wooden deck a weathered look by scrubbing it with a solution of 2 cups baking soda in 1 gallon water. Test a small area of the deck first.

Oily stains on deck wood from the grill or suntan lotion can be absorbed by sprinkling with baking soda and letting sit for 1 hour. Repeat if necessary.

PATIO FURNITURE

Clean lawn furniture at the start of the season with a solution of ¼ cup baking soda in 1 quart of warm water. Wipe, and rinse.

POOL TOYS

Remove mildew odors from plastic and vinyl pool toys with ¼ cup baking soda in 1 quart warm water.

GRILL CLEANING AND SAFETY

Loosen burned-on foods from barbecue grill racks by enclosing racks in a large plastic bag. Mix 1 cup baking soda and ½ cup ammonia, and pour over racks. Close the bag, and let it sit overnight.

Control the flames when grease drips on coals by keeping a spray bottle of 1 teaspoon baking soda mixed with 1 pint of water handy. Spray lightly onto coals when flames shoot up.

Soak a mildewed or smelly plastic tablecloth in a baking-soda solution.

ON THE OPEN ROAD

 Clean coolers and thermos bottles with a baking-soda solution to eliminate smells.

Clean canvas by scrubbing with a paste of 3 parts baking soda per 1 part water and a soft bristle brush. For stubborn stains, add a squirt of dishwashing liquid to the paste.

RV OR BOAT HOLDING TANK

To dissolve solids and control odor in toilets of recreational vehicles or boats, pour a small box of baking soda into the tank after each cleaning.

RV WATER TANK

Deodorize and help remove mineral deposits in an RV water tank by flushing periodically with 1 cup baking soda in 1 gallon warm water. Drain, and flush the tank before refilling.

FISHING TRICKS

Keep fish hooks from rusting between fishing trips by sticking them in a cork and submerging the cork in baking soda.

Add baking soda to hollow fishing lures to give them spin in the water.

CLEANING OUT THE GARAGE

A damp, musty garage can be helped by hanging grocery bags with handles from the rafters. Add ½ inch baking soda to the bottom of each bag, hang, and change every three months.

You can refresh musty old magazines found in cellars or garages if the pages aren't stuck together. Lay the magazines out in the sun for a day. Then sprinkle baking soda on the pages, and let sit for an hour or so. Brush off.

BATTERY TERMINALS

Scrub corrosion and dirt off car and lawn-mower battery terminals with a baking-soda paste. Then rub with petroleum jelly to prevent corrosion.

Neutralize acid from leaking batteries by applying baking soda to the spill. One pound of soda will neutralize 1 pint of acid.

GARAGE FLOORS

Mix equal parts baking soda and cornmeal to sprinkle on light oil spills in the garage. Let dry, then sweep or vacuum away.

For tougher spots on floors, sprinkle on baking soda, let stand, then scrub with a wet brush.

HOME MAINTENANCE

SCREENS

Dip a damp wire brush into baking soda, and scrub door and window screens clean, then rinse with a sponge or hose.

PAINTING

Soak brushes in a warm baking-soda solution to remove paint thinner.

Revive hardened paintbrush bristles by boiling them in ½ gallon water, 1 cup baking soda, and ¼ cup vinegar.

CARS

Use baking soda to safely clean lights, chrome, windows, tires, vinyl seats, and floormats in cars. Sprinkle onto a damp sponge, scrub, and rinse.

Other spots on upholstery can be cleaned with a baking-soda paste rubbed into the stain. Let dry, and vacuum.

Remove oil and grease on vinyl seats with a solution of baking soda and water, or with baking soda sprinkled on a damp sponge. Then rinse, and wipe.

Remove scuffs on bumpers from other bumpers with baking soda and a damp sponge.

REALLY ROUGHING IT

ALL-IN-ONE CAMPING TOOL

Take baking soda on camping trips to clean dishes, pots, hands, and teeth; to use as a deodorant and fire extinguisher; and to treat insect bites, sunburn, or poison ivy.

When camping season begins, deodorize sleeping bags by sprinkling in baking soda and letting them sit for 12 hours. Shake out, and set the sleeping bags in the sun.

Deodorize a smelly tent by setting it up and sprinkling in baking soda.

OUT ON THE WATER

Unlacquered brass on boats can be brightened with a paste made of baking soda and lemon juice. Rub on, and let dry. Rinse well with warm water.

Keep a large box of baking soda in your boat as a precaution in case of small oil, gas, or engine fires. Toss baking soda onto flames from a safe distance.

Clean stains on fiberglass boat bodies by scrubbing with baking soda on a damp sponge. For tough stains, leave wet baking soda on, and wipe it away when it dries.

SECOND EDITION

POLICE IN AMERICA

STEVEN G. BRANDL

Students:
Looking to improve your grades?

SAGE
Premium
Video

BOOST COMPREHENSION. BOLSTER ANALYSIS.

- SAGE Premium Video **EXCLUSIVELY CURATED FOR THIS TEXT**
- **BRIDGES BOOK CONTENT** with application & critical thinking
- Includes short, auto-graded quizzes that **DIRECTLY FEED TO YOUR LMS GRADEBOOK**
- Premium content is **ADA COMPLIANT WITH TRANSCRIPTS**
- Comprehensive media guide to help you **QUICKLY SELECT MEANINGFUL VIDEO** tied to your course objectives

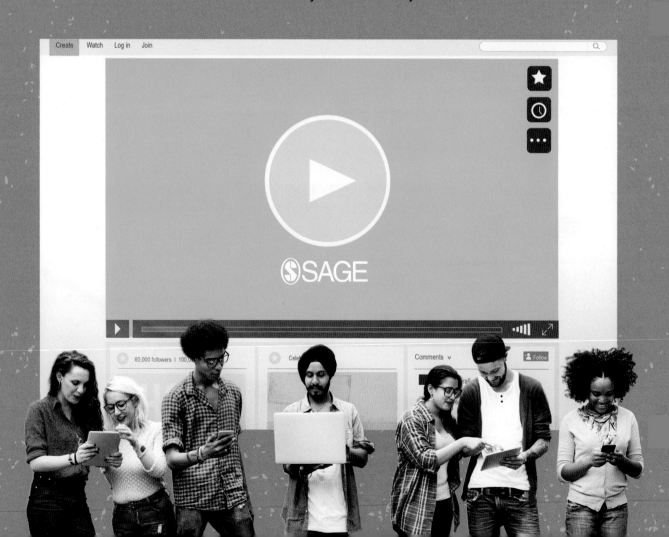

SAGE
Criminology &
Criminal Justice:
Our Story

Believing passionately in the **POWER OF EDUCATION** to transform the criminal justice system, **SAGE Criminology & Criminal Justice** offers arresting print and digital content that **UNLOCKS THE POTENTIAL** of students and instructors. With an extensive list written by renowned scholars and practitioners, we are a **RELIABLE PARTNER** in helping you bring an innovative approach to the classroom. Our focus on **CRITICAL THINKING AND APPLICATION** across the curriculum will help you prepare the next generation of criminal justice professionals.

SECOND EDITION

POLICE IN AMERICA

Sara Miller McCune founded SAGE Publishing in 1965 to support the dissemination of usable knowledge and educate a global community. SAGE publishes more than 1000 journals and over 800 new books each year, spanning a wide range of subject areas. Our growing selection of library products includes archives, data, case studies and video. SAGE remains majority owned by our founder and after her lifetime will become owned by a charitable trust that secures the company's continued independence.

Los Angeles | London | New Delhi | Singapore | Washington DC | Melbourne

SECOND EDITION

POLICE IN AMERICA

STEVEN G. BRANDL

UNIVERSITY OF WISCONSIN–MILWAUKEE

Los Angeles | London | New Delhi
Singapore | Washington DC | Melbourne

FOR INFORMATION:

SAGE Publications, Inc.
2455 Teller Road
Thousand Oaks, California 91320
E-mail: order@sagepub.com

SAGE Publications Ltd.
1 Oliver's Yard
55 City Road
London, EC1Y 1SP
United Kingdom

SAGE Publications India Pvt. Ltd.
B 1/I 1 Mohan Cooperative Industrial Area
Mathura Road, New Delhi 110 044
India

SAGE Publications Asia-Pacific Pte. Ltd.
18 Cross Street #10–10/11/12
China Square Central
Singapore 048423

Printed in Canada

Library of Congress Cataloging-in-Publication Data

Names: Brandl, Steven G. (Steven Gerard), author.

Title: Police in America / Steven G. Brandl, University of Wisconsin–Milwaukee.

Description: Second edition. | Los Angeles : SAGE, [2021] | Includes bibliographical references and index.

Identifiers: LCCN 2019030316 | ISBN 9781544375830 (paperback; alk. paper) | ISBN 9781544375816 (epub) | ISBN 9781544375809 (epub) | ISBN 9781544375793 (ebook)

Subjects: LCSH: Police–United States. | Law enforcement–United States. | Police administration–United States.

Classification: LCC HV8139 .B736 2021 | DDC 363.20973–dc23

LC record available at https://lccn.loc.gov/2019030316

Acquisitions Editor: Jessica Miller
Content Development Editor: Adeline Grout
Editorial Assistant: Sarah Manheim
Marketing Manager: Jillian Ragusa
Production Editor: Sarah Downing
Copy Editor: Karin Rathert
Typesetter: Hurix Digital
Proofreader: Theresa Kay
Indexer: Sylvia Coates
Cover Designer: Janet Kiesel

This book is printed on acid-free paper.

20 21 22 23 24 10 9 8 7 6 5 4 3 2 1

BRIEF CONTENTS

DETAILED CONTENTS

PART IV POLICE STRATEGIES AND THE FUTURE OF THE POLICE IN AMERICA   277

POLICING

is inherently controversial, and police work is extraordinarily challenging. There are higher expectations, greater scrutiny, and more calls for accountability of the police in the twenty-first century than ever before. In this environment, it is critically important that students have an accurate understanding of police in our society and be able to differentiate fact from fiction in matters relating to the police. Unfortunately, much of what we believe about policing is based on false or unsubstantiated assumptions, or misleading representations of it. These inaccuracies help fuel the controversies of policing as well as disagreements about how it can best be performed. *Police in America* addresses this issue by providing a real-world fact-based discussion of policing in the United States.

If an accurate understanding of the police in America is the goal, then a discussion of the research that has been conducted on policing is a primary means to reaching that goal. Research findings can identify and dispel the many myths, misconceptions, and false assumptions of policing. Research also can also help identify best practices in policing as well as those practices in need of improvement. An emphasis on research is also especially important given the current trends toward evidence-based policing. *Police in America* emphasizes police research. This emphasis does not mean that the text is complicated or difficult to read, however. In fact, the opposite is true: The text is easy to read and accessible to students. It is written in a straightforward and conversational manner.

Police in America provides a realistic portrayal of the police. It provides a multitude of examples of how policing is conducted in agencies across the country. It emphasizes positive aspects of policing but does so without ignoring or sugar-coating the controversies of police work. The media tend to focus on negative incidents by highlighting the bad or questionable conduct of a few officers. Although there are certainly lessons to be learned from such incidents, these images and stories can provide an inaccurate overall picture of the police. The reality is that exemplary police work is being performed by police officers and law enforcement agencies throughout the country. *Police in America* highlights some of this work.

Police in America also incorporates several other themes, including the following:

- Ethical policing: Because of the nature of the work and how the decisions of officers may affect citizens and the community, it is essential that students consider what constitutes not only a legally good decision but also a *morally* good one.
- Critical thinking: Students should be able to think critically about the complex problems and issues involved with policing.
- The impact of technology: The technological tools of policing have changed dramatically over the years, and it is important to understand how technology has fundamentally altered the nature of it.
- Diversity: To understand policing today, one must appreciate the modern-day and historical roles of race and diversity. Some of the most challenging issues of policing today are at least partly based on race.

The contributions of police research, positive aspects of policing, ethics, critical thinking, the role of technology in police work, and diversity issues are emphasized throughout *Police in America*. The text offers several features in each chapter to help establish an accurate understanding of the police in America. These features allow students and instructors to explore significant issues and questions relating to the police. The features consist of the following:

- Police Spotlight: These features introduce each chapter and discuss a particular police policy, program, or other issue that relates to the topic of that chapter.
- A Research Question: These features highlight an interesting and important research study relevant to the topic of each chapter.
- Good Policing: Each of these features includes an example of a police program, policy, or issue that relates to effective, efficient, equitable, innovative, or ethical policing.
- A Question of Ethics: The questions presented in these features relate to the topic of the chapter and require students to think critically about that particular moral or ethical issue.
- A Question to Consider: Each of these features offers a question related to the topic at hand for students to consider, answer, and/or discuss.
- Technology on the Job: These features highlight and examine a particular technology used by the police while on the job.

With regard to the unique content of *Police in America* and the issues discussed in the book, especially noteworthy are the separate chapters on police discretion and ethics (Chapter 8), the law (Chapter 9), health and safety issues in police work (Chapter 10), police use of force (Chapter 11), crime detection and investigation (Chapter 7), and intelligence-led and evidence-based policing (Chapter 14). Each of these chapters is extremely important in developing a solid understanding of the police in America, and it is through their inclusion, along with comprehensive and timely coverage of other critical topics, that *Police in America* clearly differentiates itself from other texts.

With regard to the overall content and organization of the text, the first four chapters provide a foundation for the study of the police (the history of the police, role and function of the police, characteristics of police organizations). The second section of the book includes five chapters that examine the nature of police work (police recruitment, selection, and training; police patrol; criminal investigation; discretion and ethics; and the law). The third section is devoted to the hazards of police work and provides a discussion of health and safety issues, police use of force, and police misconduct. The last section includes three chapters on the most recent strategies of policing (problem-oriented policing, evidence-based policing) as well as a discussion of the future of policing.

Police in America provides students with a substantial understanding of the role and function of police in the United States.

NEW TO THIS EDITION

All figures and statistics have been updated, and a multitude of new media and scholarly articles have been incorporated into the discussion provided in each chapter. Many discussions have been revised for purposes of clarification or succinctness. While the second edition incorporates new discussions of several critical topics, it is approximately the same length as the first edition. All existing features remain with "Research Spotlight" changed to "A Research Question." Beside these changes, listed below are the most significant changes that have been made for the second edition of the book.

Chapter 1

- New information on police-citizen contacts
- New information on body-worn cameras (BWCs)
- New information on police–citizen cooperation (and lack thereof) as a challenge of policing
- New "A Research Question" (media and citizens' attitudes toward the police)

Chapter 2

- New information on women and people of color police officers
- New "Good Policing" (good policing changes over time)
- New/revised "A Research Question" (current state of police research)
- New information about changes in policing during the community problem-solving era

Chapter 3

- New "Police Spotlight" (police officer salaries and possible consequences)
- New "Good Policing" (police as warriors versus guardians)

Chapter 4

- New "Police Spotlight" ("Scoop and Run" in the Philadelphia PD, nontraditional police task)
- New "Good Policing" (the need for new measures of police effectiveness)
- New section on police responding to people in need, people with mental illness, crisis intervention teams
- New section on police use of Narcan

Chapter 5

- New "Police Spotlight" (San Diego PD recruitment efforts)
- New information on police salaries
- New "Research Question" (how male and female college students perceive a police career)
- New information about diversity of police officers
- New "Good Policing" (Tulsa PD recruit training)

Chapter 6

- New information on foot patrol
- New "A Research Question" (CCTV and directed patrol)
- New information on traffic stops
- New information on stop, question, and frisk

Chapter 7

- New "Police Spotlight" (ancestry DNA and the Golden State Killer)
- Moved material on "proof" to Chapter 9
- New "A Research Question" (secondary transfer of DNA)
- New information on DNA and DNA analysis
- New information as video as evidence

Chapter 8

- New information on BWCs
- New information/section on race and implicit bias
- New information on de-policing and Ferguson Effect
- New "A Research Question" (BWCs)
- New "Good Policing" (BWCs)

Chapter 9

- New discussion of proof and probable cause
- Many revised discussions to shorten the chapter
- New section on juveniles and the Miranda requirement

Chapter 10

- New "Police Spotlight" (dealing with event trauma)
- New "A Research Question" (shiftwork, fatigue, and gender)
- New information on body armor

Chapter 11

- New "Police Spotlight" (de-escalation)
- New "Good Policing" (transparency in police use of force)
- New information on police use of force
- New "A Research Question" (but still suicide by cop)
- New information on police use of robots and drones
- New reorganized section on the control of police use of force
- New information/sections on de-escalation, early intervention systems, BWCs

Chapter 12

- New "Police Spotlight" (still Denver PD police complaint mediation program)
- New information on numerous chapter topics
- New information on perceptions of police honesty and ethics

Chapter 13

- Added many examples of community policing in various police departments
- New information on law enforcement use of social media
- New information on community policing versus problem-oriented policing
- New "A Research Question" (procedural justice)
- Added many examples of problem-oriented policing (with a new diagram)

Chapter 14

- New "Police Spotlight" (smart policing in Chicago)
- New information on smart policing (its meaning changed since the first edition)
- New information to clarify various strategies discussed in the chapter, including how these strategies may be controversial
- New "Good Policing" (a problem-oriented approach to CompStat)
- New "A Research Question" (crime analysis)

Chapter 15

- Condensed discussion of terrorism
- New "Research Spotlight" (far-right extremism)
- New information about future of policing including militarization of the police
- New "Good Policing" (BWCs and accountability)
- New information on the future technologies of policing

ACKNOWLEDGMENTS

MANY people helped make this book a reality, and I am very appreciative for each of them. The influence of Gary Cordner (Emeritus Professor, Kutztown University) can be traced back to the very beginnings of this book and his assistance is well reflected in this edition as well. Thanks Gary. I also acknowledge Jerry Westby, former publisher at SAGE, whose good decisions and persuasive talk led to the creation of the first edition of this book. The success of this book is due in large part to the dedication of the entire SAGE team. Jessica Miller and Adeline Grout are an author's dream come true and were always a pleasure to work with. I also would like to acknowledge the hard work of Sarah Downing and Karin Rathert. These individuals also helped make this an outstanding text.

Many law enforcement professionals assisted me either directly or indirectly with this book. In particular, I would like to acknowledge the assistance of Chief of Police Edward Flynn (retired) and Inspector Jutiki Jackson of the Milwaukee Police Department; Chief of Police Kenneth Meuler of the West Bend (Wisconsin) Police Department; Chief of Police Peter Hoell of the Germantown (Wisconsin) Police Department; and Chief of Police Peter Nimmer of the Shorewood (Wisconsin) Police Department. Also so important in shaping my world-view of the police and this book are the multitude of police officers that I have worked with and studied over the years. You expanded my reality by letting me be part of yours. Thank you.

I'd like to acknowledge the support and assistance of several colleagues on this and related projects: Professor Meghan Stroshine of Marquette University, Professor Robert Worden at the University at Albany, Professor James Frank at the University of Cincinnati, and Professor Matt Richie of the University of Wisconsin–Oshkosh. Good friends, excellent colleagues, and outstanding scholars.

I would also like to thank University of Wisconsin–Milwaukee graduate students Amber Krushas and Kevin Schlichter who assisted greatly with the second edition of *Police in America*. Thank you to Nadine Rodriguez for preparing the excellent online resources for the second edition.

Many reviewers of earlier drafts of the book also deserve acknowledgment for improving the final product.

REVIEWERS OF THE SECOND EDITION

Hadeel Al-Alosi, Western Sydney University

Paul Klenowski, Clarion University of Pennsylvania

Selye Lee, West Liberty University

Nicholas Malkov, John Jay College of Criminal Justice

Jeff O'Donnell, Community College of Allegheny County

Elizabeth Perkins, Morehead State University

Wendy Perkins, Marshall University

Carl Root, Eastern Kentucky University

Mercedes Valadez, California State University, Sacramento

REVIEWERS OF THE FIRST EDITION

Emmanuel N. Amadi, Mississippi Valley State University

James W. Beeks, University of Phoenix–Atlanta

Lt. Allen Branson, PhD, Philadelphia Police Academy

Timothy Fulk, Indiana University Kokomo

John Hamilton, Park University

Richard N. Holden, University of North Texas at Dallas

Coy Johnston, Arizona State University

Brian Kelley, Kent State University

William Kelly, Auburn University

Tristin M. Kilgallon, Ohio Northern University

Todd Lough, Western Illinois University

Marcos L. Misis, Northern Kentucky University

Thomas S. Mosley, University of Maryland Eastern Shore

Clint Osowski, Texas A&M International University

Michael D. Paquette, Middlesex County College

Jason Paynich, Quincy College

Michael S. Penrod, Kirkwood Community College

Elizabeth Perkins, Morehead State University

Michael Pittaro, PhD, American Military University

Scott Pray, Muskingum University

Melinda Roberts, University of Southern Indiana

Rafael Rojas Jr., Southern New Hampshire University

Steven Ruffatto, Harrisburg Area Community College

Kenneth Ryan, California State University, Fresno

Shawn Schwaner, Miami Dade College

Jeff Schwartz, Rowan University

Rupendra Simlot, PhD, Stockton University

Carol L. S. Trent, University of Pittsburgh

Finally, on a personal note, I gratefully acknowledge Katy, David, and Laurie. Among so many other things, you keep me grounded and provide perspective. Thank you for your love and support.

DIGITAL RESOURCES

INSTRUCTOR RESOURCE SITE

edge.sagepub.com/brandl2e
INSTRUCTORS: SAGE coursepacks and SAGE Edge online resources are included FREE with this text. For a brief demo, contact your sales representative today.

SAGE coursepacks for instructors makes it easy to import our quality content into your school's learning management system (LMS)*. Intuitive and simple to use, it allows you to

Say NO to...

- required access codes
- learning a new system

Say YES to...

- using only the content you want and need
- high-quality assessment and multimedia exercises

*For use in: Blackboard, Canvas, Brightspace by Desire2Learn (D2L), and Moodle

Don't use an LMS platform? No problem, you can still access many of the online resources for your text via SAGE Edge.

With SAGE coursepacks, you get:

- quality textbook content delivered **directly into your LMS**;
- an **intuitive, simple format** that makes it easy to integrate the material into your course with minimal effort;
- **assessment tools** that foster review, practice, and critical thinking, including:

  - **Coursepack chapter quizzes** that identify opportunities for improvement, track student progress, and ensure mastery of key learning objectives
  - **test banks** built on Bloom's Taxonomy that provide a diverse range of test items with ExamView test generation
  - **activity and quiz options** that allow you to choose only the assignments and tests you want
  - **instructions** on how to use and integrate the comprehensive assessments and resources provided;

- **assignable SAGE Premium Video** (available via the interactive eBook version, linked through SAGE coursepacks) that is tied to learning objectives, and curated and produced exclusively for this text to bring concepts to life, featuring:

  - **Corresponding multimedia assessment options** that automatically feed to your gradebook
  - Comprehensive, downloadable, easy-to-use *Media Guide in the Coursepack* for every video resource, listing the chapter to which the video content is tied, matching learning objective(s), a helpful description of the video content, and assessment questions
  - **Career videos** feature interviews with professional law enforcement discussing their day-to-day work and current issues in policing;
  - **Criminal Justice in Practice videos** feature animated, decision-making scenarios challenge students to explore how they would respond to real-world situations faced by criminal justice professionals; and
  - **SAGE News Clips** feature relevant news footage to help students apply knowledge to current events.

- **chapter-specific discussion questions** to help launch engaging classroom interaction while reinforcing important content;
- editable, chapter-specific **PowerPoint® slides** that offer flexibility when creating multimedia lectures so you don't have to start from scratch;
- **lecture notes** that summarize key concepts on a chapter-by-chapter basis to help you with preparation for lectures and class discussions;
- **integrated links to the interactive eBook** that make it easy for students to maximize their study time with this "anywhere, anytime" mobile-friendly version of the text. It also offers access to more digital tools and resources, including SAGE Premium Video; and
- **all tables and figures** from the textbook.

STUDENT STUDY SITE

edge.sagepub.com/brandl2e

SAGE Edge for students enhances learning, it's easy to use, and offers:

- an **open-access site** that makes it easy for students to maximize their study time, anywhere, anytime;
- **video and multimedia resources** that bring concepts to life;
- **eFlashcards** that strengthen understanding of key terms and concepts; and
- **eQuizzes** that allow students to practice and assess how much they've learned and where they need to focus their attention.

INTERACTIVE EBOOK VERSION

The dynamic interactive eBook version of this text goes way beyond highlighting and note-taking, giving you access to SAGE Premium Video—curated and produced specifically for *Police in America, Second Edition*. Read your mobile-friendly eBook and access SAGE Premium Video and multimedia tools anywhere, anytime across desktop, smartphone, and tablet devices. Simply click on icons in the eBook to experience a broad array of multimedia features, including:

- **VIDEO:** Boost learning and bolster analysis with SAGE Premium Video! Recapping the fundamentals in every chapter, each video activity is paired with chapter learning objectives and tied to assessment via SAGE coursepacks, offering an engaging approach that appeals to diverse learning styles.

  - **Career videos** feature interviews with professional law enforcement discussing their day-to-day work and current issues in policing;
  - **Criminal Justice in Practice videos** feature animated, decision-making scenarios challenge students to explore how they would respond to real-world situations faced by criminal justice professionals; and
  - **SAGE News Clips** feature relevant news footage to help students apply knowledge to current events.

- **OFFLINE READING:** Using the VitalSource Bookshelf® platform, download your book to a personal computer and read it offline.
- **SOCIAL SHARING AND FOLLOWING:** Share notes and highlights with instructors and classmates who are using the same eBook, and "follow" friends and instructors as they make their own notes and highlights.
- **ONLINE CONTENT:** Access more online content via links to important data, relevant background, and profiles that enrich key concepts in the text.

Steven G. Brandl (PhD, Michigan State University, 1991) is a professor in the Department of Criminal Justice and Criminology at the University of Wisconsin–Milwaukee. Professor Brandl worked in local and federal law enforcement prior to obtaining his PhD. At UW-Milwaukee, he teaches undergraduate and graduate courses, including Introduction to Policing, Criminal Investigation, and Issues in Police Practice and Policy, among others. His research interests include police use of force, criminal investigation, and health and safety issues in police work.

Professor Brandl has conducted numerous research studies and consulted with numerous national and local police departments and other state agencies on law enforcement issues. In addition to this textbook, he is the author of *Criminal Investigation* (SAGE) and many articles in professional journals. He is co-editor of *The Police in America: Classic and Contemporary Readings* and *Voices From the Field: Readings in Criminal Justice Research*.

Part I
FOUNDATIONS FOR THE STUDY OF THE POLICE

1

AN INTRODUCTION TO THE POLICE IN AMERICA

Police Spotlight: What It Takes to Be a Good Police Officer

"Policing a democracy is not an easy task. It's difficult and messy. Yet police in a democracy must *always* operate within the rule of law. They must *always* apply our shared values to the difficult daily tasks of resolving conflict, protecting unpopular people and causes, and always acting fair and respectful to those who at the time are not conducting themselves properly—those who are intoxicated, affected by other drugs, surly, disrespectful, and even violent. That's what police in a Bill of Rights do. And those who wish not to do that should not be our police.

So who can perform such a difficult task? I have said this before and I will say it again—only those who are the best of us, only those who are well-educated and well-trained, and only those who know about and can put into practice our closely-held and core values of freedom, individual rights, rule of law, fairness, and equality."[1]

—David C. Couper, former Madison (WI) police chief

THE aim of this chapter is to introduce the fundamental purposes of and controversies involving the police and to discuss how police officers are constantly dealing with ethical and moral issues in their work.

INTRODUCTION

When you think of *the police*, you most likely envision officers who work in local police agencies, such as the police in your city or county police departments. Officers who work in these agencies are the police you are most likely to see and with whom you are most likely to interact. However, there are many other law enforcement agencies, including state and federal law enforcement agencies. The focus of this book is on general service police agencies that have responsibility for

Objectives

After reading this chapter you will be able to:

1.1 Explain the challenges associated with policing a free society

1.2 Discuss the tension between citizens' rights and police power

1.3 Explain how the police are accountable to citizens

1.4 Identify and discuss the controversies and difficulties of policing

1.5 Discuss how police use of discretion and police use of force can make the police controversial

1.6 Discuss why ethical conduct of police officers is an especially serious concern

Fact or Fiction

To assess your knowledge of the police prior to reading this chapter, identify each of the following statements as fact or fiction. (See page 16 at the end of this chapter for answers.)

1. The best source of knowledge about the police is your previous interactions with them.

2. There is a trade-off between citizens' rights and police power. If there is more of one, there is less of another.

3. As long as the police avoid overpolicing, they will not be subject to criticism.

4. The use of deadly force is often considered the ultimate discretionary decision made by police officers.

5. Defining good policing is not difficult; it is simply the number of arrests made, the number of crimes solved, and the number of citizen complaints received.

6. The media tend to focus on bad police officer behavior.

7. As long as the police pursue reasonable and legitimate goals, the means used to achieve them are not a major issue.

Photo 1.1

Interpretation of inkblots may depend on a person's personality and experiences. ©iStockphoto.com/akova

crime prevention and investigation, order maintenance through patrol and other means, and the provision of other miscellaneous services. Although state and federal investigative agencies have an absolutely critical role in law enforcement efforts, in this book, limited attention is paid to the unique and specific issues associated with the operation of these organizations.

Prior to officially becoming a member of a police force, officers take a sworn oath to support the laws of the United States, their state, and their community. This is the basis for the frequent reference in this book to *sworn officers* in contrast to *civilians* who also work in police departments. Sworn officers have the authority to make arrests and to legitimately use force. As discussed throughout this book, when all the layers of complexity are stripped away, the bottom line is that it is these two fundamental powers of the police—the authority to make arrests and to use force—that can make the police controversial. Some of this controversy is reflected in the strong and varied views of citizens about the police. Some people see the police as a problem; some see the police as the solution. Some people see the police as friend; some see the police as foe. The police are, as explained decades ago by sociologist Arthur Niederhoffer, a Rorschach test in uniform.[2] Our views toward the police are shaped by our experiences with them, by other people's experiences that we see or hear about, by social and mainstream media, and by the news.[3] To one degree or another, each of these factors combine to form the basis of opinions about the police.

Of the factors that may affect your views of the police, it may be tempting to believe that your personal experiences with the police are the most valid. After all, if you've personally seen and experienced it, it must be true. However, it is important to understand that personal experience is not always a good source of knowledge from which to generalize. There are at least three reasons for this.

A Question to Consider 1.1

Why Such Strong Feelings About the Police?

Citizens tend to have strong opinions about the police. Why don't people have similarly strong opinions about other public service workers, such as firefighters, garbage collectors, or even teachers?

EXHIBIT 1.1

Contacts Between Police and the Public

The 2018 report titled "Contacts Between Police and the Public"[4] explains that in 2015, 21% (53.5 million people) of U.S. residents aged 16 or older had contact with the police during the previous 12 months. Approximately 23% of whites, 20% of blacks, and 17% of Hispanics had contact with the police. Police were equally likely to initiate contact with whites and blacks but less likely to initiate contact with Hispanics. Police had about equal contact with males and females. The most common

circumstance by which people had contact with the police was as a result of being a driver in a traffic stop. The most common reason for the traffic stop was speeding. Approximately 95% of people reported that the police behaved properly during the stop. Two percent of people who had contact with the police reported that they experienced a nonfatal threat or use of force by the police. Most of these people perceived the action by the police to be excessive.

A Question to Consider 1.2

Police Power and Crime Solving

In 2017, approximately 62% of homicides in the United States were solved, meaning that the perpetrator was identified and apprehended. Of all crimes, the police have the greatest success at solving homicides, yet this percentage is currently near a historic all-time low. There are many reasons for this. Do you think that if the police were given more power and authority to conduct investigations they would be able to solve more homicides and other crimes? If so, what would be the consequences of this? Do you think the trade-off would be worth it? Explain.

THE POLICE ARE EXPECTED TO PREVENT AND SOLVE CRIME

The police have been given a very difficult task: They are expected to prevent people from committing crimes and to solve the crimes they are unable to prevent. In doing this, the police maintain order in our society. However, many factors have been identified as contributing to criminal behavior, and the police do not control any of these factors. The police do not have any control over poverty, whether children grow up with proper role models, the weather, unemployment, or people's self-control. Furthermore, the police are primarily reactive, which means they are dependent on citizens to notify them that a crime has occurred so they can respond, and crimes are often not reported to police. The police must operate within the confines of the law, and they operate with limited resources. All of these things considered, the police are often at a disadvantage in the "game" of cops and robbers, and this helps explain the seemingly low rate at which crimes are solved (Figure 1.1). This suggests that the police have an **impossible mandate**.[20]

CITIZENS MAY NOT COOPERATE WITH THE POLICE, AND MAY EVEN DO THEM HARM

Crime prevention and crime solving depend on cooperative relationships between citizens and the police. However, for a variety of reasons, the unfortunate reality is that sometimes citizens are not interested in assisting the police. This lack of assistance can come

> **impossible mandate:** This term reflects the idea that the police have been assigned the task of crime control, but because they cannot control the factors that *cause* crime, this task is difficult—if not impossible—to accomplish.

FIGURE 1.1

Crimes Cleared (Solved) by the Police, 2017[21]

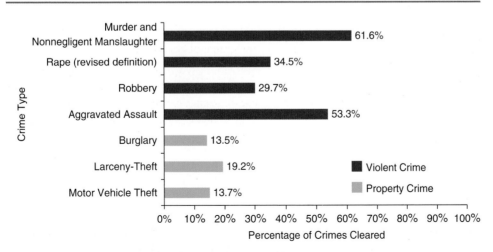

The police are expected to solve the crimes that they do not prevent, but for various reasons they have limited success in this regard.

Photo 1.5
Because of the unequal distribution of serious street crime, the police spend more time in some areas than in others. Chicago Tribune/Tribune News Service/Getty Images

in many different forms. For example, approximately 50% of crimes are not reported to the police by citizens (victims). It is difficult for the police to address crimes that they are not aware of. Further, in some places and among some people, providing information to the police that may help in a criminal investigation is severely looked down upon as demonstrated by the adage "snitches end up with stiches." Without assistance of citizens, the crime-solving abilities of the police are limited. Sometimes, as discussed more directly in Chapter 10, citizens direct violence toward the police. Ideally, citizens are a friend to the police, but in some instances, they are actually a foe. This makes the relationship between the police and the public complicated, to say the least.

THE POLICE PAY MORE ATTENTION TO SOME CRIMES, SOME PEOPLE, AND SOME AREAS THAN OTHERS

The police are not equally concerned with all types of crime. The police, local police in particular, are more oriented toward what have been referred to as *predatory* types of crime, or *street* crimes, such as murder, rape, robbery, assault, burglary, and so on. Part of the reason for the greater focus on these types of crime is that police territory *is* the streets, literally. The police patrol the streets and sidewalks. The streets are public space; the police have the most presence in public spaces, as opposed to private spaces like living rooms or business offices. When the police are in private places, it is usually only because they were invited or needed there.

Because they are responsible for crime control, officers tend to have a greater presence in areas where there is more street crime. Such areas tend to have high levels of unemployment, poverty, and population density. They are often racial minority neighborhoods. So, at least in urban settings, the police pay more attention to some areas and some people than others. Indeed, the police spend more time in some public spaces than others.[22] This can lead to criticism about **overpolicing** in some neighborhoods and **underpolicing** in others. Citizens who perceive too much police action in their neighborhoods may be just as upset as citizens who perceive too little in their neighborhoods. Either way, the police may be subject to criticism.

Law enforcement agencies other than local police departments are also more concerned with some types of crimes than others. For example, federal law enforcement agencies, such as the Federal Bureau of Investigation (FBI), devote more resources to combating predatory crimes and terrorism than other types of crimes. Although the strategies used to combat criminals differ, federal law enforcement agencies also devote more resources to certain places than others and pay more attention to some people than others.

THE POLICE HAVE OTHER RESPONSIBILITIES

overpolicing: The perception of too much police presence and action in a neighborhood.

underpolicing: The perception of too little police presence and action in a neighborhood.

Besides their important crime-related duties, local police departments have a multitude of other responsibilities. Because the police are a twenty-four-hour-a-day resource that is just a phone call away (via 911), citizens call on the police for all sorts of troubles. The local police department is often the social agency of first resort for people in need of assistance. Officers regularly deal with family members and neighbors who do not get along; they deal with homelessness issues and people with mental illness and substance abuse problems. None of these issues necessarily relate to criminal behaviors, but all require police resources.

FIGURE 1.2

Crime Rate, 1960–2017

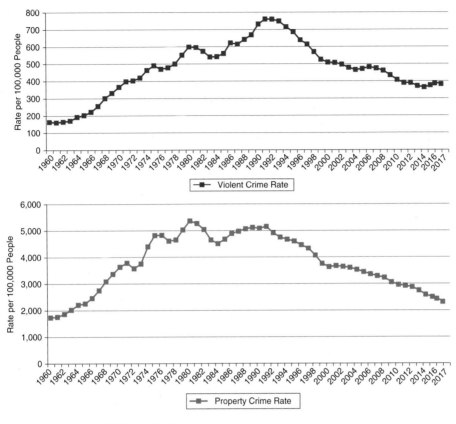

Note: 2017 was most recent data available at time of publication.

Source: Bureau of Justice Statistics

THE POLICE USE DISCRETION IN DEALING WITH PEOPLE

Police officers often must use **discretion**, or their own judgment, in making decisions about how to handle situations. This discretion can affect people's lives in dramatic ways. Whether the situation involves a decision to ticket or just warn a speeding motorist, what to do about a barking dog, how to resolve a dispute between neighbors, or whether or not to use force on a resisting suspect, the use of discretion is a critical part of the job and can raise all sorts of issues. For instance, the decision made could be an improper one. Since officers make a lot of decisions during the course of a work shift, there is the potential for many mistakes. It is very important to understand that with a critical decision comes the possibility of a critical mistake.

Another issue is that often discretion involves officers making moral or ethical judgments about who is right, who is wrong; who is the victim, who is the offender. Sometimes these distinctions are easy to make, sometimes they are not. Usually at least one of the parties involved does not like the decision that the officer has made. The reality is that officers seldom leave a situation being appreciated by all the participants. It is no wonder that citizens have strong and divergent feelings about the police, either positive or negative.

THE POLICE HAVE AUTHORITY TO USE FORCE AND ARREST CITIZENS

The ability to make an arrest is a critical but basic form of police authority. Citizen's noncompliance with the law may lead to arrests being made by officers. Further, officers have discretion about other critical actions, such as whether or not to use physical force on citizens and

> **discretion:** A police officer's personal judgment of how best to handle a situation.

what type of force to use. As discussed in more detail later in this book, many scholars argue that police authority to use force is what differentiates the occupation from all others. Workers in no other occupation can use "essentially unrestricted" force against citizens.[23] The use of force, especially deadly force, is often the most controversial discretionary decision made by police officers. Even if the force was legally justified, its use is often controversial. This fact has been repeatedly and dramatically highlighted with recent incidents in which the police have used deadly force against unarmed subjects. From these incidents have come calls for police reforms and especially for the use of body-worn cameras to provide greater transparency, accountability, and control over police actions in deadly force situations (see Technology on the Job feature on page 7).

MEASURING GOOD POLICING IS DIFFICULT

Assessing police performance is important so that corrections and improvements can be made. With police work, however, it is extremely difficult to measure good performance. For example, crime prevention is a good goal, but how do you measure crimes that do not occur? It is simply impossible for the police to accurately say that they prevented X number of crimes during the past year. Often comparisons are made to the number of crimes that occurred the previous year, but there is nothing inherently meaningful about a previous year's numbers. If there were 100 crimes last year but only 90 this year, is that a police success? Is it possible the number of crimes may decline from year to year not because the police are effective but because citizens reported crimes to the police less often? If the number of homicides went down but the number of shootings stayed the same, is that a police success? Sometimes the number of *arrests* made by the police is used as an indicator of performance; the reasoning is that making more arrests equals better performance. The problem with this reasoning is that if an arrest was made, it means that a crime was not prevented. Additionally, an arrest may not be the best or most effective way of handling a particular incident. Indeed, identifying accurate measurements of good policing has been, and remains, problematic for the police. Nevertheless, it is possible to identify specific examples of good policing practices and good qualities and actions of police officers. Such examples are provided throughout this book in the Good Policing features.

Photo 1.6

The media tend to distort the realities of policing. Some people argue that the intent of the media is more to entertain than to educate. AP Photo/Detroit News, Steve Perez

THE MEDIA DO NOT NECESSARILY ACCURATELY REPRESENT THE POLICE

Citizens often see the police through the filter of the media, including social media, entertainment media, and the news media. The problem is that the media do not necessarily accurately depict the police and their work. In particular, news media sources tend to be superficial and selective in their coverage. In some respects, they use the police as entertainment, or more precisely, "infotainment."[24] They focus on violent, random, and bizarre crimes and often call into question police abilities in controlling crime, especially when there seems to be a large amount of it occurring. The news media tend to focus on bad officer behavior, especially instances of supposed police brutality,[25] although stories about police heroism also usually make the news. Entertainment media also offer an unrealistic portrayal of the police, often exaggerating the exciting or bizarre aspects of the job (e.g., the television show *COPS*) as well as the mysterious (e.g., *Criminal Minds*). Social media are also selective in their representation and unfortunately are the sources of much misinformation about the police. As such, the media may distort citizens' views of the police.[26]

ETHICS AND MORALS IN POLICING

Discretionary decisions of police officers on the street are influenced by many factors. Ideally, one of them is ethical standards. Similarly, policy and strategy decisions in law enforcement agencies should be based on ethical principles.[29] As a result, **ethics** are important to consider when studying the police. Issues associated with ethics and **morality** are discussed in more detail in Chapter 8 and throughout the chapters via the A Question of Ethics features, but the topic is introduced here.

Ethics and morality are closely related and intertwined. Both relate to fundamental questions about what is right and what is wrong or what is good and what is bad. When a distinction is made, usually morality is concerned more with the individual and his or her internal sense of proper conduct. Ethics relate more toward the *behavior* of a person. A person whose behavior is ethical is also moral. The distinction is a fine one and is debated by philosophers.

For the police, ethical conduct is an especially serious concern because the police have extraordinary power, and "with great power comes great responsibility."[30] This axiom has great relevance for today's officers. The police have extraordinary power and authority. Nowhere is this clearer than in their ability to use discretion, to make arrests, to conduct searches, and to use force.

> **ethics:** Rules of behavior that are influenced by a person's perception of what is morally good or bad.
>
> **morality:** A person's internal beliefs about what is right or wrong conduct.

Higher Standards and Visibility

What is one of the most important things a police officer needs to know in order to do a good job? It is critically important for officers to realize that 100% of the time, on and even off duty, they are in the spotlight; they are being watched by citizens and are held to a high standard of conduct by both those citizens and by police superiors. Of course, when on duty and in uniform, people can easily recognize police officers. People notice police officers. Police officers must realize that they are *always* subject to public scrutiny, even when taking a break from their duties. An officer must also be aware that when off duty and not in uniform, some people will still know that he or she is a police officer. Any questionable conduct from an off-duty officer is still subject to concern and criticism, and the officer can be disciplined for it. Officers' off-duty use of social media is also subject to increased public and police department scrutiny and has been the basis for job sanctions. Bottom line: Police officers are held to a high standard of conduct and need to be aware of this at all times.

Officers can deprive citizens of their liberty, their property, and their *life*. In a fair and just society, the police are obligated to use their power and authority legally, responsibly, and ethically.

Many ethical issues in policing become relevant when considering the *means-ends* distinction. *Ends* are the goals to be achieved, such as apprehending criminals. *Means* are the ways in which those goals can be achieved, how things are done. When it comes to ethical concerns, usually the means are more scrutinized. Means vary from the ethical to the unethical. They are also either legal or illegal. As explained below, even if means are legal, they can still be viewed as unethical. Unethical and/or illegal means have been referred to as **dirty means**.[31]

FORMS OF UNETHICAL CONDUCT

One form of unethical conduct occurs when the police use illegal means in an attempt to accomplish **good ends**. For example, in an attempt to detect and prevent crime, an officer may stop and search citizens without the reasonable suspicion that is legally necessary. To solve a crime, an investigator may not inform suspects of their right to remain silent, which is legally required. This conduct is not legal, nor is it ethical. These actions are clearly problematic in a society that expects its police to be fair and just.

Sometimes means are legal but perhaps not ethical. An example is when the police use deception to identify and apprehend criminals. In particular, when officers go undercover and buy drugs from an unwitting citizen or lie in the interrogation room to get a suspect to confess, there is usually little debate about the goodness of the goal, but sometimes there is concern about the appropriateness of the way by which the goal was achieved. There may be ethical concerns even when conduct is legal.

Another type of situation that raises ethical concerns is when the law does not explicitly prohibit or allow certain conduct. For example, in one case,[32] the police wanted to identify associates and co-conspirators of an offender, arguably a reasonable and worthwhile goal. To do so, they used her personal information to create a fake Facebook page. At the time, these police actions were neither legal nor illegal, but they were potentially unethical nonetheless.

Finally, some ethical concerns regarding police conduct lie outside of the means-ends distinction because the goals being pursued are not appropriate. These situations relate to police corruption. Legitimate goals of policing are not present in most forms of corruption. For example, when officers seek to maximize personal gain through theft or bribery, ethical (and legal) concerns emerge. As noted earlier, one of the difficulties of policing is defining and measuring good policing. At the very least, one dimension of good policing is ethical policing. It is a worthwhile challenge to figure out how policing can be made more ethical.

dirty means: Unethical or illegal means used by police officers.

good ends: The desired goals of policing.

A Question of Ethics

Are Police Actions That Are Legal Also Ethical?

One of the first steps in understanding and controlling the unethical conduct of police officers is recognizing what it is. Are police actions that are legal also always ethical? Explain why or why not. Besides the examples already provided, identify and discuss two examples of police conduct that would help support your position.

Main Points

- Our views toward the police are shaped by our experiences with officers, by other people's experiences with officers, and by media portrayals.
- Research provides a good basis on which to develop an accurate understanding of the police.
- The United States is a free society because citizens have freedoms from the government, but this puts the police in a peculiar situation. Officers are expected not to infringe on citizens' rights and to protect citizens' rights, but at the same time, they are expected to regulate citizens' conduct.
- The Bill of Rights to the U.S. Constitution set forth the freedoms citizens have from the government.
- The system of an elected government and increasing the transparency of law enforcement agencies are supposed to provide for accountability of the police to the citizenry.
- There are many controversies and difficulties associated with policing:
  - The police are expected to control crime but they do not control many of the factors associated with criminal behavior. In addition, they are reactive, they have to follow the law, and they have limited resources.
  - Citizens may not cooperate with the police and may even do them harm.
  - The police pay more attention to some crimes, some people, and some areas than others. This can lead to criticisms about overpolicing and underpolicing.
  - The local police department is often the social agency of first resort for people with many different problems. The twenty-four-hour-a-day availability of the police compounds this issue.
  - Police very often use discretion, or their own judgment, in making decisions. Police discretion can affect people's lives in major ways. Often it involves taking action against someone in order to protect someone else. These decisions can be controversial.
  - Discretion that relates to use of force is especially controversial.
  - Measuring good police performance is problematic.
  - The news and entertainment media do not accurately depict the police and their work. This can distort the reality of policing and/or create unrealistic expectations of the police.
- In a fair and just society, the police are obligated to use their power and authority responsibly, fairly, and ethically.
- Many ethical concerns for the police relate to whether they use unethical (dirty) means to achieve good goals.

Important Terms

Review key terms with eFlashcards at **edge.sagepub.com/brandl2e**.

Bill of Rights   6

dirty means   14

discretion   11

ethics   13

free society   5

good ends   14

impossible mandate   9

morality   13

overpolicing   10

underpolicing   10

Questions for Discussion and Review

Take a practice quiz at **edge.sagepub.com/brandl2e.**

1. Why might personal experience with officers not be a good basis on which to draw accurate conclusions about the police?
2. What does it mean to say that a society is free? In general, how does policing a free society differ from policing a not-so-free society?
3. How is it that a technology, such as police body-worn cameras, may increase transparency and accountability of a police agency?
4. How is it that a system of elected government is supposed to provide accountability of the police to citizens? Does this system actually provide for accountability?
5. Do the police have an impossible mandate? Why or why not?
6. What is it about police practice that raises concerns about overpolicing and underpolicing?
7. How can police use of discretion be controversial?
8. Why is good policing difficult to measure?
9. How do the media distort the realities of policing? Why is this distortion important to recognize and understand?
10. How do ethical issues most often arise in police work?

Fact or Fiction Answers

1. Fiction
2. Fact
3. Fiction
4. Fact
5. Fiction
6. Fact
7. Fiction

Digital Resources

Get the tools you need to sharpen your study skills. SAGE Edge offers a robust online environment featuring an impressive array of free tools and resources.

Access practice quizzes, eFlashcards, video, and multimedia at **edge.sagepub.com/brandl2e.**

Media Library

View these videos and more in the interactive eBook version of this text!

Career Video
1.1: Why Is Policing so Difficult in a Democratic Society?

Criminal Justice in Practice
1.1: Crime Control v. Due Process

SAGE News Clip
1.1: Illinois—Cop Coaches

2

THE HISTORY OF THE POLICE IN AMERICA

Objectives

After reading this chapter you will be able to:

2.1 Explain why an understanding of police history is important

2.2 Identify and discuss the four eras of policing and the reasons why each era began and dissolved

2.3 Discuss the role of constables, watches, slave patrols, and sheriffs during the pre-police era

2.4 Describe how the first police departments in the country operated

2.5 Compare how the reform era of policing differed from the political era

2.6 Discuss why the 1960s were so significant for the police

2.7 Evaluate the critical concepts associated with the community problem-solving era

Police Spotlight: Policing in the Early Days

According to *The First One Hundred Years,* a publication of the Milwaukee Police Department,

> On October 4, 1855, the Milwaukee, Wisconsin, Police Department began functioning. William Beck was chosen chief of police by the mayor. His salary was set at $800 a year. Privates were to get $480 a year.
>
> Six policemen were chosen by Beck; they were picked for their size and fighting ability.
>
> To arrest a man in those days, it was nearly always necessary to whip him first. The first policemen in Milwaukee were consistently seen with black eyes, bruised cheeks, and split lips. They earned their $40 a month the hard way. Murders were reduced to practically nothing; thugs quit prowling the streets at night lying in wait for prosperous looking individuals, and citizens began writing letters of praise about the battered and bruised policemen and the fine work they were doing.[1]

As described here, formal policing in Milwaukee (and other cities) had very humble beginnings. As we will discuss in this chapter, while some aspects of policing have dramatically changed over time, others have changed very little. In particular, police use of force and the relationship between police and crime control remain central facets of the policing function.

Source: Milwaukee Police Department. 1955. *The First One Hundred Years.* Milwaukee: City of Milwaukee, p. 3.

Fact or Fiction

To assess your knowledge of police history prior to reading this chapter, identify each of the following statements as fact or fiction. (See page 35 at the end of this chapter for answers.)

1. One of the first things the colonists did when they arrived in America was set up a network of relatively sophisticated and well-run police departments.

2. To understand issues involving the police and race relations today, it is important to understand the policing of racial minorities in the past.

3. So-called black codes and Jim Crow laws were never legal, nor were they ever officially enforced by the police.

4. The first women were hired as police officers in the late 1800s and were assigned similar duties as policemen.

5. The first black officers were hired in the late 1800s and were often more educated and qualified than their white counterparts.

6. Photographs of criminals, Bertillonage, the third degree, and the dragnet roundup of suspects were common investigative strategies and tactics used during the political era of policing.

(Continued)

CHAPTER 2 chronicles the history of the police in America. It begins with a discussion of how the police first came to exist and ends with a brief discussion of modern-day policing. The chapter serves as a foundation for the rest of the chapters that focus directly on policing as it is conducted today.

(Continued)

7. The reform era of policing was an attempt to remove politics from policing and make police officers more professional.

8. Coproduction of crime prevention was the centerpiece of the reform era of policing.

9. One aspect of policing that has not changed over time is how frequently the police use batons in force incidents.

10. The terrorist attacks on September 11, 2001, immediately led to the creation of new laws and redirected law enforcement concerns, but the effects of the attacks have proven to be short lived.

WHY STUDY THE HISTORY OF THE POLICE?

There are at least three reasons why understanding the history of the police is useful and important. First, in order to better appreciate how the police in America have changed over time, it is necessary to understand what policing looked like when it first began. Some aspects of policing have changed a lot, some have changed little. Most crucially, some of the reasons why the police are controversial today are rooted in why the police were created in the first place. Overall, knowledge of police history will assist in developing a more complete understanding of the police today.

Second, police history identifies persistent policing problems and the "solutions" that were applied to those problems but did not work. Consequently, if we are aware of these ineffective measures, we can avoid duplicating them when trying to address age-old problems today. In addition, an understanding of history can help us recognize and more fully comprehend the problems that seem immune to solution, such as police misconduct.

Finally, it is useful to study police history because it may provide insight into the future. Some people say that because history is cyclical we can actually predict the future based on knowledge of the past. Although the more specific the prediction, the more likely it is to be wrong, history can be used to identify general trends and patterns that may extend into the future. This can make it possible to predict the future based on the past.

POLICING COLONIAL AMERICA

The first explorers crossed the Atlantic Ocean in the late 1400s, and the first Europeans settled permanently in America in the late 1500s and early 1600s. The most prominent settlers were the British, who created the thirteen colonies in what eventually became the United States of America. Many of the colonists had fled their homeland because they wanted religious freedom; thus, freedom became a central feature of the new government when it was created.

The colonists had a difficult time in the new land. The economy of the colonies was based almost entirely on the land and farming. Without the benefit of any modern equipment, the work was hard. Starvation and diseases were rampant. Medical care was primitive. As laborers were needed to work the land, indentured servants were first used. Many of these people were poor teenagers from England who received a free boat ride to the new colonies in exchange for years of labor.

In colonial America during the 1600s and 1700s, there were four primary policing entities: constables, watches, slave patrols, and sheriffs.

CONSTABLES AND THE WATCH

constable: The first appointed law enforcement officers in colonial America. They often organized and supervised the watch.

In the early days of colonial America, there was little need for law enforcement. The colonists were God-fearing, hard-working people who took responsibility for their own actions and the actions of their neighbors. As settlements turned into towns, **constables** were the first appointed law enforcement officers. The duties of the constable varied depending on the size of the community, but generally the post was responsible for dealing with everything from stray cattle and dogs to misbehaving children. In some towns constables even enforced church attendance.[2]

In larger villages, constables were responsible for organizing and supervising the **watch**. The watch consisted of men who would watch the town, especially at night. Looking out for fires was a major responsibility of the watch since fires had the potential to destroy entire villages. The watch was also responsible for being on the lookout for suspicious persons. In the 1630s, Boston formed a watch that consisted of one constable and six watchmen.[3] In other towns, the watch assignment rotated among the men in the village. At first, men who worked on the watch were unpaid volunteers; later they were paid, but the pay was minimal. Members of the watch patrolled on foot. When necessary, members of the watch could summon the other men in the village with what was known as the *hue and cry*. Whistles and wooden clappers or rattles were also used to alert townspeople of danger and to summon assistance.

A Question to Consider 2.1

Reflections of the Watch in Policing Today

Do you see any similarities or parallels between the watch in colonial America and policing today? Explain your answer.

SLAVE PATROLS

As the economy of the colonies continued to grow, so did the need for laborers. This led to the advent of slavery. Africans began to be transported to the colonies in the 1600s, and by 1860, approximately 450,000 had been relocated there.[4] With births far outnumbering deaths, by 1860 there were four million slaves in the country.[5] **Slave patrols** were established shortly after the mass importation of slaves began and were in place in several colonies by the mid-1700s.[6] The law typically required white landowners (slave owners) to serve on slave patrols. Because it was not a desirable duty, by the 1800s members of slave patrols usually included people who did not own slaves or land, and they were paid. The pay was about the same as that given to members of the town watch, which was about one dollar per night. In addition, when runaway slaves were captured, the slave patrol members shared the reward. The patrols typically consisted of seven men who were assigned to an area of about ten to twelve square miles.[7] In 1837 Charleston, South Carolina, had a slave patrol that consisted of more than a hundred officers.[8]

> **watch:** Men in larger villages in colonial America who were tasked with guarding the town, especially at night.
>
> **slave patrols:** Patrols tasked with looking for runaway slaves, policing the whereabouts of slaves, and making sure slaves were not in possession of weapons or property they were not allowed to have.

The purpose of the slave patrols was multifaceted. They patrolled the roads and stopped slaves to make sure the slaves had passes to be away from their plantations. They also were on the lookout for slaves who gathered for illegal worship. Members of slave patrols also had the authority to enter plantations and search the living quarters of slaves for stolen property, runaway slaves, and weapons. They also looked for books, paper, and pens, as it was illegal for slaves to learn how to read or write.[9] In some villages, the slave patrols worked alongside the watches, and some patrol members went on to serve as members of the watches. During the Civil War, the slave patrols became more active, and slaves were even more closely monitored. For example, in Atlanta, Georgia, slave patrols were authorized to arrest any blacks who were on the street after 9:00 p.m. They also prevented blacks from gathering unless members of a slave patrol or the police were present.[10]

Photo 2.1

Slave patrols provided an important form of policing in the pre-Civil War era. These patrols represented the first example of racial conflict between police and blacks in America.

@North Wind Picture Archives/Alamy Stock Photo

THE SHERIFF

The **sheriff** was another important policing figure in early America. The idea of a sheriff was borrowed from the old English system. In England, a *shire* was the American equivalent of a county; a *reeve* was an officer who functioned as a constable. A sheriff was the American version of a reeve. Normally, a sheriff was appointed by the governor and worked in a less populated area than a watch did. The primary responsibilities of the sheriff were to apprehend criminals, assist the justice of the peace, collect taxes, and supervise elections. As settlers moved west into the territory of the American Indians, the sheriff continued to have an important role. U.S. marshals employed deputies who also served as sheriffs, deputy sheriffs, or constables.[11] In some places, the sheriff could summon a posse, which was a band of armed male citizens, to assist in apprehending criminals and dealing with other violent threats. The Texas Rangers were formed as a militia to defend against Indians.

A Question to Consider 2.2

The Historical Roots of Police–Minority Conflict

Do you think the early history of the police can help explain why there are often tensions and conflict between some people of color and police in the twenty-first century? Explain why or why not.

THE FIRST AMERICAN POLICE DEPARTMENTS: THE POLITICAL ERA OF POLICING

In the early and mid-1800s, three developments converged that led to the creation of the first formal police departments in America: the **Industrial Revolution**, the rise of major cities, and the abolishment of slavery. These developments are important to consider because they provided the perfect mix of ingredients for the creation of police departments.

THE INDUSTRIAL REVOLUTION AND THE CREATION OF CITIES

With the creation of various technologies, such as electricity, the steam engine, steel, industrial equipment, and the assembly line, the focus of the economy began to move from the land more toward the production of goods. Factories were built. America was experiencing massive immigration, and these newcomers wanted jobs. Many of the jobs were in the new factories, and people tended to settle in close proximity to where they worked. As a result, new cities formed and already existing ones got much larger. For example, in 1820 Boston had a population of approximately 40,000. By 1870 it had a population of about 250,000.

Cities created a slew of new job opportunities, but they also created problems, particularly with regard to ethnic conflict, housing, sanitation, and health and medical care. Extraordinary wealth was created during this period—at least for some. Others, especially those who were unable to work, lived in poverty. Crime became a major concern, specifically among the wealthy. There were riots in many American cities, most of which were related to poor living standards, poverty, and ethnic conflict. The watch was simply no longer capable of providing the security that citizens demanded.

THE ABOLISHMENT OF SLAVERY

Slavery was officially abolished at the end of the Civil War with the ratification of the Thirteenth Amendment in 1865. The former slave owners and other pro-slavery whites now had a problem uniquely their own: a "free" black population. According to authors Jerome Skolnick and James Fyfe, "The post–Civil War South faced the enormous problem of absorbing a population of former slaves

sheriff: A police figure who typically worked in a less populated area. In early American policing, the primary responsibilities of the sheriff were to apprehend criminals, assist the justice of the peace, collect taxes, and supervise elections.

Industrial Revolution: A period during the eighteenth and nineteenth centuries marked by new manufacturing processes and a transition from rural means of production to urban ones.

while maintaining the dominance of the white caste."[12] The emergence of the Ku Klux Klan was part of the solution to this problem for the pro-slavery Southerners. The Klan's mission was to strike terror into the freed slaves and their sympathizers in order to keep them in a powerless position. Lynching was a common tool of the Klan: From 1882 to 1959, it is estimated nearly 5,000 lynchings occurred in the United States.[13] The activities of the Klan went on largely without interference from officials.

Along with the use of terror as a tool, another tactic of the pro-slavery faction was the creation of so-called **black codes**, which articulated black citizens' "rights and responsibilities." For example, blacks were prohibited from renting land in cities, and vagrancy was punishable by forced plantation labor. Other rules prohibited "insulting language," "malicious mischief," and preaching the Gospel without a license. South Carolina required that blacks be farmers or servants unless they paid a special (and unaffordable) tax.[14]

Photo 2.2

The Industrial Revolution led to the creation of cities, which in turn led to rising concern about crime and the formation of formal police departments. @iStockphoto .com/ilbusca

The black codes were made illegal as a result of the Civil Rights Act of 1866 and the Fourteenth and Fifteenth Amendments to the Constitution. In place of the black codes came **Jim Crow laws**, which mandated racial segregation in public facilities. Interestingly, Jim Crow laws actually first appeared in the North before being widely adopted in the South. These laws existed until the 1960s. The black codes and Jim Crow laws are particularly relevant when considering the history of the police because although these laws were not created by the police, the police were expected to enforce them. It is also important to remember that at this time, the police were exclusively white. It was not until the late 1800s that any blacks were appointed as officers.

THE LONDON METROPOLITAN POLICE DEPARTMENT AS A ROLE MODEL

The events that took place in the early to mid-1800s in the United States were not limited to that country. In fact, the United States trailed England in the unfolding of the Industrial Revolution. In 1829, the London Metropolitan Police Department (LMPD) was created. At this time, London had a population of approximately 1.5 million people. Londoners had made do without a formal police force as long as they could. Soon after its creation, the LMPD had 1,000 officers. It served as the model for police departments subsequently created in the United States.

The problem for the English was that although the need for a more effective means of policing was obvious, how to go about providing it was not. It was decided that the police department would exist in order to prevent crime, and this was to be accomplished through patrol. The reasoning was that by having officers patrol on foot throughout the city, their presence would deter would-be criminals from committing crimes. It was also decided that the structure of the LMPD should resemble that of the military. The person who is most often given credit for the creation of the LMPD is Sir Robert Peel. It is for this reason that British police officers are today still often referred to as "bobbies."

The founders of American police departments subscribed to the theory the English had already established: Crime prevention through patrol and using a military structure seemed reasonable. The creators of American police departments also wanted policing to be a local responsibility. With local control of police departments, there was little role for the federal government in law enforcement. Indeed, when the first police departments were created in the

> **black codes:** Codes designed to limit the rights of freed slaves in the post–Civil War South.
>
> **Jim Crow laws:** Laws that mandated racial segregation in public facilities.

mid-1800s, there were few federal laws and thus no need for federal law enforcement agencies. Early police departments operated at the local level and were controlled by citizens who lived in the towns and cities. Another important facet of the initial operations of police departments was that police power was limited by law. The Constitution made this clear.

THE CREATION OF THE FIRST AMERICAN POLICE DEPARTMENTS

As noted, the Industrial Revolution, the creation of large cities, and the abolishment of slavery set the stage for the creation of the formal police departments in America. The first police departments were created by combining the night watch with the day watch. In the South, the former slave patrols became the core of the new police departments. Although the dates are difficult to precisely pinpoint, Boston created its police department around 1838, New York City around 1845, Chicago around 1851, and New Orleans around 1857.[15]

The mid-1800s to the early 1900s is known as the political era of American policing.[16] As its label suggests, policing at the time was all about politics. Politicians, especially the mayor, controlled everything related to policing, including who got hired, who got fired, and what policemen did while they were on the job. There was little or no training. There were virtually no selection standards except for political party affiliation and connections. As politicians moved in and out of office, so did policemen and police chiefs. For example, in Hartford, Connecticut, the process went like this: Democrats and Republicans each created lists of their fellow party members that were deemed suitable as policemen. Democrats then crossed names off of the Republicans' list and vice versa. From there, the policemen were selected. The chief of the department was usually directly appointed by the mayor and was of the same political party as the mayor. Police officers were generally also of the same political affiliation as the mayor.[17]

DIVERSITY IN THE POLITICAL ERA OF POLICING

Black policemen were first hired in Selma, Alabama, in 1867 and in Houston, Texas, in 1870. In New Orleans, there were 177 black officers by 1870. Chicago hired its first black policeman in 1872. Interestingly, the black men who were appointed police officers were often better educated and qualified than their white counterparts. Nevertheless, the appointment of blacks as officers was controversial. Some cities experienced riots because of black officers taking action against white citizens.[18] Consequently, black officers were more likely not to wear uniforms and to be assigned to black neighborhoods. Some cities did not allow black officers to arrest white citizens. In Miami, blacks were called *patrolmen* while white officers were designated as *policemen*. In the late 1800s, the Civil Rights Act of 1875 was ruled unconstitutional, and later the Supreme Court upheld "separate but equal" laws. As a result, black officers lost their jobs in droves. This was most evident in New Orleans, where the number of black officers dropped from 177 in 1870, to 27 in 1880, to 5 in 1890, and to none in 1910. It was not until 1950 that another black officer was hired in the largely African American city.[19]

There were no women police officers until late in the 1800s, and even then, there were very few. Police officers were simply referred to as policemen. The first women employed in police departments were called police matrons and did not have powers of arrest. Their duties were generally limited to handling female prisoners. Although the historical record is incomplete and inconsistent with regard to the employment of women in police departments, it has been offered that the first fully sworn female police officer was hired in 1891 in the Chicago Police Department; her job was to enforce child labor laws. The first female African-American officer was appointed in 1916 in the Los Angeles (CA) Police Department.[20] Most large police departments did not employ any women until the 1920s, and it was not until the 1970s that women had the same authority as male officers. It is accurate to say that in large part the political era involved white male leaders appointing white male officers to police white and black citizens.

political era: The period from the mid-1800s to the early 1900s during which policing was heavily influenced by politics.

police matrons: Female police department employees whose duties usually involved only female prisoners.

THE ROLE OF THE POLICE DURING THE POLITICAL ERA

The capabilities of the police at the time were minimal, although officers were on duty twenty-four hours a day and police departments were better staffed than were the watches. In many cities, the first policemen wore designated hats and carried wooden clubs, but they did not wear uniforms. Not until the late 1800s did police officers begin to routinely carry firearms. In large cities, officers were assigned to extensive beats and they patrolled on foot. Some areas of cities were not patrolled at all. There was no system of communication between citizens and the police or between the police officers themselves. There were no means by which police supervisors could supervise their officers; as a result, there were very few supervisors. Needless to say, seldom were citizens able to find the police when they were needed. This situation created an environment where politicians could easy influence and control the officers' activities and the police department in general. Cities were typically divided into wards; each ward had an elected ward leader, and the police in each ward were accountable to that leader. Many police activities were political, including campaigning and, at times, assisting in rigging elections.[21] Indeed, corruption among politicians and their police forces was rampant. Payoffs and bribes were an unquestioned aspect of policing at the time. Officers accepted bribes not to enforce laws; officers paid bribes to get promoted. Chiefs and political leaders were in on the action as well, demanding a portion of the bribes accepted by officers.

Street-corner call boxes were put in place in the late 1800s. A call box was a metal box on a pole with what amounted to be a rudimentary telephone inside the box. Call boxes provided a means by which officers could communicate with supervisors at the police station, and they allowed supervisors to monitor the location of officers on the beat. However, call boxes were often vandalized by officers. They represented a first attempt to control and improve the police, if a rather unsuccessful one.

Beyond serving the interests of politicians, the police were primarily engaged in providing services to citizens: They ran soup lines, provided lodging to immigrant workers, and assisted in finding work for immigrants, all upon the direction of political leaders.[22] Something that the police did not do very often was make arrests. More than half of all arrests made at this time were for public drunkenness.[23] This was an offense that beat cops could easily discover with no investigation necessary. The police simply did not have the capability to respond to and investigate crimes. When an arrest was made, it was usually as a last resort. Making an arrest in the late 1800s presented some serious logistical difficulties; officers would literally have to "run 'em in" to the police station or, when arresting a drunk, use a wheelbarrow and wheel him into the station.[24] So-called **curbside justice** with a wooden baton often became an alternate means of dealing with drunken citizens and other law breakers.

A Question of Ethics

Changes in Ethical Standards

What historical aspects of policing could be criticized today as being ethically wrong? Why do you think ethical standards of conduct change? Do you think there is anything about policing today that is viewed as ethical (or unethical) that may be viewed differently in the future? Explain.

CRIMINAL INVESTIGATIONS DURING THE POLITICAL ERA The need to improve methods of criminal apprehension was not lost on the police of the political era. Police officers known as detectives began to appear in the late 1800s, largely in response to public concern about the increasing amount of crime. As an illustration of this increase, for most of the early to mid-1800s, there were no homicides recorded in Suffolk County (Boston), Massachusetts. Between 1860 and 1869, however, 70 homicides occurred. During the 1870s, 107 homicides

curbside justice: The use of force by police rather than arrest to deal with law breakers.

The Police Baton

Invented in the 1800s and first used in English police departments, the police baton was one of the few tools available to officers to assist them in controlling and arresting criminals. In its original form, the baton, also known as a billy club, sap, blackjack, or truncheon, was approximately twelve inches long, made of wood, and heavy. It was meant to be used as a striking instrument. There were few if any limits on how the club was to be used; most effective was when a subject was hit in the head and knocked unconscious. It was a relatively inexpensive weapon and easy to use. Over time, the baton was made longer and incorporated other features. As batons become longer, they became more difficult and cumbersome for police officers to carry. In the 1960s, the use of batons became synonymous with police brutality, and officers were frequently shown using batons on rioters.

Side handle batons, commonly known as PR-24s, were introduced in the 1970s and are still used in some police departments today. With the addition of a side handle, the baton became a more versatile tool that could also be used like a shield to protect from an attack. Collapsible batons, also called expandable or telescopic batons, were introduced to American police departments in the 1980s. In most police departments today, this type of baton is standard issue and is carried on the officer's duty belt. For most models, when collapsed the baton is less than ten inches long and can expand up to thirty-one inches. It is extended with a forceful quick swing. These batons are made of metal and are lightweight.

While batons are still standard issue in U.S. police departments, they are very seldom used by officers in force incidents as their use often results in serious injuries to subjects.[25]

Photo 2.3

The wooden baton, sometimes knows as the billy club, was one of the first tools used by the police. Over time, the police baton became more technologically advanced. Today, most officers carry an expandable baton but seldom use it against subjects.

Shawn Patrick Ouellette/Portland Press Herald via Getty Images

were reported.[26] The most important quality for detectives to possess was a familiarity with criminals and their tactics; many detectives were selected from the ranks of prison guards, and some were even reformed criminals.[27] Since they held this specialized knowledge, detectives received more pay than beat cops. Detectives also received extra compensation through witness fees or compensation for providing testimony in court. Detective work was often a clandestine activity, and detectives were sometimes considered to be members of a secret service.[28]

It was also around this time that criminal identification systems began to be developed and used in police departments. The first of these systems involved photography. By 1858, the New York City Police Department had on file a collection of photographs of known criminals called a **rogues gallery**.[29] However, photographs were extremely limited in their usefulness because the appearance of criminals could be altered either deliberately or simply by the aging process. Of course, for photographs to be useful, authorities first needed to know who committed a crime and then have a photograph of that person.

The **Bertillonage** system was considered a major improvement over the use of photographs. The system consisted of eleven measurements (e.g., length and width of the head, length of the left foot, length of the left middle and little fingers) that could be used to differentiate one person from another.[30] However, by the early 1900s, the deficiencies of the system were obvious. Besides being cumbersome and error prone, it had essentially no capabilities in identifying unknown offenders who committed crimes.

rogues gallery: A collection of photographs of known criminals.

Bertillonage: A system wherein physical measurements were used to differentiate suspects.

In addition to these identification methods, detectives during this period also used various "investigative" tactics to deal with crime and criminals. One common strategy was the **dragnet**, which involved the police "rounding up the usual suspects." The dragnet was often paired with the **third degree**—the brutal interrogation of suspects.[31] The third degree included beatings with a rubber hose,[32] placing a suspect in a sweat box for hours or days under constant questioning,[33] drilling teeth, burning flesh with lit cigars or cigarettes, and beating with blackjacks or batons.[34] Many accounts suggest that the use of the third degree to obtain confessions was commonplace into the 1930s and possibly even later.[35] However, in 1936 the Supreme Court ruled in *Brown v. Mississippi* that prolonged beatings used to extract confessions were no longer a legally acceptable police practice.

EARLY 1900s TO 1960s: THE REFORM ERA OF POLICING

Another swell of change began to sweep through American society in the early twentieth century. By 1920, automobiles were being widely used, as were radios, telephones, and other technologies. Along with advancing living standards for many, the new technology also placed increased demands on the police. Due to the use of automobiles in particular, criminals could commit crimes in one jurisdiction and easily flee to another, causing great difficulties for the police. In addition, automobiles created a need for traffic enforcement, a responsibility assigned to the police. Another element of technology that significantly affected the work demands of the police was the telephone, which turned police departments into twenty-four-hour agencies that were just a call away.

The police were also confronted with new demands unrelated to technology. Concerns about crime became a major issue. With the 1920s came a rise in serious crime—in perception if not in fact. Kidnapping, gangsters, and bombings attributed to communists were front-page news. Prohibition and the Great Depression also placed significant new demands on the police. In the face of these developments, the police were once again in the midst of a crisis.

REFORM AS ANTI-POLITICS

The new demands and technology of the early 1900s led to the **reform era** of policing.[36] Forward-thinking police leaders, such as August Vollmer and O. W. Wilson, advocated a new philosophy and methods of policing (see Exhibit 2.1). The new philosophy focused on the idea of the police as experts, police professionalism, and getting the police out from under the control of politicians. Technology was an important element of the reform era. Automobiles allowed the police to institute preventive patrol as a means of deterring criminals and to respond quickly to crime scenes in order to make more arrests. The two-way radio allowed police supervisors to be in constant communication with officers and to have supervision over them. It also allowed patrol officers to be directed to places where they were needed. With the telephone, citizens could easily summon the police when needed.

THE CREATION OF FEDERAL AND STATE LAW ENFORCEMENT AGENCIES

In the face of corrupt and ineffective municipal police agencies, state enforcement agencies were created to assist local police departments with the new demands they faced. In 1905, Pennsylvania created the first state police agency. It was designed to provide a police presence throughout the state, to assist the local police, and to provide police services in less populated rural areas.[37] In 1935, the Texas Legislature created the Texas Department of Public Safety, which consisted of the Texas Rangers and the Texas Highway Patrol. The Bureau of

dragnet: A historical reference to the process wherein when a crime occurred, the police would bring in for questioning all the suspects usually associated with that type of crime.

third degree: A historical reference to the physically brutal interrogation of suspects by police.

reform era: An era in policing that centered on removing the police from the control of politicians and making departments more professional and efficient.

EXHIBIT 2.1

August Vollmer and O. W. Wilson

August Vollmer and O. W. Wilson were two prominent police leaders who ushered in the reform era of policing.

August Vollmer was appointed police chief of the Berkeley, California, police department in 1907 (for two years prior he was the town marshal). He transformed the Berkeley department into a premiere, professional agency that was the role model for others worldwide. He hired police officers with college educations and recruited female and African American officers. He developed the country's first university criminology program at the University of California, Berkeley. Vollmer instituted automobile, motorcycle, and bicycle patrols in his department; was responsible for putting two-way radios in patrol cars; and developed the first crime laboratory in a police department. He saw policing as a profession rather than just a job, and he was extremely concerned with how police chiefs could be easily removed from office on the whims of politicians.

In 1923 Vollmer was appointed police chief of the Los Angeles Police Department (LAPD). He had limited success in reforming the LAPD given the widespread corruption that existed in the city at the time. In 1924, he returned to Berkeley. In 1931, he was the primary author of the *Wickersham Report*; the Wickersham Commission, which generated the report, studied Prohibition enforcement and related police practices and corruption. Vollmer retired in 1932 and died in 1955. He is known as the "father of American law enforcement."

O. W. Wilson had been a student of Vollmer's at Berkeley and was a police officer in Berkeley when Vollmer was chief. Wilson later became chief of the departments in Fullerton, California, and Wichita, Kansas, and during these tenures, he took many of Vollmer's ideas and extended them. He was just twenty-five years old when he was appointed chief in Fullerton in 1925. He then served as a professor of police administration and as dean of the School of Criminology at the University of California, Davis, from 1950 to 1960. He wrote the book *Police Administration*, which was widely viewed at the time as the "bible" of police administration. Wilson recognized and advocated the value of motorized patrol and rapid police response in effective policing. In 1960, he was appointed chief of the Chicago Police Department and was given wide latitude to reform and improve it. He retired from the department in 1967 and died in 1972.[38]

Photo 2.4
August Vollmer was one of the first leaders of the reform era of policing and the "police professionalism" movement. He is now referred to as the "father of American law enforcement." @Bettmann/Getty Images

Photo 2.5
A former student of Vollmer's, O. W. Wilson became chief of the Chicago Police Department in the 1960s. Wilson advocated Vollmer's beliefs and sought to institute reforms, such as motorized patrol and rapid response to calls for service. @Bettmann/Getty Images

Investigation, later known as the Federal Bureau of Investigation (FBI), was created in 1909 and quickly became a powerful law enforcement agency. The FBI led the war against communists, gangsters, and kidnappers. The FBI developed a crime laboratory, collected crime information, greatly developed the use of fingerprinting as a method of identification, and created the FBI National Police Academy to provide advanced training to police leaders across the country.

DETECTIVES AS THE ULTIMATE PROFESSIONALS

During the reform era, detectives became an important tool in the efforts of police departments to enhance their professionalism and deal with crime. Detectives were the ultimate professionals. They were well paid and trained. The entertainment media at the time portrayed detectives as efficient and effective crime solvers. As a continuing attempt to provide organizational control over officers and detectives, detective work became much more removed from interactions with criminals. Due to scientific advances made during the period, more emphasis was placed on getting information using science (and from victims and witnesses) as opposed to from criminals. In 1910, fingerprints were used for the first time as evidence in a criminal trial.[39]

A Question to Consider 2.3

The Underrepresentation of People of Color in Policing

Throughout history and even to a large extent today, people of color have been underrepresented as police officers, especially in larger cities. Why? Although this issue is discussed in detail in Chapter 5, it is worthwhile to consider the question now.

THE REFORM ERA AND (LACK OF) DIVERSITY IN POLICE DEPARTMENTS

Many police departments increased the representation of minority officers during the reform era, although the proportion of minority officers was still small and seldom approached the representation of minority citizens in the city (see Figure 2.1). However, minority officers far outnumbered female officers at the time. As in earlier decades, female officers were still most often referred to as police matrons, and their duties related primarily to women and children offenders.

THEN THE 1960S HAPPENED

Throughout the 1950s, things were going smoothly for the police. By most accounts, crime was under control. The FBI reported that over 90% of homicides were solved by the police. Then things changed. Between 1960 and 1970, the crime rate doubled. It was the time of the civil rights movement and the related demonstrations, marches, and riots. The police found themselves on the front lines of the riots and demonstrations; often it was white officers facing off against African American citizens. It did not look good on television, in the newspapers, or in person. The predominantly white police forces became viewed by many as an "occupying army" in the low-income, minority ghettos of urban cities, and suddenly the police were viewed as racists and as "pigs." American society was in turmoil. In 1963, President John F. Kennedy was assassinated. Later in the decade, senator and presidential candidate Robert Kennedy and civil rights leader Dr. Martin Luther King Jr. were

FIGURE 2.1

Representation of Minority Police Officers and Population in Select Cities, 1960s

These are estimates regarding the representation of minority officers in select police departments and the representation of minorities in those same cities late in the reform era. Minority officers were vastly underrepresented in police departments compared to their percentage of the general population.[40]

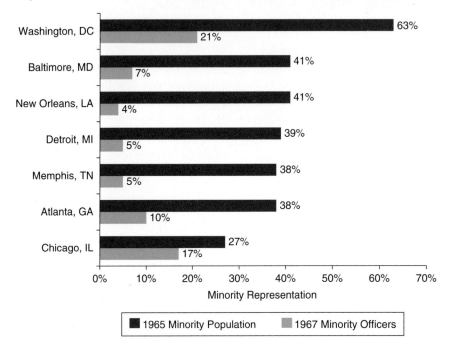

Source: Data from Advisory Commission on Civil Disorders. 1968. *Report of the Advisory Commission on Civil Disorders.* Washington, DC: National Criminal Justice Reference Service.

murdered. Helter Skelter and Charles Manson were making headlines, and fear of crime increased dramatically. America was in the grip of the Vietnam War and the attendant protests against it. Drug use, the hippie movement, and women's liberation were prominent counterculture movements. The Beatles sang "Revolution."

The police were experiencing a crisis, yet they were supposed to have the knowledge and capabilities to control crime successfully. If the situation was not bad enough for the police, the U.S. Supreme Court rendered several landmark decisions during this decade, such as *Mapp v. Ohio* and *Miranda v. Arizona*, that were seen as "handcuffing" the police. In 1967, the President's Commission on Law Enforcement and the Administration of Justice issued a report that represented the first systematic analysis of crime and how the criminal justice system could be made more effective. Especially significant was its call for the incorporation of technology, data collection and analysis, and federal resources in crime fighting.

In the late 1960s and early 1970s, several major research studies were conducted to examine the effectiveness of police operations. It was found that motorized patrols did not prevent crime,[41] detectives contributed little to solving crimes,[42] and fast police responses to crime scenes seldom led to the police making on-scene apprehensions. Given the conditions of society, many people were not surprised by these conclusions. By the end of the 1960s, it was clear that the current style of policing was not working well. The police were once again in the midst of a crisis as they struggled to deal with the demands of the new society.

Throughout this book, in the "Good Policing" feature, certain police practices and strategies are highlighted as examples of good police work. It is important to understand, however, that what was thought of as good policing years ago may not be good today and what is recognized as "good" today may not be thought of in the same way in the future. Notions of "good policing" are always changing and can be influenced by many factors. In particular, human values and morality can influence what is thought to be good policing. As an example, slave patrols used to be considered good policing, as was the practice of torture (the "third degree") to obtain confessions. As beliefs about human rights and values changed, so too did ideas about the best ways to conduct police work. Research findings can also influence notions of good policing. For instance, reliance on preventive patrol and fast police response time used to be considered the ultimate in good policing strategy. Subsequent research has strongly questioned this belief. Good policing is also dependent on technology. Use of police body-worn cameras is considered a good practice today, but because they were not yet invented, their use years ago was not even possible. Good policing may also be situational dependent; what is good policing in one community may not be considered as such in another. Good policing depends on the views and needs of citizens. As you read the "Good Policing" features in this book, remember that what is "good" is time and situation specific.

THE 1970s TO THE PRESENT: THE COMMUNITY PROBLEM-SOLVING ERA OF POLICING

In the face of these concerns, the police realized that the old ideas of professionalism no longer worked. The police needed to get closer to the community to enlist the support and assistance of its members in fighting crime. With this realization, the **community problem-solving era** of policing was born.[43] As we will discuss in Chapter 13, early (and ineffective) attempts at getting closer to the community took the form of police-community relations bureaus and team policing.[44]

COMMUNITY AND PROBLEM-ORIENTED POLICING

Community policing and problem-oriented policing have a prominent place in the community problem-solving era of policing. Community policing represents many different things to many different people, but the core idea is that the police institute policies and practices that involve citizens in policing. The intent of community policing is to foster **coproduction**. With coproduction, the idea is that police and the community coproduce crime prevention. As such, community policing is about creating cooperative relationships with citizens; having

> **community problem-solving era:**
> An era of policing that emphasizes the assistance and support of the community in fighting crime.
>
> **coproduction:**
> A concept in which the police and the community work together to prevent crime.

FIGURE 2.2

Four Eras of American Policing

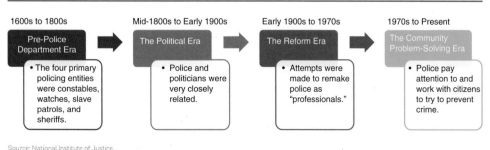

Source: National Institute of Justice.

Photo 2.6

The community problem-solving style of policing focuses on building relationships with citizens and citizens and police working together to prevent crime. @Michael B. Thomas/AFP/Getty Images

Photo 2.7

Police work today requires officers to use many different technologies and tools. @AP Photo/Ross D. Franklin

officers be in direct, day-to-day contact with citizens as much as possible; and having officers be in a position to listen to citizens and address their concerns. In areas where community policing is practiced, community meetings, community surveys, neighborhood watches, and means of patrol other than automobile (e.g., foot, horse, bicycle, and in some places even rollerblades and skateboards) have become popular.

Closely related to community policing is problem-oriented policing. With problem-oriented policing, the police become more concerned with identifying and addressing community crime problems and do so with the input and assistance of citizens. Herman Goldstein introduced the concept of problem-oriented policing when he argued that the police had succumbed to the **means over ends syndrome**, meaning that the police were more concerned with how things were done than with the goals they were supposed to achieve. He argued that the police should become more problem oriented and less incident driven.[45]

The community problem-solving era has been a time in which an extraordinary amount of research on police, crime, and criminal justice issues has been conducted. Prior to the 1970s, the number of major studies on the police could be counted on one hand. With funding from the federal government in the early 1970s to provide scholarships to individuals interested in studying police science and criminal justice, such programs began to appear in colleges, universities, and technical schools across the country. Scholars also began to receive federal funding to study police issues. Knowledge of policing has increased dramatically as a result, although gaps in knowledge still remain.

? A RESEARCH QUESTION
What Is the Current State of Research on Policing?

In spite of the mass of research on policing that has accumulated since the 1970s, fifty years later there is still much that we do not know. And since new issues continuously emerge in policing, the list of issues on which to conduct research is ever-expanding. In 2004, a committee of the National Academy of Sciences examined the current state of research in policing. The group concluded that

There are many important subjects about which there is virtually no scientific research. By any metric—whether lives lost to crime, the costs and benefits of government expenditures on law enforcement, or

the moral obligation imbedded in the use of coercive authority—police research deserves more serious attention than it has received.[46]

Much research has been conducted since the work of this committee. While we know a lot about some police strategies, it is still fair to say that we know too little about many critical police issues.[47] As discussed in this book, police practices and policies are too often based on untested assumptions and conventional wisdom. At a minimum, it is important to recognize when police practices are based on research and when they are not.

EXHIBIT 2.2

Tools of the Police in the 1980s

In the 1980s, police officers had a much more limited repertoire of tools by which to do their job. For instance, revolvers were carried instead of semiautomatic handguns. There were no Tasers or pepper spray, only a firearm, baton, and handcuffs. There were no computers, meaning that all reports were handwritten or typed with a typewriter. Large radios and pagers served as the communication link to dispatch; there were no in-squad computers or cell phones. There was no GPS for assistance in finding addresses. No in-squad computers also meant that all recorded information about calls for service and subjects had to be provided by dispatchers. There was no email; all written department communication was via telephone, letters, or memos.

The Community Problem-Solving Era of Today and Beyond

Although most scholars believe that policing today is still within the community and problem-solving era, much has changed in policing since the 1970s. For instance, many police departments have made strides in the hiring of minority and female officers. In 1970, only approximately 2% of police officers were women; by 2017, that percentage was nearly 13%.[48] In 1970, less than 10% of police officers were racial or ethnic minorities; now that percentage is close to 25%. Police departments have also greatly incorporated technology into their daily operations. This includes computers in patrol cars, DNA banks, automated fingerprint identification systems (AFIS), license plate readers (LPRs), gunshot detection systems, weaponry less likely to be lethal (e.g., Tasers), body armor vests, squad car global positioning systems, cameras in squad cars, and body-worn cameras. With regard to technology in general, policing is quite different than it was just a few decades ago (Exhibit 2.2). Some of this technology is intended to better monitor the actions of officers and increase officer accountability, not unlike how call boxes and the two-way radio were used in the past.

Community policing and the community problem-solving era in general may seem like a "kind and gentle" police orientation. However, even in the era of community policing, crime control is still controversial. The police have not shed their primary responsibilities, nor is the use of force any less significant to the role of the police than it used to be. Even with a velvet glove, there is an iron fist.[49]

Some scholars have suggested the law enforcement changes that occurred and continue to occur as a result of the terrorist attacks of September 11, 2001, represent the beginning of the end of the community problem-solving era of policing. If this is true, it is possible that the new style of policing will represent a trend already firmly in place before 2001: the increased militarization of the police and the blurring of the lines between the police and the military. Chapter 15 explores this possibility in greater detail. The remainder of this book (Chapter 15 as a noted exception) provides a detailed discussion of the current state of policing.

> **means over ends syndrome:** When police are more concerned with how things are done than with the goals they are supposed to achieve.

Main Points

- The study of police history is important for several reasons. It can be useful to understand how much or how little things have changed over time and to be aware of what solutions to problems have been tried unsuccessfully in the past. Additionally, knowledge of history can help in predicting the future of policing.
- Prior to the development of formal police departments, policing was done by constables, the watch, slave patrols, and sheriffs.

- A watch was a group of men who oversaw the security of cities and towns during the night and day and could summon others to assist when there was a disturbance. Slave patrols captured runaway slaves and monitored the conduct of slaves.
- The first police departments in America were created around the mid-1800s, at least in part because of the effects of the Industrial Revolution and the creation and rapid rise of cities. This period has been referred to as the political era of policing, as politicians controlled virtually every aspect of the practice.
- The first police departments realized the importance of criminal apprehension and used the strategies of the dragnet, the third degree, Bertillonage, and photography.
- The police baton was the first tool of the police in the mid-1800s. It was made of wood and frequently used to injure someone or to otherwise induce compliance. Batons are infrequently used today but are still standard-issue equipment. They come in many different styles.
- The reform era, which ran from the early 1900s through the 1960s, emphasized police professionalism and capabilities. This way of thinking was spearheaded by progressive police leaders such as O. W. Wilson and August Vollmer. This era began as the result of an increase in high-profile crime and additional demands on the police. These additional demands were primarily due to increased usage of the automobile.
- The 1960s represented a crisis for the police and led to a new way of thinking. The community problem-solving era of policing began at the end of this decade, and most scholars agree that it is still the current era of policing. Other scholars suggest that the 2001 terrorist attacks signaled the beginning of a new style of policing.
- The community problem-solving era of policing represents the belief that citizens have something to contribute when it comes to crime prevention. Ideally, citizens and police should coproduce crime prevention.
- Much research on policing has been conducted during the community problem-solving era, but there is still much to be learned.
- Much has changed in policing since the beginning of the community problem-solving era; most notably, the diversity of police officers and the technology used in police departments. The remainder of the book provides a discussion of the current state of policing.

Important Terms

Review key terms with eFlashcards at **edge.sagepub.com/brandl2e.**

Questions for Discussion and Review

Take a practice quiz at **edge.sagepub.com/brandl2e.**

1. What is the value of studying the history of the police?
2. Before there were police departments, policing duties were performed by constables, watches, slave patrols, and sheriffs. What was the role of each?
3. What was the political era of policing? What did policing look like in this era? What were the problems with policing during this era?
4. What were the primary crime detection and criminal identification strategies used during the political era?
5. What was the reform era of policing? What did policing look like during this era? What were the problems with policing during this era?
6. How did the police car, two-way radio, and the telephone change policing?

7. What is the community problem-solving era of policing? How is it different from the reform era?

8. Upon what factors does the definition and identification of "good policing" depend?

9. What is the means over ends syndrome and what does it have to do with the community problem-solving era of policing?

10. How are the political era and the community problem-solving era similar? How are they different?

Fact or Fiction Answers

1. Fiction
2. Fact
3. Fiction
4. Fiction
5. Fact

6. Fact
7. Fact
8. Fiction
9. Fiction
10. Fiction

⑤SAGE edge™

Digital Resources

Get the tools you need to sharpen your study skills. SAGE Edge offers a robust online environment featuring an impressive array of free tools and resources.

Access practice quizzes, eFlashcards, video, and multimedia at **edge.sagepub.com/brandl2e.**

Media Library

View these videos and more in the interactive eBook version of this text!

SAGE News Clip

2.1: Ferguson City Council
2.2: New NYPD Chief
2.3: California—Rodney King's Daughter

3

THE CHARACTERISTICS AND STRUCTURE OF POLICE ORGANIZATIONS

Objectives

After reading this chapter you will be able to:

3.1 Describe the characteristics of police organizations

3.2 Discuss the challenges of managing police organizations

3.3 Compare and contrast the structure of larger and smaller police departments

3.4 Identify the major operating units within police departments

3.5 Identify and describe the law enforcement agencies that operate at each level of government as well as those agencies with special jurisdiction

Police Spotlight: Police Salaries and the Possible Consequences

Police officer salaries vary considerably across states. The most recent statistics available from the Bureau of Labor Statistics show western states tend to provide the highest police officer salaries, southern states the lowest. For example, at the extremes are California with an average salary for officers of $96,660 and Mississippi with an average salary of $34,550. Much of the salary differences across states is a result of cost of living differences, tax base differences, as well as police labor union representation. Even within states and within counties, however, there is typically much variation in police officer salaries. For example, within St. Louis County, Missouri, police officers who work in the City of Town and County have an average annual salary of three times that provided to officers in Hillsdale.[1] In Allegheny County, Pennsylvania, there are 109 police departments. Some residents of the county are served by officers whose starting salary is $40 an hour (Castle Shannon), while others are served by part-time officers who earn $10 an hour.[2] In Wayne County, Michigan, Detroit Police Department (DPD) officers have a starting base salary of $39,545 while the City of Warren, just 15 miles away, provides a starting salary of nearly $80,000. The general pattern is that police officers who work in communities with the highest crime rates are also the lowest paid, and these officers are more likely to leave when other opportunities arise. In 2018, Warren hired 12 new officers, six were previously employed by the DPD and the other six were previously employed by other police departments in Michigan.[3]

What do you think might be the consequences of substantial differences in pay among officers in different agencies, if any? Explain.

(Continued)

Fact or Fiction

To assess your knowledge about the characteristics and structure of police departments prior to reading the chapter, identify each of the following statements as fact or fiction. (See page 61 at the end of this chapter for answers.)

1. Police departments can be accurately described as monopolies.

2. There is sometimes conflict between *street cops* and *management cops*. Street cops have very little power over management cops.

3. Much of the task of police management can be accurately thought of as the management of police discretion.

4. One of the biggest differences between small and large police departments is the amount of specialization present in the structure of the organization.

5. Of all sworn officers who work in law enforcement, the largest proportion work at the federal level.

6. While most local police departments in the country are quite small, most police officers work in larger departments.

7. State police agencies focus exclusively on highway and interstate highway traffic enforcement.

8. The Clery Act relates to local police departments and the reporting of crime to the FBI.

(Continued)

(Continued)

9. The Department of Homeland Security was created by President Richard Nixon in the 1970s.

10. There are three agencies at the federal level that have law enforcement responsibilities.

(Continued)

FIGURE 3.1

Annual Mean Wage of Police and Sheriff's Patrol Officers, by State, May 2017

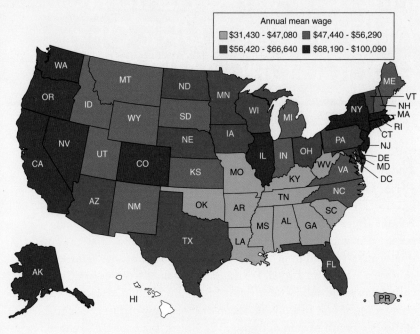

Blank areas indicate data not available.

Source: Bureau of Labor Statistics, "Occupational Employment and Wages, May 2017".

CHAPTER 3
provides an introduction to the nature and management of police departments and describes the characteristics of law enforcement agencies at each level of government.

CHARACTERISTICS OF POLICE ORGANIZATIONS

Police organizations have been described in several ways based on the characteristics they share and the means in which they operate. Specifically, police departments have been identified as being bureaucracies, quasi-military organizations, monopolies, and street-level bureaucracies. Each of these labels provides insight into the nature of police organizations and some of the challenges associated with managing them.

POLICE AGENCIES AS BUREAUCRACIES

Max Weber, a German sociologist, studied the changing nature of European society in the early 1900s and during the Industrial Revolution. Although he did not invent bureaucracies, he did identify them as the best way of structuring, managing, and operating organizations. When

Photo 3.1
Communicating through the departmental chain of command is expected and is an important rule for police to follow.
©iStockphoto.com/PGGutenbergUKLtd

people hear about bureaucracies today, they often think of red tape and the runaround; however, Weber saw bureaucracies as the solution to many managerial problems. He outlined a vision of how bureaucracies should operate in the ideal world—which is not necessarily how they *actually* operate. Police departments are still managed and structured with the five principles outlined by Weber, explored below.[4]

A DIVISION OF LABOR This principle holds that no one person or office is responsible for all of the work of an organization. Instead, labor is divided among personnel and offices so that the jobs employees are responsible for are specialized. **Specialization** is designed to allow for the development of expertise and efficiency. Many police departments divide the work of the agency among many units or divisions, such as patrol, criminal investigation, and administration, among others. Patrol may be divided among several districts and shifts. The task of criminal investigation may be divided into the investigation of homicides, property crimes, sensitive crimes, and so forth.

The amount of specialization in the division of labor of a department is largely a function of the size of the organization, as is the number of levels in the organizational hierarchy. A primary disadvantage of specialization is that with more units and divisions, effective control and coordination of the units becomes more challenging. With more divisions there is also the increased likelihood of conflict among them.

A HIERARCHY OF AUTHORITY Also known as the **chain of command**, the hierarchy of authority principle holds that every person in the organization has a supervisor, and supervisors have more authority than their subordinates. This hierarchy has direct implications for supervision and formal communication. With a chain of command, communication follows a known path from subordinate to supervisor. If a subordinate "goes over the head" of his or her supervisor (i.e., skips a level of authority in communicating with superiors), then the proper chain of command was not followed. There is no optimal number of levels of authority in an organizational hierarchy. Generally speaking, larger police departments have more levels of hierarchy than smaller ones because with more personnel comes a need for more supervisors and managers.

Within hierarchy of authority, there are some additional facets of the concept. First is *span of control*. This essentially refers to the number of people a supervisor is responsible for overseeing. *Unity of command* refers to the notion that each subordinate should only have one supervisor, thus eliminating the possibility of conflicting orders or instructions. *Delegation of authority*

specialization: A focus on certain responsibilities in order to develop expertise and efficiency in those tasks.

chain of command: This principle holds that every person in the organization has a supervisor, and supervisors have more authority than their subordinates.

means that no one person in the organization is in control of everything. With delegation of authority, the chain of command is necessary.

RULES Written rules and policies are critical to the functioning of a bureaucracy, and police departments are definitely not an exception to this. Rules outline how jobs are to be performed and what responsibilities are associated with each job. Rules also specify standards of conduct and other work processes of the organization. Police departments have written policies and procedures regarding most aspects of their operation, addressing everything from proper uniform requirements to the use of deadly force. As with the other aspects of bureaucracies, larger police departments tend to have more rules and policies than smaller ones.

IMPERSONAL RELATIONSHIPS The impersonality of relationships principle indicates that decisions must be made objectively and without emotion. In a bureaucracy, informal and formal relationships among and between employees and supervisors is not a basis upon which to make organizational decisions. For example, the fact that Officer Gunz and Officer Pistal have been patrol partners for years and are good friends should not be a major consideration if one of them needs to be reassigned to a different precinct to fulfill some organizational need. The objective of decisions is to accomplish the goals of the organization.

SELECTION AND PROMOTION BASED ON COMPETENCE When selection and promotion decisions are based on objective criteria, it helps ensure that qualified individuals will carry out the work of the organization. In the early days of police departments in the United States, this requirement was seldom followed; as a result, police officers (and police departments) did not perform well. This principle essentially requires that political considerations and relationships not be taken into account when making personnel decisions. Today, civil service requirements in police departments mandate that standard tests be used as a basis for selection and promotion decisions.

One potential drawback with civil service mandates is that they may constrain the ability of police department leaders to diversify their departments. For example, in selecting applicants for officers, only the candidates with the very highest scores may be considered for hiring. Race and gender are not part of a test score but may be valuable characteristics in selecting officers. Of course, the alternative argument would be that only those individuals most qualified for the job, as measured by a test or series of tests, should be hired, regardless of race, sex, or any other protected characteristic. Details about the selection of police officers are discussed in Chapter 5.

THE DRAWBACKS OF BUREAUCRACY Most police departments today incorporate these five principles into their operations. As noted, larger departments typically do so to a greater degree than smaller ones. While there are arguments in favor of bureaucratic management and structure, there are also unintended negative consequences. These include lack of flexibility, resistance to change, inefficiency, and insensitivity to the needs of workers. In recent decades, some police departments have slowly begun moving away from the principles of scientific bureaucratic management and closer to more open and contemporary styles of management and organizational structure.

POLICE AGENCIES AS QUASI-MILITARY ORGANIZATIONS

When the first police departments were created in the early and mid-1800s, they were fashioned after the military. Indeed, there is still a clear reflection of the military in contemporary police departments, as evidenced by command and control orientation (i.e., supervisors giving orders, subordinates carrying out those orders); an emphasis on discipline; the top-down chain of command; the rank structure; the war-like mission of the organization; an emphasis on training; uniforms; weapons; and even the tactics that are used (patrol). Today, police

bureaucracy: An organization characterized by many departments and divisions operating through a complicated structure of rules and regulations.

Photo 3.2
Police departments are designed and function as quasi-military organizations. Sometimes the reflection of the military in police operations is very clear.
Mark Boster/Los Angeles Times via Getty Images

departments are often described as being *quasi-military*. As we will explore in Chapter 15, some people argue that since the terrorist attacks of September 11, 2001, the police have become more military-like (paramilitary) and the military has become more police-like.[5]

POLICE AGENCIES AS MONOPOLIES

A monopoly is a company or agency that does not have competitors, as there are no other providers of the same services or product in a particular jurisdiction or area. With a monopoly, customers have only two options: to use the services of that agency or company or do without the product. A monopoly does not compete for clients or customers, and it has the ability to raise the cost of services without much concern about demand for those services. In the private sector, regulators are not fond of monopolies and, as a result, there are not many of them. Consider the opposite of a monopolistic industry: restaurants. In most places, there are a plethora of restaurants that customers can select to patronize. If a restaurant provides poor service, has unreasonable prices, and/or serves food that is not well liked, chances are that restaurant will eventually go out of business.

On the other hand, consider police departments. Normally, the police department that has jurisdiction over a community is the only police department that is available to citizens. If you live in Detroit and need the assistance of the police but you do not like the quality of service provided by the Detroit Police Department (DPD), you do not have any other options. You will need to call the DPD or call no police department at all. Therefore, the DPD needs not spend much time worrying about going out of business. If citizens did not report crimes to the police because of the perceived ineffectiveness of the police, the amount of reported crime would decrease. The irony is that a decrease in reported crime could be presented as evidence of a more effective police department!

However, it is important that this point not be overstated: Even though a department may be a monopoly, citizens can still demonstrate their dissatisfaction with the police. In extreme instances citizens can protest or riot in order to call attention to a poorly performing police department. And police departments are accountable to elected officials who have the power to change leadership in those departments. Also, in some instances the federal government—particularly the U.S. Department of Justice—has the authority to investigate the practices and policies of police departments and legally mandate changes in how they operate. So, while

monopoly: A company or agency that does not have competitors.

Warriors Versus Guardians

The quasi-military operations of police departments have the potential to create negative images of the police and, as such, can have negative effects on police-community relations. This is particularly the case when the police are dressed in military fatigues, armed with rifles, and deployed in armored vehicles to deal with citizen protests and riots. While such deployments may be warranted, the police need to be mindful of lessons learned during the riotous 1960s when the police became viewed as "an occupying army" in urban ghettos, which further strained police-minority relations. Especially important is that the "warrior" mindset not permeate the approach taken by officers during their more routine interactions with citizens. After all, if officers are at war, who then is the enemy? Washington State Criminal Justice Training Commission Executive Director Sue Rahr and former Philadelphia Police Commissioner Charles Ramsey have suggested that the culture of police agencies needs to be transformed to reflect more of a "guardian" approach so that officers can more fully appreciate the value and importance of showing respect to the citizens they serve.[6]

police departments are monopolies and are thus somewhat insulated from the demands and dissatisfaction of citizens, they are still accountable to citizens through the political process.

POLICE AGENCIES AS STREET-LEVEL BUREAUCRACIES

Police departments have also been described as **street-level bureaucracies**.[7] They have many features and problems in common with other agencies, such as public schools; public assistance agencies (e.g., homeless shelters, welfare agencies); municipal courts; legal service offices; public health offices; and so forth.

Specifically, these agencies all share the following characteristics:

- They process people.
- They provide services and/or sanctions.
- They are public service agencies, and most rely on tax dollars for funding.
- Most of their clientele are poor.
- A large proportion of employees in street-level bureaucracies are street-level bureaucrats. Such bureaucrats are line-level workers in the agency (e.g., police officers, teachers, social workers).
- Street-level bureaucrats use substantial discretion in processing people and providing services and/or sanctions. Their decisions are usually made on the spot.
- Clients change as a result of the decisions of street-level bureaucrats (e.g., citizens become suspects, suspects are jailed, victims may get a sense of justice).

Sometimes, there is a tendency to focus on the aspects of police departments that make them unique. In actuality, they have many features in common with other public agencies, particularly with other street-level bureaucracies.

street-level bureaucracy: A public agency that serves primarily low-income clients and whose workers have substantial discretion in processing those clients.

THE CHALLENGES OF MANAGING POLICE ORGANIZATIONS

Because police departments are bureaucratic, quasi-military, and operate at the street level, they have many management challenges. A few of the most significant are discussed here.

THE MANAGEMENT OF DISCRETION

Because officers can use their own judgment to make decisions, there is always the opportunity for that discretion to be misused or abused. Ultimately, police managers and administrators are responsible and accountable for their officers' conduct. As such, much of the task of police management can be accurately thought of as the management of police discretion.

STREET COPS VERSUS MANAGEMENT COPS

Unlike in the private sector where profit is a shared goal (especially if profits are distributed among employees), officers who work the streets and supervisors who manage the street cops do not necessarily have the same goals or priorities.[8] Police officers are most interested in completing their work with minimal interference. Their alliances and loyalties are most often to fellow officers. Managers, however, are most interested in achieving results that are in line with agency objectives. Police officers seek to maximize their autonomy; managers seek to limit that autonomy. Street cops are concerned with doing the work; management cops are concerned with how this work is represented. While there is sometimes conflict between street cops and management cops,[9] there is also mutual dependence.[10] Street officers can make work difficult for their supervisors and vice versa. For example, managers may be able to control promotions and shift assignments or other work rewards or sanctions. On the other hand, if dissatisfied with management decisions, workers can engage in work stoppages or speed-ups. Although such tactics are seldom used by officers, it illustrates the power that they can have over managers.

CONSTANT RESOURCE CONSTRAINTS AND DEMAND FOR SERVICES

In police departments, there is a never-ending demand for services but never enough resources. If there *are* more resources, there are simply more demands to consume those resources.[11] For example, if a police department has a sizeable number of patrol officers, it may have the opportunity to respond to more calls—calls that otherwise would not have been handled, such as keys locked inside of cars, burglar alarms, or missing vehicle license plates. Sometimes attempts to reduce demands result in even more demands. Some police departments use a 311 telephone number for nonemergency calls to alleviate the burden of false emergency calls to 911. The result has been even more phone calls to the police.[12] There is also never enough money to operate a police department. Departments can always benefit from having more officers, better or newer equipment, the latest technology, more training, and/or better facilities. As a result, departmental budgets are never sufficient.

Photo 3.3

Police officers and their supervisors work together to try to accomplish an agency's goals, with supervisors directing the activities of officers. On occasion, however, there is conflict between street cops and management cops.

AMBIGUOUS AND DIFFICULT-TO-ACHIEVE GOALS

The goals of police organizations are difficult to specify and even more difficult to achieve; therefore, it may also be difficult to determine what exactly constitutes good policing. The difficulty of adequately defining good policing has implications for all of police management. For example, if the goals of an organization are unclear, how does one determine who would best be able to accomplish those goals? And how does one determine good performance of employees?

When dealing with ambiguous and difficult-to-achieve goals, police departments have a tendency to focus on the *means* of reaching goals over the goals themselves or to substitute means for goals. As noted previously, this has been referred to as the means over ends syndrome.[13] For instance, making arrests is often portrayed as a goal when in actuality it is probably best considered a means to a goal. The police often present the number of calls to which they respond as a measure of performance, although this statistic says nothing about the quality of services provided. For these reasons, ambiguous and difficult-to-achieve goals represent a major challenge in managing police departments.

A Question to Consider 3.1

The Means and Ends of Policing

Police departments are good at counting and tallying activities of officers. Some even keep track of the number of miles patrolled by officers during their work shifts, and this information becomes part of performance evaluations. How might this information be legitimately useful? How might it reflect the means over ends syndrome?

THE STRUCTURE OF POLICE DEPARTMENTS

For the most part, police departments of a similar size are structured in the same way. There are also some similarities in structure between very large departments and very small departments. There are also significant differences between large and small departments, however. One of the biggest is the amount of specialization present in the organization. As noted, specialization refers to the number of distinct units within the organization as well as the number of supervisory levels in the organization.

The problems and priorities of large departments are also different than those of smaller departments. In addition, diversity among officers tends to vary by department size. Overall, approximately 27% of sworn officers are racial or ethnic minorities, although larger departments are generally much more racially diverse than smaller ones.[14] This same pattern applies to the representation of female officers (see A Research Question).

One aspect of police operations that does not vary significantly by police department size is the technology that is available to officers (see Technology on the Job feature).

THE STRUCTURE OF THREE POLICE DEPARTMENTS OF DIFFERENT SIZES

As examples of the structure of three municipal police departments of varying sizes, below are the organizational charts from the River Hills Police Department in Wisconsin (Figure 3.3), the Watertown Police Department in Massachusetts (Figure 3.4), and the Houston Police Department in Texas (Figure 3.5). Although the organizational charts of each department look quite different, notice that there are also similarities.

What Is the Relationship Between Police Department Size and the Representation of Female Officers?

Overall, approximately 12% of sworn officers are female, but the representation of female officers varies considerably by department size (Figure 3.2).

A study was conducted to identify the characteristics of police departments that were associated with greater representation of female officers.[15] Using data from approximately 4,000 agencies, the authors found that a greater representation of female officers was associated

not only with department size but also with factors such as higher education requirements, greater benefits and pay, fewer physical fitness screening criteria, no collective bargaining (police labor union), and a departmental community policing orientation. The research clearly shows that certain policies and other characteristics of police agencies have a clear impact on the degree to which they are gender diverse.

FIGURE 3.2

Full-Time Sworn Personnel by Gender and Size of Department, 2013

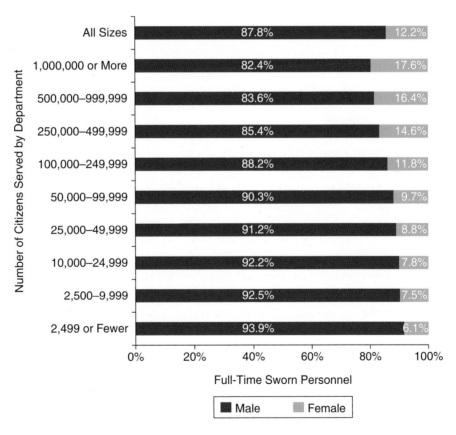

Source: Brian A. Reaves. 2015. *Local Police Departments, 2013: Personnel, Policies, and Practices.* Washington, DC: Bureau of Justice Statistics. Table 4.

Police Department Size and Technology

As we have discussed, there are structural and personnel differences between large and small police departments. Interestingly, however, as seen in Table 3.1, there are relatively few differences between such departments in terms of the technology that is used by officers, especially video technology. The biggest differences are with body-worn cameras, which tend to be used more by smaller departments, and license plate readers and drones, which larger departments are more likely to deploy.

As such, it is not accurate to say that smaller departments are less advanced technologically than larger departments. However, as mentioned in the Police Spotlight feature at the introduction to the chapter, the local tax base and the amount of money available in communities for police services will strongly influence departmental operations, including the technologies available. The local tax base may or may not be associated with the size of the department.

TABLE 3.1

Use of Video Technologies by Local Police Departments, by Size of Population Served, 2013

Population Served	Types of Video Technology						
	Any type	In-car video cameras	Body-worn cameras	Weapon-attached cameras	Cameras for surveillance of public areas	License plate readers	Unmanned aerial drones
All Sizes	76%	68%	32%	6%	49%	17%	–
1,000,000 or More	71	57	21	14	86	93	7%
500,000–999,999	80	73	30	7	87	77	3
250,000–499,999	70	63	20	9	87	87	2
100,000–249,999	75	70	19	10	76	77	1
50,000–99,999	70	63	26	11	68	55	1
25,000–49,999	79	76	22	9	67	50	0
10,000–24,999	75	71	26	9	62	24	0
2,500–9,999	80	71	34	8	51	10	0
2,499 or Fewer	72	64	35	3	35	6	0

Source: Brian A. Reaves, *Local Police Departments*, 2013: *Equipment and Technology* (Washington, D.C.: Bureau of Justice Statistics, 2015). These figures represent the most current statistics available at the time of this writing.

The village of River Hills, Wisconsin, has a population of approximately 1,600 persons, and its police department has 11 sworn officers. The organization of the department is fairly typical for its size and is discussed here as an example of the structure of a very small department. It is organized into three patrol shifts and a support division. The department is led by a chief; sergeants are in command of each of the three patrol shifts. The sergeants report to the chief. A support staff (six clerks) are managed directly by the chief. In this jurisdiction, there are very few crimes (in 2017 there was one burglary and five thefts); investigations are conducted by patrol officers.

FIGURE 3.3

River Hills, Wisconsin, Police Department Organizational Chart

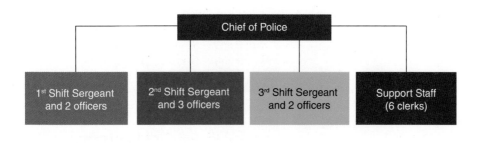

Source: Author created.

FIGURE 3.4

Watertown, Massachusetts, Police Department Organizational Chart[16]

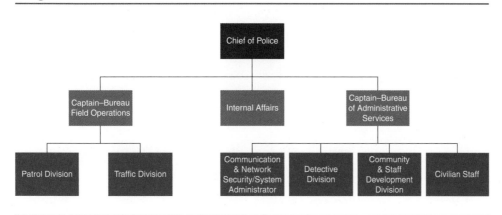

Watertown, Massachusetts, is a suburb of Boston with a population of approximately 35,000. The Watertown Police Department has sixty-three sworn officers and thirteen civilians. These officers and civilian personnel work in various divisions within the department. The department is led by a chief and has two captains; one is responsible for managing the Bureau of Field Operations, and the other is responsible for the Bureau of Administrative Services. The Bureau of Field Operations oversees all things related directly to the patrol function of the department, including the patrol and traffic division. The Bureau of Administrative Services oversees the detective division and matters that relate to the internal functioning of the department, such as the operation of computer systems. Each division is led by a lieutenant. Internal affairs is directly under control of the chief of police.

Houston, Texas, has a population of approximately 2.3 million; it is the fourth-largest city in the country. The Houston Police Department has approximately 5,200 sworn officers. Due to the size of the city it serves, this department obviously looks quite different than the others discussed here. This is primarily because of the workload demands placed on the department and its corresponding size and specialization.

The Houston Police Department has a full staff with responsibilities that relate to the varied demands and workload of the agency. To provide for reasonable supervisory control, each

FIGURE 3.5

Houston, Texas, Police Department Organizational Chart[17]

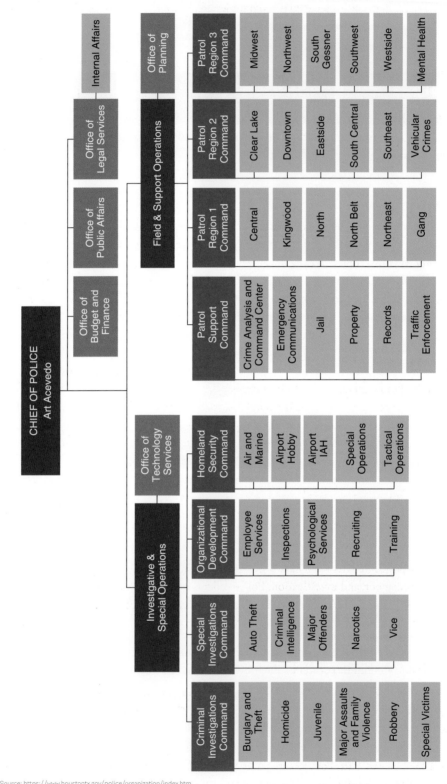

Source: https://www.houstontx.gov/police/organization/index.htm.

functional unit of the department has its own supervisors and staff. Investigative and Special Operations and Field and Support Operations are the two sections of the department. Each of these sections is further divided by division (or "command"). Especially noteworthy is the specialization in criminal investigations and the units related to recruitment, training, and psychological services. Patrol is organized by geographic organization, with further divisions in each area, along with a patrol support division.

MAJOR OPERATING UNITS IN POLICE DEPARTMENTS

Now that we have a basic understanding of variations in the organizational structure of police departments, we can turn our attention to a more detailed discussion of some of the most common operating units within departments.

PATROL

Virtually all but the smallest police departments have a designated patrol unit. Patrol officers have many responsibilities, but primary among them is responding to calls for service in the community. Typically, about 70% of officers are assigned to the patrol unit,[18] but the proportion varies by the size of the department. Typically, a greater proportion of officers is assigned to the patrol unit in smaller departments than in larger ones. Almost all local departments use automobiles for patrol, including sport utility vehicles (SUVs), trucks, or vans.[19] About 50% all local departments use foot patrol, around 33% use bicycle patrol, and about 15% use motorcycles.[20] Smaller departments often only use automobiles for patrol.[21] In larger departments, some patrol officers do not wear a uniform and are responsible for initiating activities such as pedestrian and vehicle stops, not responding to calls for service.

The patrol unit is generally staffed around the clock, although some larger departments may also staff homicide or other investigators twenty-four hours a day. Officers are assigned to shifts to provide continuous coverage over the course of the day. Different departments have different shift schedules; shifts are commonly eight hours, but some are ten hours. Uncommon but not unheard of are twelve-hour shifts. Some departments have different combinations of shift hours. Additional details about patrol shifts and shift work are provided in Chapter 6 and Chapter 10.

TRAFFIC

Some larger police departments have a designated traffic unit that is responsible for traffic control, enforcement, and investigation. Police officers assigned to the traffic unit are usually deployed in squad cars and/or motorcycles. If a department does not have a traffic unit, traffic enforcement is the responsibility of officers who work in the patrol unit.

CRIMINAL INVESTIGATION

Common among police departments is a criminal investigation unit, comprised of detectives and sometimes other police officers. Typically, about 15% of officers in a department are assigned to an investigations unit.[22] Generally speaking and as illustrated in the organizational charts presented above, the larger the department, the more specialization there is among investigators in the types of crimes investigated. Patrol officers are traditionally responsible for responding to crime scenes and conducting preliminary (or initial) investigations, including collecting evidence and interviewing victims and witnesses. Then, if warranted, detectives conduct follow-up investigations. It is becoming more common in police departments today for patrol officers to be assigned greater

Photo 3.4

Patrol officers typically conduct initial investigations of crimes and detectives conduct follow-up investigations. However, it is becoming more common for patrol officers to have increased investigative responsibilities.

Chris Pietsch/The Register-Guard via AP

responsibility for conducting follow-up investigations. As noted earlier, in the small department of River Hills, patrol officers have responsibility for all investigative activities. This is feasible because of the very small number of crimes in River Hills. Specialization is necessary in an agency such as the Houston Police Department that investigated nearly 10,000 robberies and 17,000 burglaries in 2017. Chapter 7 provides additional details about criminal investigations.

A department may have several other investigations-related units. For example, investigators and officers assigned to sensitive crimes units or special victims investigation units typically investigate sex-related crimes. Another type of investigations unit found in some departments focuses on gang-related criminal investigations. Narcotics and vice units are most often found in very large police departments. These investigators are responsible for using nontraditional investigative methods, such as sting operations, to combat drug sales, prostitution, illegal gambling, and other so-called victimless crimes.

TACTICAL ENFORCEMENT

Many police departments have a **tactical enforcement unit (TEU)** to handle high-risk situations, such as hostage or barricade situations, and to execute high-risk arrest and search warrants. In large departments, officers are assigned to the TEU on a regular, full-time basis. In smaller departments, the TEU consists mostly of officers who are assembled to handle particular situations as needed. TEU officers receive special training and equipment to handle the associated risks of their assignments.

Large police departments may also have a specially designated bomb squad. These officers are responsible for dealing with confirmed or suspected explosives in whatever situation they may be discovered. These materials can be properly and safely neutralized with special training and equipment.

YOUTH OR JUVENILE BUREAU

Officers and detectives who work in a juvenile bureau may be responsible for conducting investigations in which the victim or offender is a juvenile. School resource officers, school liaison officers, and DARE (Drug Abuse Resistance Education) officers may also be assigned to a juvenile bureau.

COMMUNICATIONS

Virtually all police departments have a communications/dispatch unit. This unit is usually staffed by civilians. Their job is to receive emergency and nonemergency calls for service and to dispatch officers to these calls.

tactical enforcement unit (TEU): A police unit that handles high-risk criminal situations, such as hostage situations and the execution of certain arrest and search warrants.

INTERNAL AFFAIRS

The **internal affairs unit** is responsible for investigating citizen complaints against officers as well as internally generated complaints (officers filing complaints against each other, supervisors filing complaints about officers, officers filing complaints against supervisors). Internal affairs investigators are often responsible for investigating officer-involved shootings, although some departments are required to have an outside law enforcement agency exercise primary responsibility for conducting such investigations.

CRIME ANALYSIS

Crime analysis units are a relatively new addition to larger police departments. Usually staffed by civilians with advanced education and training in crime analytics, these units are responsible for identifying trends and patterns in criminal incidents, locating crime hot spots, conducting analyses of offender networks and crimes, and identifying high-rate offenders. They provide this information to investigators to assist in criminal identification and apprehension and to the patrol unit to inform patrol allocation and enforcement decisions. Crime analysis is often associated with predictive policing, which is discussed in more detail in Chapter 14.

TYPES AND LEVELS OF LAW ENFORCEMENT AGENCIES

In the United States, law enforcement is provided at the local, county, state, and federal government levels. Agencies at each level are briefly described here.

LOCAL POLICE

In 2016, there were 12,267 local (e.g., city, village) police departments.[23] These departments employed approximately 468,000 full-time sworn officers and 131,000 full-time civilians.[24] Included in these totals were a few police departments operating at the county level (county sheriff's departments, not police departments, typically provide police services at the county level). About 65% of municipalities and towns in the United States do not have their own police department. Policing services in these areas are provided by the county in which the town is located (see below). The total number of local police departments in the nation is constantly changing, as some towns decide to form their own police department and others decide to dissolve theirs and rely on county sheriff's departments instead. Decisions to disband a police department are usually based on financial considerations, as policing services are expensive.

Most of the 12,267 local police departments are quite small. Approximately 51% of all local departments (6,224 of 12,267) employ fewer than ten full-time officers,[25] but in total, these departments employ only about 5% of all officers. Only 5% of local departments (665 of 12,267) employ 100 or more sworn officers. These 665 departments employ the majority of all sworn officers. The most important takeaway here is that while most police departments in the nation are quite small, most police officers work in larger departments (see Figure 3.6).

COUNTY SHERIFF'S DEPARTMENTS

The vast majority of counties in the United States have a sheriff's department; in 2016, there were 3,012 in the nation.[26] Only four states (Alaska, Connecticut, Hawaii, and Rhode Island) and Washington, D.C., do not have sheriff's departments. Sheriff's departments employ about

> **internal affairs unit:** A unit responsible for investigating citizen complaints against officers and internally generated complaints among police department members.

FIGURE 3.6

The Five Largest Police Departments in the United States

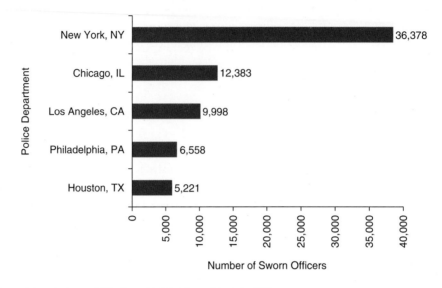

Source: Uniform Crime Report, 2017 (Washington, D.C.: Federal Bureau of Investigation, 2018).

173,000 full-time sworn officers and 186,000 civilians.[27] Compared to local and state law enforcement agencies, sheriff's departments have a large representation of civilian personnel; many of these agencies employ substantially more civilians than sworn officers.

Approximately 25% (757) of sheriff's departments employ less than ten officers.[28] These 757 departments combined account for less than 2% of all sworn officers who work in sheriff's departments. Twelve percent (364) of sheriff's departments employ 100 or more full-time sworn officers; these agencies account for approximately 65% of all sworn officers who work in sheriff's departments.[29] So, similar to local police departments, while a sizable share of county sheriff's departments are quite small, most sworn officers work in larger agencies (Figure 3.7).

As with local police departments, larger sheriff's departments are more likely to exhibit greater diversity in terms of officer gender and race. Overall, 14% of sworn officers in sheriff's departments are female, and 22% are racial minorities.[30]

Sheriff's departments have several responsibilities. First, with a few exceptions, these agencies are responsible for law enforcement in the county, on county land (e.g., parks, county roads, and highways), and in towns and villages that do not have their own police department. Second, about 75% of all sheriff's departments are responsible for operating at least one county jail. Third, they may have responsibility for providing court security, transporting prisoners to and from jail and court, and serving processes (e.g., restraining orders, court summons, eviction notices). The allocation of officers and other resources to the various functions of the sheriff's departments depends much on the characteristics of the county. Compared to urban counties, sheriff's departments of more rural counties may allocate a larger portion of resources to law enforcement responsibilities because rural counties are likely to have fewer local police departments to provide services. On the other hand, sheriff's departments in more urban counties generally allocate more resources to jail operations, if they are responsible for a jail. For example, in the Cook County Sheriff's Department (which includes the city of Chicago and the greater Chicago area), only 4% of deputies are assigned to respond to calls for service.[31]

FIGURE 3.7

The Five Largest County Sheriff's Departments

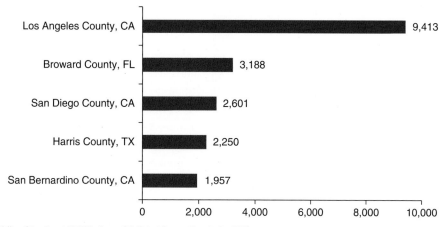

Los Angeles County, CA	9,413
Broward County, FL	3,188
San Diego County, CA	2,601
Harris County, TX	2,250
San Bernardino County, CA	1,957

Source: Uniform Crime Report, 2017 (Washington, D.C.: Federal Bureau of Investigation, 2018).

STATE LAW ENFORCEMENT AGENCIES

Each state has a primary state law enforcement agency. Depending on the state, this agency could be a highway patrol, state police, or department of public safety. Because of the variation in law enforcement agency arrangements, it is difficult to compare agencies across states. If a state has a highway patrol agency, it also has a department of public safety or its equivalent. Each of these agencies varies in its sphere of authority and jurisdiction.

State police agencies have the responsibility for providing general law enforcement services in the state. This includes traffic enforcement on state roads and highways and criminal investigations relating to drug trafficking, fraud, serial crimes, cyber and computer crimes, arson, tactical enforcement, explosives, and other matters in which local police agencies request the assistance of the state police. State police agencies also operate forensic crime laboratories, criminal information computer networks, and emergency management services, and they provide law enforcement officer training and certification. States that have this type of agency include Michigan, Illinois, New Jersey, and Colorado. The largest state police agency is the New York State Police, which numbers 5,046 sworn personnel.[32]

The jurisdiction of a state highway patrol agency is limited primarily to state and inter-state roads, highways, and interstate highways, although these agencies may assist other law enforcement agencies on other matters when requested. Highway patrol agencies are responsible for enforcement of traffic and other laws and also regulate vehicle safety through inspections and cargo movement. Patrol is provided via marked and unmarked automobiles and often motorcycles and aircraft. Wisconsin, Georgia, Florida, and Alaska are examples of states with a highway patrol (in Wisconsin and Georgia the agency is referred to as the state patrol; in Alaska it is called the state troopers). The largest highway patrol agency is the California Highway Patrol, which has 7,401 sworn personnel.[33]

As noted, if a state has a highway patrol agency, it also has a related agency that provides other state law enforcement services, including investigations and forensic support. For example, Wisconsin has the Wisconsin Department of Justice, Georgia has the Georgia Bureau of Investigation, Florida has the Florida Department of Law Enforcement, and Alaska has the Alaska Department of Public Safety. The largest of these agencies is the Texas Department of Public Safety, which employs 4,187 sworn personnel.[34]

Photo 3.5

There are about 750 college and university police departments in the United States that seek to ensure the safety of students, staff, and faculty.

©iStockphoto.com/sshepard

SPECIAL JURISDICTION LAW ENFORCEMENT AGENCIES

In addition to local, county, and state law enforcement agencies, there are approximately 1,700 other public law enforcement agencies in the United States that have special jurisdictions. In total, these agencies employ about 57,000 sworn personnel. The largest share of these agencies, about 750, consists of campus police departments serving colleges and universities.[35] For perspective, the largest university police department is Georgia State University (Atlanta), which has 126 sworn personnel (see Table 3.2).

On most campuses, the police have arrest powers not only on campus but in areas adjacent to campus and property outside the area adjacent to campus.[36] The overwhelming majority of campus police departments provide twenty-four-hour patrol coverage. About one-half of agencies serving public campuses use student patrols compared to about one-third of agencies that serve private campuses.[37] Most campus police departments also provide safety escort services for students, staff, and faculty; usually this service is provided by nonsworn personnel.[38] Almost all campus police departments have blue light phones available for emergencies. Nearly all campus police departments authorize officers to use a firearm (94%), a chemical spray (94%), and a baton (93%). Seldom are nonsworn personnel authorized to use a firearm. Most campuses have a mass notification system to alert students and others of an emergency situation; the most common of these systems are email (100%) and text messages (99%).[39]

EXHIBIT 3.1

The Clery Act and Reporting Campus Crime

The Jeanne Clery Disclosure of Campus Security Policy and Campus Crime Statistics Act was signed into law in 1990 as the Campus Security Act. The act requires institutions of higher education that participate in federal financial aid programs to keep and disclose information about crime on and near their campuses. The U.S. Department of Education monitors compliance. Violations can result in penalties of up to $35,000 per infraction and suspension from federal student financial aid programs. The Clery Act requires institutions to do the following:

- Publish an annual campus security report that documents three calendar years of specified campus

crime statistics and make it available to current and prospective students and employees.

- Maintain a timely public log of all crimes reported or otherwise known to campus law enforcement officials.
- Give timely warning of crimes that represent a threat to student or employee safety.
- Submit an annual report to the U.S. Department of Education that includes tallies of serious crimes, hate crimes, and arrests and disciplinary referrals for liquor law violations, drug law violations, and illegal weapons possession.[40]
- Clery Act statistics are available at http://ope.ed.gov/security/.

TABLE 3.2

Twenty Largest Campus Law Enforcement Agencies, Number of Full-Time Sworn Personnel, 2017

Campus Served	Full-Time Sworn Personnel
Georgia State University	126
Temple University	125
University of Pennsylvania	117
University of Alabama at Birmingham	101
University of Chicago*	100
George Washington University	100
Vanderbilt University	92
Virginia Commonwealth University*	90
Yale University	89
University of Pittsburgh	89
Michigan State University	87
University of Maryland, College Park	86
Arizona State University	84
University of Georgia	82
Tulane University	81
University of Florida	80
University of Connecticut	77

Source: Uniform Crime Report. 2017 (Washington, D.C.: Federal Bureau of Investigation, 2018)

* tally obtained via contact with the agency in January, 2019.

A Question to Consider 3.2

The Characteristics of Your Campus Police Department

Does your college or university have a police department? If yes, how many sworn officers are employed in the department? What is its geographic jurisdiction? How does your campus police department compare to typical ones as described here?

Another 250 of the special jurisdiction agencies serve public school districts, the largest of which is the School District of Philadelphia with 450 sworn officers.[41] Other law enforcement agencies include those that enforce fish and wildlife conservation laws; protect transportation systems; and serve parks, airports, state government buildings, and hospitals, among others. The largest natural resource law enforcement agency is the California Department of Parks and Recreation, which employs 523 full-time sworn officers. The largest agency with transportation-related jurisdiction is the Port Authority of New York and New Jersey, which has 1,651 sworn officers.[42]

A Question of Ethics

Do You Follow Policy That You Do Not Believe In?

As discussed earlier in this book, law enforcement agencies are controversial in large part because of the methods and strategies used to accomplish their mission. One extraordinary example of this is U.S. Customs and Border Protection. Among other responsibilities, officers of this agency have been tasked with preventing immigrants, especially from Mexico, from entering the United States illegally. In 2018, this led to some children being separated and detained away from their parents.[43] As a Border Patrol officer, if you do not morally agree with these actions, what do you do? It is your obligation to carry out the policy of the agency? Explain.

FEDERAL LAW ENFORCEMENT AGENCIES

The primary law enforcement agencies of the federal government are located in the Department of Homeland Security (DHS) and the Department of Justice (DOJ). As a result of the terrorist attacks of September 11, 2001, in 2002 federal law enforcement underwent a major reorganization. Most significant were the creation of the DHS and the removal of some agencies from the Department of the Treasury and their placement in the DHS. The major law enforcement agencies in the DHS now include the following:

- Customs and Border Protection
- Immigration and Customs Enforcement
- Secret Service
- Transportation Security Administration

The major law enforcement agencies in the DOJ include the following:

- Federal Bureau of Investigation
- Bureau of Alcohol, Tobacco, Firearms, and Explosives
- Drug Enforcement Administration
- U.S. Marshals Service

Photo 3.6

The Federal Bureau of Investigation is one of the most high-profile federal law enforcement agencies. It has authority to investigate and enforce a wide variety of federal laws.

US-Politics-Media-Police-FB/Michele Eve Sandberg/AFP/Getty Images

The primary responsibilities of each of these agencies are summarized in Table 3.3 and Table 3.4.

In addition to these well-known federal law enforcement agencies, there are a multitude of other agencies that have some law enforcement responsibilities and that employ agents, investigators, and/or police. Officers in these agencies investigate criminal offenses that relate to the jurisdiction of the agency. The other agencies with the most significant law enforcement responsibilities are listed in Table 3.5.

TABLE 3.3

Major Law Enforcement Agencies Located in the Department of Homeland Security

Department of Homeland Security (DHS)
Along with agents employed by the agencies listed below, the DHS also employs special agents and other support personnel in order to protect America and uphold public safety. In particular, Homeland Security Investigations is tasked with identifying criminal activities and eliminating vulnerabilities that pose a threat to our nation's borders, as well as enforcing economic, transportation, and infrastructure security.[44] For a complete listing of agencies within the DHS see https://www.dhs.gov/organizational-chart.

Agency	Size	Functions
U.S. Customs and Border Protection (CBP)	62,000 total employees	CBP officers protect U.S. borders at official ports of entry. Border patrol agents prevent illegal entry of people and contraband between ports of entry. Air and marine officers patrol the nation's land and sea borders to stop terrorists and drug smugglers.[45]
U.S. Immigration and Customs Enforcement (ICE)	More than 20,000 total employees	Special agents conduct investigations involving national security threats, terrorism, drug smuggling, child exploitation, human trafficking, illegal arms export, financial crimes, and fraud. Uniformed immigration enforcement agents perform functions related to the investigation, identification, arrest, prosecution, detention, and deportation of aliens, as well as the apprehension of absconders.[46]
U.S. Secret Service (USSS)	6,500 total employees (3,200 special agents; 1,300 uniformed officers)	Special agents have investigation and enforcement duties primarily related to counterfeiting, financial crimes, computer fraud, and threats against dignitaries. Uniformed division officers protect the White House complex and other presidential offices, the main Treasury building and annex, the president and vice president and their families, and foreign diplomatic missions.[47]
Transportation Security Administration (TSA)	60,000 total employees	Employees are responsible for protecting U.S. transportation systems and the traveling public. The mission of the Federal Air Marshal Service (FAMS) is to detect, deter, and defeat criminal and terrorist activities that target U.S. transportation systems.[48]

TABLE 3.4

Major Law Enforcement Agencies Located in the Department of Justice

Department of Homeland Security (DOJ)

The mission of the DOJ is to "enforce the law and defend the interests of the United States according to the law, to ensure public safety against threats foreign and domestic, to provide federal leadership in preventing and controlling crime, to seek just punishment for those guilty of unlawful behavior, and to ensure fair and impartial administration of justice for all Americans."[49] Beside the agencies listed below, the DOJ consists of more than 40 agencies (see www.justice.gov/agencies/chart), including litigation and numerous support agencies.

Agency	Size	Functions
Federal Bureau of Investigation (FBI)	35,000 total employees (14,000 special agents)	Special agents are responsible for criminal investigation and enforcement related to more than 200 categories of federal law. Criminal priorities include terrorism, public corruption, civil rights violations, organized crime, white-collar crime, violent crime, and major theft.[50]
Drug Enforcement Administration (DEA)	11,000 total employees (5,000 special agents)	Special agents investigate major narcotics violators, enforce regulations governing the manufacture and dispensing of controlled substances, and perform other functions to prevent and control drug trafficking.[51]
U.S. Marshals Service (USMS)	5,100 total employees (3,570 deputy U.S. marshals and criminal investigators)	The duties of the U.S. Marshals Service include protecting the federal judiciary, apprehending federal fugitives, managing and selling seized assets acquired by criminals through illegal activities, housing and transporting federal prisoners, and operating the Witness Security Program.[52]
Bureau of Alcohol, Tobacco, Firearms and Explosives (ATF)	5,100 total employees (2,500 special agents)	The ATF investigates and prevents crimes that involve the unlawful manufacture, sale, possession, and use of firearms and explosives; acts of arson and bombings; and illegal trafficking of alcohol and tobacco products. The ATF regulates the firearms and explosives industries from manufacture and/or importation through retail sale.[53]

TABLE 3.5

Other Federal Agencies With Law Enforcement Responsibilities

Department	Agency	Functions
Agriculture	U.S. Forest Service, Law Enforcement and Investigations Organization	Uniformed law enforcement rangers enforce federal laws and regulations governing national forest lands and resources. Special agents are criminal investigators who investigate crimes against property, visitors, and employees.
Commerce	National Oceanic and Atmospheric Administration, Office of Law Enforcement	Special agents and enforcement officers enforce laws that conserve and protect living marine resources and their natural ocean habitat in the United States and U.S. territories.
Defense	Pentagon Force Protection Agency	Officers provide law enforcement and security services for the occupants, visitors, and infrastructure of the Pentagon, Navy Annex, and other assigned Pentagon facilities.
Energy	National Nuclear Security Administration, Office of Secure Transportation	Special agents, known as nuclear materials couriers, ensure the safe and secure transport of government-owned special nuclear materials during classified shipments in the contiguous United States.
Health and Human Services	U.S. Food and Drug Administration, Office of Criminal Investigations	Special agents investigate suspected criminal violations of the Federal Food, Drug, and Cosmetic Act and other related acts; the Federal Anti-Tampering Act; and other statutes, including applicable Title 18 violations of the United States Code.

Department	Agency	Functions
Homeland Security	Federal Emergency Management Agency (FEMA), Security Branch	Officers are responsible for the protection of FEMA facilities, personnel, resources, and information.
Independent	Amtrak Police	Officers provide law enforcement and security services for the passengers, employees, and patrons of the national railroad owned by the U.S. government and operated by the National Railroad Passenger Corporation.
	U.S. Environmental Protection Agency, Criminal Enforcement	Special agents investigate suspected individual and corporate criminal violations of the nation's environmental laws.
	U.S. Postal Inspection Service	Postal inspectors conduct criminal investigations covering more than 200 federal statutes related to the postal system. Postal police officers provide security for postal facilities, employees, and assets and escort high-value mail shipments.
Interior	Bureau of Indian Affairs, Division of Law Enforcement	Officers provide law enforcement services in some tribal areas. In addition to providing direct oversight for these bureau-operated programs, the division also provides technical assistance and some oversight to tribally operated law enforcement programs.
	Bureau of Land Management, Law Enforcement	Law enforcement rangers conduct patrols, enforce federal laws and regulations, and provide for the safety of bureau employees and users of public lands. Special agents investigate illegal activity on public lands.
	National Park Service, United States Park Police	Officers provide law enforcement services to designated National Park Service areas (primarily in the Washington, D.C., New York City, and San Francisco metropolitan areas). Officers are authorized to provide services for the entire national park system.
	National Park Service, Visitor and Resource Protection Division	Park rangers, commissioned as law enforcement officers, provide law enforcement services for the national park system. Additional rangers serving seasonally are commissioned officers but are considered part-time employees and not included in the federal law enforcement officer (FLEO) census.
	U.S. Fish and Wildlife Service, Office of Law Enforcement	Special agents enforce federal laws that protect wildlife resources, including endangered species, migratory birds, and marine mammals.
Judicial	Administrative Office of the U.S. Courts	Federal probation officers supervise offenders on probation and supervised release. In seven federal judicial districts, probation officers are not authorized to carry a firearm while on duty and are excluded from FLEO officer counts.
	Federal Bureau of Prisons (BOP)	Correctional officers enforce the regulations governing the operation of BOP correctional institutions, serving as both supervisors and counsellors of inmates. They are normally not armed while on duty. Most other BOP employees have arrest and firearm authority to respond to emergencies.
Legislative	U.S. Capitol Police	Officers provide law enforcement and security services for the U.S. Capitol grounds and buildings and in the zone immediately surrounding the Capitol complex.
State	Bureau of Diplomatic Security	In the United States, special agents protect the secretary of state, the U.S. ambassador to the United Nations, and visiting foreign dignitaries below the head-of-state level. They also investigate passport and visa fraud.
Treasury	Bureau of Engraving and Printing	Police officers provide law enforcement and security services for facilities where currency, securities, and other official U.S. documents are made.
	Internal Revenue Service, Criminal Investigation Division	Special agents have investigative jurisdiction over tax, money laundering, and Bank Secrecy Act laws.
	United States Mint Police	Officers provide law enforcement and security services for employees, visitors, and government assets stored at U.S. Mint facilities.
Veterans Affairs	Veterans Health Administration, Office of Security and Law Enforcement	Officers provide law enforcement and security services for Veterans Affairs medical centers.

Source: Brian A. Reaves. 2012. *Federal Law Enforcement Officers*, 2008. Washington, DC: Bureau of Justice Statistics.

Main Points

- Most police departments today are managed and structured in accord with the five principles of bureaucracies: (1) division of labor, (2) hierarchy of authority, (3) rules, (4) impersonality of relationships, and (5) selection and promotion based on competence.

- There is a clear reflection of the military in police departments today, as evidenced by the command and control orientation, emphasis on discipline, top-down chain of command, rank structure, "war-like" mission, emphasis on training, uniforms, weapons, and even the tactics that are used. The war-like mission can have negative consequences for the police and the community.

- Police departments are monopolies but still need to be accountable to citizens for the services they provide.

- Police departments have many features and problems in common with agencies such as public schools; public assistance agencies (e.g., homeless shelters, welfare agencies); municipal courts; legal service offices; public health offices; and so forth. All of these agencies can be described as street-level bureaucracies.

- Much of the task of police management can be accurately thought of as the management of police discretion.

- Officers who work the streets and supervisors who manage the street cops do not necessarily have the same goals or priorities. However, there is also mutual dependence: Police officers can make work difficult for their supervisors and vice versa.

- There are never enough resources in police departments, and there is a never-ending demand for services.

- If the goals of an organization are difficult to determine and achieve, then it may also be difficult to specify what exactly constitutes good policing. The difficulty of adequately defining good policing has implications for all of police management.

- Large and small police departments differ in several ways, including structural specialization and diversity of sworn officers. However, large and small departments do not differ greatly in terms of the technology that is used on the job.

- Nearly all police departments have a patrol unit, although staffing levels may depend on the size of the department. Patrol is usually the largest unit in a department. To the extent that police departments are twenty-four-hour agencies, it is usually because patrol officers work both day and night shifts.

- Common among police departments is a criminal investigation unit. The larger the department, the more specialization there is among investigators in the types of crimes investigated.

- Other units within police departments may include special enforcement units (e.g., bomb squad, tactical enforcement); investigative units (e.g., sensitive crimes units); juvenile bureaus; communications units; internal affairs divisions; and crime analysis units.

- In the United States, law enforcement is provided at the local, county, state, and federal government levels.

- While most police departments in the nation are quite small, most police officers work in larger departments.

- Most sheriff's departments have more varied responsibilities than local police departments. These responsibilities include law enforcement, operation of a county jail, and court security. The amount of personnel and resources devoted to each task depends on the characteristics of the county.

- Each state has a state law enforcement agency. Some states have a highway patrol and department of public safety; other states have state police.

- There are approximately 1,700 other public law enforcement agencies in the United States that have special jurisdictions. The largest share of these agencies, about 750, consists of campus police departments serving colleges and universities.

- The Clery Act requires institutions of higher education that participate in federal financial aid programs to keep and disclose information about crime on and near their campuses.

- The primary law enforcement agencies of the federal government are located in the Department of Homeland Security (DHS) and the Department of Justice (DOJ). They include Customs and Border Protection; Immigration and Customs Enforcement; the Secret Service; the Federal Bureau of Investigation; the Transportation and Security Administration; the Bureau of Alcohol, Tobacco, Firearms and Explosives; the Drug Enforcement Administration; and the U.S. Marshals Service. There are a multitude of other federal agencies that have some law enforcement responsibilities and employ agents, investigators, and/or police.

Important Terms

Review key terms with eFlashcards at **edge.sagepub.com/brandl2e.**

bureaucracy   40
chain of command   39
internal affairs unit   51
monopoly   41

specialization   39
street-level bureaucracy   42
tactical enforcement unit (TEU)   50

Questions for Discussion and Review

Take a practice quiz at **edge.sagepub.com/brandl2e.**

1. How might police officer salary differences among police departments located in the same general area affect policing in those agencies?
2. Police are often described as bureaucracies. Explain.
3. Police departments have also been described as being monopolies. What does this mean? What are the implications of police departments being monopolies?
4. What are the most significant challenges associated with managing police organizations?
5. How are large and small police departments the same? How are they different?

6. List and briefly discuss the major operating units in police departments.
7. What are the different types of law enforcement agencies at the state level?
8. Why do you think there are law enforcement agencies at each level of government?
9. What is the Clery Act and what is its purpose?
10. Identify and discuss the law enforcement agencies located in the Department of Justice and Department of Homeland Security.

Fact or Fiction Answers

1. Fact
2. Fiction
3. Fact
4. Fact
5. Fiction

6. Fact
7. Fiction
8. Fiction
9. Fiction
10. Fiction

⑤SAGE edge™

Digital Resources

Get the tools you need to sharpen your study skills. SAGE Edge offers a robust online environment featuring an impressive array of free tools and resources.

Access practice quizzes, eFlashcards, video, and multimedia at **edge.sagepub.com/brandl2e.**

Media Library

View these videos and more in the interactive eBook version of this text!

Career Video
3.1: Communicating Effectively in a Structured Environment

SAGE News Clip
3.1: Chicago Police Get 'De-Escalation' Training
3.2: Street Gang Suspects Held After LA Raids

4
THE ROLE OF THE POLICE

Objectives

After reading this chapter you will be able to:

4.1 Discuss the various reasons why we have the police

4.2 Explain why law enforcement is important and controversial even though relatively little time is spent doing it

4.3 Identify why crime control is controversial

4.4 Explain why the police may have an impossible mandate

4.5 Describe the difficulties associated with officers deterring criminal behavior

4.6 Evaluate why the authority to use force is critical to the police role

4.7 Discuss law enforcement, order maintenance, and service as the three primary responsibilities of the police and explain why different police departments tend to emphasize different aspects of the police role

Police Spotlight: "Scoop and Run" in Philadelphia

The Philadelphia Police Department has a unique policy when it comes to rendering assistance to gunshot and stabbing victims: Instead of waiting for emergency medical services (EMS) to arrive on-scene and provide medical assistance and transportation, officers rush victims to the nearest trauma center in their squad cars, so called "scoop and run." In 2017, one-third of Philadelphia's 1,223 shooting victims were transported to one of the city's five trauma centers in this manner. While this practice is not unheard of in other major cities, Philadelphia is the only city known to have made this practice standard policy, subject to officers' consideration of proximity to a hospital, availability of EMS, and nature of victim injuries.

This practice is not without its critics. Some police officials point out that victims could be further injured during transport and typically do not receive medical care on the way to the hospital, raising liability concerns if victims do not survive. According to Philadelphia police, however, the city has never been sued over police transport of victims. Medical professionals add that less medical care during transport is made up for by faster arrival at a hospital, but they explain that this practice may not be wise if trauma centers are few and far away. Critics also highlight that police have other important responsibilities at crime scenes like finding and interviewing victims and collecting other evidence in order to solve the crime. Advocates respond that the practice may help strengthen relationships and goodwill between the community and the police and gives police officers a rare opportunity to save a life in the line of duty. As Philadelphia Police Department Captain Stephen Clark stated, "We don't join the Police Department to watch people die."[1]

Fact or Fiction

To assess your knowledge of the role of the police prior to reading this chapter, identify each of the following statements as fact or fiction. (See page 77 at the end of this chapter for answers.)

1. Police officers spend most of their time enforcing the law.

2. Crime control is relatively easy to measure accurately, and where there is less crime the police are obviously more effective.

3. The police do not deter some people from committing crimes and do not need to deter others.

4. Seldom is law enforcement controversial. It is just a matter of identifying the criminals and arresting them.

5. Media depictions of police work are similar to the realities of police work.

THIS chapter addresses the question, Why do we have the police? The answer is not as simple as it may first seem.

What is the mandate of the police? In other words, what is their purpose? What is their role in our society? Most fundamentally, why do we have the police? These seemingly simple questions have not-so-simple answers. Indeed, it takes a chapter to adequately answer them. The police have many important responsibilities; they are absolutely essential in our society. However, their mandate is difficult, if not impossible, to achieve. Given their mandate, they are also *inherently* controversial. To attain an adequate understanding of the police in America, it is necessary to examine their major responsibilities and why their mandate is so difficult and controversial.

THE POLICE ENFORCE THE LAW

Perhaps the first and most straightforward responsibility of the police is to enforce the law and identify and apprehend criminals. This is clearly an important duty. The police have been given the power and authority to arrest citizens when it is believed they have violated the law. If the police did not enforce the law and arrest offenders, then the courts would have no one to adjudicate and corrections would have no one to "correct" (or provide treatment to, administer punishment to, maintain custody of, etc.). The entire criminal justice system would cease to operate. In this light, the police, as enforcers of the law, serve as the gate keepers of the criminal justice system, determining who comes into the system and who stays out.

GIVE MEANING TO THE LAW

It is also important to understand that as enforcers of the law, it is police officers who give the law meaning. The law means little without the possibility that the police will enforce it. Consider the example of speeding enforcement. What does a speed limit of 35 mph really mean? Does it mean if you are going 36 mph you are going to get a ticket? It is not likely, because the police are usually concerned with more serious violations of speed laws. How about if you are traveling 50 mph? In this situation, it might depend on whether there is a police officer looking for speeders on that road. Without a police officer or other means to enforce the law, such as a speed camera, there are no worries about getting a ticket. This is what is meant by the police giving the law meaning.

IMPLEMENT THE LAW

Another way of saying the police enforce the law is by saying the police *implement* the law. Without the police implementing the law, laws are just words on paper. Not all laws are implemented at equal levels; some are enforced all the time while others are seldom if ever enforced. Every state and city has laws or other ordinances that are so seldom enforced that they are not even viewed as serious rules. For example, in Kennesaw, Georgia, it is required that every head of household own at least one firearm. In California, it is against the law to consume a frog that has died in a frog-jumping contest. In Aspen, Colorado, it is illegal to start or participate in a snowball fight.[2] In East Lansing, Michigan (home of Michigan State University), it is illegal to sing, hum, or play the University of Michigan (the cross-state rival) fight song. These laws (ordinances) exist with little chance that the police will actually enforce them. A more serious example is how many state legislatures have attempted to reduce the frequency of driving under the influence (DUI) by creating laws with severe sanctions. However, research has shown that the enforcement of these laws by police officers is not automatic; it depends on the priority the police chief gives to DUI enforcement, incentives to enforce the law, officer-peer influence, the availability of time, and other factors.[3]

It is also important to note that the police do not pay equal attention to all laws and all crime. Local police in particular are more oriented toward the enforcement of predatory crimes, such as murder, rape, robbery, assault, burglary, and theft, than other sorts of serious crime, such as insider trading, money laundering, investment fraud, insurance fraud, and so on. The latter are more of a priority for federal agencies and insurance companies.

THE CONTROVERSY OF LAW ENFORCEMENT

Police enforcement of the law is one reason why the police are controversial. In enforcing the law, the police use discretion. In other words, within limits, police officers can use their own

gate keepers: The role the police play in determining who comes into the criminal justice system and who stays out.

judgment in deciding the right course of action. They can decide what laws to enforce, when to enforce them, and against whom they should be enforced. These decisions can affect citizens in dramatic and negative ways. Sometimes the law enforcement decisions of officers can look more like discrimination than objective discretion. Police body-worn cameras are commonly used to make officers' decisions more visible and therefore possibly less controversial. However, with the use of body-cameras comes other controversies, such as instances when officers do not activate the cameras or when body-camera video is not released in a timely manner.[4] Indeed, it appears that in spite of attempts to reduce it, controversy is always present in law enforcement.

Photo 4.1

Laws are words on paper until police officers decide to enforce them. ©iStockphoto.com/kali9

Another controversial aspect of law enforcement is that sometimes when the police enforce the law, they have to take sides. They have to identify who is right and who is wrong and who is good and who is bad. In this sense, they are making moral judgments. As sociologist Egon Bittner put it, "Police work can . . . accomplish something for somebody only by proceeding against someone else."[5] There is no way around this issue. As a result, someone is always going to be unhappy with the actions taken by the police. This also makes the police controversial.

TIME SPENT ON LAW ENFORCEMENT ACTIVITIES

Although there is no doubt law enforcement is an important reason why we have the police, it must be understood that the police spend relatively little time on law enforcement activities. Several studies, although dated, have examined the workload of officers. Some have analyzed how officers spend their time (Exhibit 4.1); others looked at the types of calls that the police handle. All of the studies come to the same conclusion: Law enforcement makes up a relatively small portion of what the police do. This conclusion is so well established that it may explain why so little recent research has been conducted on the topic.

Similar to the Cincinnati and Wilmington studies (Exhibit 4.1), other studies that analyzed calls for service (not necessarily how time was spent) have concluded that approximately 25% of calls relate to crime; the remaining 75% involve noncrime matters.[6] A recent analysis of calls to the Portland, Oregon, Police Department also showed that more than 75% of calls were from citizens asking for the police to intervene in situations where a crime was not committed.[7] Indeed, according to criminologist James Fyfe, "Most of the people with whom the police interact need help with problems not related to crime."[8] Other studies call attention to the substantial amount of time that patrol officers spend on patrol, which ranges from 25%[9] to approximately 40%[10].

The bottom line is that although law enforcement is an important responsibility of the police, it constitutes a relatively small portion of what they do. As such, the reality of police activities is much different than what is portrayed in the news headlines or on television (Exhibit 4.2). Nevertheless, one could expect to find variation in the amount of time (or proportion of calls) devoted to law enforcement across cities, patrol beats, shifts, and officers' job assignments.

EXHIBIT 4.1

How Do Police Patrol Officers Spend Their Time?

Although these studies were conducted in different places at different times, the findings do not vary dramatically: In Cincinnati, Ohio, 18% of officers' time was spent on crime-related matters as compared to 26% of officers' time in Wilmington, Delaware. In both cities, most officer time was spent patrolling.

FIGURE 4.1

How Police Spend Their Time: Cincinnati, Ohio, Police Department

FIGURE 4.2

How Police Spend Their Time: Wilmington, Delaware, Police Department

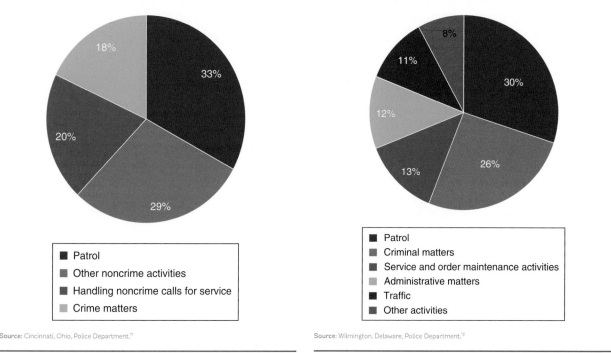

Source: Cincinnati, Ohio, Police Department.[11]

Source: Wilmington, Delaware, Police Department.[12]

THE POLICE CONTROL CRIME

Crime control is closely related to law enforcement as a reason why we have the police. We have the police so that they can control crime; law enforcement is a means by which crime may be controlled. It could be argued that the police enforce the law in order to control crime.

THE CONTROVERSY OF CRIME CONTROL

Similar to law enforcement, crime control can also be controversial. Everyone is in favor of crime control, but crime control really means that the police are seeking to control people's behavior. Unfortunately, people who engage in criminal behavior are not necessarily easy to identify. As a result, the police seek to control *everyone's* behavior. For instance, a commonly used crime control strategy in many police departments today are SQFs, which involve stopping, questioning, and frisking citizens on streets and sidewalks. For example, in Milwaukee, Wisconsin, in 2018, the police conducted approximately 7,000 subject stops (SQFs). (In 2017, more than 13,000 stops were made.) This is in a city with a population of about 600,000 people, which equates to one out of every 86 residents being stopped on the streets or sidewalks over the course of the year.[13]

EXHIBIT 4.2
"Reality" Television

The television show *COPS* is one of the longest-running shows on television, now in its thirty-first season on Spike TV. The first episode aired in 1989 and the show ran on the Fox network for twenty-five years before moving to Spike. Each episode consists of twenty-two minutes of unwritten script from police officers on the job and the citizens with whom they interact. Reportedly, to create those twenty-two minutes of program content, it takes weeks or months of filming. The show focuses on street crime and the moments of excitement and intrigue during officers' work days. The show has been criticized for not accurately portraying police work even though it is a "reality-based" program and for focusing exclusively on the crimes of the lower class and thus distorting the reality of criminal offending.

A&E's *Live PD* television show is a more recent entry into the reality-based police action genre. This show features select encounters between police and citizens in various police departments across the country shown in real-time

PD Live Archive/Bill Tompkins/Michael Ochs Archives/Getty Images

Photo 4.2
Live PD is presented as a "reality-based" police television show but it still provides a distorted view of policing.

(with a broadcast delay). As with *COPS*, *Live PD* has been criticized as focusing only on the most exciting or unusual dimensions of police work.

A RESEARCH QUESTION
Are Pedestrian Stops by the Police Effective?

Research shows that the use of street stops has increased dramatically in major cities across the United States in the past decade.[14] The popularity of the strategy has raised questions about their effectiveness as a crime control strategy and whether these stops are made in a fair manner. Studies have shown that pedestrian stops have had some success at *preventing* crime,[15] but their success at *discovering* or *interrupting* crime has been limited: From 2004 to 2010 in the New York City Police Department, 9.4% of stops resulted in an arrest, 2.9% resulted in the confiscation of drugs, and .25% of stops resulted in the discovery of an illegal gun.[16]

As for fairness of pedestrian stops, research has shown that stops have been disproportionately directed toward citizens of color.[17] A recent study also showed that black and

Hispanic citizens who were stopped were also more likely than white citizens to have force used upon them by the police.[18] These research findings have been used to support civil lawsuits against the police to stop the widespread use of the strategy. Indeed, largely as a result of such lawsuits, the use of the strategy has declined in the last few years.

Yet other research has shown that stops may lead to a deterioration in positive views of the police and make residents less likely to report crime and cooperate with the police. The outcome is a long-term negative impact on public safety.[19] As a result, there are good reasons to believe that the widespread use of pedestrian stops may cause more harm than good. The use of pedestrian and traffic stops as a police strategy is discussed in more detail in Chapter 6.

The problem is that only *after* the stop do the police determine that the vast majority of citizens are not involved in criminal behavior (see A Research Question feature). Another example of a controversial crime control tactic is when the police go undercover to identify offenders (see A Question of Ethics feature).

Simply stated, the difficulty for the police is that they are expected to control crime, but many of the tactics used to try to control crime are viewed as controversial, inappropriate, "dirty," unethical, or just plain wrong. The continuing challenge is to discover crime control strategies that actually work but are not viewed as problematic.

THE DIFFICULTY OF CRIME CONTROL

It is no easy task to control crime. The police have limited capabilities when it comes to controlling people's behaviors, especially their criminal behaviors. As such, as we discussed in Chapter 1, it has been argued that the police have an impossible mandate.[20] Some scholars have gone as far as to argue that "the presence or absence of crime has nothing to do with the police."[21]

Photo 4.3

Crime control can be controversial, often because of the methods used. In particular, sting operations raise many legal and ethical issues. Essdras M Suarez/The Boston Globe via Getty Images

POLICE LACK CONTROL OVER CONDITIONS OF CRIME One reason why crime control is difficult, if not impossible, is that the police have no control or influence over the conditions associated with criminal behavior. They do not control the rate of employment, the amount of poverty, the nature of human interactions, or whether children grow up with proper role models. They do not control the psychological or biological makeup of human beings. They do not control the demographics of the population in which they police. They do not control the weather. They do not control the policies and practices of the other components of the criminal justice system. This knowledge may lead one to the reasonable conclusion that the police can have, at best, limited impact on criminal behavior.

The Need for New Measures of Police Effectiveness

Not only is it difficult for the police to control crime, it is also difficult to *measure* the control of crime. For example, consider this likely possibility: In a retirement community, there is very little crime. In a college town or in the part of a city where a college is located, there is usually a relatively high rate of crime, especially property crime. So is it accurate to conclude the police in the retirement community are more effective at controlling crime than the police in the college town, simply based on the amount of crime in those two areas? There are many factors that may account for the difference in crime levels in the two places—factors that have nothing to do with the police. Further, the role of the police is multi-dimensional, and measures of police effectiveness should reflect these responsibilities. Therefore, for example, it is necessary for police leaders to know if citizens are being treated fairly by the police when they request services or when they are stopped by an officer, whether members of the public perceive fairness from the police in their interactions with them, if police priorities are in line with those of the community, and if services are being provided in a cost-efficient manner.[22] It is simply no longer enough to evaluate police performance in terms of crime and arrest tallies.

THE DIFFICULTIES OF DETERRING CRIMINAL BEHAVIOR Another reason why crime control is difficult is that it is hard for the police to affect people's behaviors. There are generally two approaches to controlling crime: incapacitation or deterrence. Incapacitation refers to making it impossible for people to commit crimes. For example, offenders who have been arrested and incarcerated are unable to commit additional crimes, at least outside of prison; they are incapacitated. In addition, if opportunities for criminal behavior are taken away, offenders may be incapacitated. For instance, if a would-be offender does not have a gun or access to a gun, that person is not going to be able to shoot someone, even if motivated to do so (although another weapon could be used). If a person is incapacitated in some way, he or she will not be able to commit a crime, or at least a specific type of crime. As such, the criminal behavior of that person will be controlled.

While incapacitation as a way to control criminal behavior is relatively straightforward, crime control through deterrence is not. There are many important issues that need to be considered when thinking about the ability of the police to deter people from committing criminal behaviors. In order to deter behavior, to get someone to choose one behavior over another when both behaviors are an option, there have to be consequences for the undesired behavior. In the context of criminal behavior, the consequences usually come in the form of punishment. The first thing to note is the police usually do not control the punishment associated with a crime unless an arrest, a fine, or perhaps a use of force is considered the punishment. Moreover, in order for punishment to be an effective deterrent, it has to have at least three qualities. It has to be certain, meaning that the individual must believe that if the behavior is committed, punishment will follow. Punishment also has to be swift; it has to be administered quickly. Finally, punishment has to be individually meaningful. For example, a parking ticket of $50 might be individually meaningful to a person with a moderate income but not to a millionaire. Basically, in large measure the police are not able to control the certainty of punishment, the swiftness of punishment, or if the punishment is individually meaningful to the offender. The police are responsible for identifying and apprehending offenders but have relatively limited success in doing so (e.g., 46% of violent crimes and 17% of property crimes were solved in 2017). In this light, it is difficult to see how the police can deter people who are so inclined from committing crimes.

Further and even more fundamentally, another part of the difficulty in concluding whether or not—or to what degree—the police are able to deter people from committing crimes is defining exactly what is meant by deterrence and determining how it is measured. Is a man deterred from committing a bank robbery if he decides to commit it tomorrow instead of today? What if the would-be robber sees the police at a particular bank so he decides not to rob

> **incapacitation:** Making it impossible for people to commit crimes.
>
> **deterrence:** Making someone decide not to do something.

EXHIBIT 4.3

Tendencies Toward Criminal Behavior and the Impact of the Police

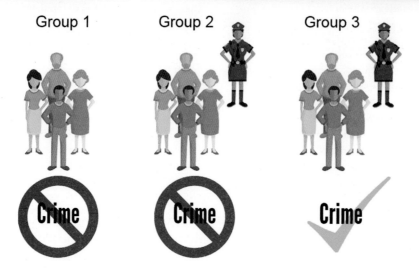

Group 1 consists of people who would obey the law even if there were no police.

Group 2 consists of people who obey the law because they fear the consequences of breaking it.

Group 3 consists of people who break the law even though they may get arrested by the police; they have no regard for the consequences. This group consists of criminals.

Some people *do not* need to be deterred; other people *will not* be deterred.

that bank but instead robs a different one down the block? Or decides to commit a bank robbery in a different city? What if instead of committing a robbery, he commits a burglary? Think about this issue in terms of the police trying to deter speeders. When you are driving down the street and see a patrol car, chances are you slow down. And then, just maybe, after you see that the officer is not behind you with lights and siren, you speed back up. Where you deterred from speeding? Deterrence is a slippery and difficult concept to define.

It is also important to recognize that not all people are equal in their criminal tendencies (see Exhibit 4.3). At some level of abstraction, there are probably people who would never commit a crime, or at least a serious crime. These people would obey the law even if there were no police, no threat of punishment (Group 1). On the other extreme are people who commit crimes with no regard for the police (Group 3). They are motivated to commit crimes regardless of the threat of police action. It is probably fair to say that no matter what the police do, these people will still commit crimes. The people in the middle are the people who would likely be affected by police actions and strategies to control crime (Group 2). Before committing a crime, they may consider the risks of getting caught by the police and punished by the criminal justice system. The point is that the police do not deter some people and do not need to deter others.

THE POLICE DEAL WITH SITUATIONS WHERE FORCE MAY NEED TO BE USED

The tools used in occupations give clues about why those occupations exist. Farmers use tractors because they cultivate crops and grow food for people to eat. Garbage collectors use trucks because people create garbage that needs to be removed. Teachers use books because people

The Continuum of Force, OC Spray, and Tasers

Although use of force is discussed in detail in Chapter 11, it is useful to introduce the concept of the continuum of force and less lethal force here.

There are many different versions of the continuum of force, but the basic idea behind all of them is that the police must only use as much force as necessary to overcome the resistance offered by a noncompliant subject. The initial steps on the continuum consist of officer presence and verbal commands. If these forms of force do not induce compliance, then it is appropriate to use *physical* force. The most common form of physical force is *bodily* force wherein officers use their bodies to gain control and compliance over subjects. If bodily force does not work, then a weapon such as a Taser or oleoresin capsicum (OC) spray is appropriately used. Finally, if none of these options produces subject compliance or is appropriate given the situation (for example, the subject is armed and poses an immediate threat to officers or citizens), then deadly force (via a firearm) would be appropriately used.

OC spray and Tasers are generally considered to be forms of force that are less likely to be lethal, at least when compared to firearms. OC is an inflammatory agent naturally found in cayenne peppers. Ideally, when a person is sprayed in the face with OC, the effects are immediate: The respiratory tract becomes inflamed; the subject experiences an intense burning sensation and swelling around the eyes, which close involuntarily; and the subject is no longer able to resist. Police departments have been using OC spray since the 1980s.

TASER (short for Thomas A. Swift Electric Rifle) is the brand name of an electronic stun device and is the most popular such device on the market. Tasers were first introduced in police departments in the 1990s. A Taser resembles a gun. When the trigger is pulled, two wired probes are discharged from the gun and fired into the body of a subject. An electrical current then runs from the weapon through the wires and into the subject's body. When both probes attach to the subject, the Taser delivers the electrical current, which overrides the central nervous system. This causes involuntary muscle contractions and the incapacitation of the resisting subject. The weapon can also be used in drive-stun mode. In this mode, the weapon itself is placed in contact with the subject's body to deliver the electrical current. Used in this way, the weapon causes localized pain that can be used to allow officers to overcome the resistance of the subject. In many police departments, the use of a Taser in drive-stun mode is discouraged or even prohibited except in extreme circumstances.

need to be educated. Physicians use stethoscopes and a variety of other instruments to diagnose illnesses so they can heal people.

Using this line of reasoning, the tools used by police include handcuffs, batons, firearms, pepper spray, and Tasers, among others. Arguably, the most important tool of the police is their ability to use force, or at least the threat of force (see Technology on the Job feature). The police have the authority to use force so that people listen to them and obey their orders. If the police are responsible for regulating the conduct of citizens, having such a tool can certainly help. Citizens might obey the orders of the police because, if they do not, the police may use force upon them. Indeed, scholars have identified police authority to use force as the central feature

A Question to Consider 4.1

Does the Authority to Use Force Really Make the Police Unique?

Although it has been argued that authority to use force differentiates the police occupation from all others, upon further reflection, it appears that people in many other occupations also have the authority to use force—for example, bouncers at a bar, football players, soldiers, and martial arts fighters. But how does police authority to use force differ from these occupations? Explain your answer. Hint: Think about the situations in which the police can use force, compared to people in these other occupations.

Photo 4.4

Many of the tools of the police, such as Tasers and firearms, relate to the use of force against subjects. Scholars argue that the authority to use force makes the police a unique and inherently controversial occupation.

AP Photo/Alex Milan Tracy

of the police function—a feature that differentiates the police occupation from all others.[23] However, it is not just the authority to use force that makes the police unique; it is that the police can potentially use force in virtually any situation or location.

Police use of force is controversial to say the least, especially when force is used in situations where it may not be justified or necessary and/or when the force being used is deadly. Police use of force, justified or not, never photographs well. Part of the issue here is that citizens and police officers (as well as prosecutors, judges, and juries) may disagree about whether the force used was actually necessary. Many of the urban riots in the 1960s began as a result of incidents in which white officers shot African American citizens. In 1991, areas of Los Angeles burned and businesses were looted after the white officers who beat Rodney King, an African American man, were found not guilty of assault by a jury (several officers were later found guilty on federal charges). In 2014, 2015, 2016, and 2017, cities across the country saw demonstrations, marches, and, in some places, riots to protest the killing of black men by white officers in New York City; Baltimore, Maryland; Ferguson, Missouri; Baton Rouge, Louisiana; and other cities. The sniper killing of five police officers in Dallas, Texas, in 2016 was apparently motivated by two high-profile police killings of black men that same year.

THE POLICE RESPOND TO PEOPLE IN NEED

Along with enforcing the law, controlling crime, and handling situations where force may need to be used, another reason for the police is to respond to people in need. In particular, the police have frequent contacts with people who are mentally ill, more now than ever before. It is estimated that between 7% to 10% of police encounters involve people with mental illness.[24] These contacts may be with people in crisis (e.g., out-of-control, bizarre behaviors, suicide attempts) or because of serious but more chronic (e.g., homelessness) mental health issues or because of other more "nuisance" or disorderly type behaviors (e.g., a person walking on a roadway). Most police contacts with persons with mental illness do not involve major crimes or violence.[25]

There are at least two reasons for the frequent contact between the police and persons with mental illness. First, even though the concerning behaviors may not be criminal, the police have responsibilities for maintaining peace and safety. Since citizens have quick and easy access to the police to deal with such situations, it is much easier and quicker to summon the police than other social services. As in other situations, the police have these responsibilities simply because they are designated as first responders.

The second reason for more contacts between the police and people with mental illness is that there are more persons with mental illness now living in communities. Decades ago, persons with serious mental illness were more likely to be institutionalized and required to receive treatment. Today, involuntary institutionalization is reserved for only the most extreme of instances. Mental illness is most frequently managed through prescribed drugs for people who live in the community. Indeed, police officers often hear from relatives of people who are experiencing a mental health crisis that "he was fine until he stopped taking his medication."

There are potential problems and difficulties when the police intervene in situations that involve persons with mental illness. First, traditionally the police have not received training on recognizing mental illness or its various symptoms. Complicating matters is that sometimes substance abuse and physical, cognitive, or other emotional problems can masquerade as mental illness. How can police officers be expected to provide "psychiatric first aid"[26] without being trained on this complex phenomenon?

Second, interactions with people with certain mental illnesses can be dangerous for officers and well as the individual, especially when mental illness is not accurately recognized or properly handled. The third problem when intervening in situations that involve persons with mental illness is that there are limited options available to the police to resolve the situation. Depending on the situation, police officers may transport the individual to a psychiatric facility for immediate treatment; try to diffuse the situation on-scene, often with a warning; or arrest the person. These options are not a solution to the problem but are simply ways for the police to manage the situation and resolve immediate concerns.

The best police practice when it comes to interactions with persons with mental illness is using crisis intervention teams (CIT).[27] Crisis intervention teams consist of police officers who have received extensive training on how to recognize mental illness and how to effectively respond to it. Officers receive training in such topics as post-traumatic stress disorder (PTSD),

Photo 4.5
Some of the many responsibilities of the police have very little to do with law enforcement or crime control. Some, such as directing traffic, are provided as a public service. ©iStockphoto.com/erreti

traumatic brain injury, psychotropic medications, suicide prevention, personality disorders, and the law relating to mental illness, to name a few. Skills such as verbal de-escalation are also a critical part of the training. The logic of CITs is that, with mental health expertise, officers will be in a better position to more effectively and safely resolve such situations. CIT also represents a more systematic and informed approach to dealing with mental health issues in the community. CIT requires the creation of relationships between the police and mental health resources and treatment options. CIT requires an entirely more systematic approach to mental health related calls for service. Police dispatchers need to send CIT officers to calls, police supervisors need to schedule adequate numbers of CIT officers to each shift, and partnerships and processes with mental health service providers need to be created, formalized, and used. Police services are recognized as the front-end of much larger process. CIT is a more holistic versus haphazard approach to effectively managing mental health needs in the community. The CIT approach is much more congruent with a service-oriented role than one focused on law enforcement.

THE POLICE HANDLE TIME-PRESSING SITUATIONS

Some scholars explain that we have the police in order to deal with a wide variety of situations where *something* is not right—situations where "something ought not to be happening, about which something ought to be done now!"[28] A gas station has been robbed, a neighbor is threatening another neighbor, traffic lights are not working, people are arguing, a dog is on the loose. Police officers may be asked to intervene in each of these situations. Some involve crime and law enforcement, but most do not. Some involve regulating citizens' behavior, others do not. Some involve the potential use of force, but not all of them do. These situations cannot wait for some future resolution; they need to be handled immediately. Cars may collide and the dog might bite someone. As this clearly illustrates, the police are needed for more than law enforcement, crime control, the use of force, or to assist persons with special needs. As criminologist James Fyfe wrote, "The police perform a variety of services that must be available seven days a week, 24 hours-a-day, that may require the use or threat of force, and that are not readily available from any other public agency or private institution."[29]

THE POLICE ENFORCE THE LAW, MAINTAIN ORDER, AND PROVIDE SERVICES

Perhaps the simplest way of explaining why we have the police—but one that still describes the breadth of the role—is to highlight the important tasks of law enforcement, order maintenance, and service. As discussed, law enforcement refers to tasks such as conducting investigations and making arrests. Order maintenance consists of activities that keep the public peace, which involves regulating citizens' conduct without resorting to citations or arrests. Service activities include all other activities of the police besides order maintenance and law enforcement, such as assisting motorists whose cars have broken down, getting keys out of locked cars, and attending to other people in need.

order maintenance: Activities of the police that involve keeping the peace without resorting to citations or arrests.

service: Other duties performed by police, such as assisting stranded motorists and attending to other people in need.

law enforcement: Tasks performed by the police that involve conducting investigations and making arrests.

EXHIBIT 4.4

Every Second Counts: Responding to Drug Overdoses

Police patrol is staffed and organized so that officers can quickly respond to emergencies, everything from crimes in progress to traffic accidents. This ability to respond to situations where "something ought not to be happening, about which something ought to be done now!" leads to the police sharing some emergency responses with emergency medical personnel such as fire and rescue. Case in point: Opioid overdoses and naloxone. Opioids most often include heroin, prescription pain killers, and fentanyl and its variants. Opioid abuse and overdoses have reached epidemic proportions. Drug overdose is now the leading cause of death for Americans under age 50; approximately 180 Americas die every day from opioid overdoses.[30] Opioids kill by depressing respiration to the point that the brain receives inadequate oxygen, a condition called hypoxia. Naloxone (or Narcan brand name) is a drug that, if delivered within minutes of overdose, can reverse the effects of opioids and restore normal breathing. It is typically delivered via a nasal spray. When administered in time, survival is likely.[31] It is not

a dangerous drug, although people who are revived by the drug may show extreme agitation. Required training to recognize overdoses and to administer the drug is not extensive. Most states provide legal authorization for police officers to carry and administer naloxone, and most states have police agencies that carry the antidote.[32] The biggest challenge is usually the cost of equipping all patrol cars with the medication, as its cost is approximately $100 per dose. While police officers armed with naloxone can save lives by being in the right place at the right time, it must be understood that these measures do very little in addressing the deeper and more complex problem of opioid abuse in our society.

Naloxone is one way by which drug abuse affects the job of police officers. Drug use also creates significant law enforcement work for the police in terms of responding to and investigating crimes committed by drug users to support an addiction, enforcement of illegal drug sale crimes, and dealing with interpersonal conflicts that result from drug abuse.

Several observations can be made regarding this classification of activities and responsibilities. First, over time police departments have emphasized some responsibilities over others. For example, during the political era, service to politicians and order maintenance were the focus of the majority of police departments. In the reform era, law enforcement was the central objective. In the community problem-solving era, service to citizens and order maintenance are probably most strongly emphasized.

Second, the lines that differentiate service, law enforcement, and order maintenance are at times rather blurry. For example, when the police ask a citizen to turn down his or her music because the neighbors are complaining, it would best be considered order maintenance. When the police return and issue a citation for loud music, it would be law enforcement. Police patrol could be considered a service to the community, order maintenance, or perhaps even law enforcement.

Photo 4.6

The police sometimes find themselves in a difficult position: They are supposed to preserve the public peace, but at the same time, they are supposed to protect the right of people to protest. Protest Against Racism and Hate in Chicago/Anadolu Agency/Anadolu Agency/Getty Images

Third, law enforcement and order maintenance may conflict. For example, citizens have the right to protest, and the police are supposed to protect that right, but the police are also supposed to preserve the public order. Sometimes the police make arrests of citizens who protest. This conflict underscores the difficult position the police sometimes find themselves in.

Fourth, some activities performed by the police, especially service activities, can be provided by other agencies. For example, citizens who lock themselves out of their cars can rely on locksmiths, not the police. Residents in some community neighborhoods rely on private security companies to provide or supplement patrol. Some departments contract with private companies to help investigate vehicle accidents.[33] The reasoning is that, by reducing work demands of the police, it allows the police to focus on other more essential responsibilities. Conversely, some departments voluntarily take on service responsibilities that are more in line with the mandate of other agencies; "scoop and run," as described in the introduction to this chapter, is an example.

Finally, different police departments emphasize different aspects of the policing role. A police department is not a "one size fits all" entity. One community may have disparate expectations of their police compared to another community. For example, in his classic book, *Varieties of Police Behavior*,[34] James Q. Wilson identified three styles of policing reflected by police departments: the legalistic, watchman, and service-oriented styles. In departments that adhere to a **legalistic style of policing**, enforcement of the law is the top priority. Not surprisingly, officers in legalistic departments make a lot of arrests and issue a lot of citations. In departments that follow a **watchman style of policing**, order maintenance and controlling disorder are the top priority. Police departments that reflect the **service-oriented style of policing** believe citizen satisfaction is the top priority. The main goal is to serve the citizens of the community; arrests are made when necessary, but police action is mostly geared toward providing services and assistance to citizens.

To illustrate the different styles of policing, consider a situation in which a police officer discovers juveniles who are in violation of a curfew ordinance. In the legalistic departments, the juveniles would most likely be issued citations. In the watchman-style departments, they would probably be brought to the police department and their parents called to pick them up. In the service-oriented departments, the juveniles would most likely be taken home to their parents, perhaps without legal consequences.

legalistic style of policing: Policing style in which enforcing the law is the top priority.

watchman style of policing: Policing style that focuses on order maintenance.

service-oriented style of policing: Policing style that maintains citizen satisfaction is the top priority.

Just like police departments may have certain priorities and orientations, so may police officers.[35] It is not a stretch of the imagination to think that police officers bring their own ideas, preferences, styles, and belief systems to the job. These attitudes relate to officers' personalities and their occupational outlooks or the way they view their role. The fact is that the role of the police varies based not only on the police department but also the officers who work within a department.

Main Points

- We have the police for various reasons:
  - To control crime
  - To deal with situations where force may be needed
  - To respond to people with special needs
  - To handle situations where something ought not to be happening and about which something needs to be done immediately
  - To enforce the law, maintain order, and provide services
- When police enforce the law, they give it meaning. The police do not pay equal attention to all laws or to all offenders.
- Although law enforcement is an important function of the police, as it is critical to the functioning of the entire criminal justice system, the police spend relatively little time on law enforcement activities.
- Crime control is a difficult, if not impossible, mandate. This means that because the police cannot control the factors associated with crime, they cannot control crime.
- Many strategies used to try to achieve the crime control mandate are controversial, such as pedestrian and vehicle stops.

- Crime control is difficult to measure. Crime statistics are flawed, and variation in crime levels is not easily attributed to police actions.
- Ideally, crime can be controlled through the incapacitation and/or deterrence of criminals. Deterrence is difficult for statisticians to measure and for the police to provide. Some people do not need to be deterred; other people cannot be deterred.
- One of the most important tools of the police is their ability and authority to use force; that is a defining feature of the police occupation. However, police use of force can be controversial.
- Not only are some functions of the police difficult or impossible to achieve, some conflict with one another.
- A police department is not a "one size fits all" entity. Different communities have different expectations of their police. As a result, departments may emphasize different aspects of the policing role (law enforcement, order maintenance, or service). The outlooks of police officers may also determine the role orientation of their departments.

Important Terms

Review key terms with eFlashcards at **edge.sagepub.com/brandl2e**.

deterrence   69
gate keepers   64
incapacitation   69
law enforcement   74
legalistic style of policing   75

order maintenance   74
service   74
service-oriented style of policing   75
watchman style of policing   75

Questions for Discussion and Review

Take a practice quiz at **edge.sagepub.com/brandl2e.**

1. Why do we have the police? Hypothetically speaking, what would happen if we did not have the police?
2. Is law enforcement the most important aspect of the police role? Why or why not?
3. Why is crime control a potentially controversial responsibility of the police?
4. How can it be argued that the police have an impossible mandate? Beside that it is impossible, why does an impossible mandate represent a problem for the police?
5. Why are many crime control strategies controversial?
6. Why is the goal of crime control difficult to measure accurately?
7. What is deterrence? Why is it difficult for the police to deter criminals?
8. Why is responding to people with mental illness an important responsibility of the police? How should the police best respond to people with mental illness?
9. Many scholars argue that police authority to use force is the defining feature of the police role. Do you agree? Why or why not?
10. Why is a police department not a "one size fits all" entity? According to James Q. Wilson, how do police departments vary in their role orientation?

Fact or Fiction Answers

1. Fiction
2. Fiction
3. Fiction
4. Fiction
5. Fiction

Digital Resources

Get the tools you need to sharpen your study skills. SAGE Edge offers a robust online environment featuring an impressive array of free tools and resources.

Access practice quizzes, eFlashcards, video, and multimedia at **edge.sagepub.com/brandl2e.**

Media Library

View these videos and more in the interactive eBook version of this text!

Career Video
4.1: Day in the Life of a Police Officer
4.2: The Most Enjoyable Aspect of Being a Police Officer

SAGE News Clip
4.1: Claim Filed to Urge Police Reform

Part II
POLICE WORK

5 POLICE RECRUITMENT, SELECTION, AND TRAINING

Police Spotlight: "Are You Ready for Your Next Challenge?"

Years past, most police departments could create a police officer recruitment advertisement, post it in the newspaper or on their website, and then just watch the applications come in. Those days are no more. Many departments today, especially large departments, report challenges in attracting adequate numbers of qualified individuals to apply for police officer positions. Part of the issue is that many police officers are at or approaching retirement age; another is the tight labor market in general as well as other more desirable career opportunities as perceived by many younger people. The San Diego (CA) Police Department is a good example of a department facing a chronic shortage of officers. As a response, the department has hired a marketing firm to create recruitment videos and a new recruitment website to explain the process and benefits of becoming a police officer in San Diego. The headline on the website is "Are you ready for your next challenge?" (see joinsdpdnow.com). In an effort to reduce turnover among officers and enhance recruitment efforts, the department also gave pay raises of 25% to officers. The department revised recruitment policies, selection exams, and procedures and standards for background checks. According to San Diego mayor Kevin Faulconer, "The best way to keep San Diego one of the safest big cities in the country is by attracting the best and the brightest recruits to the San Diego Police Department. We're going to be innovative and creative in how we do that."[1]

Objectives

After reading this chapter you will be able to:

5.1 Discuss the importance of the recruitment process in police departments

5.2 Explain the most common recruitment strategies used by police departments and how the benefits provided by particular police departments may relate to the success of the recruitment process

5.3 Discuss the importance of the selection process in police departments

5.4 Summarize the most important equal opportunity employment (EOE) laws as well as the purpose and potential criticisms associated with affirmative action policies

5.5 Describe the importance of bona fide occupational qualifications in the selection of police officers

5.6 Identify and discuss the four most common preemployment tests used by police departments as well as other tests used by some departments

5.7 Evaluate the representation of female and racial minority group officers in police departments and discuss the possible reasons for their underrepresentation

5.8 State the importance of training in police departments and identify the training stages

Fact or Fiction

To assess your knowledge of police recruitment, selection, and training, prior to reading this chapter, identify each of the following statements as fact or fiction. (See page 107 at the end of this chapter for answers.)

1. The presence of good officers in a department depends on appropriate selection standards, a thorough selection process, and adequate training. The recruitment process is not nearly as important.

2. Most police officers tend to spend their entire career in a single police department—the one in which they were first hired.

THE proper functioning of police departments depends on, at a minimum, selecting the right people for police careers from an inclusive pool of applicants and then providing those people with the best training on the skills of the job. This chapter focuses on these critical processes.

(Continued)

(Continued)

3. One of the top five reasons expressed by police recruits *and* experienced officers for wanting to become a police officer is to help people.

4. Based on equal opportunity employment laws, it is not legal to discriminate for any reason when selecting police officers to hire.

5. Because of claims of reverse discrimination, affirmative action policies are no longer used in most police departments.

6. Psychological examinations are more frequently used by police departments in selecting officer applicants than are background investigations.

7. There are more racial minority police officers than there are women police officers.

8. Most police academies are operated by police departments.

9. Field training usually consists of one year of training with a field training officer (FTO).

10. In-service training is only for officers who have been identified as requiring additional training on the basic tasks of police work.

THE RELATIONSHIP BETWEEN THE RECRUITMENT, SELECTION, AND TRAINING OF POLICE OFFICERS

In the Police Spotlight feature, San Diego mayor Kevin Faulconer makes reference to the need for "the best and brightest" police officers. Indeed, a well-performing police department needs good officers, and this depends largely on strong recruitment efforts, appropriate selection standards, a thorough selection process, and adequate training. Each process depends on the others: Recruitment of a large pool of applicants allows for a more selective process of hiring. Well-qualified applicants can be effectively trained. Well-qualified and well-trained officers may serve as good supervisors of other officers and as good leaders of the department.

DIVERSITY BEGINS WITH RECRUITMENT AND SELECTION

An important goal of recruitment efforts and the selection process is to maximize diversity among officers. Without question, diversity in the workforce is a legitimate and important goal for police departments to pursue. A police department should have a workforce that reflects the characteristics of the community it serves. If recruitment efforts result in a diverse pool of applicants, this provides an opportunity for diverse officers to be hired. Conversely, if recruitment does not lead to a diverse pool of applicants, it guarantees that the selection process will not result in diversity among hires.

Diversity in police departments is a good thing and has been loudly called for since the 1960s. People with different experiences and backgrounds bring different ideas, understandings, and skills to the job. It is argued that through the sharing of this information, police officers can become more knowledgeable of the life situations of people in the community. Diversity comes in many forms: race, ethnicity, religion, gender, age, sexual orientation, and even skills and experiences. Shared understandings may provide a basis for more appropriate and compassionate behavior in interactions with citizens. In addition, there may be less tension between the police and the community when the police reflect the characteristics of the community. Indeed, research supports this view: The public tends to associate diversity among police officers with impartiality, fairness, and trustworthiness among the police.[2] As an example of when this was not the case, as discussed earlier in the book, consider urban policing in the 1960s. Here, the predominantly white officers that policed the urban ghettos became viewed as "an occupying army" that did not have the interests of the community in mind. This created conditions where poor police–minority relations flourished.

Photo 5.1

Police departments are more diverse today than ever before, but many of them still seek to make progress on this front. Bloomberg/Bloomberg/Getty Images

THE RECRUITMENT OF POLICE OFFICERS

Recruitment, which consists of efforts to create a group of qualified applicants to hire from, sets the stage for the selection of officers. Ideally, with more applicants from which to choose, the more selective police managers can be in choosing people believed to be most qualified for the job. For example, if five police officers are to be hired and there are 100 applicants from which to choose, it allows managers to be more discerning in identifying the best applicants than if there are only ten applicants. In essence, all else equal, the more applicants there are, the more likely well-qualified and desirable individuals will be included and selected.

Recruitment is designed to create a pool of applicants that (a) includes the best and most diverse individuals that would be well suited for a law enforcement career and (b) would be interested in working in the particular police department that is conducting the recruitment. This brings up two questions: Who wants to be a police officer, and who wants to be a police officer in a particular department? A national survey conducted in 2017 found that in government employment, police officer positions were the most difficult to fill.[3] Independently, many police administrators have also highlighted the recent difficulties in recruitment efforts. Some have argued that recent high-profile controversial policing events and negative images of policing have reduced the number of people who wish to become police officers. According to Chuck Wexler, the executive director of the Police Executive Research Forum (PERF), "We go through periods where all of a sudden, being a police officer isn't as attractive as it has been in the past. And, quite frankly, I think that's what we are seeing."[4] If this is the case, logically it would follow that recruitment has become more difficult but no less important. Another important issue related to police recruitment is that individuals who are interested in police careers today are different in many important respects from those who sought such careers decades ago; they have different skills, abilities, and values.[5] Recruiting strategies, as well as the operations of police departments themselves, may need to change to attract and retain these new officers.

MOTIVATIONS FOR BEING A POLICE OFFICER

As a basis for recruitment efforts, it is important to understand why people wish to become police officers. These reasons would seem to be influential in attracting individuals to the job and important in "selling" the job to potential applicants. When asked of academy recruits, the three most common reasons given for becoming a police officer were the opportunity to help others; job benefits (e.g., health insurance, retirement benefits); and job security. There was little variation in motivations across race and gender of recruits.[6] After six years on the job, motivations for being a police officer remained stable. The top five motivations expressed by officers were job security, job benefits, early retirement, opportunities for job advancement, and opportunity to help people in the community.

Photo 5.2

The importance of the recruitment process in hiring good police officers cannot be underestimated. Jeff Greenberg/Alamy Stock Photo

At this time, there were still minimal differences in responses across race and gender.[7] A recently conducted experiment showed that efforts to solicit applications were more effective when emphasizing "the challenge" of the job and the job as a "long-term career" compared to "service" as a motivation.[8] Each of these factors may represent significant "selling points" of the occupation and, accordingly, could be emphasized in recruitment efforts.

RECRUITMENT STRATEGIES AND PLANS

At a minimum, an adequately resourced recruitment plan is necessary to create a well-qualified pool of applicants. Since recruitment efforts are so important, most police departments have a budget devoted to them. As one would expect, recruitment budgets are generally more substantial in larger departments than in smaller ones. This makes sense as larger departments hire more officers and, as a result, can benefit from larger pools of applicants.

How do police leaders go about identifying qualified individuals who are interested in a law enforcement career, and how do they get those individuals to apply for a position in their particular police department? Variation exists among departments in how this is done, most of which relates to the size of the department doing the recruiting and the resources they have available.[9] The three most commonly used recruitment methods in police agencies are newspaper advertisements, Internet advertisements, and personal contacts and referrals (sometimes providing incentives to current officers for identifying candidates). Other methods that are more likely to be used by larger agencies include job fairs; recruiting at special events (e.g., information booths at sporting or other entertainment events); and radio, magazine, and television advertisements.[10] In response to a shortage of applicants, some departments have directed their recruitment efforts to other communities and even other states.[11] Regardless of the methods used to solicit applications, research highlights the importance of timely pre-employment tests, a process that is clear and understandable from the perspective of the applicant, and the value of personalization in recruitment efforts.[12]

Community service officer (CSO) programs or police aide programs may also be used as a recruitment tool. These programs involve police departments hiring college students as nonsworn personnel. They wear uniforms that are different from those worn by sworn officers. The CSOs have varied responsibilities, depending on the department. They may conduct patrol and respond to nonemergency calls (e.g., issue parking tickets, direct traffic, and respond to animal complaints); answer phones; and/or assist with the processing of arrestees and other administrative matters. If their performance is satisfactory, these individuals may then automatically advance for hiring consideration as police officers in that department. The **Police Explorer programs** operated by some departments serve a similar purpose. Police Explorers is a voluntary program for young people aged fourteen to twenty designed to educate them about policing activities and to provide them training and hands-on experience with matters related to a law enforcement career. At the conclusion of the program, participants may be selected for police academy training and employment in that department.

Police departments may target certain groups of individuals for recruitment. The most common example of this is when agencies make special efforts to recruit applicants who possess

community service officer (CSO) programs: Programs wherein college students are hired as nonsworn police personnel. CSOs may assist in nonemergency situations and perform administrative duties.

Police Explorer programs: Voluntary programs for young people aged fourteen to twenty designed to educate them about policing activities and to provide them training and hands-on experience with matters related to a law enforcement career.

prior law enforcement experience.[13] This holds true for both large and small departments. From the perspective of the police department, this practice makes sense in that the department would then probably not have to pay for academy training for the officer or pay the officer's salary during the academy. A police officer may be willing to leave one department for another if it means a higher salary or better working conditions (e.g., a lighter workload) even though he or she will likely lose seniority (discussed below). A majority of departments with more than 500 sworn officers report they also strenuously attempt to recruit women, people of color, military veterans, multilingual persons, and graduates of four-year colleges.[14]

JOB BENEFITS AND RECRUITMENT EFFORTS

The benefits associated with the position may also have implications for how many people apply for the job. Salary is typically the most important benefit, but other factors, such as the reputation of the department, location of the department, and policies of the department, may also play a key role in an individual's decision to apply for police work.

SALARY In 2017 (the most recent year for which statistics are available), the national average salary for police and sheriff's patrol officers was $64,000.[15] However, this figure masks the significant variation that exists in police officer pay. Starting salaries and top salaries vary widely by department, even departments in the same state. Generally speaking, higher pay is provided by agencies that serve a population of 50,000 to 249,999 residents, and the lowest pay is provided by agencies that serve fewer than 2,500 residents.[16] However, even within these categories, there is major variation in salaries. For example, among the largest cities in Pennsylvania and the suburbs of Pittsburgh, the lowest starting salary is $40,000 and the highest is $80,000. Top salaries range from $60,000 to nearly $100,000.[17] In addition, overtime pay can substantially increase officers' salaries. Although exceptional, not unheard of is officers doubling their base pay as a result of overtime. Many factors may affect the pay rate of officers, not least of which is the ability of police unions in many states to negotiate police contracts.[18] Detectives and supervisors may receive a substantially higher salary than police officers, and police commanders and chiefs may receive a significantly higher salary than police detectives or supervisors. Depending on the city, most police chief salaries range from $75,000 to $200,000. However, higher-ranking positions are usually paid on a salary and therefore not eligible for overtime compensation.

Not surprisingly, salary is a major consideration among individuals who are evaluating police officer positions. To attract applicants, many police departments have substantially increased officers' pay. Nationally, since 2010, police officer salaries have increased by over 15% on average, outpacing other public service workers.[19] Some departments have increased salaries by 25% or 30% in recent years, and other departments offer signing bonuses.[20] While dramatically higher salaries may help solve the recruitment problem, it causes another: Departments that cannot afford to raise wages lose the "wage war" and, as a result, also lose officers and applicants to the higher paying departments. Some departments that pay higher salaries to officers have ended up laying off officers due to budget shortfalls.[21] Clearly, difficulties in recruitment have created challenges for many departments.

OTHER BENEFITS AND POLICIES Other common benefits that may enhance the desirability of police employment include free academy training for recruits, paid salary during academy training, and uniform and equipment allowances. Some departments, especially larger ones, offer extra pay to those with a college degree and/or provide college tuition reimbursement. Less commonly offered benefits may include flexible hours, signing bonuses, academy graduation bonuses, moving expenses, health club fee subsidies,[22] and student loan repayment.[23]

Residency requirements may also influence an applicant's decision about where to work. Most police departments have residency requirements; however, there is variation in how restrictive these requirements are. Most agencies do not require officers to live in the city in which they work. More common is that officers must live within a certain distance of the city. A few agencies are not restrictive at all regarding where officers live and only require they live in the state where their city of employment is located.

JOB REQUIREMENTS AND SELECTION STANDARDS

Job requirements and selection standards also impact recruitment efforts: All else equal, fewer or lower selection standards may produce a larger and more diverse applicant pool. However, while lower standards may solve one problem (too few applicants from which to select), it may cause another (the hiring of people who should not be police officers). Police departments must balance job requirements and selection standards with recruitment realities.

THE SELECTION OF POLICE OFFICERS

The selection process involves evaluating applicants in relation to important qualities, job-related qualifications, and test performance, and selecting the best applicants available. As discussed below, various standards, qualifications, and tests are used to try to predict who is most likely to be a well-performing police officer. The goal is to identify and hire those people.

Over the course of an officer's career, he or she may be involved in many selection processes. The first is the process to be hired as a police officer; others may involve promotions and assignments, including the selection process for the chief of the department. The discussion provided here is focused primarily, but not exclusively, on the initial selection process of being hired as a police officer.

No two police departments have the exact same selection process, and the general process followed in local police departments differs in important respects from that of other agencies, such as federal law enforcement agencies. However, regardless of the agency and the position at hand, selection processes are heavily guided by the law.

THE PERMANENCE OF SELECTION DECISIONS

Most police officers tend to spend their entire careers in a single police department—the one in which they were first hired. This is primarily for three reasons. First, once a police officer is hired and successfully completes a mandatory probationary period, it is very difficult to dismiss an officer or otherwise terminate his or her employment. This is due primarily to **civil service protections** in place in public sector employment. These protections are designed to minimize political considerations in job-related processes. They ensure that government employees are recruited, hired, and promoted based on qualifications and merit and demoted or dismissed only for cause. As a result, when officers get fired it is usually the result of a conviction for a felony or other serious infraction, which is an uncommon event (Exhibit 5.1). Officers often identify job security as an important benefit of the job; this is due at least in part to civil service protections and also because it is very rare for police officers to be laid off.

A second reason why police officers tend to spend their entire career in a single department is that officer seniority (and sometimes retirement eligibility) is usually department specific, and in most departments, seniority is a basis for job assignments and preferences. To lose seniority is to lose opportunity for preferred work assignments. For example, an officer may begin his or

civil service protections: Protections that ensure government employees are recruited, hired, and promoted based on qualifications and merit and demoted or dismissed only for cause.

EXHIBIT 5.1

EXHIBIT 5.1

Facts and Figures on Officer Separations

Approximately 11% of officers in all state and local law enforcement agencies leave (or *separate from*) their employment each year.[24] Police departments with fewer than ten sworn officers experience a greater rate of turnover than agencies with 500 or more officers—approximately 20% versus 5%. Turnover occurs when people leave a job and are replaced with others.

FIGURE 5.1

Reasons for Police Officer Employment Separations

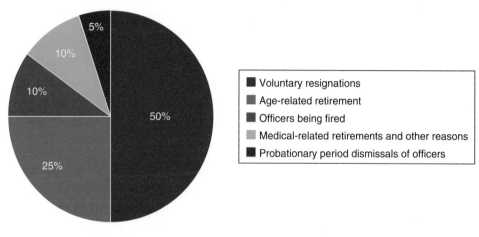

- Voluntary resignations
- Age-related retirement
- Officers being fired
- Medical-related retirements and other reasons
- Probationary period dismissals of officers

Source: Brian A. Reaves. 2012. *Hiring and Retention of State and Local Law Enforcement Officers, 2008—Statistical Tables.* Washington, DC: Bureau of Justice Statistics.

her career working on the midnight shift. After a period of time, with older officers retiring, the officer may have the opportunity to move to a different shift. The newest officers would then be assigned to the midnight shift. However, if that officer were to quit and get hired in a different police department, it is likely he or she would begin again with the lowest seniority and be assigned to the least desirable shift.

Similarly, it is relatively uncommon today that ranking police officers (such as sergeants and lieutenants) move to a different police department and retain their supervisory position; an exception would be a chief or sometimes other high-ranking command positions, such as captains, inspectors, and assistant chiefs. When officers move from one department to another and retain their position and seniority, it is known as **lateral entry**. The fact that lateral entry is not common keeps officers—especially ones who have already accrued meaningful seniority in a department—in one place.

Last, voluntary turnover among police officers (quitting the job prior to retirement) is low in most departments. For a variety of reasons, if police officers do not voluntarily quit within the first years of employment, they are unlikely to ever do so. The bottom line is that due primarily to civil service protections and seniority provisions, once a police officer is hired in a police department and completes a probationary period, that officer is likely to remain in that department until retirement. If a poor selection decision was made, police leaders may have to live with that decision for an officer's entire career.

turnover: The rate at which people leave a job and are replaced by others.

lateral entry: When officers move from one department to another and retain their position and seniority.

EQUAL EMPLOYMENT OPPORTUNITY (EEO) LAWS AND DIVERSITY

As discussed earlier in Chapter 2, the first police officers in American police departments were hired mostly because of their connections and relationships with local politicians. There were few selection standards or rules governing the process. This was true in spite of the U.S. Constitution, which prohibited discrimination generally, and the Civil Rights Acts of 1866, 1870, and 1871, which prohibited racial discrimination in employment in particular.

In 1964, the Civil Rights Act was passed. Title VII of the act reads, "It shall be an unlawful employment practice for an employer . . . to discriminate against any individual with respect to his compensation, terms, conditions, or privileges of employment, because of such individual's race, color, religion, sex, or national origin." The act prohibits discrimination in any aspect of employment, including hiring and firing, promotions, and compensation. Although progress was made in **equal employment opportunity (EEO)** with this act, it was not until passage of the Equal Opportunity Act of 1972 that discriminatory employment practices became more consequential for employers. The 1972 act contains the following provisions:

- The Equal Employment Opportunity Commission (EEOC) was provided with more power, including the power to investigate and litigate claims of violations of equal opportunity employment against employers, unions, and employment agencies.
- Educational institutions were included.
- State and local governments were no longer exempt.
- The federal government was also included under the law.
- Any employer with fifteen or more employees was included.
- People claiming employment discrimination have more time to file charges.
- Employers are prohibited from imposing mandatory leaves of absence on pregnant women or terminating women because they become pregnant. Women affected by pregnancy, childbirth, or related medical conditions shall be treated the same for all employment-related purposes, including receipt of benefits under fringe benefit programs, as other persons not so affected but similar in their ability or inability to work.[25]

Other laws also prohibit employment discrimination.[26] In particular, the Equal Pay Act of 1963 (EPA) makes it illegal to pay different wages to men and women if they perform equal work in the same workplace, except where pay is based on seniority or merit or on a system that measures earnings by quantity or quality of production. The Age Discrimination in Employment Act of 1967 (ADEA) protects people who are forty or older from discrimination because of age, except where certain retirement plans and normal retirement ages apply. Title I of the Americans with Disabilities Act of 1990 (ADA) makes it illegal to discriminate against a qualified person with a disability in the private sector and in state and local governments unless the disability prevents the person from being able to perform the job. The law also requires that employers reasonably accommodate the known physical or mental limitations of an otherwise qualified applicant or employee with a disability unless doing so would impose an undue hardship (significant difficulty or expense) on the operation of the employer's business. Information about past or current disabilities cannot be asked on an application form. An offer of employment must be made before the employer can require a medical examination, but the actual hiring of the candidate may be contingent on the results of the examination. If a psychological test is considered a medical exam and designed to identify mental disorders, it is to be administered only after an employment offer has been made. If it is not designed to be a medical test, it can be used at the preemployment stage of the selection process. Employment tests or other selection criteria that screen out individuals with a disability cannot be used.

Each law also makes it illegal to retaliate against a person because the person complained about discrimination, filed a charge of discrimination, or participated in an employment discrimination investigation or lawsuit. It is important to understand, however, that these equal opportunity employment laws are intended to stop *arbitrary* discrimination regarding

equal employment opportunity (EEO): Provision for employment regardless of employees' characteristics. Laws related to equal employment opportunity are intended to stop arbitrary discrimination.

employment. If a person is required to have certain skills, traits, or abilities to adequately perform the tasks of a job but does not, then an employer is not required to hire that person. In other words, there are no laws that prevent employers from making decisions on the basis of **bona fide occupational qualifications (BFOQs)**. It is generally acceptable to select one applicant over another applicant as long as the basis for the selection is a BFOQ. Essentially, job selection criteria must be job related. For example, if the basic demands of a job require that a woman be hired for the position, a man does not have to be hired. Such might be the case when looking to hire a corrections officer to work in a female prison or when a women's locker room attendant is needed at a gym. However, if fifty-year-olds are successfully performing the job of patrol officer, it would be illegal not to hire other fifty-year-olds to do the job, short of some other condition of employment (e.g., retirement plan requirements). If female officers are successfully able to perform the job of police officer, it would be illegal to exclude women from employment as officers.

Sometimes discrimination in hiring is indirect. An employment policy banning beards might discriminate against men who have the condition *pseudofolliculitis barbae* (commonly known as razor bumps). The question then becomes, is not having a beard a BFOQ for being a police officer? Regardless of the particular circumstances or claims, the burden is on the employer to demonstrate that the job requirement and/or selection tests are related to successfully performing the job. The burden of showing that the requirement or test had a disparate impact on hiring patterns is on the employee or complainant.

AFFIRMATIVE ACTION

An **affirmative action** policy (AAP) is intended to (a) proactively recruit, hire, and promote women, minorities, disabled individuals, and other protected groups of people and (b) correct past discrimination in order to eliminate its present-day effects. Some agency leaders freely adopt AAPs because they wish to operate without discrimination in their workplace; others adopt them because they have been threatened with legal action. Some agencies are legally required to implement AAPs based on substantiated allegations of discriminatory employment practices. In this circumstance, the AAP is likely to take the form of a **consent decree**, in which specific goals or timetables are provided to increase the employment representation of protected groups in an agency. Quotas are seldom prescribed or allowed as a basis for AAPs or consent decrees. Affirmative action often takes the form of a written policy. The inducement to adopt and follow an AAP is powerful; without an AAP in place, an agency is at risk of becoming ineligible for any federal financial support.

Affirmative action is viewed by some people as very controversial because of claims that it leads to **reverse discrimination**, in which women or minorities are hired (or promoted), not because of their qualifications but because of their sex or race. If this happens, it is troubling; however, evidence suggests that reverse discrimination does not happen frequently.[27] Nevertheless, there is little doubt that some white men have suffered discrimination as a result of attempts to prevent and correct it.

A Question of Ethics

Hiring Decisions and Affirmative Action

You are a police chief. Your department has an affirmative action policy, and you are a strong proponent of creating a diverse workforce. You are considering two final applicants to be hired for one police officer position. Both are about equally qualified on all known dimensions: They both have college educations and clean background checks, and they have similar previous work experience. They did equally well on the written test and the oral interview. One of the applicants is an African American female; the other is a white male. Who do you hire? Why?

bona fide occupational qualifications (BFOQs): Qualifications such as skills, traits, or abilities that are required to correctly perform a certain job.

affirmative action: Action intended to (a) proactively recruit, hire, and promote women, minorities, disabled individuals, and other protected groups of people and (b) correct past discrimination in order to eliminate its present-day effects.

consent decree: A decree that provides specific goals or timetables to an agency in order to increase the agency's employment representation of protected groups.

reverse discrimination: Discrimination that occurs when women or minorities are hired (or promoted), not because of their qualifications but because of their sex or race.

MINIMUM QUALIFICATIONS FOR POLICE OFFICERS

Little variation exists across departments on some factors that are considered minimum qualifications for employment as a police officer. These factors include age (applicants must be at least twenty-one), the possession of a valid driver's license, U.S. citizenship (or legal authorization to work in the United States), no felony convictions, and proportional height and weight. On other factors, however, departments vary on the qualifications applicants must possess.

EDUCATION

At the very least, police departments require officers to have a high school diploma or its equivalent. However, some *states* require that a person have at least sixty college credits or an associate's degree to be hired as a police officer. In states that do not have such a law, some police departments still have this requirement. Very few police departments formally require a four-year college degree (see Figure 5.2).

As will be discussed in Chapter 8, a multitude of studies have examined the relationship between educational level and police performance. The findings are mixed,[28] although it must be understood that the adequate and accurate measurement of police performance is always difficult. If there is not a correlation between college education and better police performance, it calls into question how necessary this requirement is in hiring officers. Because a college education requirement can hinder recruitment efforts and result in fewer applicants, some departments have dropped this requirement.

CRIMINAL RECORD AND DRUG USE

A previous felony conviction disqualifies a person from applying to be a police officer, but a misdemeanor conviction may not. In fact, a majority of agencies of all sizes are willing to consider applicants with a misdemeanor conviction as long as it is not related to domestic

FIGURE 5.2

Education Requirements for New Officers in Local Police Departments, 2013

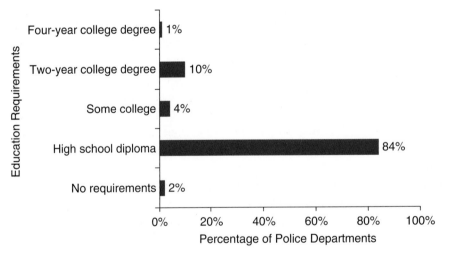

Source: Brian A. Reaves. 2015. *Local Police Departments, 2013: Personnel, Policies, and Practices*. Washington, DC: National Institute of Justice. 7.

Note: These are official minimum requirements; many police departments may have a preference for officers with higher education. Total does not sum to 100% due to rounding.

violence. The particulars of the conviction (e.g., the offense, how long ago it occurred) are important factors to evaluate. With regard to drug use, about half of all agencies allow applicants with prior marijuana use to be hired, although a drug test prior to hiring (and periodic random testing throughout employment) is common. Also about half of agencies do not disqualify applicants based on driving-related problems, such as a suspended license or citation for driving under the influence. Larger agencies tend to have more relaxed selection standards for nearly every qualification.[29]

SELECTION PROCEDURES FOR POLICE OFFICERS

Four selection procedures are most commonly used by police departments when identifying applicants to hire: written tests, background investigations, oral interviews, and medical/physical tests. The nature of these procedures and the order in which they are administered vary dramatically. For example, the process used by the Milwaukee (Wisconsin) Police Department follows these steps: (1) written test, (2) physical ability test, (3) oral interview, (4) writing sample exercise, (5) background investigation, and (6) medical exam. The Waco (Texas) Police Department uses the following process: (1) written test, (2) physical fitness test, (3) background investigation and a polygraph, (4) psychological examination, (5) oral interview, (6) oral interview with the chief of police, and (7) physical exam.

The process may also vary in terms of whether or not tests are considered compensatory or noncompensatory. With a compensatory process, a high score on one exam can compensate for a low score on another. With a noncompensatory process, a low score on an exam disqualifies an applicant from further consideration. Usually police departments use a mix of compensatory and noncompensatory tests in the process. For example, written tests, psychological tests, medical exams, and physical agility tests are likely to be pass/fail (noncompensatory), but oral interviews and background investigations are likely to be graded on a scale (although a failing score would still be possible and grounds for disqualification).

WRITTEN TESTS

Not surprisingly, the great majority of law enforcement agencies include a written test as part of the selection process for police officers. These tests are usually administered at or near the agency, although some agencies, in an effort to recruit more applicants, administer them in other cities or even different states. Each police department may have its own version of an exam; there is not a national standard police officer exam. Written tests are used to measure various aspects of applicants' aptitude, including verbal and mathematical skills, reasoning abilities, reading comprehension, memory capabilities, and/or problem-solving skills. Most exams are multiple choice, although some require applicants to provide a writing sample, such as a written response to a question. Exams are timed and usually consist of 50 to 100 questions divided into sections based on subject matter.

Most often, the written test is the first step in the selection process and is graded on a pass/fail basis. Applicants who do not achieve a minimum score (e.g., 70%) are not allowed to continue to the next stage. Exhibit 5.2 provides a sample of questions obtained from various police entrance exams.

Written tests are also often part of the process for police promotions. These exams are usually scenario based and require officers to draw upon their police and supervisory knowledge.[30]

EXHIBIT 5.2

Examples of Exam Questions on a Written Police Exam

Many different types of questions may be included on a police officer selection exam. Some examples are provided here.[31] The quality being tested is identified as well. (See page 107 at end of chapter for answers.)

1. A car was stolen downtown. Four eyewitnesses wrote down the license plate number, but each wrote down a different number. Which answer is most likely correct? (Reasoning)

   a. PLU 015
   b. RJW 005
   c. RJU 015
   d. RJW 015

2. As part of a theft prevention program, Officer Milton visits three local businesses every week—except during his four-week vacation. How many groups does he visit in one year? (Math ability)

   a. 48
   b. 75
   c. 144
   d. 156

3. Choose the correct word or phrase to complete the following sentence: As soon as they arrived, police officers decided to _____ evacuating people immediately. (Grammar)

   a. begin
   b. had began
   c. began
   d. The sentence is complete as is.

4. Which of the following best describes the Fifth Amendment? (Basic job knowledge)

   a. Police are not permitted to search your house unless they can prove that they have good reason to think you have committed a crime.
   b. When on trial or when being questioned, you can choose to remain silent rather than revealing incriminating information.
   c. You cannot vote if you have been arrested within the last year.
   d. If arrested, you have the right to a trial within a reasonable amount of time.

5. An officer stops a car for driving left of center and not stopping at a stop sign. While speaking with the driver, the officer notices that the man is a member of the city council and is driving under the influence. How should the officer proceed? (Judgment)

   a. Give the city council member a ride home.
   b. Allow the city council member to go on his way in his car.
   c. Continue as he or she would normally with an investigation and notify a supervisor.
   d. Give the city council member the necessary citations and let him go.

6. There were six _____ at the accident scene. (Spelling)

   a. witness
   b. witnesses
   c. witness's

Read the following scenario and answer questions 7–10. (Reading comprehension)

In an effort to cut down on traffic and air pollution in the center of town, an ordinance is enacted and signed by the mayor that forbids certain size trucks from entering the city limits, except for delivery trucks dropping off or picking up goods or materials from local businesses or residences. To enter the city, delivery vehicles must first obtain a special permit from the office of licenses and permits and display the permit on the front windshield. The ordinance authorizes the police to ticket any truck in the city not displaying a permit.

On May 2nd, Officer Brown saw a truck go by him in the town without a permit displayed. He pulled the truck over and questioned the driver. The driver said he was from out of town and needed to make a delivery but was not aware of the ordinance. Officer Brown gave the driver a warning and told him to either drive out of the city or, if he needed to make a delivery, to go to the office of licenses and permits to get a permit. The driver drove out of town. On May 10th, Officer Brown saw the same truck and driver in the town, again without the required permit. The driver apologized and said he did not know he was in the city limits. Officer Brown ticketed the driver.

The driver appealed his ticket, and a hearing was held in the city on May 28th. The judge was not sympathetic to the driver's excuse since there was a sign giving notice of the city limit that the driver had passed when entering the city. When Officer Brown was leaving the courthouse, he noticed the driver had parked his truck in a lot down the street. Officer Brown was surprised to see that the truck still failed to display the required permit. Officer Brown put a ticket on the front windshield of the truck and went about his day.

7. Why did Officer Brown ticket the driver the first time?

   a. The driver ignored the warning that was given the first time.
   b. The driver did not have the required permit.
   c. The officer wanted to teach the driver a lesson.
   d. The driver parked his vehicle outside the courthouse without the required permit.

8. Why was the judge not swayed by the driver's excuse?

   a. The driver had sufficient notice of the city limit.
   b. The driver ignored the first warning given by the police officer.
   c. The driver parked the truck outside the courthouse without the required permit.
   d. None of the above.

9. Based on what the driver did after the warning, what can be concluded?

   a. The driver was in the city to make a delivery.
   b. The driver was not making a delivery within the city limits.
   c. It cannot be determined from the passage whether the driver was making a delivery within the city limits.

10. On what date did the driver get his first ticket?

   a. May 2nd
   b. May 3rd
   c. May 10th
   d. May 28th

Source: *This material is a compilation of sample police test questions. The questions come from various police departments across the country, from county sheriff departments to the U.S. Capitol police department.*

ORAL INTERVIEW

Interviews of police officer candidates are also a very common part of the selection process. As with written exams, much variation occurs in the nature and process of oral interviews. Typically, they are conducted by a group of police personnel, educators, and/or other government agency employees at the police department that is hiring. The questions are generally designed to measure reasoning and judgment abilities, problem-solving skills, logical thinking, and general job-related knowledge. Other factors assessed in oral interviews may include communication abilities, maturity, and composure. Exhibit 5.3 presents a sample of the type of questions that may be asked.

BACKGROUND INVESTIGATION

Background investigations are always conducted as part of the police officer selection process;

Photo 5.3

Oral interviews, written tests, background investigations, and medical/physical tests are most commonly used to screen police officer applicants. Much variation can be found in the structure and content of oral interviews. Monica Synett/Daily Chronicle via AP

however, some are more thorough than others. At a minimum, these investigations involve a check of the applicant's criminal history and driving record. Other aspects may include a drug test, verification of education, military history, and employment history. A background investigator may interview the applicant's friends, associates, neighbors, teachers, and previous employers in order to determine the applicant's personal reputation and character. Financial records, including credit history, may be examined. A special interview may be conducted with the applicant to determine previous criminal or unethical conduct. Some police departments may conduct a polygraph examination to verify the truthfulness of information provided by the applicant. Many judgments have to be made when evaluating the employability of an individual based on background investigation information, but the importance of these investigations cannot be minimized.

EXHIBIT 5.3

Oral Interview Questions

Listed below are some typical questions asked in an interview for a police officer position. How would you answer these questions? What do you think would be good answers to these questions?

Background and motivation questions:

- Why do you want to become a police officer?
- Why do you want to work for this agency?
- Explain how your background and experiences have prepared you to become a police officer.
- Describe a challenging situation you have faced in your current or previous job and how you handled it.
- Describe yourself in one word. Explain.
- What do you like to do for fun?
- What do you like the least about your current job?
- What is your most significant weakness? What is your greatest strength?
- What is cultural diversity? Provide an example of its benefits in the workplace.

Basic and more advanced knowledge of law and law enforcement procedure:

- What is the difference between a burglary and a robbery?

- What are the Fourth and Fifth Amendments to the U.S. Constitution?
- What is the use of force continuum?

Some interviews include scenario-type questions—"What would you do if?"

- You are a police officer and while off duty you attend a party with your friends. Upon arriving at the party, you see that drugs are being openly used by people at the party. What would you do?
- You are a police officer on patrol with your partner. At 3:00 a.m., you discover the back door of a liquor store is open. Upon checking the store with your partner, it appears that a burglary may have taken place. As you are calling in the burglary, you see your partner carrying a case of beer out of the store and putting it in the trunk of the squad car. What would you do?
- You are a police officer on patrol. You notice a motorist weaving across the center line and suspect he may be driving while intoxicated. You pull over the vehicle and discover that the motorist is an off-duty police officer from a neighboring jurisdiction. His speech is slurred and he smells of alcohol. What would you do?

MEDICAL EXAMINATION

A medical exam is usually required as part of the hiring process. This exam is designed to make sure the applicant does not have any condition that would negatively affect his or her ability to perform the duties of a police officer. An eyesight and hearing exam is typically included with the medical assessment. Police departments may require a minimum eyesight ability and/or correction to 20/20.

PHYSICAL FITNESS EXAMINATION

To assess an applicant's physical ability to perform the tasks of a police officer, the majority of departments require a physical fitness or physical agility test. A great degree of variation can be found in the content of such tests across departments. Typical requirements include the ability to run a mile under a certain time limit (e.g., twelve minutes) and to navigate an obstacle course. The content of obstacle courses can also differ from agency to agency. Fairly typical of larger departments is the obstacle course used by the Denver Police Department. Within 63 seconds, participants must run around a track wearing an eight-pound utility belt, climb over a six-foot chain-link fence, crawl under a table (two times), climb through a window, identify a previously described subject, and drag a 150-pound mannequin dummy approximately eight feet (https://www.youtube.com/watch?v=mmP6KgGuL0E). The New Orleans Police Department physical agility test consists of a 1.5-mile run to be completed in a maximum of 19 minutes 50 seconds, a 300-meter sprint within 2 minutes, 14 sit-ups in a minute, and 10 push-ups with no time limit. Regardless of the test used, in most departments all participants must successfully complete the same course.

PSYCHOLOGICAL EXAMINATION

A psychological exam is designed to assess the personality characteristics of applicants and their psychological and emotional well-being, which is necessary to be a police officer. Several standardized tests may be used for this purpose. An interview with a psychologist or psychiatrist may also be part of this examination. Some studies have been critical about the value such tests add to the selection process for police officers.[32] However, given the importance of not hiring individuals who are mentally unfit to be police officers, validated psychological assessments are essential in selection processes.

Photo 5.4
Through the administration of physical fitness tests, police departments seek to identify applicants who can perform the physical tasks of the job. AP Photo/Mel Evans

ASSESSMENT CENTER

Some police departments use an **assessment center** as part of the police officer selection process, although these centers are used much more often in the promotion of officers to investigative, supervisory, or leadership positions in the department.[33] When used as part of the police officer selection process, an assessment center usually involves a role-playing exercise that is observed by officers from the department. The applicant is then rated on his or her performance in the exercise. For example, the applicant may be instructed to act as an officer and intervene in an argument between two people (police officer actors). The applicant's actions are observed and rated on the basis of predetermined criteria, such as verbal communication abilities, judgment, self-control, and problem-solving abilities. Although assessment centers are relatively expensive to operate, they offer the advantage of actually seeing how the applicant performs in potentially stressful situations.

THE OUTCOMES OF THE RECRUITMENT AND SELECTION OF POLICE OFFICERS: DIVERSITY

Because of the importance of diversity in police departments, it is necessary to discuss the outcomes of the recruitment and selection processes as they relate to the gender and race of officers. Many departments today are actively seeking to diversify their workforces.

THE REPRESENTATION OF WOMEN AS POLICE OFFICERS

With regard to gender, in 2017 approximately 12.5% of sworn officers were women.[34] In contrast, 51% of the adult population and 47% of the adult workforce were female. This clearly shows that females are underrepresented in police work. However, the representation of women in policing is much higher than it was just a few decades ago (see Figure 5.3).

Understand that the data presented in Figure 5.3 are national averages. There is substantial variation across police departments in the representation of women officers. Generally speaking, the largest police agencies employ the largest proportion of women officers. Specifically, cities with a population of one million or more had the highest representation of women officers in 2017 (18.1%),[35] while many small agencies had no female officers. Some departments clearly make the recruitment and selection of women officers a priority; as evidence, the workforce of several agencies is made up of 20% to 30% women officers. In the Madison (Wisconsin) Police

> **assessment center:**
> A center that allows officer applicants or officers seeking promotion to be judged on their performance in role-playing exercises.

Photo 5.5

Policing is a genderized occupation; only about 13% of police officers in the United States are women, although this percentage has gradually increased over the years. The underrepresentation of women as police officers may relate to recruitment, selection, and/or retention issues. simon leigh/Alamy Stock Photo

Department, approximately 35% of officers are women. Although many departments are becoming more diverse in terms of gender, policing is still clearly a **genderized occupation**. However, this does not make the police unique; most occupations are represented by either more men or more women.

There are three possible reasons why women are underrepresented in police work: (1) they are simply not applying (a possible recruitment issue), (2) they are applying for police officer positions but are not being hired (a selection issue), or (3) they are applying and getting hired but are then leaving (a retention issue). A combination of all three reasons may represent the likeliest scenario. To examine this issue further, researchers sent surveys to female police officers and all police chiefs in three counties in Pennsylvania. To different degrees, the officers and the police chiefs believed that the most likely reason for the underrepresentation of women was that "not many women apply for police positions," followed by "women apply but are not selected." "Women are selected but turnover is high" was the least frequently chosen answer (Figure 5.4).

With regard to why the recruitment of women is challenging, most women officers believed that "police agencies here do not recruit women very proactively," followed by "women here have other employment options that are more attractive."[36] As for selection, the most frequently chosen explanation was that "physical fitness tests tend to eliminate women and/or push them down the eligibility list." With regard to retention, the most frequently chosen reason for women officers leaving police employment was that "the culture in police agencies is male-dominated and not very woman friendly." Obviously some of the issues related to

> **genderized occupation:** An occupation that tends to be represented by one gender over another.

FIGURE 5.3

Representation of Women as Sworn Police Officers

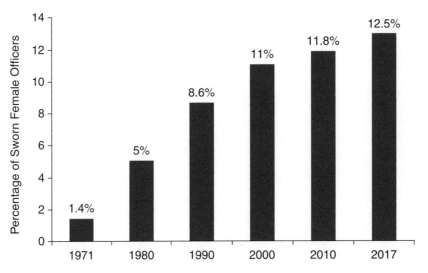

Y-axis: Percentage of Sworn Female Officers

Year	Percentage
1971	1.4%
1980	5%
1990	8.6%
2000	11%
2010	11.8%
2017	12.5%

Sources: Gary Cordner and AnnMarie Cordner, "Stuck on a Plateau? Obstacles to Recruitment, Selection, and Retention of Women Police," *Police Quarterly* 14 (2011): 207–226. 2017 figure obtained from the 2017 FBI UCR, Table 74.

FIGURE 5.4

Primary Reasons for the Low Level of Employment of Women in Selected Pennsylvania Police Departments According to Police Chiefs and Female Officers

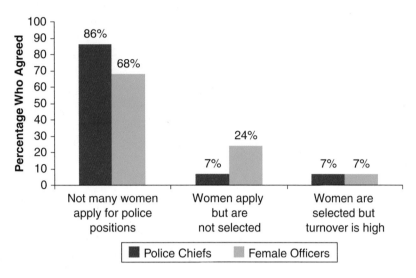

Source: Data from Gary Cordner and AnnMarie Cordner. "Stuck on a Plateau? Obstacles to Recruitment, Selection, and Retention of Women Police." *Police Quarterly* 14 (2011): 207–226.

retention may also have implications for recruitment and selection. For example, if the police occupation is viewed as not very woman friendly, it may prevent some women from applying in the first place. As for ways by which to increase the representation of women officers, the majority of female respondents believed that "more targeted recruitment of women" would be effective, as would "more family-friendly policies," "providing mentors for women officers," and "providing mentors for women applicants."[37]

? A RESEARCH QUESTION

How Do Male and Female College Students Perceive a Policing Career?

It is an undisputed fact that women are underrepresented in the policing occupation, but why this is the case is an open question. One strong possibility is that women are less interested in the occupation and therefore do not even apply. If this is the case and if police leaders can identify perceptions and beliefs that prevent women from applying to police departments in the first place, then police agencies may be in a better position to address and correct the underrepresentation problem.

A 2018 study surveyed 387 male and female undergraduate criminal justice students to compare their perceptions and beliefs about the occupation. It was found that "female students are less interested in a law enforcement career, perceive themselves as potentially less successful, and perceive less potential personal fulfilment. Women believe that current female officers receive less respect, acceptance, and opportunity."[38] In short, it appears that the police occupation has an image problem among many female students, and this may account for lower levels of applications among women. The study provides important information to police departments and recruiters who wish to increase the representation of women as police officers. Police departments must either create a more receptive work environment for female officers or correct the incorrect perceptions about the career.[39]

THE REPRESENTATION OF RACIAL MINORITIES AS POLICE OFFICERS

According to most recent national estimates, approximately 27% of police officers are members of a racial/ethnic minority group, compared to 38% of the U.S. population.[40] As such, like women, people of color are generally underrepresented as police officers today. However, there has been a substantial change in their representation in departments within the past few decades. For example, in 1967 only 5% of officers in Detroit, Michigan, were African American; in 1992, 48% were.[41] By 2017, the figure had risen to 67%.[42] Although this represents substantial change, it is also important to note that in 2017, 79% of the population of Detroit was African American.[43]

As one might expect, police departments in larger cities tend to be more racially diverse than ones serving smaller cities. This may have more to do with the demographics of city populations than the size of the city, per se. Indeed, research has shown that the representation of African Americans in the population of a city is the strongest predictor of the hiring rates of African American[44] and Hispanic[45] police officers. Generally—though not always—the more diverse a city, the more diverse the police department.[46] Research has also shown that a minority chief is associated with increased representation of minority officers in a department.[47] Research has been rather unsuccessful in identifying factors associated with racial diversity at police managerial levels.

Possible explanations for the underrepresentation of minorities in police departments are similar to those for women: people of color are simply not applying; they are applying for police officer positions but are not being hired; or they are applying and getting hired but then leaving. Studies have distinguished individual and cultural reasons for why some African Americans lack interest in a policing occupation[48] and have identified barriers to the selection of African American applicants—in particular, the initial application, the written exam, and the background investigation.[49] Having a larger representation of people of color in the community would suggest that efforts to diversify departments would be easier or more effective in these cities. That a minority police chief is associated with increased minority police representation suggests that these chiefs create a more hospitable work setting for officers of color and provide unique benefits in terms of recruitment, selection, and/or retention.

One way to enhance recruitment to diversify departments is to adjust selection criteria and standards. For example, college education requirements have an adverse impact on the recruitment of racial minorities.[50] Upper-body strength requirements tend to have a negative effect on the recruitment of women. Relaxing education and physical strength requirements would likely result in the hiring of more people of color and women. For example, the New Orleans Police Department long operated under the standard that new officers without two years of military service be required to have at least sixty college credits. However, due to difficulties in finding such applicants, the college credit requirement was dropped.[51] Police departments must balance high standards with recruitment realities.

THE REPRESENTATION OF GAY AND LESBIAN POLICE OFFICERS

There has been no research to systematically determine the representation of gay and lesbian police officers over time.[52] Studies that have examined the experiences of gay and lesbian officers usually focus specifically on those officers, without the benefit of comparison to other officers or changes over time.[53] In one study conducted in a large Midwestern police department, 3.1% of officers identified as gay, lesbian, or bisexual.[54] This study showed that the work experiences of these officers were generally similar to the work experiences of white male officers except for exposure to unwanted comments about homosexuality by other officers. Of course, part of the challenge in conducting research on lesbian, gay, or bisexual officers is that those officers may not wish to reveal their sexual orientation.

Photo 5.6
With in-service training, officers can stay current on the skills of policing. Here officers receive training on how to best conduct field sobriety tests.
AP Photo/Hans Madsen

POLICE OFFICER TRAINING

The primary goal of training is to provide officers with the knowledge, skills, and instruction so they can perform the tasks of police work with competence. Because there are many tasks associated with police work, there is a lot for officers to learn. Much of their training is oriented toward the technical aspects of police work: writing reports, communicating with hostile subjects, first aid, use of force, and so forth.

As we will discuss, training is not a one-time event; training (and retraining) is a career-long process. The initial training of officers occurs first in a police academy and then in the field. The training that occurs over the course of the officer's career is known as in-service training. Both are explored below.

ACADEMY TRAINING

Academy training, also known as basic recruit training, consists of primarily classroom lessons on the skills of police work. On average, academy training consists of approximately 840 hours completed during forty-hour work weeks. Some states and police departments require more hours, some require fewer. Successful completion of training leads to graduation and employability as a sworn police officer.

There are several ways by which police academies are administered. Most academies (47%) are operated by universities, colleges, community colleges, or two-year technical schools. After police applicants are hired by a department, the recruits are enrolled into the academy. The department is responsible for paying the cost of the academy training and usually pays the recruit a salary for the duration of training. Some states allow citizens to apply for acceptance into police academies operated by educational institutions. In this case, the citizen is responsible for the cost of the training and would not receive any salary as part of participation. Upon successful completion of training, the citizen would be certifiable as a sworn police officer upon being hired by a department.

Twenty percent of police academies are operated by police departments.[55] These academies typically serve only the department that operates it; recruits are hired by the department and then are provided training in the academy. There are 132 such academies in the United States, most of which are operated by larger police departments. Other academies are operated at the state level by the relevant state police, county police, sheriff's office, or other agency, or by a combination of regional agencies. Academies at the state level are most often operated by Peace Officer Standards and Training (POST), which is the agency typically responsible for certifying law enforcement officers in the states (see Table 5.1).

Academies vary in terms of the degree to which stress is a part of the training experience. Stress-based academy training reflects military training and includes drills, intensive physical demands, public disciplinary measures, daily inspections, value inculcation, and the withholding of privileges.[56] Non-stress academy training is more relaxed and oriented toward academic

TABLE 5.1

Types of Police Academies

Academy Agency	Number of Academies	Percentage of All Academies
All Types	664	100.0
State POST*	30	4.5
State Police	41	6.2
Sheriff's Office	66	9.9
County Police	22	3.3
Municipal Police	132	19.9
College/University	307	46.3
Multi-Agency	49	7.4
Other Types	17	2.6

Source: Brian Reaves, *State and Local Law Enforcement Training Academies, 2013* (Washington, D.C.: Bureau of Justice Statistics, U.S. Department of Justice, 2016).
* Peace Officers Standards and Training (state-operated academy).

TABLE 5.2

Major Subject Areas Included in State and Local Law Enforcement Academy Training

Topics	Percentage Academies With Training	Median Number of Hours of Instruction
Operations		
Report writing	100	20 hr
Patrol	99	40
Investigations	99	40
Basic first aid/CPR	99	24
Emergency vehicle operations	97	40
Computers/information systems	58	8
Weapons/Self-Defense		
Self-defense	99	51 hr
Firearms skills	98	60
Non-lethal weapons	98	12

Topics	Percentage Academies With Training	Median Number of Hours of Instruction
Legal		
Criminal law	100	36 hr
Constitutional law	98	12
History of law enforcement	84	4
Self-Improvement		
Ethics and integrity	100	8 hr
Health and fitness	96	46
Stress prevention/management	87	5
Basic foreign language	36	16
Community Policing		
Cultural diversity/human relations	98	11 hr
Basic strategies	92	8
Mediation skills/conflict management	88	8
Special Topics		
Domestic violence	99	14 hr
Juveniles	99	8
Domestic preparedness	88	8
Hate crimes/bias crimes	87	4

Source: Brian Reaves, *State and Local Law Enforcement Training Academies, 2013* (Washington, D.C.: Bureau of Justice Statistics, U.S. Department of Justice, 2016), Table 11.

achievement, physical training, administrative disciplinary procedures, and an instructor–trainee relationship.[57] Academies that are administered through police agencies, especially state police agencies, are more likely to incorporate stress-based elements than academies operated by colleges and two-year technical schools. The instructors in academies may be sworn full-time police officers assigned to the academy, part-time sworn officers hired to teach certain topics, and/or full- and part-time civilian instructors. Most instructors are full-time sworn officers who work at the academy part time.

Students move through the academy as a group; they begin and end training at the same time and attend most or all classes together. With regard to the actual content of the training, most time is spent learning firearm skills, self-defense skills, and fitness training. Lessons are also devoted to investigative procedures, report writing, patrolling, and emergency vehicle operations (see Table 5.2). Academies are required to provide training on topics and issues as mandated by state requirements. Additional training can be provided based on departmental needs and objectives. Academies have recently been placing more emphasis on mediation skills/conflict management, also known as **de-escalation training**.

Recruits are evaluated during academy training primarily through written tests, skill tests, and physical fitness tests. On average, approximately 86% of recruits successfully complete academy training. When considering subgroups, this figure breaks down into 87% of males, 80% of females, 86% of whites, and 79% of blacks.[58] Females tend to be more successful in academies that emphasize a community policing curriculum.[59] Completion rates are higher in non-stress academies than stress-based ones.[60]

FIELD TRAINING

Upon successful completion of academy training, the officer goes to his or her police department to begin **field training** and work as a police officer. During the first six to eighteen months of employment, the officer is in a probationary period. Field training involves the probationary officer working alongside a field training officer (FTO) to learn and perform the tasks of

de-escalation training: Training that is designed to teach officers how to resolve potentially violent situations without the use of force.

field training: This training involves the probationary officer working alongside a field training officer (FTO) to learn and perform the tasks of policing while actually on the street.

policing while actually on the street. In other occupations, this type of training is referred to as on-the-job training or job shadowing. Today, field training in most police departments is very structured, although the amount of time devoted to it depends on the agency. Field training most often consists of twelve to sixteen weeks divided into four phases that each last three to four weeks; each phase is characterized by increased responsibility for performance of police tasks (Exhibit 5.4). During each phase the officer may be assigned to a different FTO.

The FTO is an experienced police officer who has received special instructions on how to train new officers. The FTO plays several roles. He or she is a teacher who facilitates the development of the knowledge, skills, and abilities necessary for the new officer to become a successful police officer. The FTO also introduces the new officer to the subculture of the police occupation and organization—to the unwritten rules and values of the workgroup. He or she is a coach and mentor who is available to the probationary officer for consultation and is a supervisor who evaluates the performance of the new officer. In most field training programs, a critical responsibility of the FTO is the completion of daily observation assessments (see Exhibit 5.5) and weekly evaluation reports. In the reports, the FTO documents the amount of time spent on each identified activity and the strengths and weaknesses of the officer, and he or she provides overall assessment of training and performance during the week.

GOOD POLICING
Recruit Training in the Tulsa Police Department

The Tulsa (Oklahoma) Police Department (TPD) has revamped its recruit training philosophy to reflect the new demands and expectations of police officers today. According to Captain Matt McCord, the TPD has developed a more "professional" academy to replace the "boot camp-style" or militaristic approach. For example, more emphasis is placed on learning and using de-escalation techniques, and less time is spent on teaching recruits about how to anticipate exceedingly uncommon ambush situations. According to Captain McCord, the emphasis on survival training made recruits unjustifiably "believe everybody everywhere was trying to kill them." Recruits receive "plenty of survival training," but they also spend the necessary time on learning communication, empathy, and cultural competency skills. Recruits also attend classes on LGBTQ (lesbian, gay, bisexual, transgender, questioning) and Muslim culture. The overall goal is to make recruits more effective communicators and therefore better police officers.[61]

EXHIBIT 5.4
Phases of Field Training[62]

Phase 1 (four weeks): The probationary officer is assigned to an FTO and becomes accustomed to patrol. The officer learns by observing the FTO and performs simple tasks. During weeks two and three, the officer performs the most common patrol activities: using the squad radio, mobile data computer, and radar. During week four, the officer also drives the patrol vehicle, completes reports, and assists with investigations.

Phase 2 (four weeks): The officer does all of the driving, handles more serious and complicated calls for service, and initiates field activities.

Phase 3 (four weeks): The officer handles all calls with limited assistance and should have working knowledge of all tasks that are expected to be performed as a police officer. The officer receives instruction in more advanced and complex investigations.

Phase 4 (two weeks): The FTO role is to observe and evaluate the new officer's independent performance. The FTO will provide a final assessment of the new officer's performance.

The FTO is also likely to complete a summary assessment at the conclusion of each phase of field training. Here, critical knowledge areas (see Exhibit 5.6) are reviewed to ensure that the officer either performed the associated task, the FTO demonstrated the task, and/or the task was practiced or discussed. At the end of each phase, the FTO makes a recommendation that (a) the officer progress to the next training phase, (b) the training period be extended, or (c) an employment status hearing take place in order to dismiss the officer from employment. As noted earlier, it is relatively easy to legally dismiss an officer during his or her probationary period. However, for this decision to be defensible, it must be made on the basis of demonstrated job performance. The documentation of the officer's performance during field training (daily, weekly, and phase evaluations) may serve as the basis of the decision. Upon successful completion of field training, the patrol officer is assigned as needed.

EXHIBIT 5.5

Police Tasks an FTO Would Evaluate on a Daily Basis

Most field training programs require FTOs to complete assessments that evaluate a new officer's abilities during each work shift. For example, ratings would be provided on the following activities:[63]

- Ability to multitask
- Driving skills, non-stress conditions
- Orientation skills, directions, and navigation
- Report writing
- Complaint processing/investigations
- Problem-solving/decision-making ability
- Telecommunications (use of radio, mobile data computer, telephone, computers)
- Officer safety/tactics
- Self-initiated field activity
- Understanding and application of criminal laws, ordinances, traffic violations
- Knowledge of department rules and policies
- Interpersonal skills and appearance
- Understanding of chain of command

EXHIBIT 5.6

Critical Knowledge Areas of Field Training

Field training programs may require that the critical knowledge areas of policing either be performed by the officer or demonstrated or trained by the FTO. The list of critical knowledge areas is long and includes the following:[64]

- Emergency vehicle operations
- First responder duties and equipment
- Methods of arrests
- Traffic stops
- Bicycle enforcement
- Victim rights
- Crimes in progress
- Responding to a crime scene
- High-risk vehicle stops
- Domestic violence
- Temporary restraining orders
- Disturbance calls
- Noise violations
- Operating while intoxicated incidents
- Blood draw procedures
- Accident investigations
- Warrants
- Juvenile contacts
- Child custody disputes
- Landlord/tenant issues
- Fires
- Alarms
- Hazardous material incidents
- Animal bite incidents

IN-SERVICE TRAINING

An officer's training does not end with the successful completion of field training. Every year over the course of a career, all full-duty officers are required to complete additional training known as **in-service training**. The number of in-service training hours required annually varies by department and by state; typical is about forty hours per year. In many departments, in-service training is provided within the department by officers who work there or by other officers or experts who are brought in to the agency. This is referred to as in-house training. Some training may be provided through seminars and workshops offered through outside agencies or instructors that require officers to travel to the location of the training. This is known as out-of-department training.

Some in-service training is mandatory; some is voluntary. Mandatory in-service training focuses on topics that relate to critical tasks of police work (Exhibit 5.7), especially police use of force. Some topics of in-service training may be mandatory for only certain officers based on their duties and responsibilities.

Depending on funding and resource availability, officers may also volunteer for other elective training offered by local, county, state, or federal law enforcement agencies or other public or private agencies. The topics for elective training are almost limitless and many relate to criminal investigation and police management topics. Examples include interviewing and interrogation methods, managing confidential informants, detecting deception, narcotics investigations, smartphone forensics, homicide investigation, bloodstain pattern analysis, hostage negotiations, management of a K-9 unit, performance leadership, media relations, training for field training officers, and so forth.

EXHIBIT 5.7
Common In-Service Training Topics

- Departmental policy changes and legal updates
- DAAT (defense and arrest tactics)
- De-escalation techniques
- EVOC (emergency vehicle operation control)
- Handgun; training and testing ("qualifying") for skill competence
- Rifle; training and testing ("qualifying") for skill competence
- Vehicle contacts
- Tactical response

- Active shooter situations
- Taser and other less lethal weapons
- Report writing
- Cultural competence
- Professional communication
- CPR and first aid recertification
- Biohazards and bloodborne pathogens
- Operating a motor vehicle while intoxicated (OMVWI) violations; administration of and interpretations of standardized field sobriety tests (SFST)

For training to make a difference, the officer's newfound knowledge and behavior must be transferred to the job. To increase the likelihood of a successful transfer, the training environment should have identical elements to the actual circumstances in which the behavior being taught is to occur. In other words, for training to be effective, it should be realistic. With some forms of training, this is especially difficult to do (see Technology on the Job: Use of Force Training Simulators).

As discussed in this chapter, recruitment, selection, and training of officers are critical to the performance of police departments. Each process has implications for the other two: Successful recruitment improves the chances of being able to hire good officers; good officers are most likely to be effectively trained and perform well. As such, police departments should devote substantial attention to these processes.

in-service training: Training that occurs throughout an officer's career.

Use of Force Training Simulators

Providing realistic training is a major challenge when it comes to teaching use of force, particularly deadly force. Police officers have traditionally trained with firearms at well-lit shooting ranges, firing at paper targets while wearing ear and eye protection. The only thing realistic about this method is that an actual gun is being fired. Overall, the firing range does not much resemble the reality of deadly force situations, and it does not prepare officers well for the sort of conditions in which split-second deadly force decisions need to be made in the field. Computerized use of force simulators are designed to address this issue and to make force training more realistic. They are frequently used in police departments and training academies.[65] A great degree of variation exists in the sophistication and cost of use of force training simulators, but the basic principle is the same: The simulator uses video projected onto a large screen to present scenarios in which police officers are required to react. Professional actors appear in an array of different situations, such as fights, traffic stops, or confrontations with people who are intoxicated and/or mentally ill. The videos are interactive; situations can change based on the actions of the officer. With numerous videos and a multitude of possible choices and outcomes within each video, hundreds of situations are possible. The use of force is not necessary or justified in all the situations presented, and officers may be able to avoid using force through de-escalation tactics.

Photo 5.7

Even though the use of deadly force is extremely uncommon among police officers, they must be prepared to use it with skill and good judgment if necessary. Realistic training, such as that provided with a simulator, can better prepare officers to competently use force when necessary. Al Seib/Los Angeles Times/Getty Images

Officers are armed with simulated weapons equipped with lasers, and when force is used the actions of the officer and the accuracy of any weapon deployment can be recorded. Conditions of the simulations can be altered—for example, the level of lighting can be lowered—and officers may be required to engage in physical exertion prior to the exercise. This can more closely mimic the real-life situations officers may find themselves in when having to use force. After the completion of the simulation exercise, an officer's thought process can be discussed and the performance evaluated. Continuing technological advances will likely further enhance the realistic nature and effectiveness of use of force training simulators.

Main Points

- The presence of good officers in a department depends largely on strong recruitment efforts, appropriate selection standards and a thorough selection process, and adequate training. Each process depends on the others.
- Recruitment, which consists of efforts to create a group of qualified applicants to choose from, sets the stage for the selection of officers. The more applicants there are, the more likely well-qualified and desirable individuals will be included and selected.
- An important goal of recruitment efforts and the selection process in police departments is to maximize diversity among officers.
- Regardless of the recruitment methods used, it is reasonable to expect the success of recruitment efforts is directly linked to the job benefits provided by the agency.

- There are several important laws and policies that address equal employment opportunity (EEO): the Civil Rights Act of 1964, the Equal Opportunity Act of 1972, the Equal Pay Act of 1963, the Age Discrimination in Employment Act of 1967, the Americans with Disabilities Act of 1990, and affirmative action.
- EEO laws do not in any way prevent well-qualified applicants from being selected for a position. Job selection criteria must be job related.
- Four tests or procedures are most commonly used by police departments when identifying applicants to hire: written tests, background investigations, oral interviews, and medical/physical tests. The nature of these evaluations varies dramatically, as does the order in which they are administered.
- Females and people of color are underrepresented in police departments today. However, a great degree of variation is found across departments in this regard, and the representation of both is much higher than just a few decades ago. This is at least partly due to equal employment opportunity laws.
- Training occurs over the entire course of an officer's career. The purpose of training is to provide officers with knowledge, skills, and instruction so they can perform the tasks of police work with competence.
- Academy training primarily consists of classroom training on the skills of police work. Successful completion of this training leads to graduation and employability as a sworn police officer.
- Field training involves the probationary officer working alongside a field training officer (FTO) to learn and perform the tasks of policing while actually on the street.
- Over the course of a career, all full-duty officers are required to complete additional training known as in-service training. The number of in-service training hours required annually varies by department and by state.

Important Terms

Review key terms with eFlashcards at **edge.sagepub.com/brandl2e.**

affirmative action   89
assessment center   95
bona fide occupational qualification (BFOQ)   89
civil service protections   86
community service officer (CSO) programs   84
consent decree   89
de-escalation training   101
equal employment opportunity (EEO)   88
field training   101

genderized occupation   96
in-service training   104
lateral entry   87
non-stress academy training   100
Police Explorer programs   84
reverse discrimination   89
stress-based academy training   100
turnover   87

Questions for Discussion and Review

Take a practice quiz at **edge.sagepub.com/brandl2e.**

1. Why is the selection of patrol officers especially critical to the proper functioning of police departments?
2. What is the relationship between police officer pay and officer recruitment? How might pay differences among departments in the same geographical area negatively or positively affect these departments?
3. What are the advantages and disadvantages of higher selection standards in police departments? What are the advantages and disadvantages of lower standards?
4. Job selection criteria must be job related. Explain.
5. What are the possible intended and unintended consequences of affirmative action policies?
6. Why are females and people of color underrepresented in police departments? How has this situation changed over the past several decades?
7. What are the four most common tests and procedures used by police departments to select officers? Why might background investigations be particularly important?
8. What types of agencies operate police academies? What does police academy training consist of?
9. What is field training? How is field training conducted?
10. What is in-service training? Why is it important? What types of topics are covered through in-service training?

Fact or Fiction Answers

1. Fiction
2. Fact
3. Fact
4. Fiction
5. Fiction

6. Fiction
7. Fact
8. Fiction
9. Fiction
10. Fiction

Exhibit 5.2 Answers

1. d
2. c
3. a
4. b
5. c

6. b
7. b
8. a
9. c
10. c

⑤SAGE edge™

Digital Resources

Get the tools you need to sharpen your study skills. SAGE Edge offers a robust online environment featuring an impressive array of free tools and resources.

Access practice quizzes, eFlashcards, video, and multimedia at **edge.sagepub.com/brandl2e.**

Media Library

View these videos and more in the interactive eBook version of this text!

Career Video
5.1: Challenging Aspects of the Recruitment and Selection Process
5.2: Recruitment and Interview Tips for Students
5.3: Experience and Importance of Training Process

Criminal Justice in Practice
5.1: Pre-employment Character Issues

SAGE News Clip
5.1: Sheriff's Office Explorer Program

6

POLICE PATROL

Police Spotlight: The Reemergence of Foot Patrol

Many police departments across the country have taken renewed interest in foot patrol as a way to improve police–community relations. For example, the Cambridge (Massachusetts) Police Department deploys a minimum of two officers, 24 hours a day, seven days a week, in two of the city's business districts as well as in some of the city's higher crime areas. Although officers respond to calls for service in their areas, their primary responsibility is to engage and interact with citizens.[1] In the Evanston (Illinois) Police Department, two officers are assigned to patrol a one-mile square area Tuesdays through Saturdays during the afternoon and evening hours. Officers only respond to in-progress crime calls; their priority is on interacting with citizens.[2] The St. Petersburg and Cocoa (Florida) Police Departments have a program called Park, Walk and Talk.[3] As the name suggests, the program requires officers to park their cars, walk, and talk to neighborhood residents at various times during their shifts.

On the face of it, the benefits of such policies make sense. Police officers and citizens interact with each other face to face, which may improve police–community relations. And as discussed in more detail later in this chapter, recent research points to the crime reduction benefits of foot patrol. For instance, a 2018 study in San Francisco showed that foot patrol in high property crime areas led to a 17% drop in thefts, including vehicle break-ins, and a 19% drop in assaults.[4] Research conducted in Philadelphia, Pennsylvania, showed that foot patrol in high-crime areas led to a 23% drop in violent crime during the course of the study.[5] However, the Philadelphia study found cause for concern as well: The study showed that pedestrian stops (or **Terry stops**) went up 64% in the foot patrol areas.[6] The researchers question the potential negative consequences of these stops on police–community relations. If foot patrol means more intrusion into the daily lives of law-abiding citizens, it may not necessarily be good policing. It may be just another way in which the police are controversial.

Objectives

After reading this chapter you will be able to

6.1 Identify the strategies police may use to manage calls for service and discuss their actual impact

6.2 Identify the circumstances under which a fast police response is likely to have an impact on the ability to make an on-scene arrest and discuss the importance of police response time in general

6.3 Describe how the Kansas City preventive patrol experiment was conducted and discuss the results of the study

6.4 Assess why routine preventive patrol appears not to have an impact on crime but hot spot patrol does

6.5 Identify the components of a police crackdown and discuss the effectiveness of the strategy

6.6 Examine the intended and unintended consequences of police traffic stops and so-called Terry stops

6.7 Identify the advantages and disadvantages of one-officer and two-officer squads

6.8 Evaluate the benefits of foot patrol and offender-focused strategies

Fact or Fiction

To assess your knowledge of police patrol prior to reading this chapter, identify each of the following statements as fact or fiction. (See page 129 at the end of this chapter for answers.)

1. Ten codes are the universal and extensively used language of police officers' radio communications today.

2. Mobile data computers put a wealth of information in the hands of police officers while on patrol.

3. Unlike alternative phone numbers to reach the police, telephone reporting of crimes can actually reduce the workload of officers on the streets.

4. Research shows that most of officers' time spent on patrol is handling calls for service.

5. If the police respond quickly to a report of a serious crime, there is a high probability that the police will make an on-scene arrest.

(Continued)

(Continued)

6. The Kansas City preventive patrol experiment showed that increasing the amount of police patrol had a major impact on crime.

7. If patrol has an impact on crime, it will most likely happen in crime hot spots.

8. Research shows that traffic stops can reduce crime regardless of when and where they are used.

9. Studies have shown that the use of Terry stops has few, if any, effects on crime.

10. Foot patrol is an outdated and seldom used strategy in police departments today.

IN this chapter, we will discuss police patrol, which is the backbone of policing, and issues associated with it.

POLICE PATROL AND CALL PRIORITY

Police patrol involves the police providing a presence in a community in order to prevent crime. Officers on patrol also respond to calls for service. If officers respond quickly to crime scenes, they may be able to apprehend offenders before they get away. Officers may also be expected to engage in proactive activities while on patrol, such as making traffic and pedestrian stops in an attempt to discover or interrupt criminal activities. Police patrol can be conducted in several ways. Today the most common method is via vehicle patrol. Other methods include bicycle patrol, motorcycle patrol, foot patrol, and horse patrol. Some departments have officers on mopeds or even roller blades.

A Question to Consider 6.1

The Value of Police Patrol

It makes as much sense for the police to patrol randomly to prevent crime as it does for firefighters to patrol randomly to prevent fires. As you read this chapter, keep this statement in mind. When you are finished with the chapter, discuss whether this statement is mostly true or mostly false and why.

ALLOCATION OF POLICE PATROL

Patrol officers are assigned to certain sections of their jurisdiction. In small cities, this might be the entire city; in larger cities, officers might be assigned to a particular area. These patrol areas are often referred to as beats, squad areas, sectors, or reporting areas. In larger cities, a collection of patrol areas comprise a command area; command areas are often referred to as precincts, districts, or divisions. County and state law enforcement agencies may have different names for patrol areas, but the process of assigning officers to patrol those areas is the same. Patrol areas are typically created based on geographic considerations, such as natural boundaries (e.g., rivers, highways); defined neighborhoods; busy roads; or census tracts. The size of a patrol area most often depends on the number and nature of calls for service that originate from that area, particularly the area's level of crime. The number and nature of calls are likely to be related to the population characteristics of the area, including population density. For example, some patrol areas may be geographically large but have few residents and little housing. Other areas may be relatively small in size but contain a large number of people.

Patrol staffing levels also vary based on the time of the day, the week, or the year. Some areas may be geographically small but have a large population at certain times of the day, such as a downtown area. Since there are normally more calls for service at night, on weekends, and in the summer, ideally more officers would be on patrol during these times.

CALL PRIORITY

Not all calls for service are equal. Accordingly, police departments categorize calls for service into priorities. The number of **call priority** categories and their definitions vary by department.

Terry stops: Brief detentions by the police if there is suspicion of criminal activity.

call priority: A system in which a faster response and more resources are given to more critical calls for service.

For example, the Detroit Police Department in Michigan uses the following scheme to categorize calls for service:

- Priority 1 calls involve emergency situations that are still in progress, where a perpetrator is on the scene, where emergency medical service is needed, and where the preservation of evidence is of an urgent nature.
- Priority 2 calls deal with serious problems that have happened more than fifteen minutes before the call was made, where the perpetrator is still on the scene, where the likelihood of apprehension may be high or low, and where the preservation of evidence is urgent.
- Priority 3 calls are those that are not serious in nature, where the perpetrator is still on the scene, and where an incident happened less than fifteen minutes before the call.
- Priority 4 calls involve nonserious situations where the perpetrator is no longer on the scene, where the likelihood of apprehension is low, and where an incident occurred more than fifteen minutes before the call.
- Priority 5 calls are similar to Priority 4 calls but involve situations where damage or loss is less than $10,000.

A goal is often to immediately assign and dispatch officers to high-priority calls. Other calls would be categorized, placed in a queue, and then assigned to officers as officers are available to respond to them. Depending on the department, lower-priority calls, such as Priority 4 and 5 calls in Detroit, may be handled in ways other than assigning an officer for a mobile response.

> **mobile data computers (MDCs):** Computers mounted in police vehicles that are connected wirelessly to a department's computer network and computer-aided dispatch system.
>
> **computer-aided dispatch (CAD):** A computer system and database that tracks calls for service as they are received, monitors the status of patrol units, and provides various reports relating to calls for service, the activities of officers, and the calls to which officers respond.

TECHNOLOGY ON THE JOB
Mobile Data Computers (MDCs) and Computer-Aided Dispatch (CAD)

When officers are on patrol, they need information in order to make smart decisions. **Mobile data computers (MDCs)** can help fill this need. An MDC is a computer mounted in a police vehicle and connected wirelessly to the police department's computer network and computer-aided dispatch (CAD) system. MDCs provide officers with the ability to directly access vehicle information, criminal history, and warrant databases without having to rely on police dispatchers. MDCs may also provide squad-to-squad communication, email, in-field access to mug shots, and digital field lineups. MDCs allow officers to write incident and arrest reports and citations without the need to return to the police department to do so.[7] More than 90% of local police departments that serve populations of more than 25,000 people use MDCs.[8]

A **computer-aided dispatch (CAD)** system is a computer system and database that tracks calls for service as they are received, monitors the status of patrol units, and provides various reports relating to calls for service, the activities of officers, and the calls to which officers respond. Since MDCs can communicate directly with a

Photo 6.1
Mobile data computers allow officers to access a great deal of information and to be more productive while on patrol. AP Photo/Detroit News, Todd McInturf

department's CAD system and vice versa, this reduces police radio communications and allows the radios to be used for urgent situations.[9] More than 90% of all local police departments that serve populations of more than 25,000 persons use CAD.[10]

MANAGING CALLS FOR SERVICE

The police are a twenty-four-hour agency that is just a phone call away. While a relatively small proportion of calls made to the police are about crimes, a majority of them still involve matters that the police have an obligation to address. To contact the police, citizens have several options. One option is to call 911 (in some places texting 911 is also an option). The 911 system was created as a national emergency phone number in 1968, although it was not until the 1970s and 1980s that it was widely marketed and available to citizens. When a call is made to 911, it is normally received by a designated emergency dispatch center located either locally or regionally. Most 911 call systems have the unique advantage of being able to identify the caller's name, address, and phone number; most of the time cell phone location is also available. Problematic in some jurisdictions is the call volume to 911 centers. In particular, it is estimated that 20% to 25% of 911 calls are not emergencies;[11] considering accidental, prank, and hang-up calls, along with other nonemergencies, the total percentage of false emergency calls is much higher. For interest, a sample of actual 911 calls to the Los Angeles Police Department is accessible here: http://www.lapdonline.org/communications_division/content_basic_view/27361.

Another option is to call the designated police department nonemergency ten-digit phone number. These calls are usually either received at the police department itself, at the police department dispatch center, or at a regionally located dispatch center. A third option for citizens who have questions or information for specific police offices or officers is to call one of the administrative phone numbers provided in a phone book or on a police department website. A fourth option available in many larger U.S. cities is to call (or text) 311 or another designated three-digit nonemergency phone number.

311

The 311 phone number is marketed as a nonemergency number for police and/or city services. Accordingly, it is operated either by the city or the city's police department. The 311 system was created to alleviate some of the workload of overburdened 911 centers and operators. Researchers who examined the impact of 311 in Baltimore found that its introduction led to a large reduction in 911 calls (a decrease of approximately 34%). However, the actual number of calls remained about the same; most just moved to the 311 system. Further, the researchers found that there was virtually no impact on the number or nature of calls to which officers were dispatched.[12] Overall, the study showed that the benefits of the 311 phone number are limited: It may provide an easy method of contacting the police for nonemergency reasons, and it may reduce the number of 911 calls, but those 311 calls must still be answered, and 311 does not appear to affect the workload of patrol officers on the street.

EXHIBIT 6.1
Police 10-Code Radio Communications

Although the use of MDCs and CAD systems are transforming radio communications between officers and dispatchers, some police departments still use 10 codes. In most departments, these codes are not used as extensively as years ago. Ten-codes codes offer an abbreviated and standardized method of communication on police radios. Certain 10 codes are used as communication shorthand, such as 10–4 for an affirmative acknowledgment, 10–9 for out of service, and 10–50 for a vehicle accident that only includes property damage.

DIFFERENTIAL POLICE RESPONSE (DPR)

While the impact of 311 appears to be limited to the front end of call management (i.e., calls being answered by 311 operators instead of 911 operators), differential police response (DPR) may be more likely to affect the back end of call management (i.e., the number and nature of calls that patrol officers respond to). In some police departments, officers go from one call to the next for an entire shift, making it impossible to engage in any other productive activities. Some people have referred to this phenomenon as *pinball policing*. However, it may not be necessary or productive for patrol officers to respond to every call for service in which they are requested. In an attempt to do more with less and to structure patrol time, the police may be able to handle calls for service in ways other than an immediate mobile response. For example, the police might be able to handle calls by taking reports over the telephone or Internet, or police department nonsworn personnel (e.g., community service officers) could respond and take reports.

The determination as to when a DPR is appropriate is based on the type of incident being reported and the timing of the crime in relation to when it is reported. Some crimes, such as an assault, would never be eligible for an alternative response regardless of when it was reported. A study conducted in a medium-sized municipal police department showed that there were approximately 320 larceny and vandalism calls to the department per month. If just these calls were assigned an alternative response, it would translate into freeing 180 to 240 hours of patrol time per month.[13] This is time during which officers could be available to respond quickly to calls that warranted it. Not only can the use of DPR reduce the call burden on officers, it may also allow the creation of blocks of time in which certain officers could be free from responding to calls for service. During this time, officers could conduct follow-up investigations, engage in other proactive activities, or engage in other patrol activities as directed.

Before police managers develop and implement alternative responses, it is important to consider the volume of calls for service to the police department, the type of calls appropriate for alternative responses, citizen expectations of the police, and how any additional patrol officer free time created by the alternative responses may be used. Alternative responses may not be justified if call volume is not problematic, if citizens always expect responses from police officers, or if additional patrol time is not needed. Research has shown citizens are most willing to accept phone reporting as a DPR, although some people were more accepting of alternative responses than others.[14] It should also be noted that in an effort to reduce patrol officer work demands, some departments have stopped responding to certain calls, such as burglar alarms, which are often false,[15] and keys locked in cars.

PREVENTIVE PATROL

There has been a long-standing belief that having officers move around and provide a visible presence in an area will lead to the prevention of crime, particularly through the deterrence of criminal behavior. Patrol is designed to create a belief in the omnipresence of the police. If the police could be anywhere, then it may seem like they are everywhere. Therefore, patrol may lead to offenders believing that their risk of apprehension is higher, and this belief can deter them from committing crimes.

Belief in patrol as a crime control strategy took hold when it became the centerpiece of operations in the London Metropolitan Police Department in the early 1800s, and the notion grew in the early 1900s, when patrol was conducted via the automobile. Officers could cover much more territory using vehicles, thus enhancing the perception of omnipresence. Further, with officers more mobile and assigned to particular patrol areas, they could respond quickly to crime scenes and make more on-scene arrests. Or so it was thought.

Preventive patrol usually involves one or two officers riding in a marked police vehicle. In general, while patrolling, officers are on the lookout for people who are in need of assistance or

differential police response (DPR): A response to a call for service other than immediately dispatching an officer to the scene.

FIGURE 6.1

How Do Patrol Officers Spend Their Time?

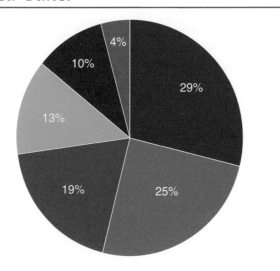

- ■ Administrative and personal activities
- ■ Patrolling in a vehicle
- ■ Handling calls for service to which they were assigned
- ■ Backing up other officers dispatched to their own calls
- ■ Other self-initiated work activity
- ■ Traffic enforcement

Source: Data from Christine N. Famega, James Frank, and Lorraine Mazerolle. 2005. "Managing Police Patrol Time: The Role of Supervisor Directives." *Justice Quarterly* 22: 540–559.

who are acting suspiciously or illegally; the officers provide a presence on the streets and in parking lots and alleys. They may visit business establishments, such as gas stations, shops, restaurants, and so forth. They look for individuals and vehicles that are known to be wanted. Patrol officers also respond to calls for service and engage in self-initiated activities (e.g., make traffic stops) and activities directed by their supervisors (e.g., attempts to locate subjects, surveillance).

Another way of thinking about patrol is that officers have *assigned time* and *unassigned time*. Assigned time consists of time spent responding to and handling calls for service. A study was conducted in Baltimore involving 1,304 hours of patrol time.[16] The authors found that officers responded to an average of five calls per shift and each call took an about twenty minutes, leaving approximately 81% of shift time unassigned (see Figure 6.1).

Based on this study, a large share of patrol officers' time is spent simply patrolling.

OUTCOMES OF PREVENTIVE PATROL

As noted earlier, police patrol is supposed to have positive effects on crime control in a community. Those effects are discussed here.

APPREHENSION THROUGH FAST POLICE RESPONSE One potential benefit of having officers on patrol is that they should be able to respond to calls for service quickly. If officers are assigned to relatively small patrol areas, when a request for service is received, the officers should be in fairly close proximity to where they are needed. Part of the rationale for patrol is that, if the police respond quickly to crime scenes, they may be more likely to apprehend the perpetrators.

For the sake of clarity, is important to define **response time**. Response time can be thought of in terms of three segments: (1) citizen reporting time (from when the incident occurs to when the police are notified); (2) police dispatch time (from when the call is received by the department to when the call is dispatched to an officer); and (3) police travel time (from when the police officer receives the call to when the officer arrives at the scene). All three segments combined represent response time; the latter two segments represent *police* response time.

The most significant study examining the relationship between response time and on-scene apprehensions found that "the traditional practice of immediate response to all reports of serious crime currently leads to on-scene arrests in only 29 of every 1,000 cases."[17] In other words, a fast police response seldom leads to an on-scene arrest. Why? The authors explain that 75% of all crimes (FBI index crimes, not including homicide or arson) are **discovery crimes**. These are crimes that are discovered only after they have been completed. By the time they are discovered, it is likely that the perpetrator has already fled the scene. No matter how quickly victims or witnesses notify the police and no matter how quickly the police respond in these situations, it is extremely unlikely that the police will find the perpetrators at or near the crime scene. The other 25% of crimes are **involvement crimes**. These are crimes where a victim or witness sees the crime as it is occurring. Robberies and assaults are good examples of involvement crimes. A burglary where a witness sees the crime as it is occurring would also best be

response time: The amount of time that elapses between when a crime occurs and when officers arrive at the scene. It includes both citizen reporting time and police response time.

discovery crimes: Crimes discovered after their completion.

involvement crimes: Crimes witnessed as they are occurring.

Photo 6.2
Ideally, preventive patrol results in crime deterrence and the apprehension of offenders through fast responses to crime scenes. However, studies have questioned the effectiveness of preventive patrol on these *outcomes.* Michael Matthews - Police Images/Alamy Stock Photo

considered an involvement crime. In these situations, a fast police response could potentially lead to an on-scene apprehension; however, the problem is that in these cases, victims (and/or witnesses) often delay in calling the police. They may instead call a friend, tend to injuries, or even chase after the offender (not a wise action) before calling the police. Any delay in notifying the police will lower the odds of a fast officer response and potential on-scene arrest. The study further reported the following:

- 13% of involvement crimes were reported while still in progress. The chance of an on-scene arrest in these cases was 35%.
- 14% of involvement crimes were reported within the first minute after the crime was committed. If reported within the first few seconds, the chance of an on-scene arrest was 18%; if reported sixty seconds after the crime, the chance of an on-scene arrest was 10%.
- 27% of involvement crimes were reported between one and five minutes after the crime occurred. In these cases, the chance of an on-scene arrest was about 7%.
- The chances of an on-scene arrest were about equal for those crimes reported after five minutes as when reported in sixty minutes.

Unfortunately, in the calls covered by the study, the average amount of time it took for involvement crimes to be reported to the police was between 4 and 5.5 minutes.[18] Clearly, seconds make a difference when it comes to police ability to make on-scene arrests. This study was conducted before there were cell phones. Although one might expect that the availability of cell phones could affect the findings, the few studies that have reexamined this issue have drawn similar conclusions. For example, in one more recent study, researchers looked specifically at the relationship between police response time, on-scene arrests, and in-progress burglaries.[19] The researchers found that on-scene arrests for even in-progress burglaries are a rare event. Of the 5,290 in-progress burglaries reported to the police over the course of a year, only 8% involved an on-scene arrest.

Since a fast police response seldom leads to an on-scene apprehension, one might be tempted to conclude that response time is not an important aspect of police patrol operations.

This conclusion would be incorrect for several reasons. First, police response time has the *potential* to make a difference in certain types of crimes (again, those that are reported to the police while they are still in progress). Efforts to minimize police response time in these situations are important.

Also, one must not forget that a fast police response may allow the police to secure and collect evidence and identify, separate, and interview victims and witnesses. Information from these sources may assist in later identification and apprehension of the offenders.

Last, citizens may believe that response time is important and expect a fast police response. Generally speaking, people expect that when they call the police, officers will respond in a timely manner. Studies have shown that it is not so much the *actual* amount of time that it takes for the police to respond that is important in ensuring citizen satisfaction, it is more the *expectations* of response time in relation to when the police actually arrive. Fifteen minutes may seem like a long time to wait for the police unless you were told that it would take thirty minutes. Thus, expectations of response time are critical.

CRIME REDUCTION THROUGH DETERRENCE: THE KANSAS CITY PREVENTIVE PATROL EXPERIMENT In addition to faster police response time, another reason for patrol is that it leads to crime prevention through deterrence. The Kansas City preventive patrol experiment (KCPPE) is the most important study that has examined this factor.[20] The KCPPE was conducted in 1972 and 1973 in Kansas City, Missouri. The intent of the study was to examine in the most rigorous way possible the impact of random preventive patrol. The KCPPE used an experimental design in order to test the effects of motorized patrol on a variety of outcomes, the most important of which was the amount of crime. To conduct this experiment was no small feat; it involved the reallocation of officers so that some areas of the city did not have any patrol during the course of the study. This was a big risk for the police department in terms of political and legal implications and explains why the study has never been replicated.

The experiment was implemented in fifteen of the twenty-five beats of the South Patrol Division of the city. Each beat consisted of thirty-two square miles and included residential and commercial areas. The beats had a total population of 148,395, with more persons per square mile than in the rest of the city as a whole. The racial makeup of the beats ranged from

78% black to 99% white, and there was substantial variation among the beats in terms of median family income.

Groups of three beats were created so that all groups were similar in terms of the amount of crime in them, geographic size, number of calls for service to the police, ethnic composition, median income, and the transiency of the population (i.e., on average, how long residents had lived in their present home). Each group of beats consisted of approximately ten square miles of land. Then one of the three beats in each group was designated as a *reactive* beat, one as a *proactive* beat, and one as a *control* beat. In the five reactive beats, there was no preventive patrol, but marked squads still responded to calls for service in these areas. When not responding to calls, the officers assigned to the reactive beats patrolled only the boundaries of their beats. In the five control beats, the normal level of patrol was maintained, with one squad car per beat. In the five proactive beats, there were two or three patrol squads per beat instead of one. In the control and proactive beats, officers were instructed to patrol in the same manner as usual. The experiment was in place twenty-four hours a day. The experiment did not apply to other police units; for unmarked squads, tactical units, and helicopter patrol, it was business as usual.

As noted, the primary outcome of interest was crime prevention. Did crime levels vary across the areas that had different levels of patrol? Before and during the experiment, the researchers collected data on crimes reported to the police and victimization data directly from citizens. They also collected a multitude of other data, including data on traffic accidents, arrests, response time, and citizens' fear of crime and their attitudes toward the police.

If the amount of patrol was to affect levels of crime, it would be expected that there would be the least crime in the proactive areas and the most crime in the reactive areas. But the researchers found there were no differences in crime across the beats (Figure 6.2), nor were there differences in any of the other outcomes of interest. There were not even variations in police response time to calls for service across the three experimental conditions. Basically, increasing or decreasing the amount of preventive patrol had no effect on crime, citizen fear of crime, community attitudes toward the police, police response time, or traffic accidents.

Are these findings valid? Should we believe them? There are several issues to consider when evaluating the veracity of the KCPPE findings:

- Although the reactive areas did not have any police patrol, they still had a police presence as a result of officers responding to calls for service and other police units in those locations.
- Citizens (and criminals) were not told of the experiment, and apparently they did not notice or care.
- Instead of one car assigned to each of the proactive beats, there were two or three. And on average, each beat consisted of approximately two square miles. One could question whether this is meaningful variation in the level of patrol.
- The experiment provided for the increased *visibility* of patrol. Is this enough to deter would-be offenders? Or might it take officers engaged in

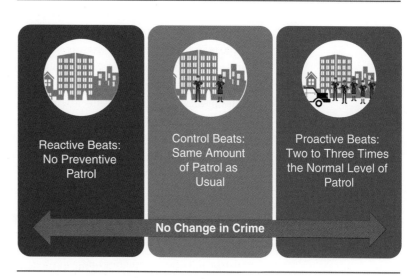

FIGURE 6.2

Three Levels of Patrol in the Kansas City Preventive Patrol Experiment

Reactive Beats: No Preventive Patrol

Control Beats: Same Amount of Patrol as Usual

Proactive Beats: Two to Three Times the Normal Level of Patrol

No Change in Crime

proactive activities with greater frequency (e.g., vehicle stops, pedestrian stops) to have an impact on crime?

- If the areas (and times) in which the experiment applied had little crime in the first place, how could patrol have prevented crime?
- At best, one might expect patrol to affect offenders' willingness to commit some types of crimes but not others. For example, might offenders be less likely to engage in drug sales than domestic assault since the former take place in public areas much more frequently than the latter?

All of these issues raise questions about the study's findings. What if patrol was greatly increased in relatively small areas that experienced a lot of crime? What if officers were instructed not just to be visible on patrol but to be active, to make traffic stops, and to stop and question people acting suspiciously on the sidewalks? This would clearly be a different sort of experiment.

In spite of its shortcomings, the findings of the KCPPE were taken seriously, and the study led to the police searching for other ways to prevent crime.

HOT SPOT POLICING

A well-established fact about crime is that it is not distributed equally; it is more likely to occur at certain times and in certain places. In other words, crime is generally not random. In a study conducted in Minneapolis, Minnesota, it was found that all robberies reported to the police over a one-year time frame occurred at only 2.2% of all locations (addresses or intersections), all auto thefts at only 2.7% of all locations, and all rapes at only 1.2% of all locations. The authors reported that 95% of all locations in the city were not involved in any of these crimes in the one-year period under examination.[21] It is likely that other types of crimes, such as vandalism or thefts from autos, are not as concentrated as the more predatory crimes. With regard to time, an examination of calls for service (crime calls and calls about other matters) to the Milwaukee Police Department showed that the frequency of calls gradually increased throughout the day, peaking between 5:00 p.m. and 6:00 p.m., between 9:00 p.m. and 10:00 p.m., and between 1:30 a.m. and 2:30 a.m. Overall, 5:00 p.m. to 10:00 p.m. saw the highest call volume. With regard to seasonal variation, the summer months had the highest call volumes.[22]

If crime is most likely to occur in certain places and at certain times, then it would follow that for the police to be most effective, they should also be concentrated in those places at those times. This is the basis for hot spot policing. Hot spot policing is sometimes known as place-based policing. This type of policing includes a variety of initiatives that focus on locations (and times) where crime is concentrated. These initiatives include increasing police patrol visibility and proactive activities as well as more problem-oriented approaches to crime prevention. It is widely accepted today that a police focus on crime hot spots is an effective approach in reducing crime; the accumulation of research on the issue supports this position. As explained by the National Research Council, "Studies that focused police resources on crime hot spots provide the strongest collective evidence of police effectiveness that is now available."[23] And since the council's report was published, the evidence has gotten even more persuasive. In essence, it has been established that if vehicle patrol is to have an impact on crime, this impact will most likely affect crime hot spots.[24] If foot patrol is to have an impact on crime, this impact will most likely affect crime hot spots.[25] If offender-focused strategies have an impact on crime, this impact will most likely affect crime hot spots.[26]

THE KANSAS CITY HOT SPOT PATROL EXPERIMENT (KCHSPE)

One of the first empirical tests of hot spot policing examined the effects of enhanced police patrol and officer activities on gun crimes in Kansas City.[27] Because it provides clear contrast to the KCPPE, the study is described here.

hot spot policing: Concentrating police in areas where crime is most likely to occur.

crime hot spots: Small geographic areas with a high concentration of crime.

Photo 6.4

Many studies have shown that increasing levels of patrol in crime hot spots has an impact on crime. Daily Camera Archives/ MediaNews Group/Boulder Daily Camera/ Getty Images

In the experiment, two additional squads were assigned to an eight-by-ten-block area from 7:00 p.m. to 1:00 a.m. for twenty-nine weeks. These officers were not assigned calls for service; their activities were oriented toward detecting and seizing illegal guns. During the course of the experiment, the officers involved issued 1,090 traffic citations, conducted 948 car checks and 532 pedestrian checks, and made 170 state/federal arrests and 446 city arrests. Seventy-six guns were seized during the experiment, twenty-nine of them by the extra patrol. In the beat covered by the additional squads, the number of guns seized increased, and the number of gun crimes went down. A similar beat without extra patrols was used as a comparison and saw no change in the number of guns seized or gun crimes committed. The researchers also did not find that gun crimes were displaced to any of the other beats that surrounded the experimental beat.

After reviewing the results of their study, the researchers asked the question, "Why should seizing 29 more guns in 29 weeks make any difference in gun crime in an 8-by-10-block area?"[28] The authors offered two explanations. First and most directly, the twenty-nine people who had their guns confiscated were at least temporarily unable to commit any gun crimes. Second and perhaps more significantly, the seizures may have had a deterrent effect among those who might otherwise have carried their illegal guns on the street. If would-be offenders perceived an increased risk of having their guns discovered and confiscated by the police, they may have left their guns at home. This would make it impossible for them to use the guns, at least outside their home. In this sense, the effect of the extra patrol was not simply limited to the additional twenty-nine guns seized but also possibly to the amount of gun crimes committed during the study period.

PREVENTIVE PATROL VERSUS HOT SPOT PATROL

It is important at this point to ask why the Kansas City preventive patrol experiment (KCPPE) did not find police patrol to impact crime but the Kansas City hot spot patrol experiment (KCHSPE) did. There are numerous differences in the studies that could account for the contrasting results. These differences include the following:

- The KCHSPE study focused on a small high-crime area, whereas the KCPPE included a larger area with relatively low levels of crime.

- The KCHSPE study provided for a greater level of police patrol in a relatively small area compared to the KCPPE.
- The KCHSPE involved increased police activities; the KCPPE provided for only an increase in police visibility.
- The KCHSPE study focused on a particular type of crime (gun crime), while the KCPPE did not.

OTHER ISSUES ASSOCIATED WITH HOT SPOT PATROL

Subsequent to the KCHSPE, many other studies have demonstrated the effectiveness of hot spot policing.[29] It is now widely accepted that this method of policing works. Not surprisingly, then, a survey of larger police departments revealed that fewer than 1% of them had not implemented some form of hot spot policing.[30] Nevertheless, some questions still remain about hot spot policing. In particular, we know too little about how it affects citizens' perceptions of the police and police legitimacy. Given the demographic characteristics of many hot spots (e.g., high rates of poverty and unemployment, people of color) and the proactive tactics often used by officers (e.g., frequent vehicle and pedestrian stops), it is not a stretch to expect that hot spot policing may lead to strained police–community relations. These communities "may begin to feel like targets rather than partners,"[31] which can lead to criticisms of overpolicing. However, research on this issue conducted in Philadelphia shows that hot spot policing did not negatively affect residents' perceptions of the police.[32] In contrast, in Baltimore, a hot spot policing strategy led to many abuse complaints and lawsuits but a significant decrease in homicides and shootings as well.[33] It is interesting that although hot spot policing is effective at crime control, it can still be controversial.

Other unanswered questions regarding hot spot policing include the following:[34]

- Is hot spot policing effective as a crime reduction strategy in smaller cities and rural areas?
- Most evaluations of hot spot policing have been limited to less than one year. What are its long-term crime reduction impacts?
- Most evaluations of hot spot policing have focused on crime reduction effects in limited areas of the community. Can hot spot policing reduce overall crime in a jurisdiction?

As these and other questions are addressed by research, additional strengths and limitations of hot spot policing will become clearer.

POLICE CRACKDOWNS

Crackdowns were the forerunner to hot spot policing and are still used by departments today. Like hot spot policing, a **crackdown** is a strategy designed to produce crime reduction effects. Crackdowns can be offense specific (e.g., focused on drunk driving, jaywalking, or street-level drug sales) or geographic specific (e.g., focused on the beach, the campus, downtown, or the highways). In both cases, a crackdown involves allocating additional police resources to the enforcement of laws with the intent of deterring illegal conduct. Crackdowns are usually temporary, at least in part because they are expensive to operate.

ELEMENTS AND OPERATION OF A CRACKDOWN

Here is how a crackdown is supposed to operate: First, the police make citizens aware of a pending increase in law enforcement efforts directed at a particular offense or location. For example, through press releases and other media notifications, the police advertise that they will

crackdown: The allocation of additional police resources to the enforcement of laws with the intent of deterring illegal conduct.

be on the highways in greater force over a holiday weekend detecting and arresting drunk drivers. Second, the police actually increase their presence and enforcement activities. Then, at some point, the increased police presence and enforcement activity is curtailed. This is referred to as the *back-off*. The back-off is not advertised and can occur gradually or suddenly.

The increased police presence and use of sanctions is supposed to affect would-be offenders' perceptions of risk associated with engaging in certain criminal behaviors or criminal behaviors at a certain place. There may be an initial deterrent effect achieved when a crime reduction occurs while the crackdown is in place. There may also be a residual deterrent effect. A residual deterrent effect, or a hangover effect, is achieved when there is a crime reduction even after the back-off. This represents the continuing effects of previous activities. If a crackdown results in an initial deterrent effect, that is good, but a residual deterrent effect is even better. A residual effect represents a "free" benefit. Researchers also refer to the concept of decay. Initial deterrence decay is when an initial deterrent effect dissipates or disappears. Residual deterrence decay is when a residual deterrent effect dissipates and the normal level of crime resumes. In both instances, would-be offenders learn that the crackdown has terminated and it is once again safer to offend.

Do crackdowns work? Do they reduce crime? A review of crackdown evaluations[36] found that most studies examined crackdowns on either drug sales or drunk driving and that, overall, drug crackdowns were less successful than those focused on drunk driving. Displacement of crime was

> **initial deterrence decay:** Decay that occurs when an initial deterrent effect dissipates or disappears.
>
> **residual deterrence decay:** Decay that occurs when a residual deterrent effect dissipates and the normal level of crime resumes.

Photo 6.5

One tactic that can be used in a crackdown is a police checkpoint at which all vehicles are stopped. ©iStockphoto.com/Sharon Dominick

found in a few studies, but many of the studies failed to rigorously examine the issue. Drug sales appeared to be most susceptible to displacement. Most studies demonstrated the crackdowns produced initial deterrence but did not have residual deterrent effects. The most important takeaway is that crackdowns represent another tool that police agencies can use when trying to achieve an elusive crime control goal, but their effects can vary.

POLICE STOPS OF CITIZENS AS A STRATEGY

Police stops of citizens are a method the police can use to make their presence on the streets obvious. Stops come in primarily two forms: traffic stops and Terry stops.

A Question of Ethics

The Ethics of "Get Tough" Tactics

Crackdowns and hot spot policing often involve the police using aggressive, "get tough" tactics in an attempt to reduce criminal behavior. Even the term *crackdown* implies a hard-nosed approach to policing. With such tactics comes the real possibility of the violation of citizens' civil rights. Is it proper to design and implement a strategy that can easily lead to or encourage such abuses? How can the potential for abuses be reduced when using aggressive police strategies?

TRAFFIC STOPS

A common activity of patrol officers is making traffic stops, which is an essential tactic in hot spot policing (but is not limited to this type of policing). Although traffic stops are common, reliable statistics on their frequency are not consistently recorded or reported. However, one study reported that in 2018 the Milwaukee Police Department conducted nearly 91,000 traffic stops.[37] As the city's total population at the time was approximately 600,000, this translated into approximately one out of every seven people being stopped over the course of the year; this figure even included members of the population who did not drive. In 2017, the Nashville Police Department conducted approximately 250,000 traffic stops (one out of three residents stopped during the year).[38]

Police officers can legally stop vehicles for moving and equipment violations in order to then take further investigative actions (so-called pretext stops), such as questioning the driver or searching for and seizing illegal guns and drugs being transported in those vehicles. Traffic stops also are a clear reminder to would-be offenders that the police are present and on the lookout. As such, the police may be able to prevent drug sales or crimes committed with a weapon either by confiscating those items or by convincing would-be offenders to leave their weapons and/or drugs at home. Traffic stops may also be a way to find individuals who are wanted on outstanding warrants. It is also expected that traffic stops will lead to a reduction in traffic accidents.

Relatively little research has examined the crime reduction effects of traffic stops. Some studies suggest traffic stops are more likely to have an effect on crime when used in high-crime (hot spot) areas,[39] although other studies show traffic stops to have no impact.[40] Perhaps part of the reason for no impact is that so few stops actually result in an arrest or discovery of drugs or weapons (2.2% of stops in Nashville).

Not surprisingly, research also suggests that citizens do not like being stopped for traffic violations, especially for reasons other than speeding.[41] Accordingly, traffic stops can have negative (and unintended) consequences. Although it should be noted these studies have numerous limitations, it has been found that (a) citizens who experience more traffic stops are less likely to contact the police to ask for assistance or to report a neighborhood problem[42] and (b) when police make traffic stops and ask to conduct a consent search of the vehicle, it negatively affects citizens' perceptions of proper and respectful police conduct.[43] Research also shows that some people are more likely to be stopped by the police than others. A study of 60 million traffic stops in twenty states showed that black drivers were stopped more often than white drivers relative to their representation among drivers, and that blacks were also more likely than whites to be ticketed, searched, and arrested after the stop compared to white drivers.[44] The authors note however that these differences may be a result of driving behavior, not bias. Other studies have also shown black drivers to be stopped more often than white drivers.[45] Similar findings have been reported with regard to Latino drivers.[46]

With regard to the impact of traffic stops on traffic safety, there is too little research on which to draw firm conclusions. In Nashville, a 50% reduction in traffic stops was associated with a 60% increase in traffic accidents between 2011 and 2017.[47]

The bottom line is that although traffic stops (and associated consent searches) may have some crime reduction benefits in some places, traffic stops may be used in a disparate manner, harm police–community relations, reduce citizen cooperation with the police, but may have the benefit of reducing traffic accidents. It should also be noted, of course, that sometimes traffic stops can turn into vehicle pursuits. This issue is discussed in detail in Chapter 10.

STOPPING, QUESTIONING, AND FRISKING

Another common proactive patrol tactic is **stopping, questioning, and frisking (SQF)** pedestrians on the streets and sidewalks. If an officer has reasonable suspicion that a subject is involved in criminal behavior, the officer may conduct a stop of that person and search the person for weapons. Chapter 9 provides a detailed discussion of the legality of stops and searches. These stops are also known as field interviews (FIs) or Terry stops (so named after the 1968 U.S. Supreme Court case, *Terry v. Ohio*).

As noted in Chapter 4, SQFs are frequently used by police departments today, although due to lawsuits against the police for the disparate overuse of the practice, SQFs are not as widespread as just a few years ago. Among other requirements, lawsuits have mandated that police officers complete detailed reports for pedestrian stops. San Francisco has debated requiring officers to provide, in writing, their name, badge number, and information about filing a complaint or commendation to citizens who are stopped by the police.[48] Such requirements may have a chilling effect on their use.

As examples of the disparate and declining use of pedestrian stops, in 2010 in New York City, blacks were stopped, questioned, and frisked by the police at more than ten times the rate of whites.[49] Data analyzed from New York City during this same timeframe also showed that black and Hispanic citizens who were stopped were also more likely to have non-weapon force used upon them.[50] After a lawsuit, the New York City Police Department (NYPD) reduced the frequency of stop-and-frisks from 23 stops per 1,000 people in 2011 to just two stops per 1,000 people in 2014.[51] In Chicago between May and August 2014, 72% of those people who were stopped were black, although African Americans represented just 32% of the city's population.[52] From 2014 to 2016, the number of pedestrian stops in Chicago declined 80% (although traffic stops increased dramatically).[53] In Milwaukee, pedestrian stops declined from 46,438 in 2015 to 6,945 in 2018.[54]

stopping, questioning, and frisking (SQF): SQFs occur when an officer has reasonable suspicion that a subject is involved in criminal behavior so the officer conducts a stop of that person and searches that person for weapons.

Anticipating the Unintended Consequences of Police Strategies

Every police strategy has intended and unintended effects. It is usually not difficult to identify and anticipate the positive effects of a strategy, which usually relate to crime reduction. However, unintended consequences are often more difficult to anticipate. They typically happen without any foresight and are unwelcome surprises. Effective management of police agencies requires leaders to study and consider all of the possible effects of law enforcement strategies prior to implementation—intended and unintended, short and long term. Possible crime control benefits should be weighed in relation to other important factors, such as community relations, legal and liability issues, citizen and officer safety, officer workload, and so forth. Research findings can play a critical role in this process.

Photo 6.6

SQFs are a controversial tactic of the police. In some cities, they have been found to be used in a disparate manner. They have also been found to have limited effects on crime.

Rafael Ben-Ari/Alamy Stock Photo

If SQFs impact crime, it would be for the same reasons as traffic stops: SQFs may lead to the confiscation of illegal weapons and/or drugs, which, in turn, would inhibit would-be offenders from using their weapons or selling their drugs. In addition, knowledge that the police are conducting frequent SQFs may increase the perception of the risk of apprehension and keep would-be offenders from possessing illegal items. SQFs could also be a way to identify (and arrest) wanted suspects. Are SQFs an effective crime reduction tool? An elaborate study published in 2014 found "no significant effects of police stops on precinct robbery and burglary rates" in New York City.[55] A follow-up study in 2017 in New York City showed that SQFs were associated with only small crime reductions.[56] Again, as with traffic stops, the discovery of weapons and contraband as a result of SQFs is very uncommon. This fact can be portrayed as a benefit of SQFs (e.g., would-be criminals are leaving their guns at home) or as a problem (innocent people are being stopped by the police). A further negative and unintended consequence of SQFs is that these stops may actually increase delinquent behavior among adolescent boys.[57] The studies on the effects of SQFs are not without limitations, so one must be cautious in drawing definitive conclusions from them. It should also be noted that sometimes SQFs can turn into foot pursuits. This issue is discussed in detail in Chapter 10.

ONE- AND TWO-OFFICER SQUADS

An ongoing debate in many police departments, especially larger ones, is the value of one officer versus two officers assigned to a squad car. Generally speaking, officers prefer two-person squads primarily because they feel safer when another officer is present to provide immediate assistance if needed. In some cases, officers also prefer having a partner because it provides the opportunity to converse during a shift and therefore avoid boredom and fatigue. Having a partner may also assist officers in making difficult decisions on the street—the "two heads are better than one" idea. An obvious drawback to two-person squads is that they are more expensive to staff. Also, one-officer squads may provide a wider distribution of officers and greater visibility per officer. Despite the importance of this issue, there has only been one major study that has examined it.[58]

In this study the researchers selected forty-four patrol units in San Diego, California, and designated half as one-officer squads and the other half as two-officer squads. The squads were monitored for one year, and the researchers collected data on squad performance, efficiency, officer safety, and officer attitudes. In conducting their analysis, they made comparisons between the one- and two-officer squads operating in similar areas under similar conditions. Among the most important findings were the following:

- The types and frequency of calls for service and officer-initiated activities were about equal for one-officer and two-officer units, although two-officer squads produced more traffic citations.
- Calls handled primarily by one-officer units were more likely to result in arrests and formal crime reports compared to two-officer units.
- Calls handled by one-officer squads were less likely to generate citizen complaints than were two-officer squads.
- More requests for backup came from one-officer squads than two-officer squads.
- One-officer units had less involvement with resisting arrest and equal involvement with assaults on officers (and corresponding injuries to officers).
- Officers had a preference for two-officer units and perceived them as having advantages in performance, efficiency, and safety.
- With regard to cost, eighteen one-officer units could be deployed for less cost than ten two-officer squads.

On the basis of their study, the authors concluded that two-officer regular patrol units do not appear to be justified in San Diego. Separate comparisons of unit performance, efficiency, and officer safety under current conditions all suggest that one-officer units are at least equal to and often more advantageous than two-person units.[59]

A Question to Consider 6.2

One-Officer Versus Two-Officer Squads

The San Diego patrol staffing study found that officers working alone were involved in fewer incidents of resisting arrest and generated fewer citizen complaints than two-officer squads, even in similar sorts of situations. What do you think could account for these findings?

FOOT PATROL

Police foot patrol is the second most common form of patrol behind car patrol; foot patrol is used in an ongoing basis in 55% of all local police departments.[60] It is most commonly used in larger departments; 90% of departments that serve a population of one million or more have foot patrols.[61] As a result of the community policing movement, in which the police seek to have more day-to-day interactions with citizens and to develop productive relationships with citizens, foot patrol has developed a newfound importance. Prior to community policing, many police departments had foot patrol but often only low-seniority officers were assigned to it and it was considered an undesirable assignment. The areas in which officers patrolled were also quite limited—mostly the downtown areas. Now, in departments committed to community policing, more officers are typically

Photo 6.7

Police foot patrol can lead to closer, more frequent, and often more friendly interactions with citizens. Bob Daemmrich/Alamy Stock Photo

assigned to foot patrol on an ongoing permanent basis and in more geographical areas. A detailed discussion of community policing and foot patrol is provided in Chapter 13.

Foot patrols are implemented for various reasons, the most common of which are to reduce crime, reduce fear of crime among citizens, and to create more personal relationships with community residents. Indeed, one could argue that a foot patrol officer creates more of a police presence than a squad car driving through a neighborhood. An officer on foot is able to more frequently and easily initiate contact with citizens than an officer in a vehicle. The police presence in a neighborhood and contact with citizens may affect perceptions of the risk of both criminal behavior and victimization, thus deterring certain criminal behaviors and reducing citizens' fear of crime.

RESEARCH ON FOOT PATROL

Given the popularity of foot patrol in police departments, it is not surprising that quite a bit of research has been carried out on the strategy. Early research examined foot patrol in neighborhoods and commercial areas; more recent research has focused on its crime control effectiveness in crime hot spots. The Newark foot patrol experiment was one of the first studies to examine the effects of foot patrol.[62] Conducted in the late 1970s in Newark, New Jersey, the study involved adding foot patrol to four beats, discontinuing it in four other beats, and continuing it as usual in four beats. The foot patrols were in place five days a week, one shift per day; each foot beat was "a small area."[63] On the basis of the analyses performed on crime data and surveys of citizens and business managers, the researchers reported the following results:

- Residents were aware of levels of foot patrol.
- Crime levels were not significantly different from each other across the three treatment groups.
- Residents of beats where foot patrol was added perceived crime problems to diminish compared to the other two areas.
- With regard to perceived safety, higher levels of safety were expressed among residents in the beats with added patrol. However, there were no differences across the beats in perceptions of safety for businesses.
- With regard to attitudes toward the police, citizens in the beats with added patrol had the most positive attitudes. This pattern was less obvious in the business sample.

Overall, this study highlights the value of adding (versus maintaining or dropping) foot patrol in improving citizens' attitudes toward the police and reducing fear of crime but not actual crime. Another study conducted in Flint, Michigan, found that added patrol in certain areas not only led to decreased fear of crime but also to less crime.[64]

As an understanding of crime hot spots emerged, researchers began to question if the impact of foot patrol would be any different if used in these small, high-crime areas. One such study was also conducted in Newark.[65] For a one-year period, between 6:00 p.m. and 2:00 a.m., twelve officers on foot patrolled a quarter-square-mile area of the city. Each officer averaged just over one arrest, two quality of life summonses, nearly four SQFs, and a total of 8.76 enforcement actions per week. In monitoring crime rates over time, the researchers found that compared to the larger precinct area in which the foot patrol beat was located, shootings and aggravated assault (but not murders or robberies) were substantially reduced in the foot patrol beat for the year that the patrol was in place. All of the violent crimes of interest decreased in frequency in the target area during the hours that foot patrol was in place. The crime reduction was also observed during the nonoperational hours of foot patrol for all of the crimes except robbery, which suggests that at least some robberies were displaced from times of the day in which foot patrol was in operation to times of the day when it was not. It was also found that robbery was spatially displaced—that at least some robberies moved from the target area to a nearby area. The authors concluded,

> This study provides support for foot patrol as a crime prevention tool, an important finding since earlier studies have predominately found foot patrol to reduce fear of crime without producing tangible crime reductions. Furthermore, this study illustrates that displacement remains a very real threat to place-based interventions.[66]

Other studies have also found foot patrol to have crime reduction effects. A study conducted in Philadelphia examined the impact of foot patrol in sixty small, very high-crime beats.[67] Foot patrol officers worked in pairs on day and night shifts, five days a week. Analysis of crime trends over a twelve-week time frame showed 23% fewer violent crimes (fifty-three actual crimes) in the foot patrol beats compared to the control beats, even after accounting for crimes that were displaced to nearby areas.

OFFENDER-FOCUSED STRATEGIES

While on patrol, officers can focus on policing places, as in hot spot policing, or they can focus on policing people, such as known high-rate offenders. Just like hot spot patrol is grounded in the logic that crime is concentrated in certain locations and times, offender-focused policing is based on the knowledge that a small percentage of offenders are responsible for a large percentage of crime. It would seem appropriate then to focus police resources on identifying, monitoring, and apprehending high-rate offenders. **Offender-focused strategies** depend heavily on criminal intelligence in order to identify serious repeat offenders. Efforts to target high-rate offenders may lead to crime reduction through incapacitation if offenders are arrested and incarcerated or through deterrence if would-be offenders perceive an increased risk of sanctions as a result of engaging in criminal behavior. Studies have shown that high-rate offenders targeted by the police are indeed more likely to be arrested and incarcerated, but the impact of this on overall crime has not been measured.[68] One study, however, did demonstrate that by identifying and paying extra attention (e.g., serving arrest warrants, conducting surveillance) to high-rate offenders over the course of seven months, violent crime was reduced by 42% and violent felonies by 50% in crime hot spots in Philadelphia, compared to control areas where offender-focused strategies were not used.[69]

Another possible advantage of offender-focused strategies is they are less likely to intrude on law-abiding citizens, unlike the widespread use of pedestrian and traffic stops. Along these same lines, if the police focus on offenders, the community may be more supportive of the police and law enforcement actions and perceive the police as being fair. Also, if police identify and focus on high-rate offenders, there is little concern for displacement effects—the police go where the offender goes. Overall, offender-focused strategies represent a more precise method of achieving a crime reduction effect.

offender-focused strategies: Strategies in which the police depend on criminal intelligence to identify high-rate offenders on whom the police then focus enforcement efforts.

Main Points

- Major considerations in police patrol allocation decisions are the number and nature of service calls in an area, population and housing characteristics, and time.
- Mobile data computers (MDCs) and computer-aided dispatch (CAD) systems are powerful tools that provide patrol officers easy access to critical information and allow for more efficient call assignment operations.
- Alternative numbers to 911 (such as 311) reduce the call volume to 911 centers but do little in reducing the overall call demands on patrol officers.
- Differential police response (DPR) strategies are strategies other than an immediate mobile response by a patrol officer that are used to respond to calls for service. DPR may reduce the call burden on patrol officers and allow officers to spend time doing things other than responding to calls for service.
- Research shows that a substantial amount of patrol officers' time is spent patrolling.
- Research shows that since citizens often delay in calling the police and offenders have usually fled before crimes are discovered, it is rare that police are able to make on-scene arrests.
- Although police are not often able to make on-scene arrests as a result of a fast response, police response time is still important from a criminal investigation and citizen satisfaction perspective.
- The Kansas City preventive patrol experiment (KCPPE) has been the most elaborate study of the effects of routine patrol on crime. The findings showed that routine preventive patrol does not prevent crime.
- Hot spot policing is built on the notion that crime is concentrated in certain locations and at certain times.

It is widely accepted that a police focus on crime hot spots is an effective approach in reducing crime, although unintended consequences of this strategy are yet to be fully explored.

- Police crackdowns involve allocating additional police resources to the enforcement of laws with the intent of deterring illegal conduct. They may result in initial deterrence and/or residual deterrence.
- Traffic stops are a commonly used strategy of crime deterrence. Research shows that traffic stops (and searches that may result from traffic stops) may have some crime reduction benefits, but they may also harm police–citizen relationships and citizen cooperation with the police.
- Stopping, questioning, and frisking (SQF) (also known as Terry stops or field interviews) is a strategy used in many police departments today. SQFs are controversial, and their crime reduction benefits are not well established.
- Although more research is needed on the issue, research has shown that one-person police squads have unique advantages and disadvantages over two-person squads.
- Foot patrol is common policing strategy, but there is much variation in its scope and goals. Research suggests that foot patrols are most likely to have an impact on citizens' fear of crime and, in crime hot spots, actual crime.
- Offender-focused strategies focus on specific offenders, not locations or times like hot spot policing does. Research suggests that these strategies are effective at reducing crime and can avoid the problems associated with widespread and unfocused pedestrian and traffic stops.

Important Terms

Review key terms with eFlashcards at **edge.sagepub.com/brandl2e.**

Questions for Discussion and Review

Take a practice quiz at **edge.sagepub.com/brandl2e.**

1. How do MDCs and CAD impact the job of patrol officers?
2. What problems are solved with the use of alternates to 911 (such as 311)? What problems are not addressed with such systems? How is this different from the effects of DPR?
3. One study found the traditional practice of an immediate response to all reports of serious crime led to on-scene arrests in only twenty-nine of every 1,000 cases. Why? With the use of cell phones today, might this finding be different? Why or why not?
4. The Kansas City preventive patrol experiment and the Kansas City hot spot patrol experiment came to different conclusions about the impact of police patrol on crime. Why?
5. How are police crackdowns supposed to work? Are they effective?
6. What is the rationale behind the use of traffic stops as a crime control strategy? What does the research say about the effects of this strategy?
7. Are SQFs an effective crime reduction strategy? Why are SQFs controversial?
8. What are the advantages and disadvantages of one-officer squads? Of two-officer squads?
9. What is the rationale behind the use of foot patrol as a crime control strategy? What does the research say about the effects of foot patrol?
10. What is the rationale for the use of offender-focused strategies? What does the research say about the effects of these strategies?

Fact or Fiction Answers

1. Fiction
2. Fact
3. Fact
4. Fiction
5. Fiction
6. Fiction
7. Fact
8. Fiction
9. Fact
10. Fiction

$SAGE edge™

Digital Resources

Get the tools you need to sharpen your study skills. SAGE Edge offers a robust online environment featuring an impressive array of free tools and resources.

Access practice quizzes, eFlashcards, video, and multimedia at **edge.sagepub.com/brandl2e.**

Media Library

View these videos and more in the interactive eBook version of this text!

Career Video
6.1: Experience of Basic Police Functions and Experience of Patrolling

Criminal Justice in Practice
6.1: Call for Service
6.2: BOL With Consensual Encounter

SAGE News Clip
6.1: Louisiana— Police App

7
CRIME DETECTION AND INVESTIGATION

Police Spotlight: Familial DNA and the Golden State Killer

Between 1974 and 1986, in southern California and the Sacramento area, an unknown person was suspected of committing multiple murders, dozens of rapes, and over 100 burglaries. The crimes began with burglaries, progressed to rape, and then to murder. The crimes were initially linked to each other and to the same unknown perpetrator by similarities in the way the crimes were committed (modus operandi).[1] Other miscellaneous information was also collected as a result of the investigations but none of it led to the perpetrator. For instance, in one of the assaults, the victim reported to the police that the perpetrator said repeatedly "I hate you Bonnie," but the victim's name was not Bonnie.[2] Eventually, the unknown perpetrator was dubbed the "Golden State Killer." In 2001, based on DNA evidence left at the crime scenes, DNA testing scientifically confirmed that most of the rapes and murders were committed by the same individual; however, the perpetrator remained unidentified. His DNA was not contained in California's DNA database or in CODIS (see Technology on the Job).

In early 2018, authorities announced the arrest of Joseph DeAngelo, 72, a former police officer, for eight counts of murder in connection with the Golden State killings. Later in 2018, he was charged with a total of 13 counts of murder. The evidence that led to his identification was his DNA recovered years ago at the crime scenes, but investigators used the DNA in a way that it had never been used before. Investigators submitted the unknown subject's DNA to the ancestry service website GEDmatch, which was able to identify numerous distant relatives of the DNA donor.[3] With this information in hand, investigators worked with a genealogist to construct a family tree. From there, with further investigation, DeAngelo was identified. Although DeAngelo's DNA profile was not in the GEDmatch file, some of his relatives were in the file. Without DeAngelo's knowledge, investigators then collected his DNA from the handle of his car and from a tissue recovered from a garbage can outside of his house.[4] The DNA print from the tissue and vehicle was consistent with the DNA left at the crime scenes. Further investigation revealed much about DeAngelo including that he lived within travel distance of where the crimes were committed, his occupations, and his relationships (including that one of his previous girlfriends was named Bonnie).

(Continued)

Objectives

After reading this chapter you will be able to

7.1 Define the criminal investigation process and the goals associated with it

7.2 Describe the reactive criminal investigation process

7.3 Identify the major types of undercover investigations and discuss why their use is often controversial

7.4 Discuss the role of evidence, circumstances, and investigative effort in solving crimes

7.5 Evaluate the role of DNA in criminal investigations

7.6 Identify and discuss the major types of evidence used to solve crimes

Fact or Fiction

To assess your knowledge of criminal investigation prior to reading this chapter, identify each of the following statements as fact or fiction. (See page 150 at the end of this chapter for answers.)

1. A crime can still be considered solved by the police even when a conviction is not obtained.

2. Most cold case investigations eventually result in an arrest.

3. When police go undercover to buy or sell drugs, it is known as a decoy operation.

4. Inculpatory evidence suggests that the suspect is actually the perpetrator.

5. Physical evidence is especially good at making associations between crime scenes, offenders, victims, and instruments.

6. Except for identical twins, no two people have the same DNA.

7. Unlike burglaries, homicides and rapes are not more likely to be solved when DNA is available.

8. Inaccurate eyewitness identifications of perpetrators by witnesses has been identified as one of the main reasons for false convictions.

(Continued)

9. In nearly all instances when an AMBER Alert is issued, the child is recovered; however, in most cases the AMBER Alert itself had nothing to do with the recovery.

10. Digital evidence is only useful in fraud investigations.

This investigation has raised questions about privacy due to the fact that investigators used personal information (DNA and family relationships) that was submitted by others for reasons unrelated to criminal identification and apprehension without their permission. Since the arrest of the Golden State Killer in 2018, GEDmatch has been successfully used more than 15 times by law enforcement authorities to identify previously unknown offenders.[5]

ALTHOUGH the police are expected to prevent crime, this is, of course, not always possible. When crimes are not prevented, the police are expected to identify and apprehend the offenders; this is the task of criminal investigation. This chapter discusses criminal investigation and its variations, the effectiveness of the police in solving crimes, and the types and roles of evidence found in investigations.

CRIMINAL INVESTIGATION DEFINED

Criminal investigation is the process of collecting crime-related information to reach certain goals. This definition has three important components: (1) the process, (2) crime-related information, and (3) goals.

As *a process*, criminal investigations usually consist of several stages in which certain activities are performed prior to other activities. The activities performed may be extensive or minimal depending on the nature of the crime being investigated. The nature and seriousness of the crime and the amount of evidence discovered determine the activities that are to be performed during the investigation. The most common activities performed in investigations are searching for and interviewing victims and witnesses and reading and writing reports. As activities are performed and evidence is collected, they are documented in the reports written by investigators.

There are many different people who may be responsible for the tasks necessary in a criminal investigation: patrol officers, detectives, crime scene technicians, even medical examiners. Although patrol officers are often thought to play a less important role in criminal investigations than detectives, studies have shown their actions can have a major impact on whether or not crimes are solved.[6] In fact, patrol officers have more responsibilities now than ever before in conducting investigations.[7] This is especially true in smaller police departments. Because of this, it is essential that patrol officers have a solid understanding of **criminal evidence** and the criminal investigation process.

Crime-related information is criminal evidence. Evidence consists of knowledge or information that relates to a particular crime or perpetrator. Criminal investigations are conducted in order to discover evidence. Evidence is used to establish (a) that a crime occurred and (b) that a particular person committed the crime. There are many different types of evidence that may play a part in criminal investigations, such as fingerprints, eyewitness identifications, confessions, psychological profiles, and so forth. Some types of evidence depend on scientific analysis in order to be made meaningful and useful. For example, blood may be analyzed in order to develop a DNA profile from it, human skeletal remains may be analyzed for clues about the cause of death, and bullets may be analyzed to determine the gun from which they were fired. These are issues that relate to the field of **forensic science**. Forensic science broadly refers to the field of science that addresses legal questions.

criminal evidence: Knowledge or information that relates to a particular crime or perpetrator.

forensic science: The field of science that addresses legal questions.

The third definitional component of a criminal investigation is that there are *goals* associated with the process. A goal is something one wishes to achieve at some point in the future. Goals also assist in giving direction to activities to be performed. Various goals have been associated with the criminal investigation process, such as solving the crime, providing evidence to support a conviction in court, and providing a level of service to satisfy victims of crime. Perhaps the most important goal, or at least the one most relevant to the police, is to solve the crime. To do this investigators must (a) determine whether a crime has been committed and ascertain the true nature of the crime, (b) identify the perpetrator, and (c) apprehend the perpetrator (Exhibit 7.1). Often these three tasks are related to each other and are tackled simultaneously.

Photo 7.1

Joseph DeAngelo was identified as the "Golden State Killer" through a controversial technique known as familial DNA testing. AP Photo/Rich Pedroncelli

The second goal often associated with the criminal investigation process is obtaining a conviction in court. Although this is not a primary responsibility of the police, a large share of investigators' time is spent on cases after they have been solved, assisting the prosecution in preparing the case for court. The prosecutor may present the evidence collected by the police in court to prove to a jury or judge that the defendant is guilty of the crime for which he or she is charged. In this sense, the police and prosecutor are on the same team, working toward the same end. It is important to understand that "solving the crime" and "convicting the defendant" are separate but related outcomes. A crime can still be considered solved by the police even when a conviction is not obtained.

EXHIBIT 7.1

The Meaning of "Crime Solved"

The FBI does not use the word *solved* to describe crimes for which perpetrators have been identified and apprehended; instead it says that crimes are *cleared by arrest*. A crime is cleared by arrest when three specific conditions have been met: At least one person has been (1) arrested, (2) charged with the commission of the offense, and (3) turned over to the court for prosecution (whether following arrest, court summons, or police notice). However, an actual conviction in court of the person arrested is not necessary for a crime to be cleared.

In its clearance calculations, the FBI counts the number of offenses that are cleared, not the number of persons arrested (see Figure 7.1 on page 137). As a result, one arrest can clear many crimes, or many arrests can clear just one crime. In addition, some clearances that an agency records in a particular calendar year may be of crimes that occurred in previous years.

In certain situations, for reasons beyond the control of the police, it is not possible to arrest, charge, or refer cases for prosecution. When this occurs, crimes can be "exceptionally" cleared. Examples of exceptional clearances include the death of the offender, the victim's refusal to cooperate with the prosecution after the offender has been identified, or the denial of extradition because the offender committed a crime in another jurisdiction and is being prosecuted for that offense. Sometimes the clearance of crimes through exceptional means is controversial.[8]

Photo 7.2
Evidence is used to solve crimes and may be presented in court to obtain convictions. These are separate but important goals of the criminal investigation process. Joe Ledford/The Kansas City Star via AP, Pool

The third goal associated with criminal investigation is victim satisfaction. This outcome has taken on increased importance during the last few decades with the movement toward community policing. The idea is that satisfying victims is a good thing and something about which the police should be directly concerned.

The *ultimate* goal of the criminal investigation process is to reduce crime through either deterrence or incapacitation. As discussed earlier in this book, it is difficult for the police to deter offenders, but if deterrence *is* to be achieved, an offender must first be identified, apprehended, and punished. Similarly, before an individual can be incapacitated by being placed in prison or some other action, that individual first needs to be identified and apprehended. At a minimum, deterrence and incapacitation depend on successful criminal investigations as investigations lead to the identification and apprehension of offenders.

TYPES OF CRIMINAL INVESTIGATIONS
REACTIVE INVESTIGATIONS

Reactive investigations are the traditional manner in which the police become involved in the investigation of crime: A crime occurs and the police respond. The police are typically in reactive mode when investigating crimes such as homicide, rape, robbery, burglary, and so forth. There are several stages to a reactive criminal investigation. After the police learn of a crime, usually through a report by a victim or witness, the first step is the initial or preliminary investigation. In this stage, patrol officers interview the victim and/or witnesses at the crime scene, canvass for other witnesses, and search the scene for evidence. For more serious crimes, patrol officers may also be responsible for maintaining the security of the crime scene, and detectives may also be involved in the initial investigation. The specific activities performed depend on the type of crime; obviously, a preliminary investigation of a homicide will look much different than a preliminary investigation of a burglary or a theft.

If an arrest is not made during the initial investigation, then a follow-up investigation may be conducted. This decision is most often based on the seriousness of the crime and the leads that are available. The follow-up investigation is conducted by a detective or a patrol officer and can involve a wide variety of tasks, including re-interviewing the victim and/or witnesses, submitting evidence to a crime lab, seeking informants, and/or questioning suspects. If a perpetrator is identified and apprehended at any time during the investigation, the case is considered to be cleared by arrest and primary responsibility for it shifts from the police department to the prosecutor's office. If a perpetrator is not identified after leads have been developed and followed, then ongoing investigative activities are terminated.

COLD CASE INVESTIGATIONS

In recent years, much attention has been paid to cold case investigations. The case described in the Police Spotlight feature at the introduction to this chapter was a cold case investigation. Cold cases are usually serious crimes that are no longer being actively investigated because more recent crimes must hold the priority. These investigations are typically reopened due to the availability of

Photo 7.3

The activities involved in a criminal investigation depend greatly on the particulars of the crime at hand. T.O.K./Alamy Stock Photo

previously untested DNA evidence or because of new witnesses.[9] Although there are certainly successes in cold case investigations, research shows substantial difficulties in making arrests for these crimes; only one case in twenty results in an arrest and one in 100 results in a conviction. Difficulties with cold cases include uncooperative witnesses, the inability to locate key witnesses, the suspect being deceased or immune from additional charges (due to a prior plea bargain deal), and/or DNA results that do not identify a perpetrator or are otherwise inconclusive.[10]

UNDERCOVER INVESTIGATIONS

In undercover or covert investigations, the police initiate action before a crime actually occurs. In these situations, the police are not easily identified as police officers, nor are their true intentions known. Covert strategies include stings, decoys, undercover fencing operations, stakeouts, and surveillance. A **sting** involves an investigator posing as someone who wishes to buy or sell illicit goods, such as drugs or sex, or to execute some other sort of illicit transaction. Once a seller or buyer is identified and the particulars of the illicit transaction are determined, police officers can execute an arrest. Undercover drug stings are often referred to as buy-bust operations.

In a **decoy operation**, an undercover police officer presents the opportunity to commit a crime to an individual. Once the crime has been attempted, officers who are standing by can arrest the would-be perpetrator. An example would be when the police setup a car that can be easily stolen and then wait for the crime to occur. Another variant of this strategy involves the investigation of Internet solicitation of minors for illicit sexual encounters. Here, an investigator poses as a minor in an online chat room. If a sexually oriented conversation develops and arrangements are made by the offender to meet with the "minor" for purposes of sexual relations, an arrest can be made when that meeting occurs.

Undercover fencing operations are another type of undercover police strategy. A fence is a business that buys and sells property that is known to be stolen. When the police go undercover and establish a fencing operation, word gets out that there is someone who is

cold case investigations: Investigations of past, serious, unsolved crimes that have been reopened, typically due to the availability of new evidence or witnesses.

sting: A police operation that involves an investigator posing as someone who wishes to buy or sell illicit goods, such as drugs or sex, or to execute some other sort of illicit transaction.

decoy operation: An operation in which an undercover police officer presents the opportunity to commit a crime to an individual. Once the crime has been attempted, officers who are standing by can arrest the would-be perpetrator.

undercover fencing operation: In this type of operation, the police put out word that someone is willing to buy stolen goods and then arrest those who come in to sell.

The television shows *NCIS* and *Criminal Minds* are among the two most popular shows in the United States. Prior to these shows, *CSI* was one of the most popular shows in the world. No matter the sophistication of the crime or the culprit, in these shows the crime is always solved. While this may make for great entertainment, it is not how things work in the real world.

The shows often feature physical evidence as the critical evidence in the investigation, and futuristic technology usually plays a major role in the crime being solved. In *Criminal Minds* in particular, through the analysis of what the perpetrator did and how he or she committed the crime and the use of a very sophisticated computer under the control of Penelope, one of the show's characters, crimes are brilliantly and quickly solved.

In reality, if investigators only have physical evidence or a crime scene profile, chances are the perpetrator will not be identified. One could argue that the entertainment media in general, and shows such as *NCIS*, *Criminal Minds*, and *CSI* in particular, raise the public's expectations unrealistically as to police capabilities in solving crimes.

willing to buy stolen goods. The police make purchases, track the origin of the merchandise, and then make arrests.

Other covert investigative methods include surveillance and stakeouts. **Surveillance** usually involves watching a person to monitor his or her activities. **Stakeouts** most often involve watching a location and monitoring activities at that location.

Undercover strategies are necessary to effectively combat certain crimes, especially drug dealing, drug trafficking, and prostitution. However, they are also controversial, particularly because of the issue of **entrapment**. Entrapment occurs when the police induce a person to commit a crime that he or she would not have committed otherwise. Essentially, the police can provide an opportunity for a person to commit a crime, but they cannot compel or induce a person to commit a crime if that person was not previously predisposed to committing such a crime. The offender's predisposition to committing the crime is critical; usually in an undercover drug buy-bust operation, the undercover officer will make several buys from the dealer before making an arrest. Multiple buys help establish predisposition. Ultimately, the details of the undercover operation, including how and why the would-be offender was targeted, are key in determining the legality of the operation.

Like other criminal investigations, undercover investigations are supposed to lead to a reduction in crime through deterrence and/or incapacitation. But do undercover investigations actually reduce crime? Some undercover operations are more effective than others in this regard. Overall, there is little research evidence to suggest that these strategies actually deter individuals from engaging in crime on a long-term basis. In fact, there is reason to believe that undercover strategies may actually *increase* crime, especially in the short term. In particular, with undercover fencing operations, if it becomes known on the street that someone is willing to buy stolen property, then a market, or a larger market if one already exists, may be created for such property. This may lead to more burglaries and robberies, at least among those who are already so criminally inclined. Decoy operations may also provide considerable opportunity for crime and may actually increase its frequency. It is especially problematic if the police provide opportunities for crime when few exist otherwise. The research is more positive regarding the impact of well-focused strategies designed to apprehend and incapacitate known high-rate criminal offenders.

surveillance:
Watching a person to monitor their activities.

stakeout:
Watching a location and monitoring activities at that location.

entrapment:
This occurs when the police induce a person to commit a crime that he or she would not have committed otherwise.

FIGURE 7.1

Percentage of Offenses Cleared by Arrest or Exceptional Means, 2017

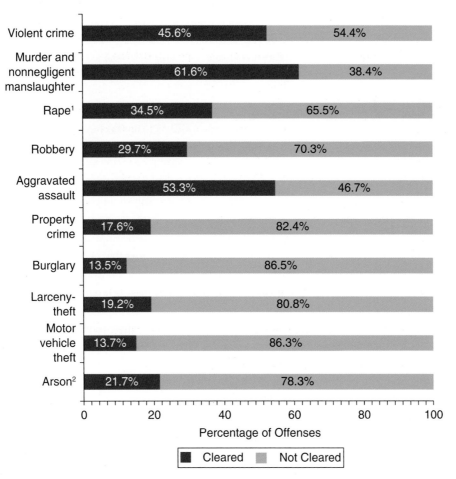

Source: Data from Federal Bureau of Investigation, "Crime in the United States, 2017." https://ucr.fbi.gov/crime-in-the-u.s/2017/crime-in-the-u.s.-2017/topic-pages/clearances.

Notes:

1. The figures shown in the rape (revised definition) column include only those reported by law enforcement agencies that used the revised Uniform Crime Reporting (UCR) definition of rape.

2. Not all agencies submit reports for arson to the FBI. As a result, the number of reports the FBI uses to compute the percentage of offenses cleared for arson is less than the number it uses to compute the percentage of offenses cleared for all other offenses.

A Question of Ethics

Deception and Miranda Rights

By law, the police must inform subjects of their Miranda rights when in custody and prior to questioning. (Miranda rights include the rights of a subject to remain silent, to not incriminate himself or herself, and to have an attorney present during questioning.) The police are required to do this so that subjects know their rights when dealing with the police. However, at other times, such as in undercover investigations, officers do not even have to tell subjects that they are dealing with the police. So the police can deceive subjects to make crimes happen, but when offenders are in police custody, the police have to inform them of their rights, including the right to remain silent. From an ethical perspective, does this strike you as being odd in any way? Explain why or why not.

HOW ARE CRIMES SOLVED?

The simple answer to this question is that crimes are solved when enough evidence is collected to show with reasonable certainty (i.e., probable cause) that (a) a crime actually occurred and (b) the person who was believed to have committed the crime is identified and arrested. However, there are a few things to keep in mind when considering this answer; doing so helps explain why such variation exists in the rate at which different crimes are solved (Figure 7.1).

First, it is important to understand that evidence is not equally available in all types of crimes. For example, the act of committing a theft usually produces much less evidence than the act of committing a rape. Rapes often occur between people who are at least acquainted with each other, which may make it possible for the victim to give investigators the identity of the perpetrator. In addition, rapes are also likely to create other evidence, such as biological evidence and DNA. Information from the victim and the DNA evidence may go a long way in solving the crime. Such evidence is not as likely to be present in a theft, however.

Second, the circumstances of the crime may have implications for the amount and type of evidence that is available and the likelihood of the crime being solved. For example, if a murder occurred in a private place, such as a residence, it is very likely that not only did the victim and offender know each other but there was a personal relationship between them, thereby reducing the pool of suspects for investigators to consider. Indeed, homicides that occur in residences are the most likely to be solved; in contrast, homicides that occur in vehicles are among the least likely to be solved. Additionally, in most solved homicides, the perpetrator is taken into custody within twenty-four hours. If the crime is not solved within this time frame, then the chances of it ever being solved fall drastically.[11] However, it is important to understand that most of the crimes solved within twenty-four hours have certain characteristics, such as considerable evidence (e.g., eyewitnesses who provide investigators with the name of the perpetrator). Therefore, it is most accurate to say that homicides *with certain characteristics* are most likely to be solved within twenty-four hours; homicides without those characteristics may not be solved within that time frame or ever.

Third, in spite of their importance, it is important to understand that the circumstances of a crime do not completely determine whether or not that crime is solved. The efforts and actions of investigators can also make a difference. More time and effort spent on an investigation may lead to more evidence, which may increase the likelihood of the perpetrator being identified and apprehended.[12] More serious crimes are the most likely to receive added investigative effort. In short, the nature of the crime, the evidence available, the circumstances in which the crime occurred, and the efforts of investigators all have implications for the type of evidence collected and whether the crime is solved.

FORMS AND TYPES OF EVIDENCE IN CRIMINAL INVESTIGATIONS

A basic understanding of criminal investigation and how crimes are solved depends on a familiarity with different forms and types of criminal evidence used in investigations.

FORMS OF CRIMINAL EVIDENCE

exculpatory evidence: Evidence that tends to exclude or eliminate someone from consideration as a suspect.

inculpatory evidence: Evidence that tends to include or incriminate a person as the perpetrator.

There are two fundamentally important distinctions in criminal evidence. First, evidence can be either inculpatory or exculpatory. **Exculpatory evidence** is evidence that tends to exclude or eliminate someone from consideration as a suspect. If a witness described the perpetrator as being five feet tall with long blond hair, that would tend to exclude a suspect who was six feet tall with black hair. On the other hand, **inculpatory evidence** is evidence that tends to include or incriminate a person as the perpetrator. For example, a lack of an alibi for a suspect may be inculpatory, as would a suspect's physical characteristics that matched a victim's description. Throughout the course of a criminal investigation, investigators will likely uncover both inculpatory and exculpatory evidence in relation to a particular suspect.

The other important distinction is the difference between direct evidence and indirect evidence. **Direct evidence** refers to crime-related information that immediately demonstrates the existence of a fact in question. As such, no inferences or presumptions are needed to draw the associated conclusion. On the other hand, **indirect evidence**, which is also known as circumstantial evidence, consists of crime-related information in which inferences and probabilities *are* needed to draw an associated conclusion. In determining whether evidence is direct or circumstantial in nature, it is necessary to consider the conclusion that is trying to be established. For example, say a police officer discovers a knife identified as the likely murder weapon, and it has the suspect's fingerprints on it. The fingerprints are direct evidence that the suspect touched or held the knife but indirect evidence that the suspect used the knife to commit the murder.

Photo 7.4

Biological evidence, including DNA, can be extremely valuable in criminal investigations, but it has limitations. Orange County Register Archive/MediaNews Group/Orange County Register/Getty Images

TYPES OF EVIDENCE USED TO SOLVE CRIMES

There are many types of evidence that can be useful in solving crimes. The most significant of these are discussed here.

PHYSICAL EVIDENCE, BIOLOGICAL EVIDENCE, AND DNA **Physical evidence** refers to tangible items that can be held or seen that are produced as a direct result of a crime having been committed. Blood, semen, fingerprints, firearms and bullets, shoe prints, tool marks, dental evidence, fibers, soil, paint, glass, and bloodstains are all best considered physical evidence.

It is important to note that when the police collect physical evidence the **chain of custody** must be maintained. The chain of custody refers to the record of individuals who maintained control (custody) over evidence from the time it was obtained by the police to when it was introduced in court. The chain of custody is to ensure the security of physical evidence. If a chain of custody is not established or if it can be questioned, the value of the evidence itself may be questioned.

Physical evidence can be used in several different ways in a criminal investigation. First, it can be used to help establish a crime actually occurred. For example, tool marks on a window can help establish that a burglary took place. Second, physical evidence can be used to make associations between crime scenes, offenders, victims, and instruments (e.g., tools); this is the most common use of physical evidence. Third, physical evidence can be used to help support other evidence. For example, semen (and DNA) can support a victim's identification of an assailant. Finally, certain forms of physical evidence may serve an identification function. For example, DNA and fingerprints can be used to determine the identity of an individual. The presence of physical evidence is also associated with increased likelihood of convictions at trial (see Figure 7.2).[13] In recent years, it has been realized that there has been too little "science" in the interpretation and analysis of certain types of physical evidence, especially bite marks, toolmarks, and writing samples. Investigators must proceed cautiously in drawing conclusions from physical evidence.[14]

BIOLOGICAL EVIDENCE AND DNA One type of physical evidence is **biological evidence**. Biological evidence has the potential to be especially useful in criminal investigations because it contains DNA, or deoxyribonucleic acid. The most common types of biological evidence recovered

direct evidence: Crime-related information that immediately demonstrates the existence of a fact in question.

indirect evidence: Crime-related information in which inferences and probabilities are needed to draw an associated conclusion.

physical evidence: Tangible items that can be held or seen that are produced as a direct result of a crime having been committed.

chain of custody: The record of individuals who maintained control (custody) over evidence from the time it was obtained by the police to when it was introduced in court.

biological evidence: Physical evidence that contains DNA, such as blood, semen, or saliva.

FIGURE 7.2

Effect of Physical Evidence on Trial Outcomes in Select U.S. Counties

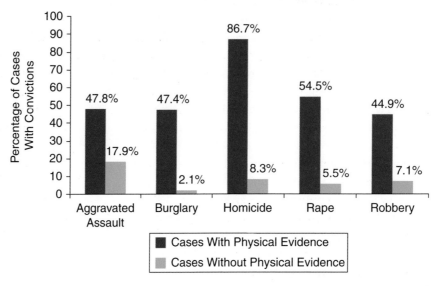

Source: Data from Joseph Peterson et al., "The Role and Impact of Forensic Evidence in the Criminal Justice Process," September 2010. https://www.ncjrs.gov/pdffiles1/nij/grants/231977.pdf.

? A RESEARCH QUESTION
What Is a Secondary Transfer of DNA and Why Is It a Problem?

With continuing technological advancements, DNA analysis is very sensitive and able to obtain profiles from tiny low-quality samples of biological material. However, a potential problem is when there is a secondary transfer of DNA. If you shake hands with your friend and then your friend handles an object (such as a murder weapon), could your DNA be recovered from that object, even though you never touched it? If yes, clearly this is an important consideration for DNA testing, as it could falsely implicate someone as a perpetrator of a crime.

Recently a study to examine the likelihood of secondary transfers of DNA was conducted.[16] Under appropriately sterile conditions, six pairs of subjects were instructed to shake hands with each other vigorously for two minutes to mimic intimate contact. One of the subjects from each pair was then instructed to handle a knife for two minutes. The entire handle of each knife was then wet-swabbed to collect DNA. Using standard accepted scientific techniques, DNA

profiles were obtained and analyzed. While taking steps to avoid cross-contamination, this procedure was repeated several times to increase the number of knives that could be analyzed.

Not surprisingly, DNA typing results showed that secondary transfer occurred in 80% of the knives (the DNA of the person who handled the knife and the DNA from the person who only shook hands with the person who handled the knife were present on the knives). However, in 20% of the samples, the secondary subject (the person who only shook the other person's hand) was identified as either the only contributor or the primary contributor of DNA on the knife. Clearly, drawing incorrect conclusions based on the secondary transfer of DNA is a serious possibility that must be guarded against when using DNA evidence in investigations. In some cases, the faulty interpretation of DNA mixtures has actually contributed to convictions of innocent people.[17]

in criminal investigations are blood, semen, saliva, and hair.[15] DNA is the genetic building block of all living organisms. Except for identical twins, no two people have the same DNA. The DNA in a person's saliva is the same as the DNA in that person's skin cells, blood, semen, and perspiration. Because of its absolute uniqueness and individual characteristics, human cells and the DNA contained in them can be a very useful form of evidence in criminal investigations. While the science of DNA can be thought of as the most significant advance in the technology of criminal investigation *ever*, it does have unique limitations (for example see A Research Question).

DNA analysis (or typing or printing, as it is sometimes known) was first used in a criminal investigation in the United States in a rape case in 1987. The defendant in the case was found guilty. Although it has been used thousands of times since then to secure convictions of defendants, it would be inaccurate to consider DNA analysis only a tool of the prosecution. DNA analysis is a tool of justice—a tool that can be used to identify and convict the guilty *and* to free the innocent.[18]

Although DNA (and physical evidence in general) can identify a culprit when one is not already known, this is still a relatively uncommon occurrence (see Exhibit 7.3). Usually evidence such as an eyewitness identification leads to the identity of the culprit. This evidence may exist due to a relationship between the victim and the offender. In such cases, DNA evidence is simply not needed to identify the perpetrator, but DNA may confirm that the already identified suspect is in fact the perpetrator. In addition, often arrests occur before DNA forensic results are even available. Not surprisingly then, homicide[19] and rape[20] cases with DNA evidence are *not* more likely to be solved than those without DNA. Alternatively, burglaries with DNA evidence *are* more likely to be solved than burglaries without such evidence.[21] In essence, DNA (and physical evidence generally) has the greatest impact in investigations when the chances of solving the crime are the smallest—when a suspect is neither named nor identified quickly after the crime. As technology allows for faster analysis and results of DNA samples, perpetrators may be able to be identified more quickly and at lower cost.[22] DNA may become more useful in identifying unknown offenders as the number of DNA profiles included in CODIS increases (see Technology on the Job: CODIS), as additional information about offenders' characteristics can be derived from DNA profiles, and as familial DNA searching (FDS) becomes more commonplace[23] (as discussed in the Police Spotlight feature discussed earlier in this chapter).

> **DNA analysis:** The analysis of DNA in order to include or exclude a particular person as the source of that DNA.
>
> **Combined DNA Index System (CODIS):** An electronic database operated by the FBI that allows federal, state, and local crime laboratories to share DNA profiles electronically.

TECHNOLOGY ON THE JOB
CODIS

The **Combined DNA Index System (CODIS)** is an electronic database operated by the FBI that allows federal, state, and local crime laboratories to share DNA profiles electronically. All states today have a DNA databank that contains DNA profiles of those who are arrested and/or convicted of certain types of crimes. These databanks are liked via CODIS; the DNA profiles represent the "offender index" of CODIS.

Investigators enter into CODIS the DNA profiles obtained from crime scene evidence, and the system scans the offender index for a match. If there is a hit, investigators can obtain a search warrant authorizing the collection of a reference sample from the identified person. The laboratory can then compare the crime scene DNA with the identified person's DNA. However, if the perpetrator's DNA is not in the system, a match will not be obtained. In this case the culprit will have to be identified through other means. CODIS can also be used to search DNA recovered from other crime scenes in an attempt to link crimes together.[24]

As of the end of 2018, CODIS contained the DNA profiles of approximately seventeen million individuals convicted of or arrested for crimes, as well as over 900,000 DNA profiles collected as a result of other investigations. CODIS has assisted in nearly 450,000 investigations.[25]

INFORMATION FROM WITNESSES AND VICTIMS

In many investigations, victims and/or witnesses provide the most critical information that leads to the identification of the perpetrator. An eyewitness may have seen the crime as it was happening, or perhaps some other witness heard about the crime or some critical aspect of it and relayed that information to the police. Witnesses can be found at the scene of the crime or through a door-to-door canvass, or they may come forward to the police of their own volition (Exhibit 7.4).

Information is obtained from witnesses through police interviews. An **investigative interview** can be defined as any questioning that is intended to produce information regarding a particular crime or a person believed to be responsible for a crime. Interviews are usually not accusatory in tone and seek to develop information to move a criminal investigation forward. There are many types of information that may be obtained from witnesses, including the actions of the perpetrator, a description of the perpetrator, and, most critically, the identification (or name) of the perpetrator. Other useful information may include descriptions of vehicles and/or stolen property.

investigative interview: Any questioning that is intended to produce information regarding a particular crime or a person believed to be responsible for a crime.

EXHIBIT 7.3
Two Examples of DNA in Investigations

Case 1: A woman was beaten, raped, and killed. The perpetrator disposed of her body in a recycling bin, and it was later discovered at a recycling center. Semen from her body was recovered, and a DNA profile was obtained. The profile was entered into the state's DNA database and produced a hit. The offender's profile was in the system due to a prior robbery conviction. It was learned that the perpetrator was currently in jail on an unrelated hit-and-run offense. In this case, DNA evidence led to the identification of a previously unknown individual.

Case 2: A young girl was abducted, taken to a wooded area, and raped. In an attempt to suffocate the child, the offender wrapped duct tape around her hands, head, nose, and mouth and then fled. The girl was able to remove the tape and summon help. She provided information to investigators for the creation of a composite sketch of the offender. Meanwhile, the offender's DNA was recovered from the duct tape (his saliva was on the tape due to his tearing of the tape with his teeth), but there was no match in the DNA database. The sketch was released to the media and several citizens identified the person in the sketch. The suspect was interviewed, and DNA was obtained from him that matched the DNA from the duct tape. In this case, the DNA confirmed the identity of an individual already believed to be the offender. Even today, this is the most common use of DNA in investigations.

EXHIBIT 7.4
The Value of Secondhand Information in an Investigation

A woman called her physician to get her oxytocin prescription filled. The receptionist informed the woman that in order to do so she would need to make an appointment with the physician. The woman then indicated that she was unable to come to the doctor's office because her child had died the night before. The woman pleaded for the prescription to be filled without an appointment, but the receptionist refused. After the phone call ended, the receptionist called the police to report this odd conversation. The police also found this to be unusual, so officers were dispatched to the woman's home and discovered the body of a dead baby. The baby had died from severe maltreatment and abuse. The woman and her boyfriend were arrested and charged with homicide. Even though the receptionist was not an eyewitness to the crime, she provided critical information in this investigation.

EYEWITNESS IDENTIFICATION There are several methods by which an eyewitness may identify a perpetrator: (a) The witness may provide information for the development of a composite picture of the perpetrator; (b) the witness may look through mug shot books (collections of photographs of previously arrested or detained persons); (c) the witness may view the suspect in a show-up situation during which the suspect is detained by the police at the scene of the crime or at another location; (d) the witness may look at photographs of the suspect and others in a photo lineup or photo array; or (e) the witness may view the suspect and others in a physical, or live, lineup. It is noteworthy that the use of mug shot books as a way of identifying offenders has become quite uncommon, especially in large agencies, because of the strong potential of false identifications.[26]

Photo 7.5

An example of a six-person photo lineup. Glendale, Wisconsin Police Department

Eyewitness identifications are extremely persuasive in establishing proof; the only evidence that is more persuasive is a confession.[27] However, eyewitness identifications are also one of the least reliable types of evidence. In fact, false eyewitness identifications are one of the most common issues present in wrongful

GOOD POLICING

Guidelines for the Proper Collection of Eyewitness Identifications

Several guidelines have been developed to improve the likelihood of collecting accurate eyewitness identification evidence.[30] These recommendations represent best practices for investigations and are not complicated, but unfortunately recent national surveys of police departments reveal that many agencies have not fully implemented them.[31]

First, "the person who conducts the lineup or photo-spread should not be aware of which member of the lineup or photo-spread is the suspect."[32] Sometimes this is known as the lineup being "double-blind." This recommendation relates to the possibility that an investigator may intentionally, or unknowingly, lead a witness to select a particular lineup member.

Second, "eyewitnesses should be told explicitly that the person in question might not be in the lineup or photo-spread and therefore should not feel that they must make an identification. They should also be told that the person administering the lineup does not know which person is the suspect in the case."[33] Eyewitnesses are less likely to identify an innocent person if they are told that the actual culprit may not be in the lineup, and in fact, this may indeed be the case. If an investigator is so certain that the culprit is

included in the lineup, then perhaps the lineup is not even necessary. In addition, if the witness is told that the administrator does not know who the suspect is in the case, it may prevent the witness from looking for cues from the administrator about which person to select.

Third, "the suspect should not stand out in the lineup or photo-spread as being different from the distractors based on the eyewitness' previous description of the culprit or based on other factors that would draw attention to the suspect."[34] In other words, the distractors (those individuals included in the lineup that are not suspects) should be selected on the basis of the witness's description of the perpetrator, not because they objectively look like the suspect. This reduces the chances of the witness selecting someone who looks similar to the perpetrator but is not the perpetrator.

The final recommendation is that "a clear statement should be taken from the eyewitness at the time of the identification and prior to any feedback as to his or her confidence that the identified person is the actual culprit."[35] The confidence expressed by the witness at the time of the identification may be the single most important factor in judging his or her credibility and the accuracy of the identification at trial.

Photo 7.6

An example of a live lineup. Glendale, Wisconsin Police Department

conviction cases.[28] Despite the error-prone nature of this type of evidence, eyewitnesses may be confident and persuasive in their inaccurate testimony, and this confidence may be quite influential on jurors and their verdicts.[29]

INFORMATION FROM PERPETRATORS: INTERROGATIONS AND CONFESSIONS

An **interrogation** can be defined as any questioning or other action that is intended to elicit incriminating information from a suspect when this information is intended to be used in a criminal prosecution. Compared to an interview, an interrogation is more of an intimidating process in which information is extracted from a typically unwilling suspect.[36] Interrogations of subjects are usually conducted when the subject is in the custody of the police (this is referred to as a custodial interrogation). *Custody* exists when the suspect is under the physical control of the police and is not free to leave. The police may also conduct a noncustodial interrogation of a suspect. This occurs when the suspect voluntarily accompanies the police and is told that he or she is not under arrest and is free to leave at any time. Miranda rights only apply to custodial interrogations.

The ultimate objective of an interrogation is to obtain a truthful statement from the suspect—ideally a confession to the crime. The police must walk a fine line in this regard, however. It is possible that the individual who is *believed* to have committed the crime may not have *actually* committed it. As a result, a confession would not of course be a desirable or appropriate outcome of the interrogation. False confessions do happen, though, and they are nearly impossible to refute. Most often, confessions are viewed as the most powerful and persuasive form of evidence—even more powerful than eyewitness identifications or DNA matches.[37]

Interrogation is a task of persuasion, of getting someone to do what he or she really does not want to do: to tell the truth and confess. To obtain a confession, investigators may rely on deceit. Indeed, deception is central to certain interrogation methods. The irony is that the "police proclaim truth as the goal of interrogation, yet interrogators regularly rely on deception and sophisticated forms of trickery" to obtain it.[38] The police do not just use random deception in the interrogation room; they use it strategically in order to make it easier for a suspect to confess. Deception can take many forms, including the following:

- Projecting a sympathetic and understanding demeanor to develop the suspect's trust.
- Overstating the evidence the police already have in the investigation (e.g., saying "An eyewitness identified you as the robber"). False claims about the evidence are limited to verbalization; the police cannot legally fabricate evidence even if it is used only in the interrogation room. In general, deception is legally permissible as long as it would not induce an innocent person to confess.
- Overstating the capabilities of technology (e.g., lie detection technology) to identify the suspect's deception.
- Misrepresentation of the seriousness of the crime (e.g., saying "I bet you didn't even intend for this to happen, I bet it was an accident").
- Providing seemingly acceptable excuses or moral justifications for the suspect's actions—again, in an attempt to make confessing psychologically easier (e.g., saying "What you did was perfectly normal, anyone in your shoes would have done the same thing").

Whatever form deception takes, its use in interrogation settings is oftentimes controversial. Deception and other high-pressure techniques of obtaining a confession have been associated with *false* confessions.

> **interrogation:** Any questioning or other action that is intended to elicit incriminating information from a suspect when this information is intended to be used in a criminal prosecution.

There are also other important aspects of an interrogation. Investigators must know the evidence in the case and what information needs to be discovered and/or tested.[39] Adequate time needs to be allotted for the questioning. Research has shown that the length of the interrogation is one of the most important factors in determining whether or not a confession is obtained.[40] Investigators must maintain control over the interrogation, both in terms of the actual questioning and the physical environment in which the questioning takes place.[41] Investigators should have a familiarity with the suspect's background and with his or her previous crimes and life situation. Investigators should build a good relationship with the suspect. The suspect has to feel like he or she can trust the investigators and that the investigators are there to help. Finally, investigators should be familiar with and comfortable using a variety of tactics, such as minimizing the seriousness of the crime, decreasing the shamefulness of the act, and appealing to the subject's hope for a positive outcome as a result of cooperation.[42] The logic behind the use of these "themes" is that they lower the psychological hurdles necessary for a person to confess to actions for which there may be significant negative consequences.

A Question to Consider 7.1

Police Deception

The police can legally lie to a person in order to get a truthful confession. Is this fair? Why or why not?

CRIME SCENE PROFILING Over the years, much media attention has been paid to the role of profiling in criminal investigations. Crime scene profiling (sometimes referred to as psychological profiling or behavioral profiling) "is a technique for identifying the major personality and behavioral characteristics of an individual based upon an analysis of the crimes he or she has committed."[43] A **crime scene profile** often includes information about an offender's race, age (or age range), employment status, type of employment, marital status, level of education, and location of residence.[44] The crime scene profiling process is oriented toward answering questions about the *type* of person who committed a particular crime. As such, crime scene profiles are most useful in focusing an investigation and reducing the number of suspects considered by the police. Geographic profiles examine the locations in which a serial offender has committed crimes in an attempt to identify where he or she lives.[45] Crime scene profiles are not capable of identifying a suspect when one is not already known; never has a profile by itself solved a crime.

INFORMATION FROM THE PUBLIC People may have knowledge about a crime simply because they saw or heard something during the course of their normal activities, even though they may not realize their observations actually relate to a crime. Some of these people may be identified as witnesses through traditional methods, such as neighborhood canvasses. At other times, however, the task for the police is to get people to realize they may have information that relates to a crime and to report that information. To identify these people and to obtain information from them, the police have several strategies at their disposal, including tip lines and special alerts.

Tip lines are designed to be an easy and convenient method for citizens to provide information to the police via a telephone or Internet. Tip lines can be created for specific crimes, or they can be oriented toward any crime on which a citizen may wish to report information. One well-known example of a tip line is Crime Stoppers.

The **AMBER (America's Missing: Broadcast Emergency Response) Alert** is an example of a special alert designed to elicit information from the public. The AMBER Alert was created in 1996 after nine-year-old Amber Hagerman was abducted and murdered while riding her bicycle in Arlington, Texas. With the AMBER plan, when a law enforcement agency is notified that a child

crime scene profile: Information about an offender that includes such details as race, age (or age range), employment status, type of employment, marital status, level of education, and location of residence.

AMBER (America's Missing: Broadcast Emergency Response) Alert: This alert is activated when a child abduction has occurred or is suspected; it includes a description and photo of the missing child and information about the suspected perpetrator, the suspected vehicle, a tip line phone number, and any other information that may assist in locating the child.

Photo 7.7

AMBER Alerts are a method of mobilizing the public in missing children cases. When an AMBER Alert is issued, the child is nearly always safely recovered. However, the alerts still have limitations, particularly in stranger abductions.

AP Photo/Mark Mitchell

abduction has occurred or is suspected, an alert is transmitted for immediate broadcast via radio, television, cell phones, and highway signs. The alert includes a description of the missing child and information about the suspected perpetrator, the suspected vehicle, a tip line phone number, and any other information that may assist in locating the child. The logic behind AMBER Alerts is that when children are abducted by strangers, harm usually comes to them quickly; thus, a quick law enforcement response is needed. In nearly all instances when an AMBER Alert is issued, the child is recovered; however, in most cases, the AMBER Alert itself had nothing to do with the recovery. Most abductions do not involve strangers, and AMBER Alerts that involve a stranger abduction are much less likely to result in the recovery of the child than those that involve a friend or family member. In cases involving stranger abductions, the child is also more likely to be harmed prior to recovery.[46] While AMBER Alerts have benefits, it is important to understand their limitations.

SOCIAL NETWORKING AND OTHER INTERNET SITES Social networking sites on the Internet, including Facebook, Twitter, and Instagram, can also be a source of evidence. These sites hold a wealth of information for investigators. Information and dialogue posted by people may provide unique and valuable evidence in criminal investigations. Photographs and videos may also be very useful, especially if they depict subjects doing illegal things. These images are evidence. Additionally, people may announce their affiliation with street gangs or terrorist groups by joining groups and interests on Facebook.

Similar to Facebook, Twitter has the potential to be a useful source of information in investigations. Users post comments and information on their activities; followers receive this information, which can be incriminating.

Many police departments also use Facebook and Twitter to disseminate information, to alert citizens to incidents, and to request information. One of the limitations of traditional media as a way of prompting citizens to provide crime tips is that the people who may be most likely to have such information may be the least likely to use traditional media. Social network sites represent a potential remedy to this problem.

Finally, YouTube is another potential source of information on crimes and criminals. YouTube is an Internet site that contains videos uploaded by people and by television networks and other media sources. There is a video on just about everything on YouTube, and some people even post videos that show illegal activity. In one case, a subject stole a Taser from a police vehicle and went home and recorded himself and his father using it on each other. The subject then uploaded the video to YouTube. Investigation by the police led to the subjects and to the YouTube video. There are many other stories where subjects recorded criminal acts while they occurred and then posted them on YouTube for later discovery by the police. Without question, as technology evolves other avenues will present themselves as sources of information in criminal investigations.

> **confidential informants:**
> Members of the public who assist law enforcement in an active and ongoing capacity; they are typically associated with the criminal underworld.

CONFIDENTIAL INFORMANTS Members of the public who assist law enforcement in an active and ongoing capacity are best considered **confidential informants**.[47] Confidential informants, also called CIs or street sources, are most often used by the police in ongoing undercover investigations, especially those that are drug related. Informants are often associated with the

EXHIBIT 7.5

Incriminated Through Facebook

The police were investigating the homicide of a man when they learned of a post on Facebook made by the man's girlfriend: "Just goin tell ya nw be4 I get caught. We was really fightn and I gt the knife and stabbed him ddnt think I would hurt him BT I did he died and I'm on the run."[48]

Source: NBC News (2016).

criminal underworld; they are in a position to have the information that the police need to "make" cases, and they are knowledgeable about the inner workings of criminal groups.

The police and investigators have regular contact with criminals and those who know of criminals and their activities. It is from these contacts that informants can be found or recruited. In particular, suspects who were recently arrested or those who are about to be arrested are particularly good candidates for being "turned" as informants.[49] However, some people come forward for other reasons, such as money. If this sounds like a controversial practice, that is because it is. The use of informants by the police is often viewed as a necessary evil.

GANG INTELLIGENCE A relatively small proportion of the population is responsible for a relatively high proportion of all crime, especially violent crime. In particular, gang members are disproportionately involved in all serious violent crime, especially murder, other nonfatal shootings (especially drive-by shootings), and robbery. Gangs are also heavily involved in drug trafficking and dealing as well as other crimes, such as human trafficking and prostitution. Some gangs also engage in white-collar crimes, such as counterfeiting, identity theft, and mortgage fraud.

Given the involvement of gangs in so many different types of crime, it is necessary for investigators to understand the operations and workings of gangs in their communities. Some police departments have a specialized gang unit designed to collect and act upon gang information. This information (or *intelligence*) can be obtained from a variety of sources, such as citizens, informants, gang members themselves, and patrol officers. In many instances, the information can be a useful source of potential leads in investigations.

CRIME ANALYSIS Broadly defined, "crime analysis involves the collection and analysis of data pertaining to a criminal incident, offender, and target."[50] Crime analysis often serves as a basis on which police managers develop strategies to confront and investigate particular crimes or patterns of crimes. Crime analysis can provide the necessary information to make informed patrol allocation decisions, to make predictions about when and where crime will occur, to identify future targets of offenders, and to track offender movements. Because crime analysis is often geographically based, maps are a commonly used and useful tool in this process. Crime analysis is discussed in more detail in Chapter 14.

ELECTRONIC DATABASES AND INFORMATION NETWORKS A multitude of electronic databases exist that investigators may use to obtain critical information needed during a criminal investigation.

For example, investigators may have access to pawnshop records. In many jurisdictions, pawnshop operators (and/or secondhand dealers) are required to electronically record all merchandise received and information about the seller. Depending on the organization of the system and how current the information is kept, such records could be particularly useful to the police in burglary and robbery investigations. As another example, some police depart-

crime analysis: The collection and analysis of data pertaining to crimes, perpetrators, and targets of crimes.

05-19-2019 Sun 00:34:15 (S)

Camera 04

Photo 7.8

Video surveillance images have become common and useful in many criminal investigations. AP Photo/Uncredited

digital evidence:
Information or data relating to a crime that is located on an electronic device.

cyberattack: Crimes in which a computer system is the target. Examples include computer viruses, hacks, and denial of service attacks.

cyberbullying: Bullying that occurs through social media sites, email, text messages, and other digital media.

identity theft: A crime that occurs when one person steals the personal information from another and uses it without permission.

ments keep electronic files on all individuals with whom the police have had contact (as victims, complainants, witnesses, suspects, etc.). This information may be helpful in developing a police contact history of individuals and their known addresses. Many other databases are available to police agencies nationwide (see Chapter 14).

ELECTRONIC DEVICES, DIGITAL EVIDENCE, AND VIDEO In this era of computerization, many of our daily activities are in some fashion recorded digitally—where we went, when we went there, what we did when we got there, who we spoke with, and what we said. This is also the case with criminals, except with criminals this information may constitute evidence. Computers, tablets, smartphones, GPS devices, digital cameras, video cameras, surveillance equipment, computer servers, video game systems, external hard drives, and other storage devices (e.g., thumb drives) serve as electronic filing cabinets that may contain evidence related to varied types of crimes.

Digital evidence is information or data relating to a crime that is located on an electronic device. It includes documents, phone numbers, location indicators of the device, emails, photos, text messages, calendars, and so forth. Digital evidence has the unique ability to cast light onto past conversations, statements, and behaviors of victims and offenders. While digital evidence often plays a central role in the investigation of cyberattacks, cyberbullying, child pornography, and identity theft, nearly every crime could have digital evidence associated with it.

Video can be an extremely useful form of evidence, and with the proliferation of CCTV (Closed-Circuit Television) cameras, it is common in criminal investigations to have video images relating to a crime. Video is especially common in business robberies and in crimes that occur in parking lots. Many businesses and homes also have video surveillance systems to monitor and record outside activity. Video is reducing reliance on accurate eyewitness descriptions of perpetrators.

Main Points

- Criminal investigation is the process of collecting crime-related information to reach certain goals. In order to solve a crime, the police must determine whether a crime has been committed and ascertain the true nature of the crime, identify the perpetrator, and apprehend the perpetrator.
- Reactive investigations are the traditional manner in which the police become involved in the investigation

of a crime. The crime occurs and the police respond. The police are typically in reactive mode when investigating crimes such as homicide, rape, robbery, burglary, and so forth.
- In undercover investigations, the police initiate action before a crime actually occurs. Covert strategies include stings, decoys, undercover fencing operations, stakeouts, and surveillance.

- Ultimately, evidence is needed to solve a crime. The circumstances of the crime and the efforts of investigators have implications for the amount of evidence available in an investigation.
- Exculpatory evidence is evidence that tends to exclude or eliminate someone from consideration as a suspect. Inculpatory evidence is evidence that tends to include or incriminate a person as the perpetrator.
- With direct evidence, no inferences or presumptions are needed to draw a conclusion. With indirect evidence, inferences and probabilities *are* needed to draw an associated conclusion.
- Physical evidence refers to tangible items that can be held or seen and is produced as a direct result of a crime having been committed. While physical evidence has several purposes in criminal investigations, it is not particularly effective at identifying a culprit when one is not already known. DNA and CODIS are changing this fact.
- Eyewitness identifications are extremely persuasive in establishing proof. However, eyewitness identifications are also one of the least reliable types of evidence. Several guidelines have been developed to improve the likelihood of collecting accurate eyewitness identification evidence.
- An interrogation refers to any questioning or other action that is intended to elicit incriminating information from a suspect when this information is intended to be used in a criminal prosecution. The ultimate objective of an interrogation is to obtain a truthful statement from the suspect, which may take the form of a confession.
- Crime scene profiling is a technique for identifying the type of person who committed a crime based upon an analysis of the crimes he or she has committed. Crime scene profiles are not capable of identifying a suspect when one is not already known.
- Tip lines and special alerts may be used to reach out to people who have information about a crime but may not realize it.
- Social networks and other Internet sites can be a source of important information in criminal investigations.
- Members of the public who assist law enforcement in an active and ongoing capacity are best considered confidential informants.
- Crime analysis involves the collection and analysis of data pertaining to a crime, perpetrator, and target.
- A multitude of electronic databases exist that investigators may use to obtain critical information needed during a criminal investigation.
- Digital evidence is information or data relating to a crime that is located on an electronic device. It includes documents, phone numbers, location indicators of the device, emails, photos, text messages, calendars, and so forth. Video evidence is valuable and common, and it reduces reliance on eyewitness descriptions of perpetrators.

Important Terms

Review key terms with eFlashcards at **edge.sagepub.com/brandl2e.**

Questions for Discussion and Review

Take a practice quiz at **edge.sagepub.com/brandl2e.**

1. What do the police need to do in order to solve a crime? Can a crime be considered solved if a conviction is not obtained?
2. Why are undercover investigations sometimes necessary? Why are they sometimes controversial?
3. Why are some crimes, such as murder and aggravated assault, more likely to be solved than other crimes, such as motor vehicle theft and burglary?
4. John is a suspect in a homicide. John's fingerprints were found at the crime scene. Are the fingerprints direct evidence or indirect evidence? Inculpatory evidence or exculpatory evidence? Explain.
5. What is DNA? What is a secondary transfer of DNA?
6. What is CODIS? What could be done to make DNA even more useful in identifying previously unknown perpetrators?
7. What are the four guidelines investigators should follow when collecting eyewitness identification evidence?
8. What are the differences between interviews and interrogations?
9. What is a crime scene profile? What purpose can it serve?
10. Explain how a suspect's Twitter and Facebook accounts may provide unique and valuable evidence in a criminal investigation.

Fact or Fiction Answers

1. Fact
2. Fiction
3. Fiction
4. Fact
5. Fact
6. Fact
7. Fact
8. Fact
9. Fact
10. Fiction

Digital Resources

Get the tools you need to sharpen your study skills. SAGE Edge offers a robust online environment featuring an impressive array of free tools and resources.

Access practice quizzes, eFlashcards, video, and multimedia at **edge.sagepub.com/brandl2e.**

Media Library

View these videos and more in the interactive eBook version of this text!

Career Video
7.1: Crime Scene Investigator: Challenges and Misconceptions

Criminal Justice in Practice
7.1: Interviewing a Suspect

SAGE News Clip
7.1: Alleged Killer Faces 13 New Rape-Related Charges

8

POLICE DISCRETION AND ITS CONTROL

Police Spotlight: Ethics, Policy, and Discretion

"The values and ethics of the agency will guide officers in their decision-making process; they cannot simply rely on rules and policy to act in encounters with the public. Good policing is more than just complying with the law. Sometimes actions are perfectly permitted by policy, but that does not always mean an officer should take those actions."[1]

Source: President's Task Force on 21st Century Policing. 2015. Final Report of the President's Task Force on 21st Century Policing. Washington, DC: Office of Community Oriented Policing Services.

Objectives

After reading this chapter you will be able to

8.1 Discuss the meaning of police discretion, why it is necessary, and why it needs to be controlled

8.2 Identify the factors that have been shown by research studies to strongly and consistently influence the decisions of officers

8.3 Identify how implicit bias may affect police discretion

8.4 Discuss the ways in which discretion may be controlled

8.5 Discuss the limitations of organizational rules and policies in controlling the discretion of officers

8.6 Describe the limitations of trying to control police discretion by enhancing the professional judgment of officers

Fact or Fiction

To assess your knowledge of police discretion prior to reading this chapter, identify each of the following statements as fact or fiction. (See page 177 at the end of this chapter for answers.)

1. Police discretion is problematic and efforts should be made to eliminate it.

2. Police discretion and selective enforcement of the law are completely different things.

3. A bad decision results in a bad outcome and/or is a decision that was made poorly.

4. Implicit bias is a phenomenon that can be easily eliminated.

5. Most studies show that suspect race has an impact on officers' arrest decisions.

6. There is overwhelming research evidence that the de-policing phenomenon is real and that police body-worn cameras contribute to it.

7. Organizational culture has a major impact on officers' decisions and behaviors.

8. Policies of police organizations are often vague, thus officers are still required to use discretion.

9. Body-worn cameras can eliminate police discretion.

10. Ethics training seminars have been shown to be effective in controlling officers' discretion.

PERHAPS no issue relating to the police has received as much attention as discretion. This chapter discusses the importance and necessity of police discretion, the factors that may influence the discretionary decisions of officers, and how and why discretion should be controlled, focusing in particular on the role of departmental policy and ethical standards of conduct.

DISCRETION DEFINED

Discretion exists when a person makes decisions based on his or her own judgment. In contrast, when a person is following directions, that person is not making discretionary decisions; he or she is just following a prescribed course of action. Police officers seldom have specific directions to follow when doing their work. Rather, they are required to use their judgment to figure things out on a constant basis, even when it comes to law enforcement situations. The police are not expected to enforce every single infraction of the law. Using their discretion, officers decide what laws will be enforced, on what occasions those laws will be enforced, against whom they will be enforced, and how they will be enforced. In essence, when police officers use discretion in law enforcement situations, they selectively enforce the law.

DISCRETION ABOUT WHAT?

Common circumstances in which patrol officers must use discretion include the following:

- Deciding whether or not to make a traffic stop and whether to warn or cite the motorist
- Deciding whether or not to stop and question a suspicious subject on the sidewalk
- Deciding where to patrol
- Deciding whether or not to pursue a fleeing suspect
- Deciding whether or not to make an arrest, particularly for a minor offense
- Deciding whether or not to use force against a resisting subject and how much and what kind of force to use

Patrol officers are not the only ones who use discretion on the job. Detectives use discretion in the following situations:

- Deciding what actions should be taken in a criminal investigation
- Deciding how much time and effort should be spent investigating a particular crime
- Deciding whether or not to seek a warrant to conduct a search
- Deciding whether to interrogate a possible suspect in the case and what tactics to use in the interrogation

Police supervisors and administrators use discretion in the following circumstances:

- Deciding who should be hired as a police officer
- Deciding whether or not to discipline or terminate a poorly performing police officer or other employee
- Deciding whether or not to create or revise policies of the department
- Deciding what priorities to give to legal infractions
- Deciding how the departmental budget should be allocated

Photo 8.1

Police officers frequently use discretion when performing their work; this need can arise in routine situations, such as traffic stops, and in extraordinary circumstances, such as incidents that may involve the use of force. ©iStockphoto.com/kali9

FIGURE 8.1

The Anatomy of a Decision

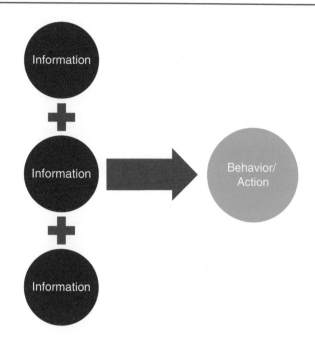

THE ANATOMY OF A DECISION

Before going any further, let's be clear about what constitutes a **decision**. A decision is when a person chooses a particular option based on consideration of the information available. Without options, there cannot be a decision. Without the consideration of information, the selection of an option is just random, like a coin flip. Decision making is a mental process, and the result of the decision is an action or a behavior. For example, you can choose to either attend class or not. This is a cognitive decision. You are likely to base this decision on information of various types and from different sources. If your decision is to attend class, then you engage in the behavior of actually going to class (see Figure 8.1).

Rightfully so or not, a good or bad decision is often determined by its outcome; bad decisions have bad outcomes, good decisions have good outcomes. However, seldom is the outcome of a decision known at the time the decision is made, although a particular result might be reasonably expected. Therefore, it is only after a decision is made that a person can tell with certainty if it was good or bad. For example, skipping class may turn out to be a bad decision because you missed a surprise quiz. A good or bad decision may also be determined by how that decision was made, especially what information was considered when making it. For example, your decision to skip class could be considered a bad one because you did not consider the professor's attendance policy or that you might be having a quiz.

POTENTIAL PROBLEMS WITH POLICE DISCRETION

To develop an understanding of police discretionary decision making, it is important to appreciate its significance and potential problems. There are many issues to consider—seven to be exact. First, police discretion is an important issue because it can affect citizens in extremely significant ways. For example, when a police officer makes a decision, a citizen might get a ticket

> **decision:** When a person chooses a particular option based on consideration of the available information.

or be arrested. An arrest may have extraordinary effects on a person's life—on his or her employment, marriage, or child custody, to name a few. In some circumstances, police discretion can mean life or death for a citizen.

Second, whenever a decision is made by a police officer, it could turn out to have a bad outcome and therefore be judged a bad decision. Police officers make decisions all day long. And, as noted above, bad outcomes in policing can be devastating for citizens and officers. For example, say a police officer decides not to make an arrest when a husband and wife are arguing, and later the husband kills the wife. Or an officer decides not to stop a jaywalker, and later that pedestrian is struck and killed by a car. In some cases, an officer uses justifiable force, but bad outcomes result. All of these decisions may have had justifiable reasons and been properly made, but they could still result in bad outcomes. Sometimes the outcomes of decisions are beyond the control of officers.

Third, police officers may make bad decisions because of the way the decision was made or because of the information that was considered when making the decision. A bad decision may be made due to lack of training, because the officer did not understand departmental policy, or simply as a result of an error in judgment. For example, a police officer could misinterpret a threat posed by a subject and use more force than necessary. In such a case, the bad decision might not be malicious in nature. Without malicious intent, the action might not be considered criminal, but that does not necessarily make the decision any less bad in terms of the harm done. A bad decision could also be made with malicious intent. A police officer may deliberately treat people with certain characteristics more harshly than others. It would be naïve to ignore this possibility. The fact is some people are racist and/or sexist, and unfortunately these individuals may find their way into the police occupation. When officers' decisions are based on these biases, it constitutes a major problem.

Fourth, good decisions made by officers are sometimes incorrectly interpreted by citizens as being improper, biased, or just plain bad. For example, if you get stopped and ticketed for speeding, you may believe you received the ticket because you are black or white, young or old, male or female, and so forth. In actuality, you received the ticket simply because the officer saw you speeding and it is his or her job to enforce speeding laws. Or maybe you *did* receive the ticket because you are black or white, young or old, male or female. Sometimes it is difficult to determine the reason for an officer's decision and whether it was properly made or based on ill-will. If citizens are treated unfairly—or even just perceive unfair treatment—it can have negative effects on police–citizen relations and on the community more generally. This important issue is discussed in more detail in Chapter 13.

Fifth, many decisions of patrol officers occur in **low-visibility situations** or those in which not many people can actually see what decisions were made. In many situations, it is only the affected citizen(s) who sees the decisions of the responding officer or officers. Supervisors may not even see these decisions. Research has shown that in half of all encounters with citizens, only one officer was present,[2] and many of these encounters occurred in private places. This can create challenges when trying to control police discretion. Of course, an important exception to the low visibility of police discretion is when officer actions (decisions) are on video, such as the case with police body-worn cameras (BWCs), and made available for police supervisors and citizens to see (see Technology on the Job: Police Body-Worn Cameras).

Sixth, most often officers' decisions are made on the spot without the benefit of time to collect information or to consider all the options available and all the potential consequences of the decision. This can contribute to bad decisions being made. Most police discretionary decisions need to be made immediately, not later in the day or next week. The speeding car, the dog that is about to attack, and the subject pointing the gun all need to be stopped *now*. Time spent collecting information and carefully considering that information prior to making a

low-visibility situations: Situations in which not many people can see the decisions of the police that are being made.

Police Body-Worn Cameras

Police officers often use discretion in situations where no one besides the officer and the citizen see what actually transpired. Body-worn cameras have the potential to change that and to make police–citizen interactions visible to others, including the general public.

Body-worn cameras are often deployed for the ultimate purpose of monitoring—and perhaps controlling—police discretion; however, ironically, the use of the cameras creates other discretionary issues for police officers, such as deciding when cameras should be on and when they should be off (Exhibit 8.1). Indeed, this issue is quickly becoming a major policy issue in the use of these cameras. For example, in response to a 2018 police shooting of a subject in Sacramento, California, where officers turned off their BWC microphones shortly after the incident, the chief of the department issued a policy whereby officers are not allowed to turn off their BWC microphone unless directed to by a supervisor.[3] Interestingly, technology is being developed to eliminate officers' discretion as to when cameras should

Photo 8.2

Police body-worn cameras show the decisions and actions of officers but their use may create other discretionary decisions such as when the camera should be turned on or off.
AP Photo/Art Gentile

be operating, such as sensors that turn the camera on and off in response to an officer's heart rate.

decision is a luxury seldom afforded to police officers. To the extent that patrol officers' decisions are subjected to review and analysis, it is usually only *after* the decision has already been made.

Finally, while police managers and executives also have discretion, their discretion is quite different than that used by patrol officers. The decisions of police executives are more likely to affect citizens in indirect sorts of ways, and those decisions are often subject to review and analysis prior to being made. In this sense, from the perspective of citizens, the decisions of police executives may be less problematic than those of officers; they may still be controversial and concerning but not to the same extent as the decisions made by officers on the streets.

THE NECESSITY OF DISCRETION

Given the potential difficulties associated with police discretion, you might think the best solution is to eliminate it. However, this is not possible for several reasons. Discretion is necessary because of limited resources. Because of time and money constraints, police officers have to pick and choose the situations in which they enforce the law. For example, the jaywalker has to be ignored so the speeder can be ticketed. Or the speeder may have to be ignored so the drunk driver can be arrested. The burglary investigation has to be set aside so

EXHIBIT 8.1

Excerpts From the 2018 Chicago Police Department's Body-Worn Camera Policy

Provided below are excerpts from the Chicago Police Department's BWC policy, which relate to circumstances in which officers must and must not activate the camera system.[4] Ironically, body-worn cameras are designed to help monitor and control police discretion, but their use creates the need to address other discretionary decisions, such as what situations should be recorded and which should not. Notice the specificity in trying to identify situations that are to be recorded.

A. Initiation of a Recording

1. The decision to electronically record a law-enforcement-related encounter is mandatory, not discretionary, except where specifically indicated.

2. The Department member will activate the system to event mode at the beginning of an incident and will record the entire incident for all law-enforcement-related activities. If circumstances prevent activating the BWC at the beginning of an incident, the member will activate the BWC as soon as practical. Law-enforcement-related activities include but are not limited to:

calls for service; investigatory stops; traffic stops; traffic control; foot and vehicle pursuits; arrests; use of force incidents; seizure of evidence; interrogations; searches, including searches of people, items, vehicles, buildings, and places; statements made by individuals in the course of an investigation; requests for consent to search; emergency driving situations; emergency vehicle responses where fleeing suspects or vehicles may be captured on video leaving the crime scene; high-risk situations; any encounter with the public that becomes adversarial after the initial contact; arrestee transports; any other instance when enforcing the law.

3. A Department member may utilize discretion to activate the BWC for non-law-enforcement-related activities in the following circumstances:

   a. in situations that the member, through training and experience, believes will serve a proper police purpose, for example, recording the processing of an uncooperative arrestee;
   b. in situations that may help document, enhance, and support the following: written reports, evidence collection, investigations, and court testimony; and
   c. when the member is engaged in community caretaking functions, unless the member has reason

to believe that the person on whose behalf the member is performing a community caretaking function has committed or is in the process of committing a crime.

4. Upon initiation of a recording, Department members will announce to the person(s) they intend to record that their BWC has been activated to record. Sworn members will not unreasonably endanger themselves or another person to conform to the provisions of this directive.

B. Deactivation of a Recording

1. The Department member will not deactivate event mode unless:

   a. the entire incident has been recorded and the member is no longer engaged in a law-enforcement-related activity;
   b. requested by a victim of a crime;
   c. requested by a witness of a crime or a community member who wishes to report a crime; or
   d. the officer is interacting with a confidential informant.

EXCEPTION:

Department members may continue or resume recording a victim or witness if exigent circumstances exist or if the officer has reasonable articulable suspicion that a victim, witness, or confidential informant has committed or is in the process of committing a crime.

2. Department members will ensure their BWC is deactivated, consistent with this directive, before providing an oral response to the public safety investigations for incidents involving a firearms discharge and/or officer-involved death.

3. The Department member will ensure that any request by a victim or witness to deactivate the camera, unless impractical or impossible, is made on the recording.

4. Justification for Deactivating a Recording

The Department member will verbally justify on the BWC when deactivating it prior to the conclusion of an incident. When a member fails to record an incident or circumstances warrant the verbal justification of a deactivation as being impractical or impossible, the member will document the reason by activating the BWC and stating the type of incident, event number, and the reason for deactivating the recording.

C. Prohibited Conduct

1. The BWC will not be activated to record:

   a. individuals in residences or other private areas not open to the public unless there is a crime in progress or other circumstances that would allow the officer to be lawfully present without a warrant.
   b. inside medical facilities, except when directly relevant and necessary to a law enforcement investigation and approved by the member's immediate supervisor.
   c. appearances at court or hearings. Members will turn off their BWC so that it is not in buffering mode after notifying the dispatcher and verbally announcing the reason for turning off the BWC.
   d. in connection with strip searches.
   e. personal activities of other Department members during routine, non-enforcement-related activities.

2. The audio recording of a private conversation is prohibited by law when obtained or made by stealth or deception or executed through secrecy or concealment.

3. According to law, no officer may hinder or prevent any non-officer from recording a law enforcement officer who is performing his or her duties in a public place or when the officer has no reasonable expectation of privacy. Violation of this law may constitute disciplinary actions consistent with the directive entitled "Complaint and Disciplinary Procedures" as well as criminal penalties such as theft or criminal damage to property. However, a member may take reasonable actions to enforce the law and perform their duties.

Source: http://directives.chicagopolice.org/directives/data/a7a57b38-151f3872-56415-1f38-89ce6c22d026d090.html.

the homicide can be investigated. Even in situations where discretionary decisions can be automated through technology, this issue is still present. For example, in instances where the police use automated traffic enforcement cameras to detect and cite motor vehicle violations, there are still limited resources for the deployment of those cameras. The cameras simply cannot be installed everywhere.

Discretion is also necessary because we believe that the police should be flexible when enforcing the law: All people should be treated equally, and yet an individual's circumstances and motivations may reasonably be considered when making decisions. For example, most of us would probably agree that the physician who is speeding to the hospital to save someone's life should probably be treated differently than the teenager who is speeding to impress his friends. The mentally ill offender should probably be treated differently than other offenders. If there was no discretion, everyone would be treated the same at all times. Discretion allows the police to provide appropriate responses to different people in different circumstances.

Finally, police officers are required to make so many complex decisions that it would be impossible to program every one of those decisions with specific instructions. It is not practical to think that the best decision in every possible situation could be dictated with specific rules. Further, even if certain discretionary decisions could be eliminated, other decisions would then be created. For example, as noted earlier, with automated traffic enforcement cameras, some discretion is eliminated, such as the decision of

Photo 8.3

Discretion is necessary because of limited police resources and the complexity of many police tasks. In addition, individuals may require different treatment by the police given their circumstances. Bob Daemmrich/Alamy Stock Photo

which motorists should be stopped and cited, but other decisions are created, such as determining where the cameras should be placed and what types of traffic infractions should be cited. It is clear that the elimination of police discretion is not practical, feasible, or even possible.

FACTORS THAT INFLUENCE THE DISCRETION OF POLICE OFFICERS

The difference between a good and bad decision depends at least partly on the factors (or information) that form the basis of the decision. In particular, selective enforcement of the law is not a problem as long as all groups of people have equal protection under the law, and from the law, as guaranteed by the Fourteenth Amendment to the U.S. Constitution. Considerable research has examined the factors that influence police decisions, and many conflicting findings have been the result. Part of the reason for the conflict is the studies were conducted at different times and in different places. Not all police departments operate in the same manner and have the same priorities and policies, and not all communities are the same in terms of their size and demographic composition. Therefore, conflicting research findings should not be surprising. Also, there is likely a complex interplay among many of the identified factors. One might expect certain factors to have an impact on officers' decisions in some situations but not others. Figure 8.2 illustrates the groups of factors that may influence the decisions of police officers.

OFFICER CHARACTERISTICS

Officer characteristics include sex, race, ethnicity, age, length of service, educational level, and attitudes. One might reasonably expect the characteristics of officers to make a difference in the decisions they make; however, perhaps surprisingly, research has generally shown officers' characteristics have minimal effects on their discretionary decisions.[5] When exceptions to this are noted, they generally have to do with the sex and age of officers: Female officers tend to make fewer arrests and use force (or different types of force) less often[6] than male officers, and younger officers tend to make more arrests than older officers. One study found female officers were less likely to make arrests in the presence of fellow officers but more likely to make arrests when being observed by supervisors.[7] Although there are many good arguments in support of college education for police officers, research has produced inconsistent findings regarding the effects of higher education on officer decision making and behavior.[8] Characteristics such as experience on the job and gender may affect the attitudes of officers;[9] however, attitudes seldom influence officers' decision-making behaviors.[10] This may be due to difficulties in accurately measuring officers' attitudes (and behaviors, for that matter) and/or because other factors (e.g., policies, expectations of coworkers) are more important and override the influence of officer attitudes. Interestingly, though, research has shown that officers bring certain styles or role orientations to their job. These styles are at least somewhat based on officers' characteristics and personalities.[11]

SUSPECT CHARACTERISTICS AND IMPLICIT BIAS

One important suspect characteristic that has consistently been shown to influence police decision making is the demeanor of the suspect; in particular, the degree of respect shown to the officer by the suspect. Informally, suspect demeanor is understood among police officers to be an important issue in how they treat suspects. In fact, some police officers have spoken of "POPO" ("pissing off a police officer") or "disturbing the police" (instead of "disturbing the peace") as a factor to consider in arrest decisions. Police officers observed and interviewed by Professor John Van Maanen identified some citizens as "assholes"—people who argue or do not cooperate with the police.[12] Research shows suspects who are defiant, uncooperative, rude, or disrespectful are much more likely to be arrested by the police or to have force used upon them regardless of other factors,[13] except for mental disorders.[14] Research has also shown black subjects are more likely to be viewed by white and black officers as disrespectful, noncompliant, and/or resistant.[15]

A Question to Consider 8.1

Why Might Education Affect Police Behavior?

In general, why might you expect that having a college education would affect an officer's decision making and behavior? Are there certain behaviors or decisions that would more likely be affected by an officers' education than others? Explain.

FIGURE 8.2

Factors That May Influence Police Discretion

The influence of respect on police decisions is not limited only to suspects but also to citizens more generally: Police officers are more likely to comply with citizen requests if citizens are respectful of the police.[16]

Beside the demeanor of the suspect, do police officers consider the demographic characteristics of suspects when making decisions about them? Particularly relevant here is the potential influence of a suspect's race, gender, age, and ethnicity on arrest decisions. A 2014 meta-analysis of forty-two studies examining these variables concluded that

> black individuals, males, and Hispanic individuals are significantly more likely to be arrested. For race and gender (not ethnicity) this effect persists regardless of whether or not a study accounts for the seriousness of the offense, the amount of evidence, whether a suspect was under the influence of drugs or alcohol, whether the suspect used a weapon, the demeanor of the suspect, and/or whether the suspect committed a crime in front of the officer during the encounter. . . . These findings are relatively consistent regardless of when and where the study takes place.[17]

However, the analyses showed that the race of the suspect was not a significant predictor of whether or not the suspect was arrested in domestic violence situations: Black and white suspects were equally likely to be arrested in these situations. This finding is interesting because the decision options for officers in domestic violence incidents are limited; often officers are required to make an arrest.

Another study, published in 2011, took a similar approach and analyzed forty research reports that examined the effects of suspect race on the probability of arrest. The study also showed "with strong consistency that minority suspects are more likely to be arrested than white suspects."[18] Indeed, most, but not all, studies of this issue have found evidence of increased enforcement against blacks. As discussed in Chapter 6, racial differences are also clear in the rates at which white citizens and people of color are subject to traffic and pedestrian stops.[19] Also noteworthy is the research that shows that police officers speak less respectfully to black than white citizens during traffic stops.[20] The disparate patterns in traffic stops have led to, among other things, use of the phrase "driving while black" to describe the racial profiling of African American motorists. To be clear, however, these patterns of decision making do not necessarily mean police officers are racist; the reason for these patterns could not be specified by the research. In addition, even if racism *does* account for these findings, that of course does not mean all officers are so biased. The issue of suspect race and police use of force is discussed in detail in Chapter 11.

RACE AND IMPLICIT BIAS

One of the ways by which suspect characteristics, especially race, may affect officers' decisions is through implicit bias. Also known as unconscious bias, it refers to "thoughts and feelings about social groups that can influence people's perceptions, decisions and actions without awareness."[21] Implicit bias is in contrast to "explicit bias," which refers to overt and conscious prejudice and racism. Most often, implicit bias relates to race, but it can also involve thoughts and actions about gender, sexual orientation, religion, and any other group identity characteristic.

Implicit bias does not just exist among police officers; it is a human phenomenon. To limit cognitive load, people take mental shortcuts. They categorize people and events based on limited information. We do this all day, every day. It is how we make sense of our world. For example, when you meet a fellow student and you learn that she is a math major, you tend to fill in missing information about her with what you assume about math majors.

demeanor: The behavior of a person.

implicit bias: Thoughts and feelings about social groups that can influence people's perceptions, decisions, and actions without awareness.

What do you assume about math majors? When you learn that another student is a dance major, what do you assume about him? These thoughts may in turn affect your actions toward these people. These assumptions are implicit biases. And these assumptions could very well be wrong.

Although implicit biases are a human phenomenon, they are especially problematic for the police. As discussed, police officers make many decisions under time pressure where they do not have the luxury of being able to search for and process all relevant information before taking action. Implicit biases may be especially prevalent and influential in these sorts of situations. Further, officers make decisions that can affect the lives of citizens in dramatic, life-changing sorts of ways, such as whether a person is arrested or is the target of deadly force. If biases affect those decisions, major injustices can result. If the stereotype of black people is that they are aggressive and dangerous,[22] black people may already be at higher risk of more punitive treatment by the police just because of their color. The research discussed above related to race and arrests suggests that implicit bias may be present. Some research on deadly force decisions shows the effects of implicit bias,[23] while other research does not.[24]

Combatting implicit bias begins with understanding that it exists and why it exists. Having exposure to members of the stereotyped group is also potentially useful in reducing implicit biases. For example, but not to keep picking on dance and math majors, if you spend time with people who are math and dance majors, you will likely find that they like doing the same sorts of things that you like to do, that you have many things in common, that they are not different from you. This exposure and these interactions will come to change your thinking about members of these groups. Same goes for different racial and gender groups.

Because implicit bias is so sneaky, often operating without even being detected, simply understanding that it exists is a good start, but other strategies also need to be used to combat its ill effects. A two-hour or even eight-hour class on "implicit bias and you" is not likely going to overpower deeply rooted human tendencies on a lasting basis. The bad news is that there are "no known, straightforward, effective intervention programs" to counter the effects of implicit bias.[25] However, other strategies may also help address the issue. In particular, police departments should provide realistic training for situations in which implicit biases may be most apparent.

Then, the training performance of officers needs to be evaluated with regard to the possible influence of implicit bias. Realistic training, simulating conditions of stressful decision making, is more likely to be transferred to the streets where police officers' decisions really matter. In addition, officers need to have ample opportunities to interact with diverse citizens to teach officers of incorrect stereotypes. Similarly, creating a diverse workforce may provide opportunities for officers to confront and dispel their incorrect preconceived notions about members of other social groups.

Photo 8.4

Most studies show that people of color are more likely to be arrested than white citizens even when the seriousness of the offense, the amount of evidence, and other factors are taken into account. This may be a result of implicit bias. Jim West/Alamy Stock Photo

VICTIM CHARACTERISTICS

When victim characteristics have an effect on police decision making, it is generally in certain situations or with certain types of crimes. In particular, when victims do not cooperate with the police, when they do not wish the police to continue an investigation, or when they do not want to pursue charges against the offender, the police are less likely to continue an investigation[26] or make an arrest.[27] An important exception to this is the more serious the crime, the more likely the police are to take action regardless of victim wishes. Such is the case with domestic violence incidents. In addition, victim credibility or believability may have an effect on police decision making, although this may have more to do with the victim's account of the crime than his or her characteristics. In essence, if there are questions about the truthfulness or credibility of a victim, the police are less likely to pursue an investigation.[28]

OFFENSE CHARACTERISTICS

The seriousness of the crime is a major factor in the decision making of officers. Regardless of any other factors, the more serious the crime, the more likely the police are to make an arrest.[29] However, it should be understood that *seriousness* is a relative concept. What is a serious offense in one city may not be so serious in another. Seriousness is typically defined in terms of injury or the potential for injury to the victim, the amount of property loss associated with the crime, or more generally the amount of harm that results (or may result) from the offense. In many instances, the seriousness of the offense can outweigh all other factors. For example, while being polite and respectful to a police officer during a traffic stop for an illegal lane change may get you a warning, being polite and respectful when stopped for going thirty miles an hour over the speed limit will probably still yield a ticket. The seriousness of the offense outweighs respect in the decision-making process.

The amount of evidence associated with the offense is also a major factor in the decision making of officers.[30] With strong evidence that an offense actually occurred and a particular person committed that offense comes a greater likelihood of an arrest or other formal action, such as the issuance of a citation. This logic makes sense; evidence is needed to establish probable cause, and probable cause is necessary for an arrest to be made.

NEIGHBORHOOD CHARACTERISTICS

The neighborhood area, or beat, in which officers work has also been shown to influence the actions of officers. Arrests, uses of force, traffic stops, traffic citations, and SQFs are all more likely to occur in low-income neighborhoods compared to higher-income areas.[31] Some offenses viewed as normal or not particularly serious in some areas are less likely to get police attention than those same offenses might in other areas.[32] For example, the police may ignore the presence of panhandlers in some areas of a city but not others.

ORGANIZATIONAL CULTURE

Many studies have found the culture of the police organization (and occupation) has a substantial impact on the decision-making behavior of officers,[33] but some officers adhere to, and are part of, this culture more than others.[34] The **organizational culture** refers to the "accepted practices, rules, and principles of conduct that are situationally applied."[35] It consists of a network of shared norms, values, attitudes, and expectations. These factors can determine everything from what is considered a good joke to the value of various police tasks. The violation of these unwritten rules carries informal and unofficial sanctions, such as being the subject of ridicule or receiving timely backup when requested. The culture of a police organization—or any agency, for that matter—cannot be learned from

> **organizational culture:** The practices, rules, and principles of conduct that govern how people in an organization behave.

a book; it is learned through experiencing the organization and the work of its members.

Several excellent and influential studies have attempted to document the culture of the police.[36] For instance, Elizabeth Reuss-Ianni developed a list of cultural "rules" that govern police officers' conduct, based on research she conducted in the New York City Police Department. These rules include the following:

Photo 8.5

Police discretion and the actions of officers may vary based on neighborhood characteristics. ©iStockphoto.com/Nuli_k

- Watch out for your partner first and then the rest of the guys working the shift.
- Don't give up on another cop.
- Show balls (courage).
- Be aggressive when you have to, but don't be too eager.
- Don't get involved in anything in another guy's sector (beat area).
- Hold up your end of the work.
- If you get caught off base (being somewhere where you are not supposed to be), don't implicate anyone else.
- Make sure other officers (but not supervisors or administrators) know if another cop is dangerous or crazy.
- Don't trust a new guy until you have checked him out.
- Don't tell anybody more than they have to know—it could be bad for you and it could be bad for them.
- Don't talk too much or too little.
- Protect your ass.
- Don't make waves.
- Don't give supervisors too much activity.
- Know your bosses.
- Don't do the bosses' work for them.
- Don't trust bosses to look out for your interests.[37]

Other scholars have focused on the critical role of danger and suspicion in the police culture.[38] It has been argued that these factors may affect the personality of police officers or the way they think and behave more generally.[39] The potential for danger is always present in police work; it is a defining feature of the job. Suspicion then becomes a survival skill for officers—a way of recognizing danger. Suspicion keeps police officers focused on detecting danger before it is too late. According to Jerome Skolnick, police officers come to identify **symbolic assailants** (people who may pose a danger) to help manage the potential for danger, and they do so with limited information. For example, when an officer sees a subject holding the waistband of his pants, the officer expects this means the subject is armed. Or the officer believes that the teenagers loitering at a bus stop are not there to get on a bus but to rob someone or sell drugs.

symbolic assailants: People who an officer believes may pose a danger.

Suspicion may also manifest itself in another way: Police officers tend to become suspicious of the motives and intentions of citizens. Given how controversial the police can be and the unique work demands of the occupation, officers learn their actions and responsibilities can easily be misunderstood by citizens. This encourages officers not to share details of their work, such as the methods used to try to control crime. The police have been known to develop an "us versus them" mentality and a so-called code of silence. When this code keeps officers from reporting on the misconduct of other officers, it becomes a barrier to the ethical operation of police departments and has serious implications for the control of police corruption. When it prevents police departments from sharing information with the community or seeking community input, it can create more distrust and misunderstanding on the part of community members. These issues are discussed further in Chapter 12.

TRAINING, SUPERVISION, AND STANDARD OPERATING PROCEDURES (SOPS)

Effective management of police officers' discretion requires that officers are accountable to the official policies and procedures of the organization, which are also known as **standard operating procedures (SOPs)**. Policies and formal rules of police departments are supported by training and enforced by supervision, and violation of rules should lead to formal sanctions. Not surprisingly, then, many studies have shown police officer discretion is influenced by official departmental policies. For example, studies show the implementation of a more restrictive shooting policy can reduce firearm discharges,[40] a restrictive high-speed pursuit policy can reduce the frequency of pursuits,[41] officers who work in agencies with more restrictive use of force policies use force less often than officers who work in agencies with more permissive policies,[42] and mandatory arrest policies can lead to more domestic violence arrests.[43]

LAW

The law shapes what the police do and how they do it. As we discuss in Chapter 9, there are many legal rules that govern how the police can conduct their duties. For example, the famous U.S. Supreme Court decision in *Miranda v. Arizona* (1966) requires police officers to inform suspects of their Miranda rights when in custody and prior to questioning. *Terry v. Ohio* (1968) allows the police to stop and frisk subjects only when an officer has reasonable suspicion a subject is engaged in illegal conduct. *Arizona v. Gant* (2009) allows the police to search an arrested individual's vehicle only if the individual could access the vehicle at the time of the search or if it is reasonable to expect the vehicle contains evidence of the offense. Perhaps the best example of how the law can affect police behavior is the Court's decision in *Tennessee v. Garner* (1985). This decision prohibits police use of deadly force against fleeing suspects, including felons, unless there is probable cause to believe the suspect poses a significant threat to officers or others. After this decision was incorporated into police department policy, it was found that shootings of fleeing suspects in Memphis, Tennessee, were reduced to zero.[44]

COMMUNITY PREFERENCES AND CRITICAL INCIDENTS

One of the primary reasons why every city has its own police department is so citizens can have more direct input into how their department operates. Citizens elect community leaders; community leaders appoint police officials. Thus, police leaders and departments should ideally be more responsive to the needs and priorities of citizens. Some police

standard operating procedures (SOPs): The official policies and procedures of an organization.

departments are most oriented toward service, some toward order maintenance, and some toward law enforcement. The orientation of the police should relate directly to the priorities and desires of community members. In this way citizens can influence the discretionary decisions of officers. Although this may seem reasonable, there is little research that has actually linked particular decisions of officers with the preferences of the community.[45]

Critical incidents may also have the potential to affect officers' decisions. For instance, in 2014 a white officer from the Ferguson (Missouri) Police Department, Darren Wilson, shot and killed an unarmed eighteen-year-old African American man named Michael Brown. The questionable circumstances of the shooting led to major disturbances and rioting in Ferguson and other U.S. cities. The incident gave birth to the Black Lives Matter movement and the slogan "Hands up, don't shoot." After an extensive investigation, a grand jury decided not to indict Officer Wilson on criminal charges, and more rioting occurred in the city. A later federal investigation of the incident concluded that Officer Wilson acted in self-defense and he was again cleared of wrongdoing. Months after the incident, Wilson resigned from the Ferguson police force, and a new chief of police was appointed to the department. Lawsuits against the officer and the police department have been filed. In response to this incident in particular, there has been speculation that police officers may try to avoid situations where force needs to be used, and therefore the police are less often enforcing the law. In turn, it has been suggested that this de-policing effect has led to an increase in crime and criminal behavior. There has also been commentary that the Ferguson incident has led to increased violence against the police. These claims have been referred to as the Ferguson Effect. To date, however, there is no solid research evidence that either of these arguments are valid on a widespread or continuing basis,[46] although beliefs of a de-policing effect have been shown to exist among some officers.[47] As another example of how critical incidents may affect police actions, a study showed that in the days after two officers were fatally shot by black suspects in New York City, officers more frequently used force against black citizens.[48]

de-policing: A phenomenon where the police withdraw from active police work and avoid interactions with citizens.

Ferguson Effect: The unsupported notion that the police enforce the law less often and, as a consequence, crime has increased and there is more violence directed against the police.

? A RESEARCH QUESTION
Do Body-Worn Cameras Cause the Police to Avoid Interactions With Citizens?

Police body-worn cameras are intended to affect officers' and citizens' behaviors so that interactions are more civil, complaints decline, and police use of force (and citizen resistance) is less likely. Research has examined whether police BWCs have the unintended effect of causing police officers to avoid interactions with citizens altogether and thus eliminate the scrutiny that may relate to their decisions during these interactions.[49] This intentional reduction in police officer activity has been referred to as de-policing. To test for this BWC effect, researchers collected data on officers' activities in Spokane, Washington, before and after the rollout of BWCs. The analyses showed that there were no differences in the number of arrests made, response

time, and time on scene before BWCs compared to while BWCs where used. Only the number of officer-initiated calls differed; officers with BWCs were more proactive than officers without BWCs. Accordingly, the researchers found no evidence of camera-induced passivity among officers in their study.[50] When interpreting their findings, the authors consider the possibility that that such passivity may be more likely in interactions with black suspects than white suspects as controversial police action more often relates to interactions with black suspects. However, given that less than 3% of the population in Spokane is black, this possibility could not be tested in the study.

HOW IS POLICE DISCRETION BEST CONTROLLED?

As discussed earlier, it is not realistic or desirable to eliminate police discretion. Many of the decisions of police officers are very complex, there are too few resources to enforce all the laws all the time, and it is considered desirable for officers to be flexible in their decision making because what is appropriate treatment for one person may not be for another. Each of these factors makes the elimination of discretion not practical. Yet because of the potential to base decisions on unwarranted factors (e.g., subject race or gender) or to otherwise make bad decisions, there is a need to guide and *control* police discretion.

To understand the need to control police discretion is one thing; to actually do it is another. This has been a management challenge since police departments were first created and the first officers walked their beats. We will now discuss the latest thinking on how best to control the discretion of officers. These methods are not an absolute solution, but they represent the most effective options available today. They include (a) the implementation of organizational rules and policy (SOPs), combined with training and effective supervision; (b) the enhancement of officers' professional judgment through personnel selection; (c) providing transparency and accountability in police operations; and (d) the incorporation of ethical standards into the culture of policing and police decision making.

ORGANIZATIONAL RULES AND STANDARD OPERATING PROCEDURES

As noted, many studies have shown organizational policies affect the decision making of officers. It logically follows then that organizational policies should be used to control police discretion. Written policies can identify the actions that must be taken in certain situations, the actions that should not be taken in certain situations, and/or the factors officers should consider when making decisions in certain situations. Essentially, policies identify how a police department expects its officers to act in certain situations. As explained by the American Bar Association,

> Police discretion can best be structured and controlled through the process of administrative rule making by police agencies. Police administrators should, therefore, give the highest priority to the formulation of administrative rules governing the exercise of discretion, particularly in the areas of selective enforcement, investigative techniques, and enforcement methods.[51]

There are many examples of policy changes affecting officers' behaviors. For example, as already discussed, policies that restrict deadly force have been shown to reduce the frequency by which officers use such force, and policies that restrict high-speed pursuits have cut down on such pursuits.

Over the years, most police departments have developed policies that address a wide range of officers' decisions. There are rules about evidence collection, searches of subjects, uniforms and equipment, incident report writing, overtime, documenting injuries on the job, and operating departmental vehicles. It is fair to say that in an attempt to control police discretion, departments today have become very rule-oriented. An officer who was interviewed by Professor Peter Manning provided a critical view to the process of rulemaking in his department:

> 140 years of fuckups. Every time something goes wrong, they make a rule about it. All the directions in the force flow from someone's mistake. You can't go eight hours on the job without breaking the disciplinary code. . . . But no one cares until something goes wrong.[52]

One must ask, however, if policies are so encompassing and effective at controlling police decisions, why are there still circumstances in which police officers misuse their discretion? This is a good question, and there are several possible explanations. First, administrative rules have limitations. Many of the most important policies of police departments do not provide *specific* direction to officers; officers are still required to make judgments, often under conditions of limited time and information. For example, part of the pursuit policy of the Seattle Police Department states that "officers will cease pursuit when the risk of pursuit driving outweighs the need to stop the eluding driver."[53] What does that mean? How exactly is the "risk of the pursuit" measured in comparison to the need to stop the driver? Clearly, even with a policy, officers must still use discretion when deciding to stop or continue a pursuit.

As another example, the use of force policy of the Seattle Police Department provides some direction regarding when use of force is *not* allowed (Exhibit 8.2), but much of the policy consists of statements that require officers to use discretion to determine when force *should* be used (Exhibit 8.3).

It is important to realize that if policies are not supported by training, they are less likely to successfully control police discretion. Training can provide an interpretation of a department's policy and serve as a basis on which police officers can model desired actions while performing their job duties. However, if a policy is vague, specific direction through training will likely be difficult.

Finally, if administrative rules are not supported with supervision and enforcement, they are less likely to control police discretion. The supervision and direction provided by sergeants, who are officers' first-line supervisors, is particularly critical.[54] However, once again, if a policy is vague, enforcement will be difficult, as conflicting interpretations of the policy may exist. In addition, sergeants are only one rank removed from patrol officers and therefore tend to have closer and stronger relationships with officers than

EXHIBIT 8.2

Seattle Police Department Use of Force Policy: When Force Is Not Allowed

An officer may not use physical force:

- To punish or retaliate.
- Against individuals who only verbally confront them unless the vocalization impedes a legitimate law enforcement function.
- On restrained subjects (e.g., including handcuffed or contained in a police vehicle) except in exceptional circumstances when the subject's actions must be immediately stopped to prevent injury, or escape, or destruction of property. All such force shall be closely and critically reviewed. Subjects who are refusing to get out of a department vehicle may be removed from the vehicle after reasonable attempts to gain

voluntary compliance have failed, subject to supervisor approval.

- To stop a subject from swallowing a substance that is already in their mouth: Officers may use reasonable force, not including hands to the neck or insertion of any objects or hands into a subject's mouth, to prevent a suspect from putting a substance in their mouth.
- In the event that a suspect swallows a harmful substance, officers shall summon medical assistance.
- To extract a substance or item from inside the body of a suspect without a warrant.[55]

Source: Seattle Police Department website, http://www.seattle.gov/police-manual/title-8.

mid- or high-level managers, such as lieutenants and captains. Their supervisory capacity may be diminished as a result. Due to the limitations of policies, training, and supervision, the challenge is to develop ways by which these factors can be made more effective in guiding police discretion.

EXHIBIT 8.3

Seattle Police Department Use of Force Policy: When Force Is Allowed

1. Use of Force: When Authorized

An officer shall use only the force reasonable, necessary, and proportionate to effectively bring an incident or person under control, while protecting the lives of the officer or others.

In other words, officers shall only use objectively reasonable force, proportional to the threat or urgency of the situation, when necessary, to achieve a law-enforcement objective. The force used must comply with federal and state law and Seattle Police Department policies, training, and rules for specific instruments and devices. Once it is safe to do so and the threat is contained, the force must stop.

When determining if the force was objectively reasonable, necessary and proportionate, and therefore authorized, the following guidelines will be applied:

Reasonable: The reasonableness of a particular use of force is based on the totality of circumstances known by the officer at the time of the use of force and weighs the actions of the officer against the rights of the subject, in light of the circumstances surrounding the event. It must be judged from the perspective of a reasonable officer on the scene, rather than with the 20/20 vision of hindsight. Factors to be considered in determining the objective reasonableness of force include, but are not limited to:

- The seriousness of the crime or suspected offense;
- The level of threat or resistance presented by the subject;
- Whether the subject was posing an immediate threat to officers or a danger to the community;
- The potential for injury to citizens, officers or subjects;
- The risk or apparent attempt by the subject to escape;
- The conduct of the subject being confronted (as reasonably perceived by the officer at the time);
- The time available to an officer to make a decision;
- The availability of other resources;
- The training and experience of the officer;

- The proximity or access of weapons to the subject;
- Officer versus subject factors such as age, size, relative strength, skill level, injury/exhaustion and number of officers versus subjects;
- The environmental factors and/or other exigent circumstances; and
- Whether the subject has any physical disability.

The assessment of reasonableness must allow for the fact that police officers are often forced to make split-second decisions—in circumstances that are tense, uncertain, and rapidly evolving—about the amount of force that is necessary in a particular situation.

The reasonableness inquiry in an excessive-force case is an objective one: whether the officers' actions are objectively reasonable in light of the facts and circumstances confronting them, without regard to their underlying intent or motivation.

Necessary: Officers will use physical force only when no reasonably effective alternative appears to exist, and only then to the degree which is reasonable to effect a lawful purpose.

Proportional: To be proportional, the level of force applied must reflect the totality of circumstances surrounding the situation at hand, including the nature and immediacy of any threats posed to officers and others. Officers must rely on training, experience, and assessment of the situation to decide an appropriate level of force to be applied. Reasonable and sound judgment will dictate the force option to be employed. Proportional force does not require officers to use the same type or amount of force as the subject. The more immediate the threat and the more likely that the threat will result in death or serious physical injury, the greater the level of force that may be proportional, objectively reasonable, and necessary to counter it.[56]

Source: Seattle Police Department website, http://www.seattle.gov/police-manual/title-8.

GOOD POLICING

Body-Worn Camera Policy

Technology is simply a tool; its real value and impact is largely based on how well it is used. Indeed, the value of BWCs in controlling police discretion and making policing more civil depends on sound policy that outlines how BWCs are to be deployed. "Police Body Worn Cameras: A Policy Scorecard" rates BWC policies from 69 "major city departments" on eight critical criteria, believed to be among the most important factors to protect the civil rights of recorded individuals.[57] According to the Scorecard, BWC policy should be evaluated on the basis of the following eight questions (yes answers indicate good policy):

1. Is the department BWC policy readily available for review such as being accessible through the department's website?
2. Does the policy limit officer discretion on when to turn-on or turn-off the camera?
3. Does the policy address citizens' reasonable personal privacy concerns? (e.g., no permission, no recording)
4. Does the policy prohibit officers from viewing the recording prior to writing reports?

5. Does the policy define limits on how long non-critical evidentiary video footage is to be retained before being deleted?
6. Does the policy provide for protections against misuse and tampering with the BWC video?
7. Does the policy make footage available to individuals when filing complaints against officers?
8. Does the policy limit the use of biometric technologies to identify individuals in the video footage?

The analysis shows that police departments score the best on Question 1 (having the BWC policy available for review) and the worst on Question 7 (access to footage) and Question 8 (biometric use). The two best policies according to this scoring system are those of the Cincinnati (Ohio) Police Department and the Montgomery County (Maryland) Police Department; each score five out of eight. Clearly, at least based on these evaluation criteria, police departments have room for improvement with regard to the formulation of BWC policy.

A Question to Consider 8.2

Important Aspects of Body-Worn Camera Policy

Besides the eight criteria listed in the Good Policing feature, are there others that should also be included that are important in a good BWC policy? Explain. Are there any of the eight criteria that should not be used to judge the BWC policies? Explain.

ENHANCING PROFESSIONAL JUDGMENT THROUGH THE SELECTION OF OFFICERS

Officers with different traits and characteristics may bring different styles of police discretion to the job. If police leaders desire a particular style of conduct from officers, perhaps it would make sense to select candidates most likely to display that orientation. For example, if police leaders most desire a department of employees who act with integrity and in which ethical decision making prevails, then the selection process should involve the identification of those applicants who have previously demonstrated in their lives high standards of character, integrity, and ethical conduct. According to the International Association of Chiefs of Police training manual, "Past interests and activities of an applicant will often prove to be a far more accurate indicator of the values, attitudes, and overall character of the applicant than any psychological screening test."[58] At the very least, if officer integrity is to be maximized, the background investigation of a candidate would seem to be one of the most important aspects of the selection process. As explained by the Independent Commission on the Los Angeles Police Department in its assessment of the LAPD,

The initial psychological evaluation is an ineffective predictor of an applicant's tendencies toward violent behavior and the background investigation pays too little attention to a candidate's history of violence. Experts agree that the best predictor of future behavior is previous behavior. Thus, the background investigation offers the best hope of screening out violence-prone applicants.[59]

Along this same line, reducing the frequency of use of force among officers was the major goal of leaders of the LAPD when it set ambitious goals of female officer recruitment in the 1990s. The idea was that female officers would be more likely to bring a "kinder and gentler" style to policing.

There are at least two problems with the selection of officers as a way to control police discretion. First, it is not entirely clear how or to what extent officer characteristics actually affect decisions. Nevertheless, it is probably fair to say there are qualities that good police officers should possess (see A Question to Consider 8.3). Second, police work can change a person's behavior. Police officers may begin their careers well suited for the occupation, but the demands of the job may change the way they think, which could result in bad on-the-job decisions. As discussed in detail in Chapter 10, this highlights the importance of monitoring officers' job experiences and performance and developing programs to help officers deal with the psychological demands of the job.

A Question to Consider 8.3

Personal Qualities of Police Officers

What three qualities would you argue are most important for police officers to possess? Why?

POLICE DEPARTMENT TRANSPARENCY AND ACCOUNTABILITY

The effective control of police discretion and a corresponding increase in accountability may be helped by departments taking steps to become more transparent in their operations. In large part, this means departments share information with citizens and allow citizens to have input into how the departments operate. This is sometimes easier said than done as it requires the "code of secrecy" that may exist at the organizational level to be dissolved and a value of openness to be instituted in its place.

While some aspects of police operations should remain confidential, departments could reasonably share a great deal of information with citizens. For example, the Seattle (Washington) Police Department provides its entire Police Department Policy Manual on-line.[60] Many police departments provide information about potentially controversial issues, such as use of force incidents and citizen complaints. For example, since 2016 the San Jose (California) Police Department has released to the public department statistics on use of force. Chief of Police Eddie Garcia was quoted as saying, "We want to get ahead of the curve. . . . There's more credibility with the community when we do this outside of

Photo 8.6

When the police share information with citizens, it may lead to greater police transparency and accountability. REUTERS/Alamy Stock Photo

crisis."[61] Similarly, the Milwaukee (Wisconsin) Police Department, through its Fire and Police Commission, provides on its website reports detailing police use of force incidents, among other issues.[62] The Chicago Police Department has released information on nearly 250,000 complaints against its officers from 1988 through 2019, although it took a court order to make this happen.[63] The New York City Police Department has released data on over five million Stop, Question, and Frisk situations from 2003 to 2017, although again it took a court order to release these data.[64]

Citizens could also have a role in reviewing complaints against the police, as is the case in many departments that have civilian review boards, and departments could make information about their budgets easily accessible to citizens. Police departments could be more open to the media. They could develop partnerships with local universities for the purpose of conducting research on important issues, problems, and strategies of the police. They could allow student internships and ride-along programs. Each of these initiatives would essentially open police departments to the public, to the people the departments are supposed to serve and protect. If police leaders and officers knew their decisions and actions might be seen by outsiders, it might lead to more careful use of police discretion and contribute to greater accountability of officers to their communities.

CULTURAL VALUES AND ETHICAL STANDARDS OF CONDUCT

Another method that may assist in the quest toward greater control of police discretion is a working understanding, appreciation, and implementation of ethics as a standard of conduct. Ideally, a commitment to ethical standards should provide officers with guidance in making difficult decisions in uncertain conditions.

Police officers' conduct should be congruent with the law, *and* it should follow departmental policy. In some instances, departmental policy is based on law. As discussed earlier, however, the problem is that law and departmental policy may not provide sufficient clarity to be useful in guiding police decisions. The law is sometimes ambiguous. For example, pedestrian stops are to be based on reasonable suspicion, but what exactly constitutes reasonable suspicion? Similarly, policy is often ambiguous as well.

Where there is ambiguity in the law and/or policy, ethical standards of conduct—doing what is right—may fill the void. Ethical standards can also be a basis for formulating and enforcing policy. But there are at least two immediate and major problems with ethics in this regard: (1) what is "right" is not necessarily easily defined or prescribed, and (2) it is difficult to agree upon which ethical standards of conduct are proper and appropriate in different situations. Police leaders understand ethics should play an important role in guiding police discretion, but probably for the two reasons identified above, most of them have hit a wall when it comes to figuring out what exactly should be done to make ethics more relevant in the decisions of officers. This was demonstrated in a survey of departments conducted by the International Association of Chiefs of Police (IACP).[65] The survey found much concern about ethical issues in police work and about the need for ethics training in policing. In particular, respondents were asked to identify what they thought were the most important ethical issues in law enforcement today. A multitude of issues were identified, including the following:

- Cultural diversity/racism/sexism
- Corruption/gratuities
- Public trust
- Morals/personal values of officers/lack of values in new officers
- Honesty
- Abuse of force/abuse of authority
- Decision making

ethical standards of conduct: Doing what is right based on moral standards.

- Code of silence
- Off-duty issues/behavior
- Poor work ethic of new recruits
- Lack of a sense of responsibility
- Lack of role models

Respondents were also asked to provide a "working definition of ethics" used in their organization. Most simply provided the police department's code of conduct or mission/vision statement (see A Question of Ethics feature).

A Question of Ethics

What Is the Value of a Law Enforcement Code of Ethics?

In most police departments today, an understanding of the role of ethics is often limited to a discussion about the law enforcement code of ethics. The code of ethics of the Los Angeles Police Department reads as follows:

As a law enforcement officer, my fundamental duty is to serve mankind—to safeguard lives and property, to protect the innocent against deception, the weak against oppression or intimidation, and the peaceful against violence or disorder, and to respect the Constitutional rights of all men to liberty, equality and justice.

I will keep my private life unsullied as an example to all; maintain courageous calm in the face of danger, scorn, or ridicule; develop self-restraint; and be constantly mindful of the welfare of others. Honest in thought and deed in both my personal and official life, I will be exemplary in obeying the laws of the land and the regulations of my department. Whatever I see or hear of a confidential nature or that is confided to me in my official capacity will be kept ever secret unless revelation is necessary in the performance of my duty.

I will never act officiously or permit personal feelings, prejudices, animosities, or friendships to influence my decisions. With no compromise for crime and with relentless prosecution of criminals, I will enforce the law courteously and appropriately without fear or favor, malice or ill will, never employing unnecessary force or violence and never accepting gratuities.

I recognize the badge of my office as a symbol of public faith, and I accept it as a public trust to be held so long as I am true to the ethics of the police service. I will constantly strive to achieve these objectives and ideals, dedicating myself before God to my chosen profession . . . law enforcement.[66]

Based on this statement

- What is your general impression of this code of ethics?
- If you were to summarize the code in three sentences or less, what would you say?
- What do you think is the main purpose of this code?
- Do you think this code actually serves this purpose? Explain.

Source: Los Angeles Police Department website, http://www.lapdonline.org/lapd_manual/code_of_ethics.htm.

To begin, it must be understood that ethics is more than an understanding of a department's code of ethics or the code of conduct. For ethics to become a basis of police action, efforts must be made far beyond putting words on paper or holding a training class or seminar. For ethics to impact decisions, police agencies must first establish an understanding among officers that displaying integrity and honesty and acting in a moral way is *the* way policing is conducted in that agency. Therefore, a police department needs to specify the values by which its members operate (e.g., respect, integrity, accountability, cooperation).

The **value statement** of a department should not be limited to the thoughts and ideas of the police chief or his or her command staff. Rather, to the extent possible, it should represent the ideas and beliefs of the entire police department; this should provide more support and acceptance of the values as a basis for police conduct.

The department's statement of values should be the basis for *everything* the department and its officers do. It must become part of the culture by which the department operates. Selection criteria and processes, departmental policies, supervision, performance evaluation, discipline, officer assistance and support, and training should be built on an understanding of agency values. For instance, officer integrity should be a primary consideration in personnel selection decisions. Ethical standards should be clearly

Photo 8.7

New police officers swear to uphold the code of ethics of their department. tom carter/Alamy Stock Photo

reflected in police department policies and procedures, and supervisors and field training officers should believe in and adhere to departmental values. A well-defined process of discipline is important as this is one way by which officers are held accountable to the policies and values of the department, and ultimately, to the community they serve. This system of discipline should be reasonable and fairly administered and should include not only punishment for violations but also rewards for positive behavior. Violators should be disciplined; officers who perform well (in relation to the values of the department) should be recognized. The code of silence among officers should be understood as the enemy of proper conduct, especially when silence translates into protecting officers known to have engaged in improper behavior in terms of the law, departmental policy, or the values of the department. The code of silence is so strong among some officers it can be a major hurdle in advancing a properly functioning department. Clear policies regarding conduct, proper supervision, and appropriate discipline can have a meaningful effect on this code. The code of silence as a feature of the police culture is also discussed in Chapter 12.

Instruction regarding ethical standards typically receives minimal attention in training curriculum, often because there is uncertainty about what exactly to train and how to conduct the training. Sometimes the argument is made that there is no one single standard of ethics and therefore ethics cannot be trained. This is not true. As explained in a report of the International Association of Chiefs of Police,

> Every police recruit should leave the training academy with a clear understanding of what is expected of him or her in terms of professional ethics and personal morals. . . . An academy graduate who knows exactly what conduct is or is not acceptable is far less likely to make ethical mistakes and is one who understands clearly where the line in the sand is drawn and that there is no tolerance for those who cross it. . . . Ethics training in police recruit classes must be reality-based and must involve more than just a simple discussion of integrity. The training must be candid and involve free discussion of the potential problems and pitfalls that challenge police officers on the job. It must include discussion of the temptations that they will face, the stresses of police work, the effects of a career in law enforcement on personal life, and related matters.[67]

This instruction should also involve a discussion of the value statement of the department as well as the policies and procedures relating to misconduct and the reporting of the misconduct of other officers. Difficult ethical situations become a little less difficult when it is known

value statement: A set of values important for an agency's employees to consider when performing their duties.

by officers that the reporting of misconduct is required by policy and the failure to do so is subject to appropriate discipline. Instruction regarding ethical conduct should not be limited to just the academy; it should be career long and suited specifically to the nature of officers' job assignments.

Police discretion is necessary, but sometimes it may be misused. As a result, it is important that efforts be made to control it. There are many methods used to attempt this, although none of them are completely new and no one method is the definitive solution to the problem. Nevertheless, by implementing these strategies, police departments may incrementally move toward the more effective control of police discretion and, in the process, make officers and departments more accountable to the citizens they serve.

A Question of Ethics

What to Do in a Difficult Ethical Situation?

The scenario below is a situation officers could confront at some point in their careers. How would you respond to this situation? Would it make a difference to know your department has a policy that requires the reporting of officer misconduct and explains how such reports are to be made? How could this situation have been prevented? Are there any other factors that should be considered when deciding what to do in this situation? Or are there any other factors that would make a difference in deciding what response would be most appropriate?

Scenario: You are a police officer on patrol when another officer, a friend of yours, reports over the radio that he is in pursuit of a motorist. Shortly thereafter, the officer reports that the fleeing motorist crashed into a parked car. You are requested to respond to the scene. When you arrive, you see the officer arresting the motorist, who has his hands cuffed behind his back and who is bleeding from a cut to his face. You then see the motorist spit toward the officer. The officer immediately reacts by punching the subject in the face. The officer then throws the subject to the ground and spits in the subject's face. After lifting the subject to his feet and pushing him into the back of the squad car, the officer looks at you and says, "Sometimes you gotta teach scum a lesson." What would you do?

Main Points

- Discretion exists when a person makes a decision based on his or her own judgment. Police officers are required to use their judgment on a constant basis, especially when it comes to law enforcement situations.
- A decision occurs when a person chooses a particular option based on the consideration of available information. Decision making is a mental process, and the result of the decision is an action or a behavior.
- Police discretion is an important issue because it can affect citizens in extraordinarily significant ways. Patrol officers generally have more discretion than police executives. Many decisions of patrol officers are not seen by people other than those directly involved in an incident.
- A bad decision may be considered bad because of its outcomes, the way the decision was made, or because

of the information that was considered when making the decision.
- Most situations in which police must use discretion need to be resolved immediately.
- Discretion is necessary because of limited resources, the need for flexibility when enforcing the law, and because many decisions of the police are too complex to program with rules and instructions.
- Many factors have been shown to strongly and consistently influence the decisions of officers, including certain suspect characteristics, offense characteristics, and organizational factors.
- Implicit bias is an important phenomenon to understand, especially when interpreting patterns in police discretion and suspect race.

- The concepts of danger and suspicion play a critical role in the police culture.
- Because of the potential to base decisions on unwarranted factors (e.g., citizen race or gender) or to otherwise make bad decisions, there is a need to *control* police discretion.
- Strategies to control police discretion consist of (a) the implementation of organizational rules and policy (SOPs), combined with training and effective supervision; (b) attempts to choose officers who display good professional judgment when hiring; (c) providing transparency and accountability in police operations; and (d) the incorporation of ethical standards in police decision making. Used together, these factors have the best chance of effectively controlling police discretion.

- Each of the strategies listed above has limitations. Perhaps that is why there are still instances of police misuse of discretion.
- Ethics is more than an understanding of a department's code of ethics or code of conduct. For ethics to become a basis of police action, the effort goes far beyond a training session on the topic.
- For ethics to impact decisions, police agencies must first establish an understanding among officers that displaying integrity and honesty and acting in a moral way is *the* way policing is conducted in that agency. Therefore, police departments need to specify the values by which its members should operate. All organizational processes should be based on the specified organizational values.

Important Terms

Review key terms with eFlashcards at **edge.sagepub.com/brandl2e.**

decision   155
demeanor   162
de-policing   167
ethical standards of conduct   173
Ferguson Effect   167
implicit bias   162

low-visibility situations   156
organizational culture   164
standard operating procedures (SOPs)   166
symbolic assailant   165
value statement   175

Questions for Discussion and Review

Take a practice quiz at **edge.sagepub.com/brandl2e.**

1. What is police discretion? Why is it necessary? Why does it need to be controlled?
2. Under what conditions and circumstances do police officers use discretion? Why is this important?
3. What factors strongly and consistently influence the decisions of officers?
4. What is implicit bias and how can the prevalence of it be reduced?
5. What is the most significant limitation of SOPs as a way to control police discretion? Explain.

6. How do the concepts of danger and suspicion play a role in police culture? How do these concepts relate to each other? What significance do they have for police officers?
7. What is meant by police department transparency? How could it aid in the control of police discretion?
8. Do police body-worn cameras cause de-policing? Explain.
9. How can ethical values be created in a department?

Fact or Fiction Answers

1. Fiction
2. Fiction
3. Fact
4. Fiction
5. Fact

6. Fiction
7. Fact
8. Fact
9. Fiction
10. Fiction

$SAGE edge™

Digital Resources

Get the tools you need to sharpen your study skills. SAGE Edge offers a robust online environment featuring an impressive array of free tools and resources.

Access practice quizzes, eFlashcards, video, and multimedia at **edge.sagepub.com/brandl2e.**

Media Library

View these videos and more in the interactive eBook version of this text!

Career Video

8.1: A Police Department's Experience With the Use of Discretion and Media Relations

Criminal Justice in Practice

8.1: Use of Miranda

SAGE News Clip

8.1: Missouri—Ferguson Response

9

THE LAW OF SEARCH, SEIZURE, AND SELF-INCRIMINATION

Police Spotlight: Arizona v. Gant (2009)

On August 25, 1999, police in Tucson, Arizona, went to investigate a house suspected of illegal drug activity after receiving an anonymous complaint. Rodney Gant answered the door, identified himself, and explained to the officers that the owner of the house was away but would be returning later. The officers left the house and performed a warrant check on Gant; they found he had an outstanding warrant for driving with a suspended license. Officers returned later in the day and saw a woman in a car parked in front of the house. The woman consented to a vehicle search, during which the officers found drug paraphernalia. The woman was arrested. A man in the back of the house was questioned by officers and arrested for providing false information. At about this time, Gant drove up to the house and was arrested for driving with a suspended driver's license. Officers handcuffed Gant and placed him in a squad car. They then searched Gant's vehicle and found a plastic bag of cocaine and a gun. Gant was subsequently convicted of cocaine possession.[1]

Gant appealed his conviction. The Arizona Court of Appeals reversed the conviction, and the Arizona Supreme Court agreed with this ruling. The Arizona Supreme Court ruled that the police can legally search the passenger compartment and containers in a vehicle without a warrant as a result of a motorist's arrest in order to protect officers' safety or to preserve evidence, but the search of Gant's vehicle when Gant was away from it was not reasonable.

In a 5-4 vote, the U.S. Supreme Court affirmed the decision of the Arizona Supreme Court. It ruled that the police may search the passenger compartment of a vehicle when the occupant of the vehicle is arrested only if (a) it is reasonable to believe that the arrestee might have access to the vehicle at the time of the search or (b) if the vehicle contains evidence associated with the arrest. Since Gant was secured in a police vehicle at the time of the search, he did not have access to anything in the vehicle. In addition, his arrest was related to a suspended driver's license, and the police could not reasonably expect to find evidence of that offense in the vehicle. Therefore, the search of the vehicle without a warrant was unreasonable.[2]

Objectives

After reading this chapter you will be able to

9.1 Discuss standards of proof and the relevance of probable cause

9.2 Differentiate between arrests, custody, stops, and encounters

9.3 Discuss the importance and intent of the Fourth Amendment to the U.S. Constitution

9.4 Identify and discuss the situations in which a search warrant is not necessary in order to conduct a search

9.5 Differentiate between a frisk and a search incident to arrest

9.6 Discuss the impact of the exclusionary rule on criminal investigations and the criminal justice process

9.7 Evaluate the importance of the Fifth and Sixth Amendments to the U.S. Constitution

9.8 Identify the Miranda warnings and the circumstances under which the police must notify suspects of their Miranda rights

9.9 Assess the impact of the Miranda requirement

Fact or Fiction

To assess your knowledge of police and the law prior to reading this chapter, identify each of the following statements as fact or fiction. (See page 201 at the end of this chapter for answers.)

1. In order for the police to make a valid arrest, they need reasonable suspicion that a crime occurred and that the person they are about to arrest committed it.

2. If the police put handcuffs on a person, then that person is automatically under arrest.

3. The police need either the owner's consent or a search warrant in order to conduct a search of a vehicle.

4. The most common circumstance in which the police conduct a search without a warrant is the crime scene exception.

5. Pretext traffic stops are not legal according to the U.S. Supreme Court.

(Continued)

(Continued)

6. In a traffic stop, the police need probable cause to search occupants for evidence.

7. Pat-down searches are conducted to discover weapons. If nonthreatening contraband is discovered in the process, it must be ignored.

8. One of the primary purposes of the exclusionary rule is to deter unlawful police conduct in search and seizure cases.

9. There are circumstances in which the police do not have to provide subjects their Miranda warnings even when subjects are in custody of the police and prior to questioning.

10. Research has shown that telling suspects they have the right to remain silent keeps most suspects from answering questions of the police.

proof: Something, such as evidence, that proves something else is true.

probable cause: A standard of proof that is generally required in order for police to justify a search or arrest.

reasonable suspicion: A standard of proof that is required to legally stop and frisk a person. It relates to the likelihood of a person's involvement in a criminal act. Reasonable suspicion is less of a hurdle than probable cause.

IN the previous chapter, we examined police discretion and learned that, while it is unrealistic to eliminate officer discretion, it is important that it be controlled. Several methods of controlling officer decisions were discussed, including departmental rules and standards of ethical conduct. Another important way discretion may be controlled is through laws, which is the topic of this chapter. This discussion focuses on issues related directly to the identification and apprehension of offenders—the goal of crime control. Other legal issues relating to such topics as police use of force, entrapment, and equal employment opportunity are discussed in related chapters.

BASIC LEGAL TERMINOLOGY AND CONCEPTS

PROOF, STANDARDS OF PROOF, AND PROBABLE CAUSE

To prove something is to eliminate uncertainty or at least some degree of uncertainty regarding the truthfulness of a conclusion. Evidence (or information) is used to establish **proof**. Proof is not a one-dimensional phenomenon; there are various levels, or standards, of proof (see Table 9.1). For instance, the police usually need **probable cause** to justify a search and always need probable cause to justify an arrest (*Manuel v. City of Joliet*, 2017). Probable cause, then, is a standard of proof. Probable cause exists when it is more likely than not that a particular conclusion is true (e.g., that a particular house contains evidence relating to a homicide or that a particular person committed a homicide). Generally speaking, probable cause means that the degree of certainty is greater than 50%.

Another standard of proof is **reasonable suspicion**. In order for police to legally stop and frisk a person, they must have a reasonable suspicion about that person's involvement in or association with a criminal act. Reasonable suspicion is less of a hurdle than probable cause.

A third standard of proof is *beyond a reasonable doubt*. Beyond a reasonable doubt is a high level of proof that needs to be established by a prosecutor in a court trial in order to convince a judge or jury that a defendant is guilty of a crime.

A fourth major level of proof is *preponderance of the evidence*. Preponderance of the evidence is the degree of certainty needed to prove and win a civil claim. It is essentially the functional equivalent of probable cause but applies only to civil matters.

TABLE 9.1

Standards of Proof in Criminal Matters

Standard	Critical Question	Situations of Relevance
Reasonable suspicion	Is there reason to believe that a particular circumstance exists?	To stop and frisk
Probable cause	Is it more likely than not that a conclusion is true?	To make an arrest
Beyond a reasonable doubt	Is the doubt about the defendant's guilt meaningful or significant?	To obtain a conviction

It is important to understand that all levels of proof are subjective in nature. The determination of what constitutes proof depends on the judgments of people. As a result, when presented with the same evidence, one judge may see probable cause, but another may not. One jury may find proof beyond a reasonable doubt, but another may not. The weight and value of evidence in establishing proof is an individual determination.

ARREST, CUSTODY, STOPS, AND ENCOUNTERS

There is often confusion about what constitutes an *arrest*, what is meant by *custody*, and what *encounters* and *stops* are. An **arrest** occurs when the police take a person into custody for the purposes of criminal prosecution and interrogation (*Dunaway v. New York*, 1979). When a person is under arrest, that person is always in *custody* of the police. However, custody and arrest are not synonymous. A person may be in custody of the police even if that person is not under arrest. A person is in custody of the police if that person's detention is not temporary or brief. If a person is handcuffed, that person may or may not be under arrest or in custody. If handcuffs are used only to briefly detain a person and to protect officer safety, that person is not under arrest or in custody. If a person is handcuffed because there is probable cause to believe that a crime occurred and that he or she committed it, then that person is under arrest and in custody. This distinction matters because only persons who are in custody are required to be informed of their Miranda rights (see discussion later in this chapter).

To make things even more complicated, a person who is *stopped* by the police is also not free to leave but is not necessarily under arrest *or* in custody. And, along with arrests and stops, there are also *encounters*, or nonstops. A nonstop is an encounter, confrontation, or questioning of a subject by a police officer that requires no justification. However, during a nonstop, the subject is legally free to leave.

ARREST WARRANT

In some cases, an **arrest warrant** is requested by the police. The arrest warrant must name the accused or provide a specific description of the person so that his or her identity is not in question. An arrest warrant typically specifies the crime committed, the evidence of the crime, and the evidence pointing to the person named in the warrant as the perpetrator of the crime. The arrest warrant must be approved by a neutral and detached magistrate, most often a judge. As with an arrest, the standard of proof necessary to justify the issuance of an arrest warrant is probable cause. The overwhelming majority of arrests made by the police are made without an arrest warrant because they are made in public. An arrest warrant is required when the police must enter a home to make an arrest, unless there are exigent circumstances (*Payton v. New York*, 1980) or **consent** is given (*Steagald v. United States*, 1981).

SEARCH

A **search** can be defined as a governmental infringement into a person's reasonable expectation of privacy for the purpose of discovering things that could be used as evidence in a criminal prosecution (*Katz v. United States*, 1967). A reasonable expectation of privacy exists when a person believes that his or her activity will be private and that belief is reasonable (*Katz v. United States*, 1967). A **seizure** involves the police taking control of a person or thing because of a violation of a law. What is seized may constitute evidence and could include items such as contraband (e.g., drugs), fruits of a crime (e.g., stolen goods), instruments of the crime

arrest: An arrest occurs when the police take a person into custody for the purposes of criminal prosecution.

arrest warrant: A document issued by a magistrate that authorizes the arrest of an individual.

consent: The granting of permission for police to take requested action, such as to enter a home or conduct a search.

search: A governmental infringement into a person's reasonable expectation of privacy for the purpose of discovering things that could be used as evidence in a criminal prosecution.

seizure: When the police take control of a person or thing because of a violation of the law.

Photo 9.1

An arrest warrant is required when the police enter a home to make an arrest unless it is an emergency situation or the police have consent to enter. Mikael Karlsson/Alamy Stock Photo

(e.g., weapons), or evidence of the crime (e.g., bloodstained clothing). Depending on the circumstances, searches may or may not need to be based on probable cause.

SEARCH WARRANT

In the absence of an exception to the search warrant requirement (see below), a **search warrant** may be required when the police wish to seize evidence. A search warrant specifies the person, place, or vehicle to be searched and the types of items to be seized by the police. It must be based on probable cause, issued by a judge or magistrate, and served immediately. Most searches are actually conducted without a warrant because an exception to the warrant requirement applies; this will be discussed in detail later in the chapter.

THE LAW OF SEARCH AND SEIZURE: THE FOURTH AMENDMENT

In order for evidence to be admissible in court, the police have to follow certain legal rules in collecting it. These laws are intended to protect citizens from unwarranted governmental intrusion into their lives; they represent the civil liberties of citizens and relate to the protections offered by the Fourth, Fifth, and Sixth Amendments to the U.S. Constitution (see appendix). The procedures associated with arrests, searches, and seizures relate to the Fourth Amendment and various legal interpretations of it. The Fourth Amendment reads as follows:

> The right of the people to be secure in their persons, houses, papers, and effects, against unreasonable searches and seizures, shall not be violated, and no warrants shall issue but upon probable cause, supported by oath or affirmation, and particularly describing the place to be searched, and the person or things to be seized.

Over the years, a multitude of legal cases have defined and redefined the meaning of the Fourth Amendment. In essence, the intent of the Fourth Amendment is to protect individuals'

search warrant:
A document that authorizes the search and seizure of an individual's property; it specifies the person, place, or vehicle to be searched and the types of items to be seized by the police.

TECHNOLOGY ON THE JOB
GPS and *United States v. Jones* (2013)

The global positioning system (GPS) has many applications, including criminal identification, apprehension, and evidence collection. Using satellite-to-ground communication, GPS can monitor the location and movements of suspects, suspects' vehicles, and contraband. However, the use of GPS or of any technology oriented toward crime control can raise questions about reasonable expectations of privacy. One example of this can be found in the U.S. Supreme Court case of *United States v. Jones* (2013).

Antoine Jones was being investigated for narcotics offenses by the FBI and the Washington, D.C., Metropolitan Police Department. During the investigation, a GPS device was installed on Jones's vehicle that tracked the vehicle around the clock for four weeks. A warrant was not obtained for the GPS tracking. Jones was eventually arrested, convicted of drug trafficking, and sentenced to life in prison.

On appeal it was argued the GPS evidence was collected in violation of Jones's Fourth Amendment rights and, as a result, should be excluded from the proceedings. The appeals court agreed and overturned the conviction. The case was then brought to the U.S. Supreme Court. The Court agreed that the installation of a GPS device on a vehicle to monitor that vehicle's movements constituted a search and required a warrant. The Court ruled the GPS evidence was illegally collected and should be excluded from trial.

In a retrial of the case, prosecutors used cell site data for which no warrant was required instead of the illegally collected GPS tracking information. The case resulted in a mistrial, and Jones accepted a plea bargain of fifteen years to avoid yet another trial.

privacy and protect against arbitrary intrusions into that privacy by government officials. As such, as interpreted by the courts, the Fourth Amendment offers protection in a variety of situations.

REASONABLE EXPECTATION OF PRIVACY

According to *Katz v. United States* (1967), searches are restricted wherever individuals have a reasonable expectation of privacy. If there is not a reasonable expectation of privacy, then there is no need for the police to restrict a search and no need for a warrant. In the case of *Katz*, a reasonable expectation of privacy was found to exist in a public telephone booth used by the defendant. Other cases have ruled that a reasonable expectation of privacy exists in a defendant's desk and filing cabinets (*O'Conner v. Ortega*, 1987), that surgery to recover evidence is a search and seizure (*Winston v. Lee*, 1985), as is the use of a thermal-imaging device to detect criminal activity in a home (*Kyllo v. United States*, 2001). Attaching a global positioning system (GPS) device on the undercarriage of a car also constitutes a search and requires a warrant (*United States v. Jones*, 2013; see Technology on the Job feature).

GPS monitoring of sex offenders also constitutes a search under the Fourth Amendment when such a device is attached to a person's body without consent (*Grady v. North Carolina*, 2015). A police dog sniffing for drugs on a subject's front porch is considered a search (*Florida v. Jardines*, 2013). However, a dog sniff of the outside of an automobile during a valid traffic stop is not a search that requires a warrant or consent (*Illinois v. Caballes*, 2005) as long as that action does not "prolong the stop, absent the reasonable suspicion ordinarily demanded to justify detaining an individual" (*Rodriguez v. United States*, 2015).

A Question to Consider 9.1

The Value of Privacy

The Fourth Amendment and all of the court cases associated with it relate in some manner to the expectation of privacy. Some people think that privacy is disappearing and are deeply troubled. The question is this: What is so great about privacy? Why do people care about privacy and about losing it? Is it only criminals who should be concerned about their privacy? Explain your answer.

THE SEARCH WARRANT REQUIREMENT AND ITS EXCEPTIONS

The general rule is that when and where citizens have a reasonable expectation of privacy, the police need probable cause and a search warrant to conduct a legal and valid search. However, there are many exceptions to this rule. In fact, most searches conducted by the police are conducted without a search warrant, just as most arrests are made without an arrest warrant.[3] Generally speaking, probable cause (or reasonable suspicion, in some cases) is required in nearly all searches, regardless if conducted with or without a warrant, unless the search is conducted with consent. When there is consent to do a search, probable cause or reasonable suspicion is not necessary. When a search is conducted without a warrant, the burden is on the police to establish a valid and lawful reason for the search. Specifically, when a search is conducted without a warrant, police actions must relate to one of the exceptions to the search warrant requirement. These exceptions can be grouped into the following categories:

- Exigent circumstances
- Vehicles
- Other places/things not covered by the Fourth Amendment
- Hot pursuit
- Incident to arrest
- Stop and frisk
- Plain view
- Consent

Photo 9.2

Even if a house is a crime scene, the police must still have a warrant to conduct a search or one of the exceptions to the search warrant requirement must apply. There is not a "crime scene exception" to the search warrant requirement. Connecticut State Police via Getty Images

Notice that there is *not* a crime scene exception to the search warrant requirement. For the police to conduct a search of a crime scene, such as a house, they either need a warrant or their actions must relate to one of the exceptions to the search warrant requirement listed above.

EXIGENT CIRCUMSTANCES Exigent circumstances, or emergency situations, allow the police to conduct a search without first obtaining a warrant. In general, the rationale for the **exigent circumstances exception** is that without immediate police action, the suspect may destroy evidence or may pose danger to herself or himself, the police, or the public, or someone else may be in further danger of harm.[4] Several Supreme Court cases define the exigent circumstances exception. For example, consider the case of *Schrember v. California* (1966). Schrember was hospitalized as the result of an automobile accident during which he had apparently been driving. A police officer smelled alcohol on Schrember's breath and noticed symptoms of intoxication at the scene of the accident as well as at the hospital. Schrember was placed under arrest and informed of his rights. On the officer's direction and despite Schrember's refusal hospital medical staff took a blood sample. An analysis of the blood revealed a blood-alcohol level indicative of intoxication, and this evidence was admitted at trial. On appeal, the Supreme Court ruled exigent circumstances existed in this situation because the alcohol in a person's bloodstream may disappear in the time required to obtain a warrant. Thus, obtaining evidence in this manner, under these circumstances, and without a warrant did not constitute a violation of a defendant's constitutional rights. *Missouri v. McNeely* (2013) challenged the *Schrember* decision. This case also involved a blood draw from a motorist suspected of being intoxicated. The motorist was taken to a hospital, where an involuntary blood draw was performed without a warrant. The Supreme Court ruled exigency must be determined on the totality of the circumstances and a warrant should be required for blood draws in routine DUI situations. In some circumstances, such as the ones present in the *McNeely* case, technology has provided the potential for officers to obtain warrants without delay. Therefore, exigency in blood draws is not automatic or present in all cases.

A more recent case that relates to the prevention of the destruction of evidence is *Kentucky v. King* (2011). Here, the Supreme Court ruled if the police reasonably believe a subject is destroying evidence, officers can take immediate action without a warrant. In this case, the police had kicked in the door to an apartment after pursuing a subject into the apartment building, smelling marijuana outside of the apartment door, and announcing their intent to forcibly enter the apartment. The search was ruled as an exigent circumstances exception and the seized evidence was admissible.

Along with the prevention of the destruction of evidence, the exigent circumstances exception can also be used to enter a home or other place in order to give emergency aid where a subject is injured, may be about to be injured, or is in need of aid, even if the police do not have "iron-clad proof" a subject has life-threatening injuries (*Michigan v. Fisher*, 2009; also see *Brigham City v. Stuart*, 2006).

In *Payton v. New York* (1980), the Supreme Court ruled there were no exigent circumstances and, correspondingly, the warrantless search in question was unconstitutional. In this case, police intended to arrest Payton for murder and went to his apartment without a warrant. After knocking on the door and receiving no answer, they used crowbars to gain entry into the

exigent circumstances exception: Emergency situations that allow the police to conduct a search without a warrant.

TABLE 9.2
Guidelines for Stopping and Searching Vehicles

Type of Search/Action	Minimum Level of Proof Necessary
To stop a vehicle	Reasonable suspicion/violation of traffic law
To order occupants out of the vehicle	Reasonable suspicion of a crime
To search occupants for weapons	Reasonable suspicion/fear of safety
To search occupants for evidence	Probable cause of a crime
To search the passenger compartment for weapons	Reasonable suspicion/fear of safety
To search the passenger compartment for evidence	Probable cause to arrest
To search closed containers in vehicle	Probable cause or an inventory search

apartment. No one was there. In plain view was a shell casing that was seized and later admitted into evidence at Payton's murder trial. Payton was convicted and appealed. The Supreme Court ruled that in the absence of consent or exigent circumstances, the police may not enter a suspect's home to make a routine felony arrest or to conduct a search without a warrant. As a result, the evidence seized from the search was not admissible.

VEHICLE EXCEPTION People have a lesser expectation of privacy in vehicles (including motor homes; see *California v. Carney*, 1985) than in their homes. Moreover, vehicles are mobile and it is therefore more difficult for the police to collect evidence contained in them. Searches of vehicles may also be conducted to minimize the dangers to officers associated with vehicle stops. Several cases have defined this **vehicle exception** to the search warrant requirement and provided guidelines for stopping and searching vehicles (see Table 9.2).

In *Chambers v. Maroney* (1970), the police stopped the car of Chambers and three other men for an armed robbery that had just occurred at a service station. The men were arrested and the car was driven to the police station, where it was searched. During the course of the search, the police found concealed in a compartment under the dashboard two .38-caliber revolvers, a glove containing change, and cards bearing the name of a different service station attendant who had been robbed a week earlier. In conducting a warrant-authorized search of Chambers's home the day after the arrest, police found and seized .38-caliber ammunition. At the trial, the evidence found in the car and the bullets seized from the home were introduced, and Chambers was convicted of the robbery of both service stations. On appeal, the Supreme Court held that if probable cause exists that a vehicle contains evidence and if that vehicle is mobile, an officer may search the vehicle at the scene or at the police station without a warrant. The search was valid and the evidence admissible.

When the police tow and impound a vehicle, even for a parking violation, a routine inventory search is reasonable without a warrant or probable cause. This procedure protects the owner's property, protects the police against claims that the owner's property was stolen while the car was impounded, and protects the police

> **vehicle exception:**
> A warrant exception that allows police to search a vehicle if there is probable cause to believe the vehicle contains evidence or contraband.

Photo 9.3
A warrant is not required for a dog to sniff the outside of a car. Photo by Steve Osman/ Los Angeles Times via Getty Images

from potential danger (*South Dakota v. Opperman*, 1976). In addition, during an inventory search, it is reasonable for the police to search closed containers, such as a backpack, without a warrant (*Colorado v. Bertine*, 1987). However, inventory searches conducted solely for the purpose of discovering evidence are illegal regardless of what is discovered during the course of the search.

In *Michigan v. Long* (1983), the Supreme Court spoke of the dangers associated with roadside encounters with suspects and stated this can justify searches of vehicles. In this case, Long was driving erratically and swerved into a ditch. Officers stopped and were met by Long at the rear of the car. Long was the only occupant of the car. Long did not respond to the officers' initial requests for his license and registration. When he began walking toward the open door of the car, the officers followed him and saw a knife on the floorboard of the driver's side of the car. At that time, the officers subjected Long to a pat-down search, but no weapons were found. One of the officers shone a light into the car and saw something protruding from under the armrest of the front seat. Upon lifting the armrest, the officer saw an open pouch that contained what appeared to be marijuana. Long was then arrested for possession of marijuana. A further search of the car revealed no additional contraband, but the officers decided to impound the vehicle. As a result of the subsequent search, more marijuana was found in the trunk. The marijuana was introduced at trial, and Long was convicted of possession of marijuana. On appeal, the Supreme Court held that if an officer has reasonable suspicion that a motorist who has been stopped is dangerous and may be able to gain control of a weapon in the car, the officer may conduct a brief warrantless search of the passenger compartment even if the motorist is no longer inside the car. Such a search should be limited to areas in the passenger compartment where a weapon might be found or hidden. If contraband is discovered in the process of looking for a weapon, the officer is not required to ignore it. However, in order to look inside a closed container in a vehicle without a warrant for reasons other than an inventory search, there must be probable cause to suggest evidence is present in the container (*California v. Acevedo*, 1991).

Important differences between the *Long* case and *Arizona v. Gant* (2009), which was discussed in the introduction to the chapter, involve *when* the suspect was arrested—before or after the search—as well as the reason *why* the subject was arrested. *Gant* relates most closely to the searches of vehicles after an arrest is made. Thus, this case is discussed below in the section on search incident to arrest.

With regard to traffic stops, in the case of *Whren v. United States* (1996), the Court ruled that any traffic offense committed by a driver provides a legal basis for a traffic stop. In this particular case, the traffic offense was a right turn without the use of a directional light. The traffic stop led to the discovery and seizure of drugs from the occupants. According to the Court, a traffic stop is legal even if it is a **pretext traffic stop** for some other law enforcement action, such as a criminal investigation.

Other cases have further defined the law with regard to traffic stops. For example, the Court has ruled that a stop is legal even if the officer made a reasonable mistake in concluding a traffic violation occurred (*Heien v. North Carolina*, 2014) and that an anonymous 911 call can provide reasonable suspicion to make a traffic stop (*Navarette v. California*, 2014). However, the police may not stop a vehicle to check the motorist's driver's license and car registration without reasonable suspicion the driver does not have a license, the vehicle is not registered, or the law is somehow being violated (*Delaware v. Prouse*, 1979), or without consideration of the totality of the circumstances that illegal actions are afoot (*United States v. Arvizu*, 2002). Checkpoints where all vehicles are stopped by the police for the purpose of locating witnesses or to collect other information are permissible (*Illinois v. Lidster*, 2004). Searches of randomly stopped vehicles are not legally permissible.

Additionally, when a traffic stop is made to issue a traffic citation, officers must have at least a reasonable suspicion of danger or a crime in order to justify a search of the vehicle (*Knowles v. Iowa*, 1998; *Michigan v. Long*, 1983). A search of the occupants depends on either reasonable

pretext traffic stop:
A traffic stop made for any traffic offense that may then allow for other law enforcement action.

suspicion, fear or safety, or probable cause (see *Brendlin v. California*, 2007). However, a dog can sniff the outside of a vehicle in a traffic stop without reasonable suspicion (*Illinois v. Caballes*, 2005) so long as this action does not prolong the stop (*Rodriguez v. United States*, 2015).

OTHER PLACES AND THINGS EXCEPTION The third exception to the search warrant requirement, the other places exception, applies to places and things not afforded Fourth Amendment protections. For example, in the case of *Oliver v. United States* (1984), the "other place" was an open field of marijuana, in spite of a "No Trespassing" sign. Other decisions of the Supreme Court have held that there is no reasonable expectation of privacy in garbage left for collection outside a house (*California v. Greenwood*, 1988), in greenhouses viewed from the sky (*Florida v. Riley*, 1989), or in bank records obtained via a subpoena

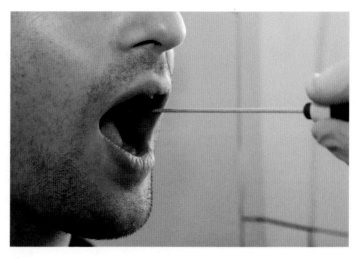

Photo 9.4

A warrant is not required to collect DNA from a subject who has been arrested, but one is required when a subject is not under arrest, unless the DNA is collected via consent or in some other way. Anton Luhr imageBROKER/Newscom

(*United States v. Miller*, 1976). In *Maryland v. King* (2013), the Court ruled that the defendant's Fourth Amendment rights were not violated when a DNA cheek swab was taken as part of arrest and booking procedures. However, police need a warrant to collect DNA from a subject who is not under arrest unless the DNA is provided with consent or collected in some other indirect way. Firefighters do not need a warrant to enter a building to extinguish a fire or to conduct an investigation of the cause of a fire (*Michigan v. Tyler*, 1978).

HOT-PURSUIT EXCEPTION Sometimes hot pursuit is considered an exigent circumstance. Indeed, the rationale for a hot-pursuit search is the same as for other exigent circumstances: to prevent harm to people or to prevent the destruction of evidence. The Supreme Court case of *Warden v. Hayden* (1967) created the hot-pursuit exception to the search warrant requirement. When pursuing an armed robbery suspect, the police arrived at the house they believed the suspect had entered. An officer knocked and announced his presence. He asked for permission to search the house, and Mrs. Hayden offered no objection. The officers found Hayden upstairs pretending to be asleep and arrested him. Another officer discovered a shotgun and a pistol. The pistol; a clip of ammunition for the pistol; and a cap, jacket, and pants that matched the description of the clothing worn by the perpetrator were admitted as evidence. Hayden was convicted. On appeal, the Supreme Court ruled the police may make a warrantless search and seizure when they are in "hot pursuit" of a suspect. The scope of the search may be as extensive as reasonably necessary to prevent the suspect from resisting or escaping. Officers do not need to delay an arrest if doing so would endanger their lives or the lives of others or allow for the destruction of evidence. However, the warrantless entry still requires probable cause that the suspect being pursued committed a crime and is in the premises to be entered. If a suspect enters a house to avoid an arrest when police are present, the police may enter the house without a warrant in order to make an arrest (*United States v. Santana*, 1976). The hot pursuit exception applies only to serious offences, felonies, and some misdemeanors (*Welsh v. Wisconsin*, 1984).

SEARCH INCIDENT TO ARREST EXCEPTION The search incident to arrest exception to the search warrant requirement applies to situations in which the police conduct a search of an individual as a result of that person's arrest. Over the years, numerous cases have addressed this exception. As with the other exceptions, the rationale is to prevent harm to the officer and/or to prevent the destruction of evidence.

other places exception: A warrant exception that allows police to search certain places and things not afforded Fourth Amendment protection.

hot-pursuit exception: If the police are in pursuit of a subject and have probable cause to believe that subject committed a crime and is in a home, then the police may enter the home to make an arrest and conduct a search.

search incident to arrest exception: This exception allows a warrantless search of an individual as a result of that person's arrest.

In the case of *Chimel v. California* (1969), police officers with an arrest warrant but not a search warrant were admitted into Chimel's home by his wife. When he arrived home, Chimel was served with the arrest warrant. Although he denied the officers' request to "look around," the officers conducted a search of the entire house, including the attic, garage, and workshop. At his trial on burglary charges, items seized from Chimel's home were admitted over the objection they had been unconstitutionally seized. The Supreme Court agreed and held that the search of Chimel's home went far beyond his person and the area within which he might have harbored either a weapon or something that could have been used as evidence against him. There was no justification for extending the search beyond the area within his immediate control—the area covered by the spread of the suspect's arms and hands.

In *Maryland v. Buie* (1990), the Supreme Court ruled a larger search was justified because of the potential for danger to officers. Specifically, the court held that a "protective sweep" is justified when there is reasonable belief a person who poses a danger to those at the scene is at the scene. However, a protective sweep by the police is not allowed every time an arrest is made, and it must be limited in scope.

If an arrest occurs outside a house, the police may not search inside the house as a search incident to lawful arrest (*Vale v. Louisiana*, 1970). However, the police may monitor the movements of a person who has been arrested. If the person who has been arrested proceeds into a private place (e.g., a dorm room, a house), the police may accompany him or her. If evidence is then observed in plain view, it may be seized (*Washington v. Chrisman*, 1982). In addition, any lawful arrest justifies the police to conduct a full-scale search of that person even without officer fear for safety or belief evidence may be found (*Gustafson v. Florida*, 1973).

If an occupant of a vehicle is arrested in or near the vehicle, the scope of the search can include a search of the passenger compartment of that automobile, including containers found within the passenger compartment, "for if the passenger compartment is within reach of the arrestee, so also will containers in it be within his reach" (*New York v. Belton*, 1981).

The Supreme Court case of *Arizona v. Gant* (2009) further clarifies the ability of the police to search vehicles incident to arrest (see the Police Spotlight feature at the introduction to this chapter). In this case, Gant was arrested for driving with a suspended license and was handcuffed and placed in the back seat of a police car. It was only then that the police searched his vehicle. The search was ruled unreasonable. As explained by the Court, "the police may search a vehicle incident to a recent occupant's arrest only if the arrestee is within reaching distance of the passenger compartment at the time of the search or it is reasonable to believe the vehicle contains evidence of the offense of the arrest." Gant had no access to the vehicle when the search was conducted, and there was no reason to suspect that the vehicle contained evidence relating to driving with a suspended license.[5]

An emerging issue involves the retrieval of information from cell phones incident to arrest. In *Riley v. California* (2014), the Court ruled that the police may not, without a warrant, search a cell phone seized from an individual who has been arrested, unless exigent circumstances or consent would allow it.

STOP AND FRISK EXCEPTION Due to the stop and frisk exception, the police may conduct a *search* of a person even though an *arrest* of that person may not be justified. Many court decisions have clarified and defined the intricacies of this exception to the search warrant requirement, and most of them note the importance of ensuring officers' safety in justifying stop and frisk searches. The most famous of these cases was *Terry v. Ohio* (1968). The facts of the case are as follows: While patrolling a downtown beat, Cleveland police officer McFadden observed two strangers on a street corner. It appeared to the officer that the two men were casing a store. Each man walked up and down the street, peering into the store window, then returned to the corner to confer with the other. At one point, they were joined by a third man, who left abruptly. Officer McFadden followed the original two men

stop and frisk exception: If there is reasonable suspicion criminal activity is afoot and the subject may be armed, the police may conduct a search of the outer clothing of a person without a warrant even though an arrest of that person may not be justified.

for a couple of blocks until they were rejoined by the third man. The officer then approached the men, identified himself, and asked for their identification. The men mumbled something, whereupon McFadden frisked all three of them. Terry and one other man were carrying handguns. Both were tried and convicted of carrying concealed weapons. On appeal the Supreme Court held that

> where a police officer observes unusual conduct which leads him to reasonably conclude in light of his experience that criminal activity may be afoot and that the persons with whom he is dealing may be armed and presently dangerous . . . he is entitled for the protection of himself and others in the area to conduct a carefully limited search of the outer clothing of such persons in an attempt to discover weapons which might be used to assault him.

The practice of stop and frisk is thus legal.

Many cases relate to the question of what constitutes reasonable suspicion that criminal activity is afoot—the prerequisite for a legal stop and frisk. In *Brown v. Texas* (1979), the Supreme Court ruled that just because an individual *looks* suspicious and has never been seen in the area before, the police do not have reasonable suspicion criminal activity is afoot. In *Illinois v. Wardlow* (2000), reasonable suspicion was determined to have been present when a suspect in a high narcotics trafficking area fled from the police once he saw them. However, a stop and frisk based on an anonymous tip is not legally permissible (*Florida v. J. L.*, 2002).[6]

When conducting a pat-down search, nonthreatening contraband (e.g., drugs) may be seized only if it is immediately apparent and it is not found as a result of squeezing, sliding, or otherwise manipulating the contents of the defendant's pockets (*Minnesota v. Dickerson*, 1993).

Photo 9.5

Stop and frisk searches are conducted to protect officers and to discover weapons. However, if evidence is discovered in the process of the search, the police are not required to ignore it. AP Photo/Ross D. Franklin, File

PLAIN VIEW EXCEPTION When the police conduct a search with a warrant or when the police are legally present at a particular place and evidence is observed, that evidence may be seized under the provisions of the plain view exception to the search warrant requirement. Consider the case of *Texas v. Brown* (1983). In Fort Worth, Texas, Brown's car was stopped at a routine checkpoint at night by a police officer. The officer shone his flashlight into the car and saw an opaque party balloon, knotted near the tip, fall from Brown's hand to the rear seat. Based on the officer's experience in drug offense arrests, he was aware narcotics are often stored in these types of balloons. He shifted his position to obtain a better view and noticed drug paraphernalia and loose white powder in the glove compartment. After he failed to produce a driver's license, Brown was placed under arrest. At trial, Brown was convicted of narcotics offenses. The Supreme Court held the officer's initial stop of the car was valid and the officer shining his flashlight into the car and changing position did not violate Brown's Fourth Amendment rights. The officer had probable cause to believe the balloon contained narcotics, so the seizure was also justified.

In the case of *Horton v. California* (1990), the police were conducting a warrant search for the proceeds of a robbery and in the process inadvertently discovered weapons in plain view. The Supreme Court ruled the seizure of items not listed in a warrant is permissible as long as those items are in plain view. In other cases, however, additional actions with regard to items found in plain view have been ruled by the Court as unacceptable. For example, in *Arizona v. Hicks* (1987), the Supreme Court held that moving a stereo in plain view to record its serial number constituted a search and was not permissible without a warrant.

plain view exception: If the police conduct a search with a warrant or are legally present at a particular place and evidence is observed in plain sight, that evidence may be seized.

CONSENT SEARCH EXCEPTION The vast majority of searches conducted by police without a warrant occur when a person provides the police consent for the search. One reason why consent searches are used so often is that probable cause or even reasonable suspicion are not needed in order to justify the search.

Consider *Schneckloth v. Bustamonte* (1973). A car containing six men was stopped for a traffic violation by a California police officer. Bustamonte, the driver, and three of the other men could not provide a driver's license. The man who did provide a license, Alcala, explained his brother owned the vehicle. The officer asked Alcala if he could search the car. Alcala gave consent and helped the officer open the trunk and glove compartment. Under the rear seat, the officer found several checks that had previously been stolen from a car wash. The checks were admitted as evidence in trial, and Bustamonte was convicted. The Supreme Court held that after validly stopping a car, an officer may ask the person in control of the car for permission to search it. If consent is given, the officer may conduct a search even if there is not probable cause or reasonable suspicion. The voluntariness of the consent is to be determined by the totality of the circumstances, and consent need not be in writing. The police do not have to inform subjects of their Fourth Amendment rights prior to receiving valid consent; however, the burden lies on the officer to prove the consent was valid.

In a related case, the Supreme Court ruled consent is valid if received from a third person believed to have common authority over the premises (*Illinois v. Rodriguez*, 1990). In *Stoner v. California* (1964), however, a search of a hotel room was deemed not valid when consent was received from the hotel night clerk, as the clerk did not have common authority over that room. If two people with common authority (e.g., a husband and wife who share a home) are present and one gives the police consent to search and the other objects, then a consent search is not justified; however, if one of those objecting parties is legally removed (e.g., arrested) and not present where the search is to occur, then the search is valid (*Fernandez v. California*, 2014).[7]

In some law enforcement agencies, so-called knock and talk searches are a frequently used investigative strategy. With a **knock and talk search**, the police approach a house, knock, talk with the occupant, and seek consent to enter and search the house. Depending on the manner

knock and talk search: A search in which the police talk with the occupant of a home in an attempt to get consent to conduct a search.

GOOD POLICING
Legal Knock and Talk Searches

A knock and talk search is used to obtain consent to conduct a legal search of a home without a warrant. Four issues need to be considered in judging the legality of a knock and talk search.[8] First is the walk. The police should approach the premises using open and accessible areas, such as driveways, sidewalks, and front doors. Second is the encounter. The police actions should be conducted in such a way that a person could feel free to decline the request to search. For instance, officers should not order persons to open the door or be unreasonably persistent in attempting to gain access/consent. Third is the knock. Again, officers should not be unreasonably persistent in summoning the occupants of the house. Fourth is the talk. Officers should be polite and ask questions (e.g., "Can you come to the door please?") versus issuing commands (e.g., "Police, open the door!").[9] With these considerations in mind, a knock and talk can be a valuable and legally justified approach to obtaining consent to conduct searches.

Photo 9.6
The legality of a knock and talk search depends on the methods used. Carlos Chavez/Los Angeles Times via Getty Images

in which they are conducted, knock and talk searches can be a legal and useful strategy of obtaining evidence (see Good Policing feature).

Finally, a consent search is limited in several ways. It is limited by the statements and actions of officers; the officer must limit the scope of the search to that which was represented to the subject (e.g., the statement "I'm only interested in looking around in the bedroom" would preclude the officer from searching other rooms). Also, the search is limited by the actions and statements of the subject; the officer may not exceed the parameters of the search as stated by the subject (e.g., "You can't search the bedroom" would prohibit an officer from searching the bedroom). Finally, the search must be reasonable. For example, consent to do a pat-down does not represent consent to do a strip search; consent to search does not allow an officer to break open things.[10]

THE EXCLUSIONARY RULE

If a search is determined to be unreasonable, the evidence obtained must be excluded from trial. This basic principle is known as the **exclusionary rule**. The exclusionary rule relates specifically to unreasonable searches and seizures. As discussed later, however, evidence collected in violation of other constitutional rights is also excluded from trial, although not technically as a result of the exclusionary rule.[11]

The case most often associated with the exclusionary rule is *Mapp v. Ohio* (1961). In this case, Cleveland police officers arrived at the Mapp residence as a result of information they received that a person wanted for a recent bombing was in the home. The officers knocked on the door and demanded entrance. Mapp telephoned her attorney and refused to let the officers in without a warrant. Three hours later, additional officers arrived at the scene. Mapp's attorney also arrived, but the police would not allow him to enter the house or to see Mapp. Mapp demanded to see a search warrant, so a paper, claimed by the police to be a warrant, was held up by one of the officers. Mapp grabbed the warrant and a struggle ensued. Mapp was handcuffed and the police searched her entire house, including dresser drawers, suitcases, and closets. A trunk in the basement was searched and obscene material was discovered inside. Mapp was charged and convicted of possession of these materials, but no search warrant was produced at the trial. On appeal the Supreme Court ruled that in state criminal proceedings the exclusionary rule prohibits the use of evidence resulting from unreasonable searches and seizures. The evidence in *Mapp* was not admissible.

There are exceptions to the exclusionary rule that occur when something may have made the search and seizure technically illegal, but the evidence is still admissible in court. Briefly, these exceptions include situations in which the police make an unintentional error in conducting the search (good faith exception to the exclusionary rule); the police would have found the evidence without the illegal search (inevitable discovery); the voluntary actions of the suspect nullify the illegal actions of the police (purged taint exception); or the evidence is obtained independent of the illegal police action (independent source exception).

THE IMPACT OF THE EXCLUSIONARY RULE The discussion of the exclusionary rule and its exceptions raises an important question: What is the purpose of the exclusionary rule? The most obvious purpose is to prevent illegally obtained evidence from being used in court against a defendant. The other purpose of the exclusionary rule is to deter unlawful police conduct in search and seizure cases. The reasoning is that if the police know illegally seized evidence cannot be used in court to prove the suspect's guilt, then the police will not seize the evidence illegally—they will follow the law in collecting the evidence. The police will not violate citizens' rights, which is the fundamental aim of the constitutional protections.

But does the exclusionary rule *really* deter police misconduct in conducting searches and seizures? The answer to this question is no, it does not—at least not as much as what many would hope or expect. The reason for this is that there are ways of "getting around the Fourth Amendment."[12] Indeed, the police can use several strategies to circumvent the rule. If these

exclusionary rule:
If a search is determined to be unreasonable, the evidence obtained must be excluded from trial.

strategies do not break the law, they at least bend it. Although these actions may not be ethical, it would be naïve to ignore the fact that they sometimes occur. For example, the police may stretch the boundaries of the consent exception and conduct searches in situations when "consent" may not actually have been legally obtained. The police may conduct an illegal search with full knowledge the evidence seized will not be admissible and the case will not be prosecuted. Or officers could conduct an illegal search and seize evidence to harass a suspect.[13] Does this really happen? In a survey of Illinois police officers, 25% stated they had witnessed at least one illegal search of a subject during the previous year.[14] Additionally, police officers may engage in "judge-shopping," which occurs when officers seek out particular judges to review applications for warrants.[15] It is interesting to consider that warrant applications are rarely rejected by judges or magistrates.[16]

A Question of Ethics

Necessary Means to Achieve the Desired Ends?

An important goal of the police is to identify and apprehend criminals, and we expect them to work hard to do so. A lot of money is allocated to law enforcement agencies so they can accomplish this goal. We also expect the police to obey the law and not violate citizens' rights. However, some criminals are crafty; they know how to avoid detection. If the police engage in conduct that is not legal, but their actions result in criminals being identified and apprehended, is there really anything wrong with that? Why or why not?

Finally, unfortunately, sometimes officers simply lie. For a variety of reasons, officers may misrepresent the facts of a case to a judge and, as a result, the "fruits" of an otherwise illegal search may be admitted into trial and considered in determining the guilt or innocence of the accused.[17] One study found that 4% of officers knew of other officers who had provided false testimony in traffic cases, 3% knew of false testimony given in criminal cases, and 7% knew of arrest reports written in a false manner.[18] Given the sensitive nature of this area of inquiry, one might expect these illegal behaviors are underreported by officers. The most common reason given for officers lying is that they view it as a necessary means to achieve the desired ends.[19] The bottom line is that the exclusionary rule may not prevent the police from engaging in these questionable, unethical, or illegal actions.

Another issue related to the exclusionary rule is whether or not potentially guilty suspects are freed because of the rule. The research on this is mixed and dated; the general conclusion appears to be that the exclusionary rule affects only a very small percentage of cases (.4% to 1.4%).[20] Although the prevailing wisdom is that the exclusionary rule has little impact on cases *after* a suspect has been identified and apprehended, it is important to realize these studies did not examine the impact of the exclusionary rule on crimes being solved. The exclusionary rule may very well prevent some crimes from being solved.

THE LAW OF SELF-INCRIMINATION: THE FIFTH AND SIXTH AMENDMENTS

The Fifth Amendment to the U.S. Constitution protects citizens against self-incrimination. It reads, in part, "No person shall be compelled in any criminal case to be a witness against himself, nor be deprived of life, liberty or property, without due process of law."

The Sixth Amendment includes several rights. The most important for the police is the right of individuals to be represented by an attorney in legal proceedings. The amendment states, in part, "In all criminal prosecutions the accused shall enjoy the right to . . . have the assistance of counsel for his defense."

The protections offered in the Fifth and Sixth Amendments are relevant when determining the admissibility of incriminating statements obtained from suspects. If information is obtained from suspects illegally, then that potentially valuable information is inadmissible in court because it violates the due process rights of the accused. The basic question is, when are incriminating statements made by a suspect admissible in court and when are they not?

The most famous and widely applied case associated with the Fifth and Sixth Amendment protections is *Miranda v. Arizona* (1966). This case involved the rape of an eighteen-year-old developmentally disabled woman. At the time the crime was reported, the woman was able to provide a description of the perpetrator and details about his car. About a week after the incident, the police were informed by a relative of the victim that she had again seen the car driven by the perpetrator, and a license plate number for the vehicle was provided. The police were eventually able to locate the vehicle, and it matched the description provided by the victim.

The police found Ernesto Miranda asleep in the house where the vehicle was parked. The police arrested him, transported him to police headquarters, and placed him in a lineup with three other Mexican Americans to be viewed by the victim. She was not able to positively identify Miranda as the perpetrator, but in the interrogation room, police told Miranda that he *had* been identified by the victim. After two hours of questioning, he confessed to the kidnapping and rape as well as two other recent crimes—a robbery and an attempted rape. Police then provided a sheet of paper to Miranda on which to provide a handwritten confession. The confession provided by Miranda was similar to the account provided by his victim.

At the trial, Miranda's written confession was presented as evidence in spite of the objections of Miranda's attorney, who argued the confession was coerced and therefore inadmissible. The judge ruled that the case of *Gideon v. Wainwright* (1963) offered the benefit of defense counsel at trial, not at the arrest, and therefore the confession was legally obtained and admissible. The jury found Miranda guilty of rape and kidnapping, and he was subsequently sentenced to prison.

Miranda's attorney appealed the conviction to the Arizona Supreme Court with the argument that the confession was not voluntarily offered. Meanwhile, during this time and while Miranda was serving his prison sentence, the U.S. Supreme Court ruled in the case of *Escobedo v. Illinois* (1964). In this case, the Court held defendants have the right to an attorney at the interrogation stage of criminal proceedings. However, because Miranda had not requested an attorney at the time he was questioned by the police, the Arizona Supreme Court ruled the *Escobedo* decision did not apply to Miranda's case. The court upheld the conviction. As a result of this decision, the conviction of Miranda was overturned, but Miranda did not go free. Prosecutors retried him for the rape and kidnapping without the original confession as evidence. At the new trial, Miranda's common-law wife provided testimony that Miranda had earlier confessed to her about the rape. Miranda was convicted again and returned to prison.

Photo 9.7
The U.S. Supreme Court case of Miranda v. Arizona *is responsible for the Miranda warnings.* ASSOCIATED PRESS

THE CONTENT AND WAIVER OF MIRANDA WARNINGS

Miranda warnings consist of the following (see Exhibit 9.1):

- The right to remain silent
- The advisory that anything the suspect says can be used against her or him in court
- The right to have an attorney present during the interrogation
- The provision that if the suspect cannot afford an attorney, one will be appointed at no cost

Suspects must be informed of these rights when in custody and prior to questioning; in other words, prior to a custodial interrogation.

> **Miranda warnings:** A list of rights that must be provided to an individual when in custody and prior to questioning.
>
> **custodial interrogation:** An interrogation that takes place when a subject is in the custody of the police.

EXHIBIT 9.1

Example of a Miranda Waiver Form

As a result of the decision in *Miranda,* police departments usually require investigators to complete a Miranda waiver form like the one shown here prior to questioning a suspect.

Incident Number: _____ Defendant: _____

Address: _____ Charge: _____

Constitutional Rights Miranda Warnings

_____ You have the right to remain silent. Anything you say can and will be used against you in a court of law.

_____ You have the right to talk to a lawyer and have him/her present with you while you are being questioned.

_____ You can decide at any time to exercise these rights and not answer any questions or make any statements.

_____ At this time, I, _____, wish to waive my constitutional rights and agree to voluntarily provide a written statement to the Glendale Police Department. This statement is given voluntarily of my own free will and there have been no promises or threats made to me.

Signature _____ Date _____ Time _____

Witness _____ Date _____ Time _____ Title _____

Statement:

Signature _____

Page _____ of _____

There is leeway in how suspects can be informed of their rights. For example, the police do not have to give verbatim warnings as long as the suspect is advised of his or her rights and no limitations are placed on those rights (*California v. Prysock*, 1981). Most important is that it is communicated to the defendant that she or he has the opportunity to consult an attorney prior to or during the interrogation. In addition, the order in which the warnings are read does not matter (*Florida v. Powell*, 2010). If a defendant's request for an attorney is clear even if the waiver of the other rights is not, all questioning must stop until an attorney has been provided (*Smith v. Illinois*, 1984). Further, a suspect cannot be questioned again for the same offense after invoking his right to remain silent unless the suspect has consulted with a lawyer or initiates further communication, exchanges, or conversations with the police (*Edwards v. Arizona*, 1981). Relatedly, once a subject invokes his or her Miranda rights not to answer the questions of the police, the subject cannot be asked about other offenses or questioned by different law enforcement authorities (*Arizona v. Robertson*, 1988). However, interestingly enough, the Supreme Court ruled that if a subject is out of police custody for fourteen or more days, the police can provide new Miranda warnings to the subject in an attempt to re-initiate questioning (*Maryland v. Shatzer*, 2010).

THE MEANING OF AN INTERROGATION AND CUSTODY

In *Rhode Island v. Innis* (1980), the Providence, Rhode Island, police arrested Innis as a suspect in the murder of a taxicab driver based on an eyewitness identification. Innis was advised of his Miranda rights. He said that he understood his rights and wanted to speak with an attorney. He was then placed in a car and driven to the station. During the drive, one of the officers commented that there were "a lot of handicapped children in the area" because a school for such children was nearby. He further stated how horrible it would be if one of the children found the gun (used in the murder) and something happened. Innis then proceeded to tell the officers where the gun could be found. The Supreme Court ruled that the respondent was not interrogated in violation of his rights. The statements the officer made did not constitute express questioning or its functional equivalent, and the officer had no reason to believe his statements would lead to a self-incriminating response from the suspect. Subtle compulsion does not constitute an interrogation.

However, in the case of *Brewer v. Williams* (1977), the Supreme Court ruled the police had explicitly sought to obtain incriminating evidence from Williams with regard to his involvement in a kidnapping/murder. Knowing that he was a former mental patient and deeply religious, the officer called Williams "Reverend"

Photo 9.8
Suspects must be told their Miranda rights when in custody and prior to interrogation, although sometimes "custody" and "interrogation" are difficult to precisely define. © iStockphoto.com/KatarzynaBialasiewicz

and suggested the missing girl's parents should be entitled to a Christian burial for their daughter, who was taken from them on Christmas Eve. Williams then showed the police where to find the girl's body. The Court held that the "Christian burial speech" was an interrogation. The statements were not admissible as evidence, but the body of the girl *was* admissible (under the inevitable discovery exception to the exclusionary rule).

There are occasions in which the police speak to prisoners in jail or in prison. The Supreme Court has ruled an inmate is not in custody for Miranda purposes simply by being incarcerated. As in other situations, if the subject is free to end the questioning and leave the interview, the subject is not in custody, even if he or she is incarcerated (*Howes v. Fields*, 2012). As a result, incarcerated defendants do not need to be informed of their Miranda rights prior to questioning. The Supreme Court has also ruled that when

> a suspect has been indicted and engaged an attorney, the police can no longer question him or her, even through a third party, such as a coconspirator who records a conversation. (*Massiah v. United States*, 1964)

As ruled in *Arizona v. Fulminante* (1991), statements obtained as the result of implied duress and without the benefit of the Miranda warnings are not admissible. In this case, a paid police prison informant promised Fulminante he would provide protection from the other prisoners if Fulminante would tell him the truth about the abduction/murder of a child victim. Fulminante then confessed to the crime. He was convicted of murder, partially on the basis of this confession. The Supreme Court ruled Fulminante's confession was involuntary because it was motivated by fear of physical violence if he did not receive protection. As a result, the confession was not admissible at trial.

THE IMPLICATIONS OF SILENCE

If a suspect remains silent after being informed of the Miranda rights, the police may continue to interrogate the suspect and incriminating statements can be used against him or her (*Berghuis v. Thompkins*, 2010).

But does silence on the part of a suspect constitute evidence? It depends on the meaning, context, and form of silence. In *Griffin v. California* (1965), the Court ruled a prosecutor cannot comment to the jury that the defendant not testifying constitutes evidence (of guilt). Similarly, the *police* cannot comment to the jury that the defendant invoked her or his Miranda rights to

Exceptions to the Miranda Requirement

The police must inform suspects of their Miranda rights when suspects are in custody of the police and prior to questioning, except when certain circumstances apply. These circumstances include the following:

- During roadside questioning of motorists
- When limited questions are asked in an attempt to protect public safety

- During undercover operations
- When statements are provided voluntarily by the suspect
- During stop and frisk ("Terry stop") searches
- When general on-the-scene questions are asked (e.g., "What happened?")
- During routine booking

remain silent. However, if the defendant simply remains silent without invoking his or her right to remain silent, the police *can* testify about the defendant's silence (*Salinas v. Texas*, 2013). Interestingly, the right to remain silent does not mean that a subject can refuse to submit to a blood alcohol test; such a refusal is admissible in court (*South Dakota v. Neville*, 1983).

JUVENILES AND THE MIRANDA REQUIREMENT

The Miranda requirement applies to juveniles and adults. If a juvenile is in custody and is about to be interrogated, Miranda warnings must be provided to the juvenile. There are continuing questions however about juveniles' ability to understand the Miranda warnings and the serious implications of self-incrimination. These concerns are magnified given the techniques that may be used by interrogators to elicit a confession. Agencies sometimes modify the wording of the warnings to make them easier to understand. For example, instead of saying, "You have the right to remain silent," the warning would be "you don't have to answer my questions." Most states do not require a parent to be notified or present for the interrogation of a child, but it may be a factor when considering whether the child waived his or her rights knowingly, voluntarily, and intelligently. Some states require that an attorney be notified and/or that a parent make the waiver on behalf of the juvenile who is to be interrogated.

EXCEPTIONS TO THE MIRANDA WARNINGS

Certain circumstances exist under which it is legal not to inform suspects of their Miranda warnings. These circumstances may be thought of as exceptions to the Miranda requirement. The police do not have to provide Miranda warnings prior to the roadside questioning of a motorist because this does not constitute a custodial interrogation (*Berkemer v. McCarty*, 1984). As a general rule, if a suspect is not in custody, Miranda warning need not be given. The Court created the public safety exception to the Miranda warnings when it ruled in *New York v. Quarles* (1984) that the police can ask a suspect about the location of a weapon without first informing the suspect of the Miranda rights. In the *Quarles* case, the gun posed a possible immediate danger to the public; the potential danger justified the officer's failure to provide the Miranda warnings prior to questioning. The questioning was limited and the statements were provided voluntarily.[21] When a suspect is not aware he or she is speaking to a law enforcement officer (e.g., during undercover operations), the police are not required to provide the Miranda warnings to the suspect unless that suspect has previously invoked the warnings. Finally, any voluntary statements made by a suspect are admissible (*Illinois v. Perkins*, 1990; *Michigan v. Mosley*, 1975).

THE IMPACT OF *MIRANDA V. ARIZONA* ON SUSPECT CONFESSIONS

As a result of the Supreme Court ruling in *Miranda v. Arizona*, the police thought confessions were a thing of the past. It was believed if suspects were told they did not have to talk to the police and that anything they said could be used against them, no one of sound mind would ever confess. However, most of the research that has examined the impact of *Miranda* on police ability to obtain confessions has shown that the decision has had minimal impact or at least much less impact than what was originally feared.

To understand the impact of *Miranda*, at least two questions need to be considered. First, to what extent are confessions obtained by the police subsequently ruled to be inadmissible? Although research has not provided a precise estimate, the short answer seems to be very few.[22] If a suspect was informed of her or his rights and those rights were voluntarily and knowingly waived, then the requirements of *Miranda* were satisfied. Only if the methods used to obtain confessions are deemed coercive would a confession be ruled inadmissible; *Miranda* does not protect against coercive interrogation methods.

Second, to what extent are confessions (and convictions) not obtained by the police because of *Miranda*? This question is more complicated than the first because it must be understood that perhaps a confession was not obtained because there was no confession to give (i.e., the person who was interrogated did not commit the crime). It is not possible to determine with certainty whether or not a person who is interrogated but does not confess is actually guilty or innocent of the crime in question, especially if the crime remains unsolved. Nevertheless, studies have shown that most subjects (78% to 84%) agree to answer the questions of the police even after being told that they do not have to.[23] The Miranda warnings do not appear to prevent subjects from answering the questions of the police.

? RESEARCH SPOTLIGHT
Why People Waive Their Miranda Rights: The Power of Innocence

A laboratory experiment was conducted with 144 students as subjects in order to discover what proportions of guilty and innocent subjects agree to answer the questions of the police and why suspects answer questions of the police.[24]

Half of the students were instructed to commit a theft in a staged and controlled setting (taking $100 out of a drawer). The other half went through the same motions but did not take the money. Then each student subject was taken alone to a room where he or she met a "detective" (an actor in the experiment). The detective explained to each subject that he wished to ask questions about a theft that occurred. The subject was then provided a waiver form and asked if he or she was willing to answer questions or not. No interrogation or questioning by the detective actually occurred for any subject in the experiment. Subjects were then asked questions by the researcher about their decision to answer questions or not.

It was found that, overall, 58% of the subjects agreed to answer the questions of the detective, but innocent subjects were much more likely to agree than guilty subjects: 81% compared to 36%. The most common reason expressed by guilty subjects for being willing to answer the questions of the detective was either "If I didn't, he'd figure I was guilty" or "I would've looked suspicious if I chose not to talk." The most common reasons expressed by innocent subjects for agreeing to answer questions were "I did nothing wrong" and "I didn't have anything to hide."

The authors explain that only innocent subjects are at risk of false confessing, and the only way in which this can happen is if these subjects waive their Miranda rights and answer the questions of the police. Other studies and DNA exoneration cases show that false confessions do happen. As demonstrated in this study, Miranda warnings do not prevent innocent subjects from allowing interrogations to proceed. According to the researchers, "These results indicate that people have a naïve faith in the power of their own innocence to set them free. Our results suggest the ironic conclusion that during the early stages of a criminal investigation, innocence may put innocents at risk."[25]

Main Points

- Evidence (information) is used to establish proof. There are various standards of proof. For example, probable cause is necessary to justify an arrest. Reasonable suspicion of a person's involvement in a crime is necessary to justify a stop and frisk search.
- An arrest occurs when the police take a person into custody for the purposes of criminal prosecution. A person who is in custody of the police is not free to leave. If a subject has been arrested, they are in custody, but a person who is in custody may not be under arrest.
- A search is a governmental infringement into a person's reasonable expectation of privacy for the purpose of discovering evidence for use in a criminal prosecution.
- The intent of the Fourth Amendment is to protect individuals' privacy and protect against arbitrary intrusions into that privacy by government officials.
- The general rule is that the police need a search warrant to conduct a legal and valid search and that warrant is to be based on probable cause. However, there are many exceptions to this rule. In fact, most searches conducted by the police are conducted without a warrant; the most common of these searches are consent searches.
- The exclusionary rule holds that if the police collect evidence illegally, that evidence is to be excluded from court proceedings. However, there are exceptions to the rule, the most common of which is the good faith exception.
- The Fifth Amendment to the Constitution protects against self-incrimination; the Sixth Amendment provides the right of the accused to be represented by an attorney in criminal proceedings.
- The police must inform a suspect of his or her Miranda rights when the suspect is in custody of the police and prior to interrogation by the police. There are several circumstances in which the police do not need to inform suspects of their Miranda rights.
- Research shows that the Miranda warnings do not prevent people from answering the questions of the police or from obtaining confessions.

Important Terms

Review key terms with eFlashcards at **edge.sagepub.com/brandl2e.**

Questions for Discussion and Review

Take a practice quiz at **edge.sagepub.com/brandl2e.**

1. Explain why a person could be placed in handcuffs by the police but not be under arrest.
2. What is a search? What is necessary for the police to obtain a search warrant?
3. The police need to conduct a search of a suspected crime scene inside a house. Under what circumstances could they do so without a warrant?
4. Under what circumstances is a search warrant not required to conduct a search? What is the reason or rationale for each exception to the search warrant requirement?
5. What is the purpose of the exclusionary rule? Is the exclusionary rule effective in this regard?
6. What is an *interrogation* from the perspective of the Fifth Amendment?
7. What are the Miranda warnings? When must they be provided to suspects?

8. Under what circumstances do Miranda warnings not need to be provided to suspects?

9. Do the police need to inform juvenile suspects of their Miranda rights when in custody and prior to interrogation? What are the special considerations with regard to juvenile suspects and the Miranda warnings?

10. What has been the impact of *Miranda* on criminal investigations and the criminal justice process? Why is this the case?

Fact or Fiction Answers

1. Fiction
2. Fiction
3. Fiction
4. Fiction
5. Fiction

6. Fact
7. Fiction
8. Fact
9. Fact
10. Fiction

$\circledS$ SAGE edge™

Digital Resources

Get the tools you need to sharpen your study skills. SAGE Edge offers a robust online environment featuring an impressive array of free tools and resources.

Access practice quizzes, eFlashcards, video, and multimedia at **edge.sagepub.com/brandl2e.**

Media Library

View these videos and more in the interactive eBook version of this text!

Criminal Justice in Practice
9.1: Search and Seizure
9.2: Traffic Stop

SAGE News Clip
9.1: Supreme Court Cell Phones

Part III
THE HAZARDS OF POLICE WORK

10

HEALTH AND SAFETY ISSUES IN POLICE WORK

Police Spotlight: Combatting Post-Traumatic Stress in the Tampa Police Department

In recent years, post-traumatic stress has become recognized as a serious risk for police officers who have experienced traumatic incidents such as killing someone in the line of duty, being shot or shot at, being assaulted, and seeing violent crime scenes and traffic accidents. As Chief Brian Manley of the Austin (Texas) Police Department explains: "This is a career where you can't unsee things that you have seen. Every officer walks around with visions of things that they have experienced during their career."[1] Many departments provide assistance to officers who have experienced traumatic events. This assistance commonly comes in the form of trainings on how to deal with emotions, peer-support programs, and opportunities to talk with a psychologist after the event.

The Tampa Bay (Florida) Police Department has gone further and has developed the Franciscan Center Post-Trauma Education and Retreat Program. According to Julia Hill and her colleagues:

> The Franciscan Center Post-Trauma Education and Retreat Program was borne out of the desire to more effectively treat post-traumatic symptoms, promote healing, and provide post-trauma education for police officers over the course of a confidential, intensive four-night, five-day retreat.

> Officers attending the retreat are in the company of peers experiencing similar crises and professionals who are trained in crisis stress management and, in some instances, have gone through traumatic events themselves. The Franciscan Center focuses on helping participants understand post-trauma symptoms and provides them with the tools and resiliency to address those symptoms. Key components of the retreat program include a trauma symptom inventory used to identify specific symptoms; eye movement desensitization and reprocessing therapy; reviews of critical incidents; and meetings with chaplains, counselors, and peers. Educational sessions are of significant importance to the program's success and focus on topics like drug/alcohol abuse prevention, post-trauma brain function, emotional survival, suicide prevention, forgiveness, cognitive behavioral techniques, and spiritual responses to post-

(Continued)

Objectives

After reading this chapter you will be able to

10.1 Discuss the difficulties of accurately measuring police stress and drawing valid conclusions about the causes and consequences of it

10.2 Identify the workplace problems and other features of police work that are most often identified by officers as stressful

10.3 Assess why shift work is potentially stressful for officers and identify what could be done to minimize its negative effects

10.4 Define burnout and post-traumatic stress disorder as potential problems for police officers and discuss what can be done to minimize their negative effects

10.5 Identify and discuss the four categories of physical hazards to police officers

10.6 Evaluate trends in officers feloniously killed and assaulted

10.7 Discuss why arresting suspects, foot pursuits, and police vehicle accidents and pursuits are potentially risky activities for officers and assess what could be done to reduce those risks

Fact or Fiction

To assess your knowledge of health and safety issues in police work prior to reading this chapter, identify each of the following statements as fact or fiction. (See page 225 at the end of this chapter for answers.)

1. Research has identified dealing with bosses and dealing with coworkers as two of the most frequently experienced stressors of police work.

2. Workplace stress is not difficult to measure accurately.

3. Burnout, post-traumatic stress disorder, and stress have the same causes and consequences.

4. The research is clear: Police officers do not live as long as members of the general population.

(Continued)

(Continued)

5. Police work is dangerous because more police officers die while performing the job than do workers in any other occupation.

6. Typically, more police officers die as a result of accidents per year than because they are murdered.

7. Research has shown that of all injuries police officers sustain, about 10% are the result of assaults on officers.

8. Most police officer motor vehicle crashes occur while officers are on routine patrol, not when responding with lights and sirens in emergency situations.

9. In police vehicle pursuits, fleeing motorists are much more likely to crash than the police.

10. Technology and training can lessen the actual risks and dangers of police work.

(Continued)

trauma stress. Participants leave with a personalized 90-day action plan to use after the program. Ninety days following completion of the program, participants complete a survey designed to gauge progress. More than 76% of participants report a positive increase in post-trauma healing during the course of the retreat. Another 69% notice increased healing from completing the program to the 90-day follow-up. The Franciscan Center hosts retreats five times annually, and each retreat usually involves four to six participants. The $3,000 cost per participant includes renting the entire center to ensure complete confidentiality for the participants. Funding comes from the department or individuals, and the center seeks other sources to reduce the financial burden.[2]

Source: Hill, Julia, Sean Whitcomb, Paul Patterson, Darrel W. Stephens, and Brian Hill. 2014. *Making Officer Safety and Wellness Priority One: A Guide to Educational Campaigns.* Washington, DC: Office of Community Oriented Policing Services.

THERE are risks and hazards associated with every occupation.

For example, exposure to communicable diseases is common among health-care workers. Military soldiers are at risk for post-traumatic stress. Furniture movers are at risk for back injuries. Exterminators may be exposed to harmful chemicals. Football players are at risk for concussions. Given the unique tasks of policing, there are also unique risks and hazards associated with the police occupation. This chapter provides a discussion of some of the more prevalent health and safety risks associated with police work, including stress and its effects. The causes of police officer injuries and deaths are also discussed.

WHAT IS STRESS?

Over the years, a multitude of studies have examined stress, and stress in the workplace in particular. These studies have used various definitions of stress. For the purposes of this book, stress occurs when a person experiences any physiological and/or psychological demand that causes that person to have to adjust to that demand, physiologically or psychologically. As a simple example, cold temperature may cause a person to shiver, which is a **physiological response**. In this case, the cold temperature is a stressor. Personal problems (a stressor) may cause an individual to lose sleep. Being confronted by someone shouting profanities (a stressor) may cause a person to experience a surge of adrenaline. Although stress is usually considered negative, some stress, and the reactions to that stress, may be positive and desired. For example, aerobic exercise may lead to increased cardiovascular abilities and better health.

HOW IS STRESS MEASURED?

An important aspect of defining stress is determining how it is to be measured. Most studies have defined stress in terms of an array of questions or statements to which the research subject responds. For example, in one study, police officers were asked (a) if they had experienced unwanted stress

physiological response: A reaction that triggers a bodily response to a stimulus.

from their job, (b) whether the amount of unwanted stress from their job had had a negative effect on their physical well-being, (c) whether the amount of unwanted stress from their job had had a negative effect on their emotional well-being, and (d) whether unwanted stress had had a negative effect on their job performance. Responses ranged from "strongly agree" to "strongly disagree."[3] Another study listed possible stressors and asked officers to score these factors on a seven-point scale from causing "no stress at all" to "a lot of stress."[4] Other studies have provided short vignettes (stories) depicting stressful situations and asked respondents how they would most likely respond.[5] Some studies defined the existence of stress by the outcomes experienced by the respondents, such as drinking, suicidal thoughts, or divorce.[6]

Photo 10.1

Stress is commonly associated with the police occupation. Although stress is difficult to measure accurately, you generally know it when you feel it. Carol Guzy/The Washington Post via Getty Images

While a great deal of research has examined police stress, one of the problems associated with this research is that there is not standard measure of job stress.[7] Further, "The breadth and degree of missing data is staggering, the inconsistency of measurement substantial, and the shortage of methodologically rigorous studies disappointing."[8] As a result, there are few conclusions that can be drawn with certainty regarding police stress. It is also critically important to realize the experience of being a police officer varies across communities. Being a police officer in Detroit is probably a much different experience than being a police officer in Beverly Hills. Although this may seem to be common sense, the current research on the topic has not sufficiently accounted for differences in stress levels, or correlates of stress, across different police organizations with different characteristics.

One of the primary reasons why stress is usually measured perceptually (e.g., by asking, "How often do you feel stress at work?") is that measuring it in any other way is difficult. If measured psychologically, brain waves would need to be measured. If measured physiologically, heart rate, blood pressure, or hormonal levels in blood would need to be measured. It is obviously easier just to ask people about their perceived stress levels. Nevertheless, a few studies have adopted a physiological approach to measuring stress levels among police officers. One study involved the measurement of heart rate and cortisol levels during firearms training;[9] another involved taking salivary cortisol samples from police officers over a three-day period.[10] Some studies have used monitors to examine the heart rates of officers over the course of entire shifts and then correlated tasks performed during the shift with heart rate increases and decreases (and presumably stress levels).[11] Although the studies that have used these methodologies have shown them to be feasible for measuring physiological stress, for practical reasons, they were limited to testing over a relatively short amount of time and with few officers. Additionally, numerous issues need to be taken into account when measuring stress in this way, including baseline heart rate measures, personality and anxiety traits, feelings of anger, medications, and caffeine intake for each officer studied. The bottom line is the accurate measurement of stress among police officers is difficult.

THE CAUSES OF POLICE STRESS

Policing has long been cited as a stressful occupation. Several studies have attempted to identify the most common causes of stress among police officers. While many of these studies have focused on task-related factors, such as being engaged in high-speed vehicle pursuits and exposure to violence, others have also included organizational or administrative factors, such as dealing with coworkers and supervisors.

In one notable study, researchers asked 365 officers in a police department to identify their top five work stressors.[12] The identified stressors consisted of the following:

- Exposure to battered or dead children
- Killing someone in the line of duty
- Having a fellow officer killed in the line of duty
- Situations requiring the use of force
- Experiencing a physical attack

In this study and in others that identify the most stressful aspects of police work, it must be realized these are all uncommon events and, if experienced, are likely to be experienced vicariously—that is, officers may recognize and internalize the stress being experienced by other officers.

Other studies that have examined the causes of stress have often focused on two particular issues: workplace experiences and shift work. These two issues are discussed in more detail here.

WORKPLACE PROBLEMS The majority of recent studies on police stress have found organizational stressors, such as inadequate support from supervisors and conflict with coworkers, to be the most frequent for police officers;[13] interactions and confrontations with citizens are rated lower.[14] Research has also shown stressors are not equally experienced by all officers, even in the same department, nor are the consequences of stress. For instance, one study of over 1,900 officers from a large police department revealed that, in general, issues such as lack of support at work and a lack of opportunity for promotion were more commonly experienced by some officers than others, as were sexually offensive behaviors and vulgar jokes and language. Further, African American females were more likely to express negative workplace experience and higher levels of stress compared to white male officers.[15]

One workplace problem, or stressor, that has received a lot of research attention is sexual harassment (Exhibit 10.1). It is understood to occur most commonly in genderized occupations, especially ones in which women are a minority, such as police work. Not surprisingly, then, research on sexual harassment in the police occupation has focused on women.

> **sexual harassment:** Unwelcome sexual advances, requests for sexual favors, verbal or physical harassment of a sexual nature, and/or offensive remarks about a person's gender.

EXHIBIT 10.1
The Meaning of Sexual Harassment

According to the U.S. Equal Opportunity Commission (EEOC)

1. Sexual harassment refers to unwelcome sexual advances, requests for sexual favors, and other verbal or physical harassment of a sexual nature.

2. Harassment does not have to be of a sexual nature, however, and can include offensive remarks about a person's sex. For example, it is illegal to harass a woman by making offensive comments about women in general.

3. Both victim and the harasser can be either a woman or a man, and the victim and harasser can be the same sex.

4. Although the law doesn't prohibit simple teasing, offhand comments, or isolated incidents that are not very serious, harassment is illegal when it is so frequent or severe that it creates a hostile or offensive work environment or when it results in an adverse employment decision (such as the victim being fired or demoted).

5. The harasser can be the victim's supervisor, a supervisor in another area, a coworker, or someone who is not an employee of the employer, such as a client or customer.[16]

Source: Equal Opportunity Employment Commission website, http://www.eeoc.gov/laws/types/sexual_harassment.cfm

In a review of the prior research on the topic, it has been reported that between 53% and 77% of female officers have experienced at least one sexually harassing behavior in their policing career,[17] although other studies have found that similar rates of sexual harassment occur over the course of a much shorter time frame, such as the last six months or year.[18] Coworkers are the most common perpetrators of sexual harassment, not supervisors. The most common form of sexual harassment experienced by male and female police officers is verbal—the telling of dirty stories or jokes followed by the making of sexually suggestive remarks toward the respondent. Female police officers are much more likely to experience sexual harassment than male officers. The majority of sexual harassment goes unreported; the most common reason given for this has been the "situation was not serious enough to warrant a formal complaint."

Perhaps surprisingly, researchers have found that in spite of female officers' greater exposure to sexual harassment, male and female officers express similar levels of workplace stress.[19] Other studies point to the highly negative effects of sexual harassment, including poor physical and mental health and symptoms of **post-traumatic stress disorder (PTSD)**.[20] Also, as discussed in Chapter 5, the perceived existence of sexual harassment in the workplace may have negative effects on the recruitment and retention of officers, especially female officers.

Photo 10.2
Estimates vary as to the extent of sexual harassment in police departments. However, whenever it occurs, it is a serious issue that needs to be addressed.
©iStockphoto.com/svarshik

SHIFT WORK Shift work is a reality of the police occupation as well as many other occupations, including medical care (e.g., nurses and doctors), transportation (e.g., pilots and truck drivers), and hospitality (e.g., waitresses and cooks). Although police departments are twenty-four-hour agencies, normally it is only patrol officers and supervisors, and in some agencies, detectives, who work in shifts to provide continuous coverage and availability. Different departments have different shift schedules: Shifts are commonly eight hours, although some are ten hours, and twelve-hour shifts, though uncommon, are not unheard of. On an eight-hour schedule, the exact hours of shifts can vary, but the "day" shift is typically 8:00 a.m. to 4:00 p.m., the "afternoon" shift is from 4:00 p.m. to 12:00 a.m., and the "night" shift is from 12:00 a.m. to 8:00 a.m.

From a scheduling standpoint, ten-hour shifts are generally most desirable[21] as they allow for a four-day work week (four consecutive days on, three consecutive days off) and two weekends off every seven weeks. With eight-hour shifts, common is five days on, two days off, five days on, three days off. Even twelve-hour shifts have advantages: Approximately half of all work weeks consist of just three days. With twelve-hour shifts, there are 182 scheduled days off per year, compared to just ninety-one with eight-hour shifts. The major problem with twelve-hour shifts is that, well, they are twelve hours.

Especially problematic and stressful for officers is when they are expected to rotate shifts on a frequent basis. Fortunately, this practice is uncommon. Also, working on weekends and holidays can add another dimension of stress to the work schedule, regardless of the officer's particular shift or the length of the shift. Overtime may also compound the negative effects of shiftwork. Sometimes, overtime is unavoidable as officers may need to complete reports at the end of the shift or attend court proceedings when normally off-duty. Many departments do not place limits on officers' overtime hours.[22]

post-traumatic stress disorder (PTSD): Mental health condition triggered by witnessing or experiencing an extremely upsetting event. Symptoms may include flashbacks, sleep problems, and anxiety.

Police officers who have held their rank the longest typically have the ability to select the shift they would like to work. The day shift is usually most desirable, as it resembles the work schedule of most other people and is least disruptive to the officers' life outside of work, although even day shift officers may work holidays and weekends. For most officers, the afternoon or night shift is least desirable. Each shift can be stressful for its own reasons. Some officers find the day shift to be stressful because it is the shift most supervisors and command staff work, and officers may receive closer supervision as a result. The afternoon shift is especially stressful from a social or family perspective. If an officer has a working spouse and school-aged children, the officer is usually starting work as the spouse and children are finishing work or school, and when the officer is done with work, the spouse and children are normally asleep. Evenings are also a common time for social and family activities, which are difficult to attend when working the afternoon shift.

The night shift is most problematic with regard to sleep. Physiologically, human beings are meant to sleep at night; being awake disrupts the circadian rhythm and can lead to chronic sleep deprivation and fatigue, which, in turn, can result in many other serious negative health and performance effects. Sleep deprivation and stress represent a vicious circle: Lack of sleep limits one's ability to deal with stress, and increased stress decreases one's ability to obtain adequate sleep. A survey conducted by Brian Vila, a leading researcher on police fatigue, revealed the following:

- 53% of officers get less than 6.5 hours of sleep daily (compared to 30% of the general population).
- 91% report feeling fatigued "routinely."
- 14% are tired when they start their work shift.
- 85% drive while "drowsy."
- 39% have fallen asleep at the wheel.[23]

With regard to negative health effects of shift work, especially working at night, studies have associated sleep deprivation with some relatively minor and short-term conditions, such as upset stomach, diarrhea, and constipation, but also with more serious long-term effects, such

---

GOOD POLICING
Managing Shift Work

While shift work is required in police work, it can be disruptive, stressful, and unhealthy for officers. Particularly at risk are those officers who work through the night because being awake at night conflicts with the body's natural circadian rhythm. As Professor Vila explains,

> In most people, there tends to be a gradual decrease in alertness after 10 or 11 o'clock at night, hitting bottom between 3 and 6 a.m. From about 6 a.m. onward light rays from the sun trigger cells in your brain that promote a renewed cycle of alertness. The longer your shift is in darkness, the more at risk of fatigue you are. If you've been up for 12 hours, you're more at risk at 4 a.m. than if you've been up for 12 hours and it's 4 p.m.[24]

Therefore, with some consideration and planning, departments could structure shifts so night shift officers could start earlier and end sooner (and be in bed earlier). Even a minor improvement would be night shift officers working 10:00 p.m. to 6:00 a.m. instead of midnight to 8:00 a.m.

Day shift officers could then start at 6:00 a.m. and be done at 2:00 p.m. In addition, departments could use shifts of different lengths during the day (twelve hours) and night (eight hours). Scheduling night shift officers to no more than three night shifts in a row could also lessen the difficulties of working nights. As Vila notes, "After about 3 consecutive night shifts, you'll start to see a substantial problem and you need time off so you can catch up on your sleep."[25] He further notes that "all this may be a bit of a pain for administrators, but it's smart in terms of risk management. Departments will end up getting better work out of their people while keeping them safer."[26]

as cardiovascular disease, diabetes, high blood pressure, high blood sugar, obesity (due to poor diet, lack of exercise, and hormonal imbalances), depression and mood disorders, ulcers, problems with fertility and pregnancy, and even cancer.[27]

As for sleep deprivation and its relationship to job performance, studies have shown that fatigue contributes to officer irritability with the public and an inability to maintain calm in situations, and it also impairs physical and cognitive abilities. One study showed that officers with less sleep were more likely to associate people of color with weapons (implicit bias).[28] Another study showed that officers who worked longer shifts (13 hours versus 10 hours) had decreased hours of sleep, concentration, cognitive processing, and overall quality of life, and increased fatigue, reaction time, and citizen complaints.[29] No question about it, shift work can have significant effects on the well-being of police officers and the quality of policing.

? A RESEARCH QUESTION
What Is the Relationship Between Shift Work, Fatigue, and Gender?

A recent study[30] examined the relationship between shift work, fatigue, and gender among police officers. The authors discuss how fatigue may have many negative effects on one's personal and work life but that little research has examined how male and female officers may differ on this issue. Participants in the study were 308 officers from the Buffalo (New York) Police Department. Officers were asked a series of questions about their levels of fatigue, their primary work shift, and their background characteristics (gender, education, physical activity, etc.). The researchers found that men were least tired on the day shift and most tired on the afternoon shift. The opposite pattern was found among women: Female officers who worked the day shift were most tired, and those who worked the afternoon shift were the least tired. The authors suggest that this pattern may relate to the timing of childcare and family obligations for men and women and these demands may affect tiredness at work. The authors explain that their findings highlight the need for sleep education among officers as well as well-informed shift work management strategies in police departments.

THE EFFECTS OF STRESS

Many studies have examined the presumed health-related effects of stress among police officers. Sleep disorders and fatigue,[31] post-traumatic stress disorder,[32] depression,[33] alcohol abuse,[34] and early death[35] have all been examined. Other studies have focused on serious physical health problems among officers, including coronary artery disease and heart disease.[36]

Although there is a strong link between stress levels and negative health outcomes and police officers experience many of these negative health outcomes, it is not clear from the research if police officers differ from demographically similar people who work in other occupations in terms of these health issues. For example, some research shows police have higher rates of alcohol abuse than members of other occupational groups or the general population,[37] but other research suggests they do not.[38] Some studies show police officers die younger than members of other occupational groups,[39] yet other research does not find this pattern.[40] Some studies show high rates of hypertension among the police,[41] but others do not.[42] As noted, some of these conflicting findings are likely due to the research being conducted in departments with varied characteristics and among officers with different assignments and characteristics. In other words, what is true in one department may not be true in another. What might be true for patrol officers may not be true among supervisors. Further, comparisons across occupations do not necessarily take into account the differences among people in those occupations. For example, a comparison of police officers with elementary school teachers is not valid as it is really largely a comparison of men with women. Observed health-related differences between the two groups may have more to do with gender than the occupation.

A Question of Ethics

Nap Time?

Imagine you are a patrol officer in a small department and you are assigned to the night shift, working 11:00 p.m. to 7:00 a.m. There are only two other squad cars on the streets at this time. This is your fourth night in a row at work. It is now 4:00 a.m., it is completely dark, there are very few motorists on the streets, and the radio is silent—it has been over forty-five minutes since the last call came in. You are seriously struggling to stay awake while you are driving. You have written some parking tickets to have something to do, but that has not helped you feel any less tired. You begin to think about how great it would be to be a firefighter—if it is the middle of the night and they are not at a call, firefighters are sleeping at the fire station. With this in mind, you drive into a nearby park and park your squad car. You turn up the volume on the radio in case a call comes in, leave the car running, and fall asleep. You wake up after about twenty minutes feeling a little better; soon the sun will rise and you only have a couple of hours left on your shift. You then have three days off during which you can catch up on sleep. The question is, is there anything wrong with taking a little nap while on duty, as described here? Why or why not?

SUICIDE

An extreme response to stress is suicide. The prevailing research suggests that police officers are no more likely to commit suicide than demographically comparable people in other occupations,[43] although there are studies that report police suicide rates to be higher than those of the general population. Valid research on this issue is very difficult to conduct in that reliable data on suicides by occupation are not available, especially on a national scale. Also, as noted above, comparisons to the general population are unwarranted as the police officer population does not resemble the characteristics of the general population (e.g., the police occupation is disproportionate comprised of men, and men are more at risk of suicide than women). Further, it is hazardous to generalize to police officers in general based on an analysis of officers in only certain departments.

On rare occasions, police officers do commit ("complete") suicide; it has been reported that there are about 140 police suicides a year out of a police population of approximately 600,000. This total approximates the number of officers who are killed in the line of duty each year (see below). Most police suicides are committed with a firearm and occur while the officer is off duty at home.[44] Other research shows that the factors associated with suicide among police officers are similar to those associated with suicide among those who are not police: access to a firearm, alcohol or drug abuse, mood disorders, life event stressors, and a minimal support network.[45] One recent study

Photo 10.3
Working during the night can place significant physical demands on police officers and can result in negative physiological and job performance outcomes. As a result, the proper management of shift work is important. ©iStockphoto.com/kali9

involved an in-depth analysis of eight police officers who had completed suicide.[46] The study found that all of the officers had experienced risk factors for suicide completion prior to being hired, including mental health diagnoses, substance abuse, traumatic experiences, and life struggles.[47] This finding clearly points to the importance of psychological and drug testing for police officers prior to hiring. It also highlights the importance of intervention from coworkers and supervisors.[48]

BURNOUT

Burnout is another possible effect of stress. It has been defined as "a prolonged response to chronic and interpersonal stressors on the job."[49] Emotional exhaustion is often identified as a key element of burnout and is symptomatic of when individuals feel "overwhelmed and hopeless without the energy to cope or respond."[50] In some instances, stress and burnout are considered the same phenomenon; in others, stress is thought to cause burnout. Like stress, burnout is typically measured perceptually. For example, in one study,[51] burnout was measured based on how frequent certain feelings were experienced by the officer or the degree to which the officer agreed with certain statements, such as "I feel used up at the end of the workday," "I feel burned out from my work," "I feel frustrated by my job," "This job is hardening me emotionally," and "I have become more callous toward people since I took this job." Burnout among police officers has been attributed to greater work–life conflict, less social support, and greater perceptions of unfairness at work.[52] A recent study showed that about 19% of a national survey of 13,000 officers experienced severe levels of emotional exhaustion.[53] Females and African Americans have been found more likely to experience burnout.[54]

POST-TRAUMATIC STRESS DISORDER (PTSD)

While workplace stress and burnout are typically a result of ongoing conditions and circumstances, post-traumatic stress is more focused on a specific incident or incidents. It has been defined as "direct exposure to an event involving death, threatened death, serious personal injury or sexual violence."[55] The diagnosis is made on the basis of seven dimensions:

1. The person has experienced, witnessed, or been confronted with an event or events that involved actual or threatened death or serious injury, or a threat to the physical integrity of oneself or others.
2. The response to the event involved intense fear, helplessness, or horror.
3. The individual reexperiences the traumatic event.
4. The individual experiences avoidance/numbing symptoms.
5. The individual experiences symptoms of hyperarousal.
6. The symptoms persist for a defined period of time.
7. The symptoms result in significant distress or functional impairment.[56]

Many stressful events may also come together in memory to increase the risk of PTSD.[57] Nightmares or flashbacks are common with PTSD, as is loss of memory relating to the event. Sleep disorders and difficulty controlling emotions are also frequently experienced. Symptoms of PTSD may develop in some people but not others, even when faced with the same traumatic situation. What is traumatic for one person may not be as traumatic for another.

Estimates vary on the prevalence of PTSD in police officers and range from 7% to 19%, depending on the study.[58] PTSD is often found to be more common among police officers than among members of other occupations not typically exposed to traumatic situations. Whether a person will develop PTSD depends on many factors, including prior history of trauma, coping strategies, work hours, and social supports at work and at home.[59]

> burnout: A prolonged response to chronic and interpersonal stressors on the job.

Photo 10.4

PTSD is a serious risk for police officers who are exposed to threatening and traumatic events. Professional mental health counseling is imperative for the effective treatment of this disorder. AP Photo/Don Ryan

Numerous studies have associated PTSD with negative physical outcomes, such as heart disease, hypertension, stroke, diabetes, sleep problems, chronic fatigue, migraine headaches, gastrointestinal disorders, and autoimmune disorders. However, although associations have been established between PTSD and these outcomes, a causal relationship between many of these conditions and PTSD *has not* been established. For example, just because someone has PTSD and hypertension does not mean that the PTSD caused the hypertension.[60] Research has also found links between PTSD and depression, panic attacks, agoraphobia, social phobia, and substance abuse[61] as well as increased use of sick leave, more medical appointments, and generally poorer health.[62]

PTSD is a serious condition, and its treatment depends on expert care. Research speaks to the effectiveness of psychotherapy, cognitive behavior therapy in particular, which trains subjects basically to think differently about the traumatic experiences and foster a sense of control, safety, ability to cope, and anxiety reduction in subjects. Drug therapy with or without psychotherapy has also been used to control the symptoms of PTSD, as have other therapies.[63] It is imperative that affected officers seek and obtain professional counseling, but this is often easier said than done because of peer group pressure to avoid the stigma associated with obtaining mental health help.[64] In some departments, mental health questions could have implications for fitness for duty, preferred assignments, or promotions. It is extremely critical, however, that the police culture changes in order not just to allow but also encourage officers to seek psychological services, that the stigma associated with mental health services be dissolved.[65] It is only in this way that the effects of PTSD can be minimized and officer effectiveness maximized.

A Question to Consider 10.1

Stress, Counseling, and the Police Culture

All sorts of factors contribute to police officer stress, and all sorts of negative outcomes are associated with this stress. Police agencies can help their officers deal with the demands of the job by making professional and peer-counseling programs available. However, the police culture, which rewards "toughness" and discourages "weakness," may dissuade officers from participating in such programs even when they could do much good. The question is, how can the police culture be changed so officers feel more comfortable seeking help for their stressful experiences?

EARLY DEATH

Four studies have specifically examined the mortality rates and age at death of police officers in the United States. Each study was conducted in a single police agency. Three of the studies showed police officers die younger than a comparable group,[66] but one study found police officers live longer than the average person.[67] Clearly, there are inconsistent findings with regard to this issue. Once again, these findings may highlight the fact that officers who work in different departments may experience their jobs differently. Being a police officer in some places may be more hazardous than being a police officer in other places.

WHAT CAN MEDIATE THE EFFECTS OF STRESS?

Studies have examined factors that mediate stress, such as friends and social supports, as well as individual characteristics, such as time on the job and personality. A recent review of studies on police stress found that one of the most consistent findings in the research is that "increased levels of social support are consistently related to decreased levels of perceived stress."[68] In other words, having friends, family, and other social supports can lessen the perceptions of stress and perhaps even the negative effects of stress. In addition, studies have highlighted the value of formal peer-counseling and related programs to officers in need of such support.[69] For more serious mental health issues, such as PTSD, professional counseling and other services would be most appropriate (see Police Spotlight at the introduction to this chapter). Most departments need to pay more attention to the health and well-being of officers.

PHYSICAL HAZARDS OF POLICE WORK

Danger is an important dimension of the police job. Police officers are constantly vigilant for threats that may cause them harm. While the *potential* for harm to officers is constant, and there are certainly serious risks involved with being a police officer, is the police job actually *dangerous*? This is not an easy question to answer for at least two reasons. First, everything is relative. Danger can only be determined in relation to some comparison or standard. Second, it depends largely on how *danger* is defined. There are many possibilities. The dangers of the job may take the form of the following:

- Incidents in which officers are actually physically harmed. These situations would include the following:
  - Fatal Injuries
    - As a result of accidents and/or illnesses
    - As a result of homicides
  - Nonfatal Injuries
    - As a result of assaults
    - As a result of accidents and/or illnesses

- Incidents in which there is risk of harm to officers but no actual physical harm—for example, responding to "subject with gun" calls where a suspect is not located
- Incidents where there is unseen emotional or psychological harm to officers—for example, gruesome crime scenes, tragic accidents, and assaults on officers, as discussed earlier in this chapter

DEATHS ON THE JOB: ACCIDENTS AND HOMICIDES

If danger is defined exclusively in terms of fatal injuries on the job, the police occupation does *not* rank among the top ten most dangerous jobs. This is according to U.S. Bureau of Labor Statistics figures from 2017, the most recent year in which data are available (see Figure 10.1).

In nearly all of the occupations listed in Figure 10.1, deaths that occurred on the job were as a result of accidents, not homicides. If *danger* was defined only in terms of homicides, the police occupation would score quite high.

The FBI *Law Enforcement Officers Killed and Assaulted* (*LEOKA*) report is the definitive source on injuries and deaths to police officers in the United States. The report provides tallies of officers

FIGURE 10.1

Top Ten Most Dangerous Jobs in 2017

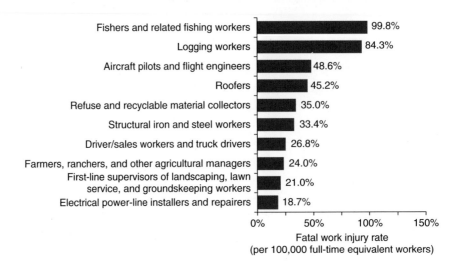

Source: Data from U.S. Department of Labor, Bureau of Labor Statistics, "2017 Census of Fatal Occupational Injuries."
Note: Does not include the military occupation.

FIGURE 10.2

Felonious and Accidental Police Officer Deaths, 2004–2018

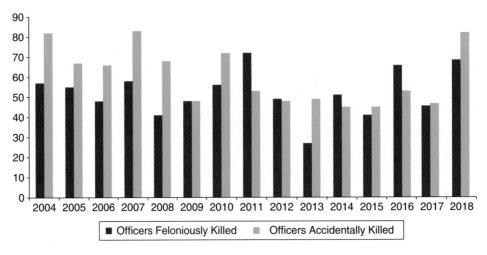

Source: FBI, *Law Enforcement Officers Killed and Assaulted*, https://ucr.fbi.gov/leoka/2017; 2018 statistics obtained from the Officer Down Memorial Page https://www.odmp.org/search/year/2018

killed per year and how those officers were killed. As seen in Figure 10.2, in most years, more officers are accidentally killed than are murdered. This might be surprising, as the media tend to pay much more attention to incidents in which officers are murdered compared to ones in which they are accidentally killed.

A seen in Figure 10.2, there was no clear trend in accidental deaths or murders from 2004 to 2018. The year 2011 had the highest number of *felonious* deaths of officers (seventy-two), and 2007 had the highest number of *accidental* deaths (eighty-three). Based on increased media attention alone, it may seem that more officers are being murdered in recent years. In particular, the tragic mass shooting that left five police officers dead in Dallas, Texas, in July 2016 was the deadliest day for police officers since September 11, 2001.

FIGURE 10.3

Police Officers Assaulted, 2004–2017

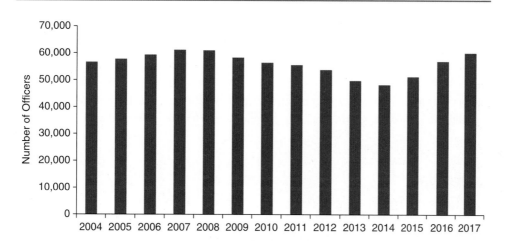

Source: Federal Bureau of Investigation, "FBI Releases 2015 Statistics on Law Enforcement Officers Killed and Assaulted," October 18, 2016, https://www.fbi.gov/news/pressrel/press-releases/fbi-releases-2015-statistics-on-law-enforcement-officers-killed-and-assaulted

INJURIES ON THE JOB: ACCIDENTS AND ASSAULTS

The FBI *LEOKA* report also tallies the number of **assaults** that occur among officers, as reported by agencies that submit data to the bureau (see Figure 10.3). According to the FBI,

> Law enforcement agencies report . . . the number of assaults resulting in injuries to their officers or instances in which an offender used a weapon that could have caused injury or death. Law enforcement agencies report other assaults (i.e., those not causing injury) if they involved more than verbal abuse or minor resistance to an arrest.[70] In 2017, twenty-nine percent of assaults involved injuries to officers.[71]

Trends in assaults on officers are clear from these data: Assaults on officers peaked in 2007 and 2008 and then began slowly to decline through 2014, then in 2015 through 2017 assaults on officers increased (2017 is the most recent year for which data are available).

assault: A physical attack.

The *LEOKA* assault data have been criticized for not being uniformly or accurately reported to the FBI. As a result, one should be cautious about the accuracy of patterns in the data. A study that used an alternative source of data on assaults of police officers, injuries treated at a national sample of hospital emergency departments from 2003 to 2014, showed that assault-related injury rates for officers increased from 2008 to 2012 and then decreased until 2014.[72]

A study that examined all injuries to officers showed assaults are the "tip of the iceberg" in terms of injuries to officers.[73] As shown in Figure 10.4, of all injury incidents reported by officers, approximately 10% occurred as a result of an assault on the officer (e.g., officer was intentionally hit, bit, shot, stabbed, kicked, or spat upon), approximately 35% occurred as a result of subject actions but did not constitute an assault (e.g., suspect resisting arrest, officer injured while chasing suspect on foot), and the remaining 55% occurred as a result of an accident (e.g., vehicle accident, slip and fall).[74]

In another study, all officer injury incidents that occurred over a period of six years (approximately 5,000) in one large police department were analyzed; among these incidents there was one fatality—the murder of an officer.[75] The most common injuries among officers were abrasions and lacerations. Serious injuries, such as broken bones, dislocated joints, torn ligaments/tendons, serious burns, gunshot wounds, knife wounds, and human bites, were relatively uncommon, accounting for approximately 5% of all injuries (Table 10.1). It was also found that most of the serious injuries, medical treatment, and time taken off work were due to accidents. Even the three gunshot wounds sustained by officers were the result of accidents. Clearly, when weighing the risks of the police job, accidents cannot be ignored.

It is important to remember incidents caused by the deliberate actions of another person may have both physiological consequences (physical injuries) *and* psychological or emotional ones. Therefore, one might expect injuries that occur as the result of one person doing harm to another are more psychologically significant than injuries that occur as the result of an accident. Indeed, as discussed earlier, research has identified "experiencing a physical attack" as a major

FIGURE 10.4

Causes of Injury Incidents to Police Officers

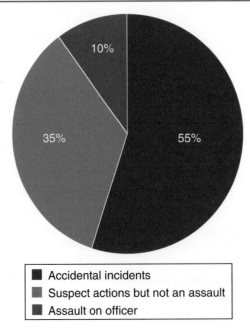

- ■ Accidental incidents
- ■ Suspect actions but not an assault
- ■ Assault on officer

TABLE 10.1

Types of Injuries Sustained by Police Officers[76]

Type of Injury	Percentage of All Injuries
Laceration/abrasion	19.4
Sprain/strain	18.9
Contact with infectious disease	18.3
Contusion	16.6
Broken bones/dislocated joints	2.7
Human bite	2.1
Gunshot wounds	.3
Burns	.2
Other	21.5

Source: Steven G. Brandl and Meghan S. Stroshine." 2012. The Physical Hazards of Police Work Revisited." *Police Quarterly* 15: 262–282.

stressor associated with police work. Thus, when considering the risks (or dangers) of the job, one must consider not only accidents and physical harm but also the psychological consequences of injury incidents.

SO IS POLICE WORK DANGEROUS?

It depends. Although this not a very satisfying answer, it is the truth. It depends on how danger is defined and the comparison used. At the very least, it is important to realize that risks to police officers come in many forms: homicides, assaults, fatal accidents, nonfatal accidents, and psychological. Workers in few other occupations face the potential for this wide array of risks. Certainly, to the extent policing is a dangerous occupation, it is not just because of murders and assaults, as these incidents are relatively uncommon. Training and equipment (such as body armor; see Technology on the Job feature) may reduce the actual dangers of the job; however, there is still constant *potential* for danger, and therefore the job may constantly *feel* dangerous.

A related question is whether police work is becoming more or less dangerous. Recent trends in FBI *LEOKA* data would tend to suggest the rates of murders, assaults, and accidental deaths of officers are generally on the decline over the long term, with some exceptions year to year, indicating perhaps police work is becoming less dangerous. As noted, the decline in injuries may have to do with increased training, use of less lethal technology, and the utilization of protective equipment.

RISKS OF POLICE WORK

There are several noteworthy tasks and activities of the police that increase the risk of injuries to officers, including making arrests of suspects, foot pursuits, vehicle accidents, and vehicle pursuits. These issues are discussed here.

ARRESTING SUSPECTS AND USING FORCE Police officers are at increased risk of sustaining injuries when making arrests, especially arrests of resisting suspects. When suspects resist arrest, almost always some type of force is used upon the suspect; suspect resistance and officer use of force are closely related. Of all injuries officers sustain, approximately 40% occur

Police Body Armor

Body-worn devices to prevent injury and death have been in existence for a long time—for example, gladiators wore metal chest plates to prevent being impaled by an opponent's sword or spear. Armor to prevent gunshot wounds has been a more recent development. As there are some misunderstandings about the capabilities of police **body armor**, clarification is provided here.

Police body armor is in the form of a vest. Most models are constructed of strong fibers that are designed to stop a bullet on contact. Sometimes body armor vests are referred to as Kevlar vests. Kevlar is one particular type of material that can be used in vests, although several other materials are also available. Body armor vests are also sometimes referred to as *bullet-proof* vests. It is more accurate to say body armor is *bullet resistant*. Different models of body armor have different ratings that refer to their effectiveness in stopping bullets fired from various firearms. Some vests are designed to only stop bullets from short barreled handguns, some are constructed to stop armor piercing rifle ammunition. Some vests are designed to be stab resistant, and are most often worn by correctional officers. High risk activities (e.g., hostage situations) may necessitate a higher level of protection from a vest than would be the case with regular patrol duty. The vests are designed to absorb and distribute the force of the bullet throughout the vest. Officers wearing vests when struck by gunfire often experience bruising, blunt trauma, or other related injuries at the point of contact. Once a vest has been struck by gunfire, it is no longer effective.

During the last thirty years, the lives of more than 3,000 police officers have been saved as a result of body armor.[77] However, it is still possible to sustain serious or fatal gunshot wounds when wearing bullet resistant body armor. For instance, in 2018, 51 officers were feloniously shot and

Photo 10.6
Equipment such as bullet-resistant vests can help mitigate some of the most serious physical risks of police work, but it cannot eliminate them.
John Moore/Getty Images

killed. Of these 51 officers, 43 (84%) were wearing body armor at the time. The year 2018 was not dissimilar to other recent years in this regard: During the past ten years, 72% of officers shot and killed were wearing vests at the time. Most of these officers (56%) sustained gunshot wounds to the head. Other officers were also shot in areas not protected by the vest (neck, upper torso, armpit, lower torso).[78] According to the FBI, in 2017 there were 268 officers who were non-fatally injured with a firearm.[79] It is unknown how or to what extent body armor saved the lives of these officers.

Most police departments today require officers to wear bullet-resistant vests. The most common reasons expressed by officers for *not* wearing a vest is that it is uncomfortable.[80] Most body armor vests cost between $200 and $400. Often federal funds are available to cover the cost of at least some of the vest as long as the agency has policy that requires uniformed officers to wear vests while on duty.

---

body armor: Protective vest worn by officers to help prevent injury or death, particularly from gunshot wounds.

in arrest situations,[81] and nearly all assaults on officers occur when officers are making arrests.[82] However, to keep things in perspective, it is important to understand that only about 3% of arrests the police make involve force, and only in about 20% of the situations where force is used is an officer injured.[83] In other words, a large proportion of injures to officers occur in arrest situations, but officers actually being injured in arrest situations is quite uncommon.

Injury to officers (and to subjects, as discussed in Chapter 11) is highly dependent on the type of force that officers use. Specifically, officers (and subjects) are most likely to be injured when officers use bodily force, such as decentralizations and focused strikes and kicks, to overcome the resistance of suspects.[84] Officers are least likely to be injured when using weapon-based force, such as electronic control weapons (ECWs; e.g., Tasers).[85] Weapons like Tasers allow the police to subdue resisting suspects while maintaining a safe distance from them. While weapon-based technologies have reduced the risks of injuries to officers, police use of bodily force is still the most common type of physical force used by police officers today.[86]

FOOT PURSUITS Very little research has been conducted on police foot pursuits, but the studies that have been done suggest foot pursuits are risky for both officers and for the subjects being chased. In six years of injury incidents analyzed in a large police department, nearly 5,000 injury incidents were reported by officers, 13% of those injuries occurred as a result of foot pursuits. One incident involved the death of an officer shot and killed while engaged in a foot pursuit of a suspect.[87] Officer use of force is also associated with foot pursuits.[88] It is not known what proportion of all foot pursuits actually result in injuries to officers or subjects.

It is a balancing act to weigh the need for officer (and suspect) safety with the need for criminal apprehension. While serious injury as a result of a foot pursuit is a relatively uncommon event, it does happen. As a result, some departments have developed policies and training with regard to foot pursuits. For example, in the Dallas Police Department, officers are directed not to initiate or continue a foot pursuit if they are acting alone and would be chasing two or more suspects at the same time, if they lose their weapon, or if they and other officers split up to chase multiple suspects. Other police department policies require officers to terminate a foot pursuit if the subject enters a building.[89] The intent of this provision is to prevent an opportunity for suspects to ambush officers. Research has also called attention to the risks confronted by officers as a result of engaging in foot pursuits at night with low visibility, running on uneven terrain, and climbing fences, especially chain link fences.[90]

VEHICLE ACCIDENTS AND VEHICLE PURSUITS Police officers, especially those who are assigned to patrol, spend much of their time in motor vehicles, usually squad cars. Therefore, it is not surprising police officers are involved in motor vehicle accidents at a higher rate than other U.S. workers. Indeed, research has shown the fatality rate for police officers and sheriff's deputies in transportation-related incidents is four times greater and the injury rate seventeen times greater than the rate for all other U.S. workers.[91] In addition, of all injuries police officers sustain, approximately 9% are due to vehicle accidents,[92] and most accidental police fatalities are the result of motor vehicle accidents.[93] While other transportation workers, such as truck drivers, are also involved in vehicle accidents at a high rate, the police are at increased risk because they also operate vehicles in emergency situations and have other tasks to attend to while driving.

In spite of the risk vehicle accidents pose to police officers, not until recently has research been conducted on the nature, extent, and causes of them. Most of these studies have been conducted on high-speed pursuits. However, only a small fraction of all police accidents result from police pursuits—approximately 5%.[94] Therefore, it is problematic to draw conclusions about *police vehicle accidents* based on an analysis of *police pursuit accidents*. Furthermore, although vehicle pursuits are risky, most often harm comes to the fleeing motorists and innocent bystanders, not to police officers: Of all people killed as a result of police vehicle pursuits, about 1% are police officers.[95]

POLICE VEHICLE ACCIDENTS There is no national reporting system of police vehicle accidents. Most of the available studies on police vehicle accidents simply describe the accidents and, unfortunately, include only a subset of all accidents (e.g., they include only accidents that result in injury and/or only accidents from one state). The common characteristics of the accidents in these studies included the following:

- The most common police vehicle accidents involved a police car being rear-ended by another car.[96]
- 11% of all police vehicle accidents resulted in some injuries to officers, and 31% of "lights and siren" accidents resulted in injuries to officers.[97] However, the overwhelming majority of vehicle accidents occurred while officers were on routine patrol; only 8% of all crashes involved police vehicles that were traveling with lights and siren. Therefore, although lights and siren crashes were much more likely to result in officer injuries, they were relatively uncommon.[98]
- In multiple-vehicle collisions, citizens were usually at fault; in single-vehicle collisions, officers were usually at fault.[99]
- In all types of accidents, sprains, strains, or soft tissue tears were the most common types of injuries reported, accounting for approximately 50% of all injuries. The most common body parts injured were the neck, back, and leg.[100] The most common cause of injuries was striking something within the police vehicle, most commonly the mobile data terminal.[101]
- Although only 6% of accidents involved police motorcycles, police motorcycle accidents were significantly more likely than car crashes to result in officer injuries or fatalities.[102]
- Officers wearing seatbelts at the time of the accident were less likely to sustain serious or fatal injuries.[103]
- Officers were more likely to be injured when alone in the vehicle compared to when there was another officer in the vehicle; this may relate to officer distraction.[104]
- Finally, when the officer vehicle was at fault in a crash, driver distraction was the most common reason, and the primary source of distraction was use of the mobile data terminal.[105] Other research has found police officers drive significantly worse while distracted; distraction leads to greater lane deviations, more instances of unintentionally leaving the driving lane, and slower reaction time for breaking, all of which increase the likelihood of collisions.[106]

In summary, these studies call attention to the importance of seatbelt use, the likelihood of injuries in crashes involving emergency calls (lights and siren), the role of officer distraction in causing accidents, and the risks of motorcycle accidents.

POLICE VEHICLE PURSUITS A motor vehicle pursuit can be defined as

an event involving one or more law enforcement officers attempting to apprehend a suspected or actual violator of the law in a motor vehicle while the driver is using evasive tactics, such as high speed driving, driving off a highway, turning suddenly, or driving in a legal manner but failing to yield to the officer's signal to stop.[107]

On the one hand, of course the police have a legitimate interest in apprehending law violators, even when those violators seek to escape capture. On the other hand, vehicle pursuits create potentially dangerous circumstances for not only the fleeing subjects but also for the police and innocent bystanders. Most pursuits end without incident, but some result in crashes, injuries, and property damage. Research shows fleeing vehicles are more likely to crash than are the involved police vehicles, and the fleeing subjects and innocent bystanders are more likely than police officers to be injured in crashes. Due to the serious potential consequences of police vehicle pursuits, substantial research has been conducted on how best to manage such pursuits and develop effective policies to reduce their frequency.

Many issues are relevant when considering police vehicle pursuits, and many factors need to be considered in judging the reasonableness of the pursuit. One of the most important considerations is the reason for the pursuit. Clearly, a pursuit of an armed robber may be better justified than a pursuit of a subject who made an illegal turn. Many other factors need to be considered when deciding to initiate, or terminate, a pursuit, including the following:

- Public safety
- Officer safety
- Passenger in officer's vehicle (e.g., citizen, witness, prisoner)
- Pedestrian and vehicular traffic patterns and volume
- Other persons in or on pursued vehicle (e.g., passengers, co-offenders, hostages)
- Location of pursuit (e.g., school zone, playground, residential neighborhood, downtown)
- Time of day
- Speed of fleeing suspect
- Weather, viability, and road conditions
- Identity of the offenders (if known)/can offender be located at a later time?
- Capabilities of law enforcement vehicle(s) (e.g., speed, maneuverability)
- Ability of officer(s) driving
- Availability of additional resources
- Whether supervisory approval is required
- Officer's/supervisor's familiarity with the area of the pursuit
- Quality of radio communication (e.g., out of range, garbled, none)[108]

Photo 10.7

Research shows that police pursuits are risky especially to the fleeing motorist and innocent bystanders. Adwo/Alamy Stock Photo

Other important considerations relate to the number of law enforcement vehicles involved in the pursuit. The more units involved, the more likely a collision. Policy should also specify the number and types of police vehicles authorized to engage in a particular pursuit.[109] It is also important to consider (a) the role and responsibilities of each police unit involved in the pursuit, (b) the need and nature of communication among and between units involved in the pursuit and the communications center, (c) supervisory responsibilities in monitoring the pursuit, and (d) procedures relating to interjurisdictional coordination of pursuits, among other factors.[110]

Based on the research on police vehicle pursuits, some facts about them include the following:

- Research findings often vary based on the jurisdiction being studied and the pursuit policies in place in that jurisdiction.
- The trend is for police departments to restrict, not expand, officer discretion in pursuit situations, becoming more common are policies that restrict pursuits to violent crimes only.
- There is little research evidence that more restrictive vehicle pursuit policies will lead to more offenders fleeing, cause crime rates to increase, or cause clearance rates to decrease.
- Most vehicle pursuits begin as the result of a traffic violation.
- Studies that have asked offenders the reasons why they fled the police have found the most common reason was that the car the suspect was driving was stolen. As a side note, police technology, such as automated license plate readers (LPRs), may increase the

likelihood officers will identify stolen vehicles while they are being driven by offenders. Thus, LPRs could potentially lead to increased opportunity for vehicle pursuits.

- Once a pursuit begins, most offenders continue to flee until the police terminate the pursuit or the offender crashes his or her vehicle.
- Approximately 30% of reported pursuits result in a crash, although estimates vary by study. When a pursuit involves a crash, it is most often of the fleeing offender's vehicle. Approximately 30% of vehicle crashes result in injury to one or more persons, most often the fleeing offender(s) or bystander(s).
- Being young, male, intoxicated, and without a driver's license increases the odds of engaging the police in a pursuit.
- The use of police helicopters reduces the likelihood of a pursuit-related crash.
- Emerging technology can reduce the need for pursuits yet allow the police to make apprehensions. For example, StarChase technology involves firing a projectile with a GPS tracker onto a fleeing vehicle. The technology then allows the police to identify the location of the vehicle and offender in real time without the need for a pursuit.[111]

Police officers are confronted with many potential health and safety issues, such as stressful work environments, assaults, foot pursuits, and vehicle accidents. This chapter has provided details regarding these potential hazards. As discussed, technology and well-informed policy may reduce many of these risks.

Main Points

- Stress occurs when a person experiences any physiological and/or psychological demand that requires that person to adjust to that demand. Most studies have measured stress perceptually using an array of questions or statements posed to the respondent.
- The majority of recent studies on police stress have found organizational stressors, such as inadequate support from supervisors and conflict with coworkers, to be the most frequent stressors for officers. Research has also shown stressors and their consequences are not equally experienced by all officers, or even by all officers in the same department.
- Shift work can be stressful for officers; each shift is stressful for its own reasons. Fatigue is associated with shift work and can have negative health and job performance consequences for officers.
- Although there is a strong link between stress levels and negative health outcomes, and police officers experience many of these negative health outcomes, it is not clear from the research if police officers differ from people who work in other occupations in terms of these health issues.
- When studying the causes and consequences of stress, it is important to realize there may be important differences between officers with different assignments and characteristics and between departments in different settings.
- The prevailing research suggests police officers are no more likely to commit suicide than demographically comparable people in other occupations. Factors associated with suicide among police officers are similar to those associated with suicide among other occupations.
- Burnout is a possible effect of stress. Burnout is a prolonged response to chronic and interpersonal stressors on the job. Emotional exhaustion is a symptom of burnout.
- Post-traumatic stress is defined as a response to an extraordinarily threatening or traumatic event. Estimates vary on the prevalence of PTSD in police officers.
- Having friends, family, and other social supports can lessen the perceptions of stress and perhaps even lessen the negative effects of stress.
- The danger of the police occupation is difficult to determine. Danger can only be determined in relation to some comparison or standard. Police work often feels dangerous. The risks to police officers come in many forms: homicides, assaults, fatal accidents, and nonfatal accidents. Training and equipment may reduce the actual dangers of the job.

- During recent years, more officers have been killed as a result of accidents than because they were murdered.
- While there is little research on the issue, the studies that have been done suggest foot pursuits are risky for officers and for the subjects being chased.
- Police vehicle accidents are a significant risk for police officers. Research has shown the fatality rate for police officers and sheriff's deputies in transportation-related incidents is much higher than the rate for all other U.S. workers.
- While only a small fraction of police vehicle accidents result from pursuits, pursuits create potentially dangerous circumstances for not only the fleeing subjects and innocent bystanders but also for police officers.

Important Terms

Review key terms with eFlashcards at **edge.sagepub.com/brandl2e.**

assault   217
body armor   220
burnout   213
circadian rhythm   210

physiological response   206
post-traumatic stress disorder (PTSD)   209
sexual harassment   208

Questions for Discussion and Review

Take a practice quiz at **edge.sagepub.com/brandl2e.**

1. What is police stress? How is it different from burnout? How can each of these conditions be prevented and/or treated?
2. What are the most frequent workplace problems and other features of police work identified by officers as being stressful?
3. Why is shift work potentially stressful? What can be done to help officers reduce the negative consequences of shift work?
4. There are many inconsistent research findings regarding the extent and consequences of stress among police officers. What might help explain these different findings?
5. What is post-traumatic stress disorder and what can be done to help officers who have symptoms of it?
6. Is police work dangerous? Why or why not?
7. Why should accidental deaths and injuries to police officers be a major consideration when drawing conclusions about the risks of the police occupation?
8. How can the risks of police vehicle pursuits be reduced?
9. What are the research findings with regard to the characteristics and patterns of police vehicle accidents?
10. What are the research findings with regard to the characteristics and patterns of police vehicle pursuits?

Fact or Fiction Answers

1. Fact
2. Fiction
3. Fiction
4. Fiction
5. Fiction
6. Fact
7. Fact
8. Fact
9. Fact
10. Fact

$SAGE edge™

Digital Resources

Get the tools you need to sharpen your study skills. SAGE Edge offers a robust online environment featuring an impressive array of free tools and resources.

Access practice quizzes, eFlashcards, video, and multimedia at **edge.sagepub.com/brandl2e.**

Media Library

View these videos and more in the interactive eBook version of this text!

Career Video
10.1: Identifying and Handling Police Stress and Managing Relationships With Family and Friends

SAGE News Clip
10.1: Dallas Police Chief: "We Are Hurting"

PRACTICE AND APPLY WHAT YOU'VE LEARNED

▶ edge.sagepub.com/brandl2e

⑤SAGE edge™

WANT A BETTER GRADE ON YOUR NEXT TEST?

Head to the study site where you'll find:

- **eFlashcards** to strengthen your understanding of key terms.

- **Practice quizzes** to test your comprehension of key concepts.

- **Videos and multimedia content** to enhance your exploration of key topics.

11
POLICE USE OF FORCE

Police Spotlight: De-escalation Versus Use of Force

The Seattle (Washington) Police Department and other departments across the country are in the midst of a fundamental but controversial shift in how officers are supposed to handle potentially violent situations. The old-school way of thinking is that the police job is to make arrests, the more arrests the better. Arrests were the answer regardless of the question. Unfortunately, it is during arrests that force is most likely to be used, so an arrest can actually make a situation violent, where police and citizens literally fight each other. Charles Ramsey, former police commissioner in Philadelphia, asks important questions related to the value of this approach: "What is the collateral damage after that policing strategy? Have we alienated people? Yeah . . . but if you've alienated people, have you served your purpose?"[2]

A different approach to dealing with citizens in potential arrest situations can reduce the frequency of force and help improve citizen's trust in the police. This approach is known as de-escalation. The de-escalation approach to interactions with citizens is not complicated, but it does require a change in the traditional way of doing things in police departments. As discussed in more detail later in this chapter, de-escalation involves the use of various techniques to defuse heated situations. For example, in the Seattle PD, officers are trained to talk and behave calmly even with angry people, to ask questions instead of issuing commands, paraphrase what a person said in order to demonstrate to the person that he or she is being heard, and to make statements to the citizen that show empathy. The goal is to resolve the situation without having to make an arrest or use force. Kathleen O'Toole, a former Seattle police chief, explains, "I was trained to fight the war on crime, and we were measured by the number of arrests we made . . . But over time I realized that policing went well beyond that, and we are really making an effort here to engage with people, not just enforce the law."[3] De-escalation has its critics. Some officers argue that the approach is dangerous for officers, that sometimes the use of force is a preferred and most effective way of handling a situation, and that the approach is just political correctness in the extreme. Nevertheless, it is difficult to argue against an approach that seeks to make policing a little more civil and a little less controversial.

Objectives

After reading this chapter you will be able to

11.1 Define reasonable force, unnecessary force, and brutality in relation to the continuum of force

11.2 Discuss the meaning and importance of threat assessment in situations where officers make decisions about the use of force

11.3 Identify the various types of force that can be used by the police, the circumstances in which they can be used, their frequency, and their effects

11.4 Explain the phenomenon of suicide by cop, how frequently it occurs, and the characteristics of such incidents

11.5 Identify the factors that lead to police use of force

11.6 Explain how police use of force may be controlled

Fact or Fiction

To assess your knowledge of police use of force prior to reading this chapter, identify each of the following statements as fact or fiction. (See page 250 at the end of this chapter for answers.)

1. The continuum of force consists of specific instructions that eliminate police discretion in use of force situations.

2. By federal law, every police department in the United States operates under the same continuum of force policy.

3. The twenty-one-foot rule eliminates police discretion in deadly force situations.

4. Bodily force is the most common type of physical force used by police officers but also is most likely to lead to officer and subject injuries.

5. Data show that most people who are shot and killed by the police are male and black.

6. The phenomenon known as *suicide by cop* is basically a media construction; in reality, it does not actually exist.

(Continued)

(Continued)

7. If de-escalation is effective, then there is likely to be fewer instances of police use of force compared to when de-escalation is not used.

8. Research has shown that subjects' risk of death in a Taser-related use of force incident is less than 0.25%.

9. Early intervention systems are designed to identify officers that represent a potential problem and correct their behavior before they engage in more serious conduct.

10. The best predictor of how often officers use force is the number of arrests that they make.

AS discussed in Chapter 4, the ability to use force is a fundamental responsibility of the police; to use force only when necessary is a major legal and moral obligation of the police. The right to use force can be thought of as a powerful tool of the police: With this ability, officers can intervene in situations in which order needs to be maintained and offenders need to be arrested. However, it is also the use of force that makes the police inherently controversial.[1] Simply stated, the police protect the rights of people by limiting the actions of other people, and in order to limit the actions of other people, sometimes force must be used.

Because of the central role that force plays in policing, the use of force by officers may influence citizens' attitudes and behaviors toward the police, and the misuse of force may erode confidence in the police and the entire criminal justice system. Even when it is necessary and justified, the use of force never photographs well. In particular, as seen in numerous well-publicized incidents, when it appears police use of **deadly force** is racially motivated, this can lead to riots and community destruction.

There is no doubt that an understanding of the complexities, controversies, and dilemmas of police work depends on an understanding of police use of force. This chapter is devoted to this critical issue.

REASONABLE FORCE AND USE OF FORCE GUIDELINES

Although the police are authorized to use force and can use it in virtually any situation, they are legally not allowed to use force in an arbitrary or excessive manner. As explained by the U.S. Supreme Court in *Graham v. Conner* (1989), force must be "objectively reason-

Photo 11.1
A police officer uses OC spray on protesters at the University of California, Davis. Justified or not, the use of force by police never photographs well.
Wayne Tilcock/The Enterprise via AP, File

deadly force: Forms of force used by the police that have a high likelihood of resulting in the death of a subject.

able in view of all the facts and circumstances of each particular case."[4] Accordingly, the obvious question is, what makes force "objectively reasonable"? To define **reasonable force**, most police departments provide guidelines for officers in the form of a **continuum of force**.[5]

While there are many variations in continua of force across police departments,[6] and there are major limitations to the continuum of force principle, the basic idea is the police can and should only use as much force as necessary to overcome the resistance offered by the subject (see Figure 11.1 for an example of a police department's continuum of force). In this sense, the use of force by the police is a stepwise progression in which the officer uses the next more severe form of force only if the previous form of force did not induce subject compliance. In a typical force continuum, the first step is the mere presence of the police. In many situations, when order needs to be restored or subjects controlled, the simple presence of the police at the scene is enough to resolve the situation. If mere presence of the police does work, then verbalization would be appropriate. The characteristics and tone of the verbalization may depend on the nature of the situation. Verbalization can take several different forms:

- Search talk (e.g., "Good afternoon, sir. May I please see your driver's license?")
- Persuasion (e.g., "Sir, would you mind stepping over here?")
- Light control talk (e.g., "Show me your hands.")
- Heavy control talk (e.g., "Drop the knife or I will shoot!")

Sometimes "stun" words are used in an attempt to gain the attention of the subject and to emphasize the seriousness of the command. Stun words are typically swear words.

If verbalization does not induce compliance, then the use of hands-on techniques is often the next step on the continuum. These techniques can include putting hands on the subject or using escort holds or compliance holds to control subject movements. If these actions are not effective and resistance continues or escalates, then **bodily force** would be appropriate. Bodily force includes vertical stuns (moving the subject forcefully into a vertical structure such as a wall), takedowns (decentralizations), and punches or kicks (focused strikes). If bodily force is ineffective at inducing compliance, then impact weapons would be appropriate. **Impact weapons** include batons; flashlights; chemical sprays (e.g., OC spray, otherwise known as pepper spray); and conducted energy weapons (CEWs; e.g., Tasers). Finally, the last option in the continuum of force is the use of deadly force, which usually involves the use of a firearm.

VARIATIONS AND LIMITATIONS OF THE CONTINUUM OF FORCE

As noted above, different police departments have different guidelines for continua of force. One study found 123 different variations of

> **reasonable force:** The minimum amount of force necessary to overcome the resistance offered by the subject.
>
> **continuum of force:** The principle that police can and should only use as much force as necessary to overcome the resistance offered by the subject.
>
> **bodily force:** Force that involves physical restraining maneuvers, such as vertical stuns, takedowns, and punches or kicks.
>
> **impact weapons:** Weapons used to induce compliance; they include batons, flashlights, chemical sprays, and Tasers.

FIGURE 11.1

Example of a Use of Force Continuum

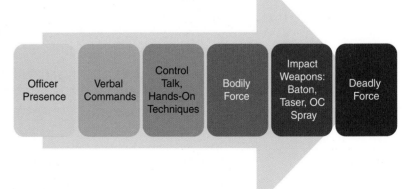

| Officer Presence | Verbal Commands | Control Talk, Hands-On Techniques | Bodily Force | Impact Weapons: Baton, Taser, OC Spray | Deadly Force |

continua in a survey of 662 police and sheriff's agencies.[7] Much of the variation relates to weapons such as OC spray and Tasers, which are less likely to be lethal. While OC spray and Tasers are usually placed at the same level in force continua,[8] there is little agreement between and among departments at which level this should be. In some departments, they are placed at the lower end of the continuum, which authorizes their use against passive resisters, but other departments place them closer to lethal force on the continuum, thus authorizing their use only against active resisters. Where OC spray and Tasers are located on the continuum of force matters when understanding the circumstances in which these weapons are used[9] and their effectiveness in inducing compliance among subjects.[10] Some police departments do not equip their officers with Tasers, and in some departments that do have them, not all officers are authorized to carry them. Obviously, the weapon is not even an option in a continuum of force if it is not available.

There are at least three major limitations to using continuum of force as a guide for police action. First, although the idea of a continuum is to apply just as much force as necessary to overcome **subject resistance**, most agencies do not actually incorporate elements of citizen resistance in the continuum of force.[11] In other words, it is not clear what type of force is appropriate given different forms of resistance, only that some forms of force are less severe than other forms. For example, it is not clear from the continuum of force itself if it is appropriate to use a Taser on a subject who continues to offer verbal resistance, or if it is acceptable to use a baton on a subject who is resisting being handcuffed. When citizen resistance is part of the continuum of force (see Table 11.1), most agencies make distinctions between (a) passive resistance, (b) verbal resistance, (c) physically defensive resistance, (d) physically active or aggressive resistance, and (e) deadly resistance (see Table 11.2).

Police agencies may not include subject resistance as part of the force continuum because there may be circumstances in which it would be necessary to use more force against a subject who is presenting less resistance. For example, if an officer encounters a subject who is making threats (verbal resistance), hands-on or even intermediate weapons may be most appropriately used, even though the force continuum may indicate that verbal commands and **control talk** would be justified. Even though a higher level of force may actually be appropriate given the circumstances of the situation, it would not be congruent with the force continuum and therefore could be judged as unnecessary force according to department policy. Less specificity allows for more flexibility and discretion.

subject resistance: Subject resistance to police; can involve (a) passive resistance, (b) verbal resistance, (c) physically defensive resistance, (d) physically active or aggressive resistance, and/or (e) deadly resistance.

control talk: A police officer's verbalization to induce compliance from a subject.

TABLE 11.1

Example of a Use of Force Continuum That Incorporates Subject Resistance

Subject Resistance	Officer Use of Force
No resistance	Officer presence
Verbal resistance	Verbal commands, control talk
Passive resistance	Verbal commands, control talk, hands-on techniques
Defensive resistance	Hands-on techniques, intermediate weapons: baton, Taser, OC spray
Aggressive resistance	Intermediate weapons, intensified techniques
Deadly resistance	Deadly force

TABLE 11.2

Subject Resistance Defined

Type of Subject Resistance	Examples of Subject Behavior
No resistance	Compliant
Verbal resistance	Verbally rejecting verbal communication or direction
Passive resistance	Unresponsive to verbal commands or directions
Defensive resistance	Efforts to evade police attempts at control; fleeing, resisting
Aggressive resistance	Attacking or striking officer, preparing to fight officer
Deadly resistance	Attempted or actual attacks on officer or another citizen that could cause death

The second limitation is that many factors need to be considered in assessing the threat posed by a subject, but these factors are not explicitly incorporated into force continua. These factors may include the following:

- Age: A significant age difference between the officer and the subject may affect the officer's threat assessment. A twenty-five-year-old officer facing a nine-year-old child would probably assess the threat differently than if he or she were facing another adult.
- Size: If the subject is much bigger or smaller than the officer, it will change the threat assessment.
- Strength: If the subject is much stronger or weaker than the officer, it will change the threat assessment.
- Skill level: If the subject is a skilled fighter, that fact may change the threat assessment (if the officer is aware of it). On the other hand, if the officer is highly skilled, that will also affect the threat assessment.[12]
- Availability of backup: If backup is far away or not available, the threat assessment may be higher than if immediate help is available.
- Degree of stabilization: Stabilization refers to the subject's ability to use force. An unrestrained, standing subject is in a much better position to be able to deliver force against an officer than a subject who is prone on the ground and in handcuffs. However, an officer should never assume that just because a subject is restrained he or she no longer poses a threat.[13]
- Weapon: If the subject is known to be in possession of a weapon or even suspected of being in possession of a weapon, the threat assessment may be substantially different.

A third limitation is that use of force situations often unfold in a matter of seconds, making a stepwise progression of force unrealistic. For example, research shows that most of the time the police use deadly force it is within two minutes of arriving at the scene.[14] A continuum appears neat and orderly on paper, but use of force incidents are seldom neat or orderly. Officers may find themselves in situations where their mere presence escalates events to the use of impact techniques or deadly force very quickly.

It must be understood a continuum of force is simply a general guideline; it does not specify or dictate appropriate officer action in every use of force situation. Training and educating officers on the appropriate interpretation and application of the continuum of force, along with appropriate **threat assessment**, is necessary. Even then officers must use their discretion in determining the appropriate type of force to be used.

threat assessment: An officer's assessment of the degree of danger posed by a subject; factors involved in the assessment can be physical or situational.

Photo 11.2

Justified use of force by the police depends on the amount of resistance offered by the subject and the overall threat presented by the subject. Antonio Perez/Chicago Tribune/TNS via Getty Images

THE TWENTY-ONE-FOOT RULE AND ITS LIMITATIONS

Another guideline sometimes associated with police use of force is the **twenty-one-foot rule,**[15] also known as *edged-weapon defense*. This concept was created thirty years ago but is still discussed and/or trained in some police departments today. The concept should not be considered a rule but a guideline at best. This principle holds a person can pose a significant threat to a police officer when that person is within a twenty-one-foot boundary of the officer; this is particularly true for someone who possesses a knife. (Obviously, this principle does not apply to a subject with a firearm as a firearm can present a threat to an officer well beyond twenty-one feet.) The reasoning holds that it takes 1.5 to 2.0 seconds for a police officer to unholster, aim, and fire a gun. In that amount of time, a person with a knife can cover about twenty-one feet and could attack the officer.

There are several problematic issues with this principle. Overall, its validity depends on many variables and, as such, is subject to misinterpretation. For example, it does not explicitly take into account any other important characteristics of the situation besides the distance between the subject and the officer. A *potential* threat does not mean the subject is an *actual* threat. Clearly, not every subject who is within twenty-one feet of an officer poses grave harm to the officer—for example, when a subject is moving away from the officer and no other people are in the area. A subject may not be an actual threat until the subject faces the officer or movements are made toward the officer. The actions of the subject need to be considered. Also, as presented, the principle does not take into account the characteristics of the subject. Is the subject intoxicated, or is the subject coherent and verbally threatening? Is the subject a seven-year-old child, an eighty-five-year-old lady, or a twenty-four-year-old man? It also does not take into account the capabilities of the officer or the environmental conditions of the situation. For instance, a subject who is on ice and snow may not present the same threat as someone on a dry surface. Another obvious limitation of the principle is the judgment as to the actual distance between the subject and the officer. Finally, the twenty-one-foot rule does not necessarily indicate when a firearm should be used; it may be appropriate for an officer to unholster the firearm, but actually shooting may not be justified. Similarly, other, less lethal weapons could be appropriately deployed when the situation allows for it. The bottom line is the twenty-one-foot rule should be carefully interpreted as a threat assessment technique.

twenty-one-foot rule: A guideline used by some departments stating a subject can pose a significant threat to a police officer when that person is within a twenty-one-foot boundary of the officer.

Making Use of Force Incidents More Transparent

Police departments are sometimes criticized and distrusted by community residents because they do not share important information with citizens. Citizens may interpret this unwillingness to share information as mismanagement or the police hiding misconduct. Indeed, sometimes police departments are reluctant to share information with the public for fear citizens may criticize police actions and operations due to misinterpretation of the information or otherwise. Nowhere is this more the case than when it comes to releasing information about citizen complaints against police officers (and the disposition of those complaints; see Chapter 12) and police use of force. However, some police departments routinely collect, analyze, and make available such information for public consumption. Arguably, this openness leads to a greater degree of accountability for departments and their officers, which is a good thing.

For example, the New York City Police Department publishes an annual report that provides descriptive information on each incident in which an NYPD officer used force (e.g., used a Taser, discharged a firearm, used other physical force).[16] The year 2016 was the first year a report on all use of force incidents was made available to the public; incidents of police firearm force have been reported since 2007. Among many other details, in 2017 the department reported that there were 23 incidents where NYPD officers shot

(or shot at) citizens. In 19 of these 23 incidents, a subject was struck by gunfire. Of the 19 subjects who were shot, nine subjects were injured and ten subjects were killed. Although this information is not necessarily flattering for the police department, its release by the police shows that the public deserves to know this important information about police activities.

Interestingly and as discussed later in this chapter, from 2003 to 2009 the federal government attempted to collect data from local agencies on "arrest related deaths" of citizens but stopped because agency submission of data was voluntary and incomplete as a result. In 2015, the U.S. Department of Justice announced a program whereby these statistics would be collected via media reports and disseminated through the Bureau of Justice Statistics.[17] Media reports were to be used as voluntary submission of data was problematic. The initiative was expected to be fully operational at the start of 2016, but the project never materialized. In 2018, the FBI announced that it had developed a different system to provide information about police use of force on a national scale. Beginning in January 2019, law enforcement agencies can report use of force data to the FBI via a web portal. Submission of data to the FBI is voluntary.[18] Meanwhile, in recent years, citizen deaths by police have been tallied and reported by various news outlets.

DEVIATIONS IN USE OF FORCE: UNNECESSARY FORCE VERSUS BRUTALITY

When force is used beyond what is prescribed in the continuum of force, it is deemed unjustified, unnecessary, or excessive, or sometimes it is referred to simply as police brutality. However, because use of force situations are often complex and use of force continua do not specify the exact actions officers should take in each use of force situation, the determination of what is actually "unnecessary" or "excessive" is often a matter of judgment. A situation may be viewed differently by different people: The subject upon whom force was used may see the use of force differently than the officer who actually used the force. Citizens, other officers, police administrators, the media, the prosecuting attorney, the judge, and the jury may all have different perspectives on whether the force used in a particular incident was excessive or not, which can help make police action controversial. Sometimes it appears to citizens that the police "got away with murder" if the use of force is ruled as justified.

Some scholars have made a distinction between **unnecessary force** and force that constitutes brutality.[19] In this case, **brutality** "is a conscious and venal act committed by officers who usually take great pains to conceal their misconduct."[20] According to Jerome Skolnick and James Fyfe,

> Unnecessary force, by contrast, is usually a training problem—the result of ineptitude
> or insensitivity, as, for instance, when well-meaning officers unwisely charge into

unnecessary force: Force that violates the principles of the continuum of force due to training errors or good faith mistakes.

brutality: Use of force that is consciously and purposefully cruel and harsh.

situations from which they can then extricate themselves only by using force. Hasty cops who force confrontations with emotionally disturbed persons and who consequently must shoot them to escape uninjured have used unnecessary force.[21]

They also maintain that "unnecessary force may be a good-faith police mistake. Good faith plays no part in brutality."[22] It may be difficult to differentiate one form of force from the other as it depends on the officer's intent, but the distinction is potentially an important one in that unnecessary force and brutality have different causes and, as a result, different remedies.

TYPES OF FORCE

As noted, there are many types of force that can be used by the police as part of the continuum of force. Several of these options are discussed in more detail here.

BODILY FORCE

Bodily force can take many different forms. Included here are takedowns or decentralizations (where officers get a resisting subject off of his or her feet) and punches and/or kicks, designed to make the subject stop resisting. The primary goal is to minimize the subject's overall movements and limit resistance by securing his or her arms; the use of handcuffs is particularly useful in this regard. Bodily force is the most common type of physical force used by the police. Analyses have shown that approximately 70% of all use of force incidents involve the use of bodily force alone. Another 10% involve bodily force along with some other form of force.[23]

A Question to Consider 11.1

Unnecessary Force Versus Brutality

Presuming there is a meaningful difference between police use of unnecessary force and police brutality, how can a manager actually go about determining what is an instance of unnecessary force and what is an act of brutality?

Of all types of force, bodily force is most likely to result in injuries to officers and subjects. Studies show when police use bodily force on a subject, approximately 60% of the incidents result in at least minor injuries to the subject, and 20% result in injuries to the officer. The most common injuries sustained by officers and subjects in these situations are abrasions or contusions; serious injuries as a result of bodily force are very uncommon.[24]

When controlling resisting subjects, officers must be aware of what is known as **positional asphyxia** (Exhibit 11.1).

DEADLY FORCE

Deadly force typically refers to an officer's use of a firearm to shoot a subject, although other forms of force could also potentially cause fatal injuries to a subject. Officers may use deadly force in defense of their own lives or the lives of others. Although this statement may sound straightforward, it is not because its interpretation depends somewhat on the situation at hand and the officers involved. In some situations, an officer may feel that his or her life is in jeopardy, whereas a different officer in a similar situation may not. Prior to 1985, officers in many states were authorized to use deadly force to apprehend all fleeing felony suspects, but the U.S. Supreme Court decision in *Tennessee v. Garner* (1985) changed that.

positional asphyxia: A dangerous condition that occurs when a person's body position prevents normal and adequate breathing.

EXHIBIT 11.1

Positional Asphyxia

Subjects in police custody have died as a result of positional asphyxia. Positional asphyxia occurs when a person's body position prevents normal and adequate breathing. Usually this relates to when the subject is face down with hands secured behind the back. Positional asphyxia may also result when a subject's hands and feet/legs are secured via a specially designed belt (e.g., "hobble" belt or restraint) or the person is hog-tied.

The following factors contribute to positional asphyxia deaths:

- Obesity: A large abdomen or "beer belly" means that when the person is prone, the contents of the abdomen can be forced upward under the diaphragm, restricting breathing.
- Psychosis: Stimulant drugs (amphetamines, speed, "ICE," ecstasy) can create an "excited delirium" in which the person is paranoid, overexcited, and potentially violent. The stimulation of the heart can produce cardiac rhythm disturbances that can be fatal. In this situation, any difficulty breathing can result in sudden deterioration in condition and death.
- Preexisting physical conditions: Any condition that impairs breathing under normal circumstances will put a person at a higher risk when they are physically restrained. Examples are heart disease, asthma, emphysema, bronchitis and other chronic lung diseases.
- Pressure on the abdomen: Even a thin person will have difficulty breathing if there is pressure on the abdomen.

The more [police] officers there are holding a person down in a prone position, the greater the risk that there will [be] pressure on the person's abdomen, making it difficult to breathe.

The following actions will reduce the likelihood of a positional asphyxia death occurring:

- Identify persons at risk: Knowledge of the risk factors will help identify potential situations.
- Avoid prone restraint unless absolutely necessary: Consider alternative methods for resolution. The person should be repositioned from the face down/prone position as soon as practical.
- Do not sit or lean on the abdomen EVER.
- Identify danger signs of asphyxia: [Police] officers must remember that some restraints put the subject in danger and they should avoid tactics that are associated with deaths.
- Constantly monitor the person: Continuously monitor a restrained person and where possible utilize a person not involved in the restraint to monitor the restrained person's condition.
- Seek medical attention: Immediate medical attention should be obtained where there is any concern over the health of a person who has been actively restrained.[25]

As part of arrest tactics training, police officers are educated about the risks associated with such actions and on proper procedures to avoid their potentially serious consequences.

Source: Victoria Police. 2012. Positional/Restraint Asphyxia. Available at http://www.police.vic.gov.au/content.asp? Document_ID=37779.

As a result of this case, officers are only authorized to shoot fleeing suspects who are demonstrably dangerous, such as a suspect with a gun. Most police departments have written rules and guidelines relating to the situations where deadly force can, and cannot, be used (Exhibit 11.2).

Sometimes, questions arise about the appropriateness and necessity of police officers pointing their firearms or threatening to use their firearms without actually firing them. The most informed policy and training on this issue provides for officers "to draw their guns only when circumstances present a reasonable expectation that they will encounter life-threatening violence."[26] In larger cities where officers encounter potentially dangerous situations more often, it is not uncommon for them to unholster and hold their weapons. It is less common that they point or threaten to use those weapons. Rarely do officers actually fire their firearms at a suspect. For example, in Milwaukee, Wisconsin, from 2009 to 2018, there were no more than fifteen incidents per year in which police officers fired at another person. To put that number in context, officers had hundreds of thousands of contacts with citizens and arrested thousands of violent criminals during each of these years. In 2018, there were nearly 700 incidents in which officers used some type of physical force against a subject, five involved the police shooting at a subject.

EXHIBIT 11.2

New York City Police Department Policy Regarding Use of Deadly Force

To ensure that officers use only the minimal amount of force necessary, the department has nine rules that guide a New York City police officer in the use of deadly physical force. They are as follows:

1. Police officers shall not use deadly physical force against another person unless they have probable cause to believe they must protect themselves or another person present from imminent death or serious physical injury.

2. Police officers shall not discharge their weapons when, in their professional judgment, doing so will unnecessarily endanger innocent persons.

3. Police officers shall not discharge their weapons in defense of property.

4. Police officers shall not discharge their weapons to subdue a fleeing felon who presents no threat of imminent death or serious physical injury to themselves or another person present.

5. Police officers shall not fire warning shots.

6. Police officers shall not discharge their firearms to summon assistance except in emergency situations when someone's personal safety is endangered and unless no other reasonable means is available.

7. Police officers shall not discharge their firearms at or from a moving vehicle unless deadly physical force is being used against the police officer or another person present, by means other than a moving vehicle.

Photo 11.3

Most police departments have well-developed policies regarding the use of deadly force and devote much officer training to this critical issue. Nevertheless, police officer discretion in such incidents is necessary. Rick Loomis/Los Angeles Times via Getty Images

8. Police officers shall not discharge their firearms at a dog or other animal except to protect themselves or another person from physical injury and there is no other reasonable means to eliminate the threat.

9. Police officers shall not, under any circumstances, cock a firearm. Firearms must be fired double action at all times.[27]

Source: New York City Police Department. 2016. *2015 Annual Firearm Discharges Report.* New York: City of New York Police Department, 4.

In incidents involving police use of firearms, injuries to subjects are usually either catastrophic (fatal) or the subject is uninjured. If a subject is killed as a result of the use of force, it is nearly always as a result of police use of a firearm. Still, using the Milwaukee Police Department in 2018 as an example, of the five subjects who were shot at, two died, two were not struck, and one subject was shot but did not die. In other words, 40% of subjects who were the intended target of the firearm were uninjured and 40% died.[28] In New York City in 2017, four of twenty-three (17%) subjects who were shot at were uninjured and ten of twenty-three (43%) were killed.[29] An analysis of Miami-Dade (Florida) Police Department data found that 52% of subjects who were shot at were uninjured.[30]

As discussed earlier in this chapter, since January 2019, the federal government has been attempting to track incidents in which police officers shoot and kill subjects. Such efforts were made in the past, but the data were incomplete and invalid because many police departments chose not to submit information. It remains to be seen if the current initiative has more success. Since 2015, however, the *Washington Post* has tallied these incidents based on publication and review of online media reports (search *Washington Post police shooting count*). In 2018, 998 citizens were shot and killed by the police.[31] In 2019, as of July 23, 505 people had been shot and killed by the police.[32] The *Washington Post* data also provide information on the characteristics of deadly force incidents (see Figure 11.2).

It is important to be clear that, unlike in some media depictions, police officers are not expected to be, or trained to be, heroes. Being a hero is not part of the continuum of force or the police position description. Police officers are not expected to risk life or limb; they are not expected to jump in front of bullets. Police officers are also not trained to wrestle guns away from suspects, shoot to wound, or use martial arts techniques to disarm subjects with knifes. Officers are expected and trained to use as much force as necessary—and only as much force as necessary—to stop the threat with which they are confronted. If a police officer is confronted with a situation in which she or he needs to use a firearm, the goal is to stop the threat. When facing a subject armed with a firearm, an officer who tries to shoot the subject in the hand is not achieving the goal of stopping the deadly threat, of keeping the subject from trying to kill an officer or someone else. It bears repeating: An officer is trained to fire his or her weapon as many times as necessary to stop the threat before harm can come to officers and/or citizens.

FIGURE 11.2

Characteristics of Subjects in Deadly Force Incidents, 2018

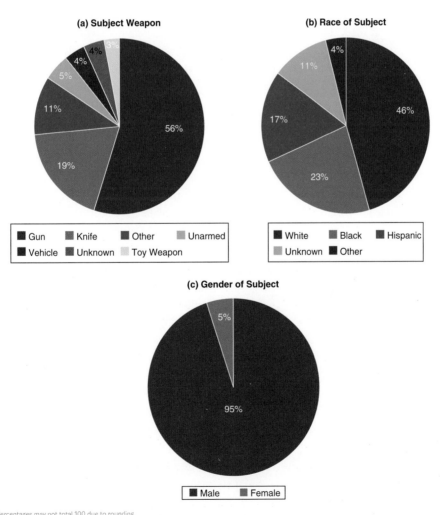

Note: Percentages may not total 100 due to rounding.

Source: Data from *Washington Post* https://www.washingtonpost.com/graphics/2018/national/police-shootings-2018/? utm_term=.743f27051b7e

The Value of Police Restraint in Deadly Force Situations

Police officers often find themselves in situations in which it would be legal and justified to use deadly force, yet it is rare the police actually shoot a subject. Because of this, the police are sometimes said to act with "great restraint." Why do you think police restraint in deadly force situations is a good or bad thing?

SUICIDE BY COP Suicide by cop (SBC) is evidenced by a subject's intentional life-threatening behaviors displayed "to coerce a law enforcement officer to respond with lethal force."[33] The SBC subject has the goal of being killed by officers, and a highly motivated SBC subject may do whatever it takes to reach this goal. The subject's actions usually involve jeopardizing the lives of officers and/or other citizens by threatening them with a deadly weapon. Generally speaking, SBC situations are much more dangerous for police officers and bystanders than are conventional suicides or suicide attempts in which the person seeks to harm only himself or herself. In fact, SBC situations are just as dangerous as other situations that may require police deadly force against a subject.

Accurately determining the extent to which SBC occurs is difficult for at least three reasons. First, the critical factor that determines SBC is the intent of the subject. Did the subject intend to be killed by the police or display threatening behavior for some other reason that led to his or her death? In some SBC cases, the intention is clear: The subject may actually beg officers to shoot. In one SBC case, the subject left a suicide note on his phone addressed to the police. It read, "Dear Officer(s), you ended the life of a man who was too much of a coward to do it himself. Please don't blame yourself. I used you. I took advantage of you."[34] The subject had failed to comply with the orders of the police, pulled a gun from his waistband (which turned out to be a BB gun), and was fatally shot. More often than not, however, SBC is not so easy to discern.

Another reason it is difficult to accurately tally SBC situations is all of these incidents may not end up involving deadly force. A subject may have SBC intentions, but the police resolve the situation without shooting or even causing injury to the subject. Analyses designed to identify SBC situations may miss those that do not involve firearms or any significant force, as would be the case if officers were able to de-escalate the incident. Of all incidents in which a subject intended to commit SBC, only a fraction may actually have been successful.

Finally, different researchers use different samples of cases (e.g., only officer-involved shooting cases versus all use of force cases) and different criteria for classifying SBC intent. Sometimes SBC is measured by degrees of intent,[35] whereas other times subjects are classified as having SBC intent or not.[36] Simply stated, there is no agreed-upon measure of what constitutes SBC ideation or intent; what might constitute SBC in one study may not constitute it in another. Therefore, estimates of frequency can vary dramatically depending on different research approaches—in fact, an article in *Archives of Suicide Research* reported estimates of officer-involved shootings involving SBC have varied between 11%[37] and 80%.[38]

While suicide calls are primarily about suicide prevention, SBC situations are about

> **suicide by cop (SBC):** Intentional subject behavior that seeks to coerce police officers to respond with lethal force.

Photo 11.4

Suicide by cop subjects may go to extraordinary lengths to get police officers to use deadly force. These situations are especially risky for police officers. OJO Images Ltd/Alamy Stock Photo

What Are the Characteristics of Suicide by Cop Incidents?

A 2016 study analyzed 18 previous studies conducted over the past 20 years in order to describe suicide by cop (SBC) incidents. The authors explain that in summarizing the research findings they place more emphasis on more recent studies. As defined in this study, SBC incidents involved a subject displaying "intentionally life-threatening behaviors in order to coerce a law enforcement officer to respond with lethal force." The results include the following:

- SBC subjects are usually white males between the ages of 31 to 35 who are under the influence of alcohol and/or other drugs (cocaine or methamphetamines).
- SBC subjects typically have significant mental health and criminal histories.
- Clear evidence of suicide intent is usually lacking; typically it is shown by either refusing to follow police orders or advancing toward police with a deadly weapon.
- SBC subjects most often possess a loaded functional firearm.

- SBC incidents are most often preceded by domestic or interpersonal disputes.
- De-escalation of such incidents is often attempted but are seldom effective.
- The rate at which subjects are killed by the police in SBC situations is high.

Although there may be some benefit for the police in recognizing SBC situations quickly, the fact of the matter is that any subject with a weapon may pose grave danger to officers or others, regardless if that person has suicide motivations. Motivations of the subject may simply be unknown. As such, given the possibility for immediate danger, it may not be appropriate for officers to treat SBC incidents different than other similar incidents. In any such situation, the goal is to end the situation with as little violence and harm as possible.[39]

Source: Christina L. Patton and William J. Fremouw. 2016. "Examining 'Suicide by Cop': A Critical Review of the Literature." *Aggression and Violent Behavior* 27: 107–120.

suicide prevention *and* public safety. SBC incidents pose a serious risk to police officers and other people involved in the situation. The challenge for the police in SBC incidents is that people who are highly motivated to commit SBC are as dangerous or even more dangerous than other subjects in deadly force situations. SBC situations also often have the added challenge of involving subjects with serious mental health issues.

FORCE LESS LIKELY TO BE LETHAL

In between bodily force and deadly force exist some less lethal options for inducing compliance among resisting or threatening subjects. Most common among these options are oleoresin capsicum (OC) spray and Tasers. As discussed below, these weapons can be deadly, but they are less likely to be so and for that reason they are broadly described as **less lethal weapons**.

OLEORESIN CAPSICUM (OC) SPRAY Oleoresin capsicum (OC) spray, otherwise known as pepper spray, was introduced to law enforcement in the 1980s. Today, approximately 94% of police departments authorize officers to use OC spray.[40] Oleoresin capsicum is an inflammatory agent naturally found in cayenne peppers. Ideally, when a person is sprayed with OC spray, the effects are immediate: The respiratory tract becomes inflamed, the subject experiences a burning sensation and swelling around the eyes, and the subject's eyes close involuntarily.[41] OC spray has the potential to render a resistive suspect passive and compliant so the officer may be able to take the suspect into custody without the need for additional force. However, although the subject may be in extreme discomfort, he or she may still be able to resist. OC spray is used in approximately 10% of all use of force incidents.[42] With other less lethal options available, such as Tasers, the use of OC spray has declined over time.

less lethal weapons: Weapons less likely to cause death to a subject. They commonly include OC spray and Tasers.

oleoresin capsicum (OC) spray: Spray containing an inflammatory agent that causes an inflammation of the respiratory tract and a burning and swelling of the eyes; also known as pepper spray.

Police Robots

Robots are no longer just a science fiction trope. While robots are being more frequently being used in police operations, their tactical use originated on the battlefield. In particular, TALON robots are used to identify and detonate improvised explosive devices (IEDs), and drones are used to conduct surveillance. Drones with smart bombs have also been used to kill terrorists in Iraq, Afghanistan, and Syria. Several types of robots have been adapted for use by law enforcement. Robots are routinely used to defuse bombs. "Throwbots," which are thrown into position and then controlled remotely, have been used to obtain visual data from closed or dangerous settings.[43] Drones are being used to respond to shooting incidents, to provide a very quick visual of the area, including possible getaway vehicles and suspects.[44] Some robot deployments are especially noteworthy. For example, in 2016 the police used a robot to deliver an explosive device to a barricaded sniper who had killed five officers at a demonstration in Dallas, Texas. Police Chief David Brown told reporters, "We saw no other option but to use our bomb robot and place a device on its extension for it to detonate where the suspect was."[45] In 2015, the San Jose Police Department used a robot to deliver a phone and pizza to an armed, suicidal man on a freeway.[46]

The use of robots for dangerous missions has clear advantages but also raises questions and concerns. Where will the trend toward the greater deployment of such technology end? Could the use of robots for police work become routine—for instance, could robots be used to handle high-risk traffic stops? If robots are a means by which force is delivered, might the use of force become "easier" and thus perhaps overused? Although the days of *RoboCop* are probably not near, it would appear they are certainly closer on the horizon.

Photo 11.5

The use of robots in police work is becoming more common. Initially they were used to help diffuse bombs, now they are also sometimes used to perform other potentially hazardous tasks. AP Photo/Matt Rourke

Once introduced, OC spray immediately demonstrated advantages over other forms of force. In particular, the effects of OC spray, while generally immediate and dramatic, are more temporary than other forms of chemical gasses used previously by police,[47] and OC spray is less likely to cause serious or lasting injury than bodily force, batons, and flashlights.[48] OC spray can also be used effectively on multiple subjects simultaneously.[49] However, OC spray also has numerous disadvantages.[50] These include the following:

- When using the spray, innocent bystanders (including other police officers) can be affected.
- Officers have to be in relatively close proximity to the resisting subject they seek to spray.
- OC spray must be aimed at a subject's face for it to be effective (i.e., the target is limited).
- OC spray can be affected by environmental conditions, such as wind.
- People who are impaired by alcohol or drugs, mentally ill, emotionally disturbed, or highly motivated to resist or escape may not be affected by exposure.[51]
- Most research concludes that OC spray is not as effective in incapacitating subjects as Tasers.[52] Perhaps as a result of this, research shows OC spray is more likely to be used along with some other type of force (e.g., bodily force) than by itself.[53]

- Safety concerns have been raised with regard to OC spray because in rare instances people have died after being exposed to it. However, research has shown the deaths involving OC spray were actually the result of positional asphyxia or preexisting health conditions, or they were drug related.[54]

TASERS Taser (short for the Thomas A. Swift Electric Rifle) is the brand name of the most popular electronic stun device on the market. These devices are also known as conducted energy devices (CEDs), electronic control devices (ECDs), or electronic control weapons (ECWs). This technology was first introduced in police departments in the 1990s; today 81% of departments have CEDs such as Tasers. All police departments that serve populations of one million or more have CEDs, although not all officers in these departments may actually carry or be authorized to use them.[55]

Photo 11.6

OC spray leads to extreme discomfort mostly in the eyes and, if inhaled, the nose, mouth, and sinuses. Although OC spray can incapacitate a resisting subject and its effects are reversible, some concerns have been expressed over its use. Jabin Botsford/ The Washington Post via Getty Images

Tasers can be used in either probe mode or drive-stun mode. A Taser resembles a gun. In **probe mode**, nitrogen cartridges are discharged to fire two probes into the body of a subject. As long as both probes attach to the body, an electrical current then runs from the Taser through the wires into the body. The electrical current overrides the central nervous system, causing involuntary muscle contractions and incapacitation of the resisting subject.[56] In **drive-stun mode** the weapon itself is placed in contact with the subject's body and the electrical current is delivered directly. A Taser in drive-stun mode is a pain compliance weapon; it causes localized pain that can allow officers to overcome the resistance of the subject. However, the weapon is not necessarily incapacitating and in fact may lead to additional subject resistance. In many police departments, the use of a Taser in drive-stun mode is discouraged or even prohibited except in extreme circumstances.

In probe mode, the Taser has several potential advantages over other less lethal alternatives, such as OC spray, including the following:

- Tasers can be used from greater distances and allow officers to keep more space between themselves and combative or assaultive subjects. As a result, Taser incidents seldom involve injuries to officers.
- Tasers provide for a relatively quick recovery time for the subject.
- Tasers are effective at inducing suspect compliance.[57] Research indicates a significantly higher level of effectiveness with Tasers compared to OC spray.[58]
- Because in probe mode Tasers do not rely on pain to induce compliance, ideally they are more effective than OC spray on persons who have a higher tolerance for pain. This can include people under the influence of drugs or alcohol or who have a mental illness.[59]

In spite of their potential advantages, serious questions exist about the safety and potential overuse of Tasers.[60] There are concerns Tasers are used in a disparate fashion against members of minority groups.[61] A related worry is that Tasers are rated too low in continuums of force and consequently are being used against passive (versus active) resisters.[62] Finally, there are concerns about the use of Tasers (and OC spray, for that matter) on the elderly, children, pregnant women, and persons with medical conditions that put them at greater risk of experiencing dangerous medical side effects.[63]

Taser: An electronic stun device that resembles a gun. It delivers an electrical current to incapacitate a subject.

probe mode: Taser mode in which two probes are fired from the Taser into the body of a subject; an electrical current then flows from the Taser through the wires attached to the probes and into the subject.

drive-stun mode: Taser mode in which the weapon itself is placed in contact with the subject's body, delivering the electrical current directly.

excited delirium: A potentially deadly medical condition involving psychotic behavior, an elevated temperature, and an extreme fight-or-flight response by the nervous system.

With regard to the potential overuse of Tasers, much depends on the department being studied. Research has shown Taser use accounts for 10%[64] to 50%[65] of all use of force incidents. And concerning disparate use, most studies examining the issue have not found racial disparities in the use of Tasers on subjects compared to other forms of force; Tasers are equally likely to be used on white, Hispanic, and black subjects.[66]

As previously noted, it is very uncommon for officers to be injured when force is limited to Taser use alone. If Tasers work effectively, however, they *will* cause injuries to subjects, if only because of the probes that stick into the person. While the probes seldom cause more than minor physical injuries, it is unknown if there are lasting physical or emotional consequences of Taser injuries. There is also the potential for injuries to subjects as they become incapacitated and fall to the ground, although these types of injuries are surprisingly rare. The most serious issue is that Tasers have been linked to in-custody deaths.[67] Between January 2015 and November 2015 in the United States, forty-eight people died in incidents in which police used Tasers.[68] However, to put this statistic in perspective, it is estimated Tasers are used on citizens more than 900 times *per day*. The reasons for Taser-related deaths are varied; most of the individuals who died soon after a Taser incident suffered from hypertensive heart disease, coronary heart disease, drug intoxication (especially methamphetamines or PCP), cocaine toxicity, or **excited delirium** (Exhibit 11.3).[69]

Overall, more than half of the citizens who died after Taser use suffered from mental illness or were under the influence of drugs,[71] which may reflect the sample upon which Tasers are most commonly used.[72] A National Institute of Justice study that examined Taser-related deaths concluded, "The risk of death in a CED-related use of force incident is less than 0.25%, and it is reasonable to conclude that CEDs do not cause or contribute to death in the large majority of those cases."[73]

EXHIBIT 11.3
Excited Delirium

According to an article in the *FBI Law Enforcement Bulletin* titled "Excited Delirium and the Dual Response: Preventing In-Custody Deaths,"

Excited delirium syndrome (ExDS) is a serious and potentially deadly medical condition involving psychotic behavior, elevated temperature, and an extreme fight-or-flight response by the nervous system. Failure to recognize the symptoms and involve emergency medical services (EMS) to provide appropriate medical treatment may lead to death. Fatality rates of up to 10% in ExDS cases have been reported. In addition to the significant mortality associated with unrecognized ExDS, a substantial risk for police liability and litigation exists. These patients often die within 1 hour of police involvement. One study showed 75% of deaths from ExDS occurred at the scene or during transport.

ExDS subjects typically are males around the age of 30, and most have a history of psychostimulant use or mental illness. Law enforcement agents or EMS personnel often are called to the scene because of public disturbances, agitation, or bizarre behaviors. Subjects are usually violent and combative with hallucinations, paranoia, or fear. Additionally, subjects may demonstrate profound levels of strength, resist painful stimuli or physical restraint, and seem impervious to self-inflicted injuries. This information becomes particularly important to law enforcement personnel who may use techniques intended to gain control and custody of subjects through physical means, chemical agents, or ECDs. During initial assessment patients often are noted to have elevated body temperatures, fast heart rates, rapid breathing, elevated blood pressures, and sweaty skin.[70]

Source: Brian Roach, Kelsey Echols, and Aaron Burnett. July, 2014. Excited Delirium and the Dual Response: Preventing In-Custody Deaths. *FBI Law Enforcement Bulletin.* https://leb.fbi.gov/2014/july/excited-delirium-and-the-dual-response-preventing-in-custody-deaths.

THE CAUSES OF POLICE USE OF FORCE

It is well established that a small percentage of officers account for a relatively large proportion of force incidents in any given police department.[74] For example, one study found 58 out of 1,084 (5.4%) patrol officers accounted for 40% of all use of force incidents in the department under examination.[75] That same study found 71% of all officers in the department were not involved in any use of force incidents over the course of the year.[76] According to the Christopher Commission report of the Los Angeles Police Department, 5% of the officers accounted for 20% of use of force reports.[77] This is a potentially useful research finding because it suggests if a department wishes to address the use of force issue, it is necessary to focus on the behavior of a small group of officers.

OFFICER CHARACTERISTICS AND USE OF FORCE

The fact that a small group of officers accounts for a relatively large proportion of use of force incidents raises an important question: How do officers who frequently use force differ from officers who do not? Studies examining background or demographic characteristics of officers and their relationship to use of force have produced many inconsistent findings. Most (but not all) suggest that officer gender does not influence the frequency with which officers use force.[78] Similarly, studies also suggest officer race does not influence the likelihood of using force.[79] Research does show, however, that younger officers are more likely to use force than older ones,[80] and less experienced officers are more likely to use force than more experienced ones.[81] It would be useful to know if there is a relationship between officer characteristics and the specific type of force used, but few studies have examined this issue. For example, perhaps female officers are less likely to use bodily force but more likely to use less lethal weapons. Research regarding the impact of officer characteristics has also been limited because much of it has failed to consider the job assignments and arrest activity of the officers under study—important variables that might attenuate the relationships between individual characteristics and police use of force.

OFFICER ASSIGNMENT, ARRESTS, AND USE OF FORCE

One might reasonably expect location/type of patrol assignment to be related to use of force.[82] For example, high-crime areas may foster conditions in which use of force is more frequently necessary. Also, like officer assignment to a high-crime area, an officer's assignment to a high-crime *time* could affect the likelihood of involvement in force incidents. Serious crime and arrests are more likely to occur during the late evening and early morning hours. As such, an officer's shift (particularly one that covers the hours of 9:00 p.m. to 3:00 a.m.) may be related to the frequency with which the officer uses force.[83]

If there is one pattern that is most clear with regard to police use of force, it is that most instances of force occur during the course of arrests.[84] Not surprisingly, then, research has found arrest activity is the strongest predictor of use of force. Officers who make more arrests are significantly more likely to be involved in more force incidents.[85] However, some officers make a lot of arrests and are *not* involved in a lot of force incidents. Also, the time ordering of the arrest-force relationship is not entirely clear. Some officers may make more arrests simply to justify the use of force.

POLICE CULTURE

The police culture is another important factor that needs to be considered when understanding and controlling police use of force. As discussed earlier in this book, one of the primary values of the traditional police culture is "toughness." Tough cops are seen as good cops, and this may

Photo 11.7
Tasers are commonly carried by police officers and are considered a less lethal form of force. They have many advantages over other forms of force, but their use is still sometimes controversial. ©iStockphoto.com/seanfboggs

have direct implications for the tendency of officers to use force. Professor Dennis Rosenbaum from the University of Illinois at Chicago explains,

By the time officers leave their academy training, they're already more prone to want to use force to resolve any kind of situation rather than talk to people. They're less likely to want to engage in active listening and more machismo about how to interact with people. They haven't even hit the road yet. So I think it is part of the police culture that is being instilled at our training academies and something to think about.[86]

In addition, one might expect police departments oriented more toward law enforcement would have higher rates of use of force than departments oriented more toward service. The values espoused by departmental leaders and supervisors could potentially affect officers' tendencies toward using force. However, little research has examined these issues.

THE CONTROL OF POLICE USE OF FORCE

Use of force is associated with officer and subject injury, may lead to lawsuits against the police department, and has seriously negative consequences for police–community relations. As a result, there is good reason for the police to avoid the use of force if at all possible. However, it is important not to lose sight of the obvious: Police use of force is sometimes necessary. The goal is only to use force when it is the best option. As discussed in detail in Chapter 8 and earlier in this chapter, numerous studies have shown that organizational policies and rules can substantially reduce the frequency of use of force incidents.[87] The careful assignment of certain officers to certain shifts and patrol areas may be another way by which to reduce the frequency of force. In particular, when young male officers are assigned to high-crime areas during high-crime shifts, it creates situations in which force may more likely be used. Other important practices have been offered as ways to reduce the frequency of force such as the use of de-escalation techniques, early intervention systems, and police body-worn cameras. Each is discussed below.

DE-ESCALATION TECHNIQUES

Potentially violent situations do not necessarily have to be resolved with violence. As discussed in the Police Spotlight earlier in this chapter, police departments today are increasingly providing de-escalation training to help officers resolve situations peacefully without force being used by either the officer or the subject. De-escalation can be attempted in many situations that involve angry, agitated, threatening, or uncooperative subjects.[88] It must be emphasized however that de-escalation is not a method to eliminate the use of force by the police. Police use of force is sometimes necessary, especially when subject resistance is high. In some situations, if force is not used, the situation may become more dangerous to officers and citizens.

De-escalation involves the use of certain techniques to move a person from a high state of tension to a reduced state of tension.[89] With de-escalation, there are two important concepts to keep in mind.[90] First, reasoning with an enraged person is not possible. The first and most important step in de-escalation is to reduce the subject's level of anger so that discussion is possible. Communication, the exchange of useful information, is the basis of de-escalation. If communication is not possible, neither is de-escalation. Second, de-escalation techniques are not natural, which highlights the need for training, if police departments want officers to effectively deploy these techniques. People are driven to fight, flight, or freeze in emotionally charged situations; however, this cannot be the case in de-escalation situations. Officers must appear calm and learn to "back-off."

Active listening skills and effective verbal communication are the legs on which de-escalation stands. Active listening is demonstrated through reflective statements (e.g., "I see that you are upset"), minimal encouragers (e.g., "Okay," "I understand"), and other techniques (i.e., introducing oneself, using "I" statements, restating the subject's statements, summarizing the subject's statements). Strategic use of open- and close-ended questions is also important (e.g., "Tell me what happened?" "Will you please let me help you?"). It is also important to not argue with subjects, to not get defensive, and to use a soft and slow tone of voice. Officers' control over the conversation may remove the dysfunctional emotion from the situation and allow for a more empathetic connection with the subject.[91] Nonverbal behaviors, such as limiting eye contact with the subject, appearing calm, and maintaining a neutral facial expression, are also critical to de-escalation.[92] Although there are many calls for the use of de-escalation by the police, there are relatively few studies that examine the effectiveness of it , and most focus on fields other than policing, such as nursing and social work.[93] An exception is a 2018 study that showed when police officers simply talked with subjects in a respectful tone and avoided condescending comments, subjects were more likely to be calm at the end of the encounter.[94] De-escalation is most likely to be a feasible and effective option when subject resistance is low, in situations where police use of bodily force would otherwise be most likely used.

There is nothing magical about de-escalation although it does require a shift in how officers approach uncooperative and disrespectful subjects. The goal is to transfer a sense of calm and respect to the subject so that he or she responds favorably and use of force can be avoided. This outcome is in the best interest of everyone involved.

EARLY INTERVENTION SYSTEMS

An **early intervention system (EIS or EI system)** is a data management system designed to record and monitor certain behaviors of officers with the objective of identifying potentially problematic patterns of behaviors while they can still be corrected. The ultimate goal is to prevent more serious behaviors. These systems are used in most major police departments today. EIS have much promise, but they are more complicated than what they may first seem to be. Because these systems also have implications for the control of a wide range of police misconduct, they are discussed in more detail in Chapter 12.

To control police use of force, EIS requires the continuous collection of data on incidents where officers used force in the line of duty. And herein lies the first question: What types of force are considered problematic and deserve to be monitored and controlled? Pointing a gun? Discharging a weapon? When a suspect is injured? Subjective judgments may need to be made. The second question is, How many times must an officer use force to indicate "problematic" performance? Is two a year too many? Is ten too many? Is ten too many if the officer made 300 arrests? More subjective judgments are required. The third question is what is to be done to address the identified problem? Should the officer be re-trained? Counseled? Fired? Reassigned to a different job? Transferred to a different area? And whatever the intervention may be, will it be effective? The bottom line is that officers' use of force should be monitored, but EI systems by themselves should not be considered a solution to the use of force issue. Research on the effectiveness of EI systems supports this conclusion: "For all the stock that has been placed in EI systems, social science has not provided much or very strong evidence on their effectiveness or on their unintended effects."[95]

early intervention system (EIS or EI system): Data management system designed to record and monitor certain behaviors of officers with the objective of identifying potentially problematic patterns of behaviors while they can still be corrected.

POLICE BODY-WORN CAMERAS

Police body-worn cameras (BWCs) have many expected and documented benefits. As discussed in other places in this book, BWCs may provide valuable evidence in criminal prosecutions and lawsuits, enhance the transparency of police actions and improve police officer accountability, provide protection against baseless citizen complaints and lawsuits, and provide more accurate portrayals of police actions, to mention just a few benefits. Another critically important intended consequence of BWCs is that they may cause a reduction in the frequency by which officers use force, presumably force that was not necessary. When this effect is found, it may be because of two reasons: (1) officers avoid using force as a result of the possibility of "being caught on camera" violating department rules and policy, and/or (2) citizens cooperate with the police knowing that their actions are being recorded and therefore the police do not need to use force to control the citizen. This second reason relates to BWCs having a "civilizing effect" on citizens' behavior, or the "demeanor hypothesis." Research suggests BWCs affect citizen and police behaviors.[96]

BWCs have been shown to reduce the frequency by which the police use force. In a study conducted in Rialto (California), researchers found that officers who did not have BWCs were involved in more than two times as many force incidents compared to officers who did have cameras, even while taking into account other important factors. In addition, in separate analyses, it was shown that the number of force incidents declined significantly in the twelve months after BWCs were deployed in the department, compared to the twelve months before they were used. The authors conclude that BWCs may "substantially reduce force responses . . . and noticeably affect police-public encounters."[97] Another study that examined the impact of BWCs on use of force was conducted in the Spokane (Washington) Police Department.[98] Here the researchers also found that following BWC deployment, officers who were equipped with BWCs were less likely to be involved in a use of force incident than officers who did not have cameras, although for unexplained reasons this effect disappeared after six months. The authors suggest that after six months officers with BWCs may have become more comfortable with the technology and the potential for supervisory review, and BWC-generated restraint may have dissipated over time as a result. Yet another study showed that BWCs led to a decrease in use of force incidents only when cameras were activated and when officers did not have discretion about when to turn on the cameras.[99] Once again, it must be emphasized that police use of force is a relatively rare event; that BWCs have been shown to reduce these rare events is an important benefit of the technology.

Police use of force is a complicated and controversial issue. Much still needs to be learned about its causes and how it can best be controlled.

Main Points

- The right to use force can be thought of as a powerful tool of the police. With the ability to use force, the police can intervene in situations in which order needs to be maintained and offenders need to be arrested. However, it is also the use of force that makes the police inherently controversial.

- The continuum of force principle holds that police can and should only use as much force as necessary to overcome the resistance offered by the subject. The use of force is a stepwise progression in which the officer uses the next more severe form of force only if the previous form of force did not induce subject compliance. It is a general guideline that does not eliminate police discretion of force situations.

- Much variation in continua of force can be found across police departments, and all continua have limitations as a guide for police decision making. The twenty-one-foot rule is a guideline indicating a subject within a twenty-one-foot boundary of an officer can pose a significant threat to that officer; this is particularly true for a subject who possesses a knife. This "rule" also has many limitations as a useful guide for police decision making.

- When force is used beyond what is prescribed in the continuum of force, it is deemed unjustified, unnecessary, or excessive, or sometimes it is referred to simply as police brutality. However, because use of force situations are often complex and continua do not specify the exact actions officers should take in each situation, the determination of what is actually "unnecessary" or "excessive" is often a matter of judgment.
- Bodily force can take many different forms and is the most common type of physical force used by the police. Of all types of force, bodily force is most likely to result in injuries to officers and subjects.
- Deadly force typically refers to an officer's use of a firearm to shoot a subject, although other forms of force could potentially cause fatal injuries to a subject. Officers may use deadly force in defense of their own lives or the lives of others, although this determination is subjective and may depend on many factors.
- Suicide by cop (SBC) is a method of suicide that occurs when a subject engages in threatening behavior in an attempt to be killed by law enforcement. SBC situations are potentially very dangerous for police officers and other people involved in the situation.

- Oleoresin capsicum (OC) spray and Tasers are examples of weapons that are less likely to be lethal when used. Both have unique advantages and disadvantages, and both have also been criticized for overuse and safety concerns.
- Excited delirium syndrome is a serious and potentially deadly medical condition involving psychotic behavior, an elevated temperature, and an extreme fight-or-flight response by the nervous system.
- It is well established that a small percentage of officers account for a relatively large proportion of force incidents in any given police department. Officers who make more arrests are significantly more likely to be involved in more force incidents.
- Even though police use of force is a relatively uncommon event, its consequences are potentially serious, so reasonable efforts should be made to reduce the frequency of it. De-escalation skills can be used by officers to resolve situations peacefully without force being used. Early intervention systems (EIS) also hold potential in this regard. Police body-worn cameras have been shown to reduce the frequency by which officers use force.

Important Terms

Review key terms with eFlashcards at **edge.sagepub.com/brandl2e.**

Questions for Discussion and Review

Take a practice quiz at **edge.sagepub.com/brandl2e.**

1. What is a continuum of force and what are its major limitations? Given these limitations, is a continuum of force of any value? Explain.
2. What is the difference between unnecessary force and brutality? How can one be differentiated from the other? Why is this distinction potentially important?
3. What is bodily force? How might de-escalation techniques affect the frequency by which officers use bodily force? Why?
4. What is positional asphyxia? What is excited delirium? Under what conditions are they likely to occur?

5. What were the rulings in the U.S. Supreme Court cases of *Graham v. Conner* (1989) and *Tennessee v. Garner* (1985)? How did these cases affect police use of force?
6. Why have previous government-sponsored mechanisms of reporting use of force by the police failed?
7. What is suicide by cop (SBC)? Why are these situations especially problematic for police officers?

8. What are the two most common weapons that are less likely to be lethal to subjects? How do they work? What are the advantages and disadvantages of each?
9. Research shows that a small proportion of officers in a police department account for a relatively large proportion of use of force incidents. Why?
10. How might the frequency by which officers use force be reduced? Can police use of force be eliminated? Why or why not?

Fact or Fiction Answers

1. Fiction
2. Fiction
3. Fiction
4. Fact
5. Fiction

6. Fiction
7. Fact
8. Fact
9. Fact
10. Fact

ⓈSAGE edge™

Digital Resources

Get the tools you need to sharpen your study skills. SAGE Edge offers a robust online environment featuring an impressive array of free tools and resources.

Access practice quizzes, eFlashcards, video, and multimedia at **edge.sagepub.com/brandl2e.**

Media Library

View these videos and more in the interactive eBook version of this text!

Criminal Justice in Practice
11.1: Use of Force (Non-Lethal)
11.2: Use of Force (Lethal)

SAGE News Clip
11.1: Peaceful Protesters React to Police Shooting Video

12

POLICE MISCONDUCT AND CORRUPTION

Police Spotlight: Denver's Citizen/Police Complaint Mediation Program

When a citizen believes a police officer has acted improperly, the citizen may decide to file a complaint against the officer. In Denver, Colorado, a complaint can be filed with the Internal Affairs Bureau (IAB) of the Denver Police Department (DPD) or with the City of Denver Office of the Independent Monitor (OIM; an oversight agency responsible for monitoring the DPD).[1] The traditional way of processing complaints is for police department personnel—usually persons assigned to the IAB—to investigate them. Citizens generally have little faith in this type of process as investigations are reviewed only from the perspective of the police.

An alternative method of processing complaints was created in the DPD in 2005. The new program involves a face-to-face meeting between the citizen who filed the complaint and the officer who received it. According to the website of the Denver Independent Monitor:

> Community-Police Mediation is an alternative to the traditional way of handling police complaints. Complainants have the opportunity to sit down with the officer in a neutral and confidential setting, with the assistance of a professional mediator. This completely voluntary process allows both sides to be heard—the complainant talks to the officer about the behavior s/he felt was inappropriate, harmful, scary, or discourteous and helps the officer see the incident from his/her perspective. The officer has an opportunity to explain what happened from his/her perspective as well, and often shares what kind of information s/he had going into the situation as well as relevant policies and procedures that may have impacted his/her decisions or actions. The mediator helps both sides to feel safe and comfortable in getting all of the issues out on the table and working through them.[2]

A study that evaluated the operations of the program explains that most of the mediated cases involve allegations of discourtesy (80%) and/or policy or procedural issues (16%), although some relate to racially disparate treatment or how force was used (5% and 6% respectively). If the complaint involves an allegation of serious misconduct that could result in an officer's dismissal, demotion, or a substantial suspension, it is not eligible for mediation. Both the officer and the complainant must agree to participate in mediation, or the complaint is not eligible. Details relating to the mediation are

(Continued)

Objectives

After reading this chapter you will be able to

12.1 Give examples of police corruption and misconduct and explain why serious efforts need to continue to be made to prevent such behaviors

12.2 Identify the sources of data commonly used to study the nature and extent of police corruption and misconduct and discuss their limitations

12.3 Describe how characteristics of officers, the job, and the organization can lead to police corruption and misconduct

12.4 Discuss promising approaches to controlling police corruption and misconduct

12.5 Explain why officer honesty is such a difficult but important issue in police work

Fact or Fiction

To assess your knowledge of police misconduct and corruption prior to reading this chapter, identify each of the following statements as fact or fiction. (See page 276 at the end of this chapter for answers.)

1. Police use of excessive force is a good example of police corruption.

2. Some types of police corruption can be good.

3. Unauthorized use of sick time can be considered a form of police misconduct.

4. Self-report surveys are a good way of accurately estimating the extent of misconduct among officers.

5. According to research, of all citizens who report officers did not behave appropriately during a street stop or traffic stop, fewer than 5% actually file a complaint.

6. Citizen complaints are most likely to be received by officers early in their careers.

7. The number of citizen complaints received by a police department may have more to do with what is required to file a complaint than it does the quality of police services provided in that community.

(Continued)

(Continued)

8. There is a strong argument to be made that an officer who has been sanctioned for lying should actually be dismissed.

9. When it comes to combatting police corruption and misconduct, there is no greater enemy than the code of silence.

10. The code of silence is unique to the police. It seldom exists in other social groups, occupations, or organizations.

(Continued)

confidential. Typically, mediations are held in a municipal conference room, and the officers are allowed to be on duty and in uniform. If a complainant drops out of the mediation process, the case against the officer is declined. If an officer drops out of the process, then the complaint is returned to the regular complaint process for investigation.[3]

Evaluating the success of such a program can be difficult for several reasons. For example, different types of complaints are handled through mediation compared to a traditional complaint process. Also, citizens and officers generally have to volunteer to be included in the program. The evaluation of the program showed that citizens who participated in mediation were more likely to be satisfied with the complaint process compared to citizens whose complaints were handled in the traditional way.[4] Female officers and Latino complainants were most likely to be satisfied with the process. Despite some issues with program assessment, mediation as a way of resolving relatively minor complaints of misconduct shows promise.

BECAUSE of the unique nature of the job and the environment in which the police work, there are many opportunities for officers to engage in misconduct. This chapter discusses the various forms of police misconduct and corruption, the factors that contribute to these behaviors, and the tactics and challenges associated with controlling them.

THE IMPORTANCE OF UNDERSTANDING POLICE MISCONDUCT AND CORRUPTION

When police officers and departments are in the news, often it is because of something bad, such as questionable use of force by the police or other claims of misconduct. By the nature of their job, the police are given power and authority ordinary citizens do not have. In particular, because officers have the authority to enforce laws, in some respects the police *are* the law. It is a tragic irony when officers violate the law. In such circumstances, the police violate the public trust. The police can work hard to develop a positive relationship with the community and create goodwill, but a well-publicized incident of police misconduct can threaten or even destroy the reputation they worked so hard to achieve. Because police misconduct can have such negative and far-reaching consequences, it is important to understand the complexities of this issue.

As a starting point, it is necessary to define the phenomenon of interest. While it could be argued definitions do not matter, that the conduct is wrong no matter what it is called, it may still be useful to make distinctions in these types of bad behaviors. Different behaviors have different causes and may have different remedies. Police corruption and police misconduct are related but are not the same. The concept of police integrity is also sometimes part of the corruption/misconduct conversation. It is defined here as well.

POLICE CORRUPTION

The term **police corruption** is often used to describe many types of illegal conduct on the part of officers. In actuality, however, police corruption refers to a relatively narrow set of behaviors. Most definitions of corruption have three elements in common: Police corruption refers to (1) illegal behaviors, (2) committed by police officers who use their authority as police officers to commit crimes, and (3) that are motivated by personal gain. As such, if a police officer engaged in illegal conduct but did not use his or her authority to assist or facilitate the illegal behavior, then that conduct would not technically be considered corruption. Using this definition, most forms of criminal activity committed by *off-duty* officers would not be considered police corruption. However, if, for example, an off-duty officer sold drugs that were taken from dealers while the officer *was* on duty, the behavior would fit the definition.

With regard to the third dimension of the definition, if the officer's illegal conduct was not motivated by personal gain, which is usually thought of in terms of financial gain, then the conduct would not represent corruption. For instance, the intentional use of excessive force on citizens by officers seldom leads to personal gain and would typically not be considered corruption. The most common form of corruption is **theft**. If the acceptance of **gratuities** is considered corruption, then gratuities would be the most common form of corruption (see A Question of Ethics feature). **Bribery** (accepting money in exchange for some consideration, such as not issuing a speeding ticket) and **extortion** (using threats to obtain something, usually money) are also good examples of corrupt actions.

A Google search of *police corruption stories* produces thousands of news media accounts relating to police corruption. Some examples include the following:

- In exchange for money, a police officer provided information to a gas station owner that investigators were about to execute a search warrant relating to a food stamp fraud investigation.

police corruption: Illegal behaviors committed by police officers who use their authority as police officers to commit crimes motivated by personal gain.

theft: The taking of property without permission.

gratuities: Small gifts provided for services.

bribery: The acceptance of money, services, or goods in exchange for some consideration, such as not issuing a speeding ticket.

extortion: The use of threats to obtain something, usually money.

- Police officers stole hundreds of thousands of dollars in cash, drugs, guns, and other items while pretending to be seizing the goods for legal reasons.
- A police officer allegedly embezzled money from a police athletic program he managed.
- An officer sold and helped distribute narcotics and provided protection for drug dealers for cash.
- A police officer allegedly stole drugs from the police department evidence room and then used the drugs herself.
- A police officer stole tools from a home construction site and then deleted a video recording of the theft that was to be used as evidence.
- Six officers were arrested on a range of charges, including conspiracy, robbery, extortion, kidnapping, and drug dealing.

In some cases, reference has been made to what is known as **noble cause corruption**.[5] The idea here is that in some circumstances, the police may take illegal action to achieve a greater good; in other words, the end justifies the means. For example, to make sure known criminals are convicted, a police officer might illegally plant evidence, lie on a report, or lie on the witness stand. Such actions are wrong regardless of the justification and even if the behavior is not committed for personal gain. It must be clear that noble cause corruption is an oxymoron; it is impossible for corruption to be noble.

A Question of Ethics

Corruption in the Form of Free Coffee?

While theft, bribery, and extortion are clearly wrong, and officers who engage in these crimes must be sanctioned, there is sometimes debate about whether the acceptance of gratuities constitutes a form of corruption. Gratuities are small gifts provided for services. For example, on-duty police officers may receive free coffee at 7–11 stores or a discount on meals at certain restaurants. Do you think it is wrong for police officers to accept gratuities? Explain why or why not.

POLICE MISCONDUCT

There are many wrongful actions of the police that fall outside of the narrow definition of police corruption. **Police misconduct** is more general and inclusive; it refers to corruption but also to other inappropriate or illegal behaviors of police officers. It can include on-duty as well as off-duty conduct and may have motivations besides personal or financial gain. It may involve violating departmental rules and/or violating actual laws. It includes relatively minor and very serious conduct. For example, everything from an officer's unjustified use of sick time to an officer committing a sexual assault would be considered police misconduct. An officer's intentional use of excessive force could also be considered a form of police misconduct. Sometimes police misconduct is referred to as **police deviance**. These two concepts are essentially overlapping.

A Google search of *police misconduct* provides many examples of serious forms of police misconduct. These include the following:

noble cause corruption: A belief the police may take illegal action to achieve a greater good.

police misconduct: Inappropriate or illegal behaviors of police officers; also sometimes referred to as police deviance.

- Officers were suspended after being seen on a body-worn camera video planting evidence at a crime scene.
- An officer allegedly sexually assaulted thirteen women over the course of three years. Most incidents began with a traffic stop. The women were illegally searched and threatened prior to the assaults.
- A police officer was fired after he mocked a subject on a social media post.
- An officer was arrested for the domestic assault of his girlfriend.
- Officers were charged with fraud after being paid for overtime shifts that they never showed up to work.

Photo 12.2
Gratuities are often provided to officers in the form of free or discounted meals. There are reasons to argue that the acceptance of gratuities is wrong and is a form of corruption, and there are arguments that gratuities are a reasonable expression of gratitude for officers.

@age fotostock/Alamy Stock Photo

EXHIBIT 12.1
Dirty Harry and Noble Cause Corruption

The notion of noble cause corruption was captured well in the classic 1971 film *Dirty Harry*. In the film, Inspector "Dirty" Harry Callahan is a San Francisco detective on the trail of a psychopathic serial killer named Scorpio. After killing several people, including a police officer, Scorpio kidnaps a young girl and buries her alive with only enough air for her to breathe for a short time. Dirty Harry tracks down Scorpio and shoots him in the leg.

With Scorpio in excruciating pain and asking for a lawyer, Dirty Harry stands on Scorpio's injured leg in order to get him to reveal the location of the girl. But the police are too late, and the girl has already died. Dirty Harry searches Scorpio's nearby home and takes Scorpio's gun as evidence. But Dirty Harry conducts this search without a warrant, so the gun cannot be used as evidence, and Scorpio is released.

(Continued)

Photo 12.3

The fictional character Dirty Harry often violated the law in order to apprehend offenders and solve crimes. Violating the law is wrong regardless of the justification.

Source: AF archive/Alamy Stock Photo

Because Dirty Harry violated Scorpio's rights, his supervisor orders him off the case. However, Dirty Harry knows that Scorpio will kill again, so he continues to pursue the criminal. Sure enough, with Dirty Harry in pursuit, Scorpio steals a gun and takes a bus full of children hostage. The bus crashes and Scorpio flees, but he finds a young boy and holds him at gunpoint. Dirty Harry closes in. He shoots Scorpio in the shoulder, and the boy gets away. With Scorpio injured and unarmed, Dirty Harry stands over Scorpio, points his gun at him, and intones, "Ask yourself the question, do I feel lucky? . . . Well, do you punk?" Scorpio makes a grab for his gun, and Dirty Harry shoots him dead. After a long pause, Dirty Harry takes out his police badge and throws it into the water where Scorpio's body is floating. The movie ends.

POLICE INTEGRITY

Police integrity can be defined as the "inclination among police to resist temptations to abuse the rights and privileges of their occupation."[6] Police integrity can also be thought of as the opposite side of the misconduct coin. If a police force has integrity, it has little or no corruption or misconduct among its officers. If a police officer has integrity, then the officer does not engage in corrupt behaviors or other misconduct. Integrity is good, and corruption and misconduct are bad.

Much discussion also occurs among police leaders and scholars about what is known as the culture of integrity. As discussed in more detail below, this concept involves the importance of the police culture in either supporting or preventing misconduct. If the culture of a police department emphasizes the value of integrity, then that department is more likely to be free of corruption and misconduct.

THE NATURE AND EXTENT OF POLICE MISCONDUCT AND CORRUPTION

Because there is no national reporting system for instances of police misconduct, it is difficult to come to an accurate conclusion about the extent of the problem in departments today. Hypothetically, it would not be surprising to learn that some departments have more of a problem with police corruption and misconduct than others. However, unless major and widespread scandals come to light, it is simply impossible to accurately judge the extent of misconduct in any given department.

Discussed below are five common sources of data that have been used to try to provide insight into the nature and extent of police misconduct. Each of these sources has major limitations, however.

SELF-REPORT SURVEYS

Self-report surveys have been used to estimate officers' involvement in misconduct or to estimate coworker involvement in misconduct. For example, one study surveyed over a thousand officers from twenty-one states about their knowledge of their coworkers'

police integrity: The inclination among police to resist temptations to abuse the rights and privileges of their occupation.

EXHIBIT 12.2

Sledding on Duty

In one particularly unusual instance of police misconduct, six officers decided to go snow sledding while on duty. Of all places to go sledding, they chose a cemetery. One of the officers ended up fracturing several ribs and needed medical attention. To cover their conduct, one of the officers made up a story about how they noticed a robbery about to happen and gave chase to the would-be robbers. The officers told their supervisors the injured officer had fallen while chasing the suspects. The police chief was so impressed with the injured officer's brave conduct that she gave the officer a meritorious conduct award. However, shortly thereafter the officers' story began to unravel, and the truth was discovered. The officer who came up with the false story was fired, charged with a felony, and sentenced to four months in jail. The injured officer and two other officers were fired, and another was given a twenty-day suspension. An officer who cooperated with investigators was not punished.

misconduct and whether they reported that misconduct to their supervisor.[7] Another study surveyed and interviewed forty officers from fourteen police departments in the St. Louis metropolitan area.[8] One of the major problems with this approach is it relies entirely on the honesty of officers, and obviously there may be reason for officers not to be honest when it comes to questions about misconduct. Another serious problem is that it is difficult to put the number of reported incidents in perspective when using a sample of officers, especially a sample from more than one department, given that experiences with misconduct may vary considerably. Some have gone so far as to say that self-report surveys of police misconduct are completely ineffective.[9]

A related approach is to ask officers about their perceptions of the wrongfulness of various forms of misconduct described in hypothetical scenarios (as in the Question of Ethics feature).[10] This method can provide insight into how officers perceive potentially wrongful actions and can serve as a basis for instruction about acceptable conduct in the department. However, once again, the results of such surveys are highly dependent on officer willingness to participate and respond honestly. This method also does not provide an indicator of the extent of *actual* misconduct among officers in the department, only officer perceptions of it.

A Question of Ethics

How Wrong Are These Police Actions?

Listed below are ten examples of what could be considered police misconduct. Identify the three most serious and the three least serious forms of misconduct. If you believe there is no misconduct described, identify those scenarios as well. Provide reasons for your judgments.

Case 1: A police officer knowingly conducts an illegal search of a subject's vehicle and in the process finds drugs. She orders the suspect to dispose of the drugs in the sewer.

Case 2: Many on-duty police officers eat dinner at a particular restaurant because the manager never charges them for their meals. This arrangement has been in place for years.

Case 3: An on-duty police officer goes through the drive-through at a restaurant and pays for an ice cream cone. As soon as he gets the ice cream cone, the officer receives a radio call to respond to a crime in progress. The officer finishes the cone before responding lights and siren to the call for service.

Case 4: A police officer is using radar for speed enforcement. White motorists who are going ten to fifteen miles an hour over the speed limit receive warnings. Everyone else gets a ticket.

Case 5: An on-duty police officer finds a wallet in a shopping mall parking lot. It has $400 in it. He takes the money for himself and reports the wallet as found property.

Case 6: While on patrol, a police officer discovers a subject trying to break into a car. When the officer attempts to investigate, the subject runs away. After a short chase, the police officer apprehends the subject and places him in handcuffs. The officer knowingly tightens the handcuffs to the point where it causes pain and disregards the subject's complaints.

Case 7: A police officer discovers a burglary at a shoe store. While investigating, the officer finds four of her favorite pairs of shoes and puts them in the trunk of the squad car. She then calls in the burglary. The shoes taken by the officer are included in the list of items stolen in the burglary.

Case 8: A male police officer makes a traffic stop at 2:00 a.m. and discovers the motorist is an intoxicated woman. The officer demands the woman perform a sex act or he will arrest her for driving while under the influence.

Case 9: A police officer makes a traffic stop at 2:00 a.m. and discovers the motorist is an intoxicated off-duty police officer from a neighboring jurisdiction. She gives the officer a ride home.

Case 10: A police officer plays an important role in solving the burglary of an electronics store. As an expression of gratitude, the owner of the store offers the officer a television for his own use—a value of about $400. The police officer accepts the television.

CITIZEN COMPLAINTS

Citizen complaints are a second source of data used to study and document the extent of police misconduct. To allow increased transparency, some police departments provide public access to complaints filed by citizens; however, easy access to such data is not common. As a notable exception, as a result of a lawsuit filed against the Chicago Police Department, complaints filed against Chicago police officers are published online ("Citizens Police Data Project"; https://cpdp.co/). As with self-report surveys, there are many potential problems with this source, including the following:

- The misconduct is limited to what citizens see or experience in an interaction with officers. Most citizen complaints relate to claims of discourtesy.
- Citizen complaints are uncommon. Research has shown that of all citizens who reported the police did not behave properly during a street stop or a traffic stop, only about 3% actually filed a complaint.[11]
- The number of complaints received by a police department is determined not only by the quality of services provided by its officers but also the nature and requirements of the complaint process in that department. In departments where it is easier for citizens to file complaints, more complaints will be filed.
- Citizen complaints may not actually relate to police misconduct but rather to the citizen's *perception* of misconduct. The officer may have displayed proper conduct but that conduct was interpreted by the citizen as improper.
- Officers may learn how to avoid citizen complaints without actually changing their problematic behaviors. For example, officers may learn certain types of people are less likely to complain than others, so improper behaviors may be directed toward those people.[12]

Studies that examine police misconduct from a citizen complaint perspective offer several findings, including the following:

- Every study that has examined the issue found a small number of officers account for a disproportionate amount of citizen complaints.
- The officers who account for a relatively large proportion of citizen complaints tend to be male, young, and inexperienced.[13]
- Complaints are most likely to be received by officers early in their careers (years two through five) and then steadily decline thereafter as officers "master their craft."[14] (See Figure 12.1.)
- Research has generally found officers with a college education receive fewer complaints compared with officers without a college degree.[15]
- Patrol officers are more likely to receive citizen complaints than supervisors. Patrol officers are also most likely to receive *internal* complaints *from* supervisors.[16]
- An officer's rate of arrests and citations affects his or her likelihood of receiving complaints from citizens, with more productive officers receiving more citizen complaints.[17] However, it is not clear if complaints are a by-product of officer activity (e.g., arrests) or whether more productive officers are also more likely to act improperly in police–citizen encounters.[18]

LAWSUITS AGAINST THE POLICE

A third source of data relating to police misconduct is lawsuits filed against the police. Some of the same limitations that apply to citizen complaint data also apply to lawsuits, but those limitations are even more pronounced with lawsuits. Lawsuits are filed in only a small fraction of citizen complaints and generally for only the most serious allegations of misconduct. As a result, lawsuits do not provide a complete and accurate representation of police misconduct. For example, in 2018, thousands of citizen *complaints* were filed against Chicago Police Department officers, but only a few hundred involved *lawsuits*. Approximately 200 lawsuits led to payouts to citizens, which involved more than $113 million.[19]

An analysis of 655 settled lawsuits against the Chicago Police Department between 2012 and 2015 discovered the following:

- From 2012 to 2015, the city of Chicago spent $210 million to settle police misconduct lawsuits. This figure does not include all attorney fees. A portion of police department annual budgets and long-term loans were used to cover the costs of settlements.
- In approximately half of all cases, the city paid out $36,000 or less to settle the lawsuit.
- Approximately one-half of the lawsuits claimed officers filed false reports and/or committed perjury to cover up their misconduct.
- Most misconduct lawsuits related to routine encounters between police and citizens. The most common allegations included false arrest and excessive force.[20]

Depending on one's perspective, lawsuits are either an inescapable by-product of controversial police work or an indicator of a problem that needs to be addressed. Some police departments, including those in New York City and Philadelphia, routinely analyze lawsuits and recommend policy changes based on the findings of those analyses.

FIGURE 12.1

Average Citizen Complaints by Officers' Years of Work Experience for Late 1980s Cohort

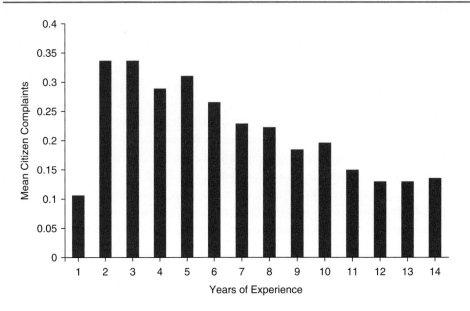

Source: Christopher J. Harris. 2009. "Exploring the Relationship Between Experience and Problem Behaviors: A Longitudinal Analysis of Officers From a Large Cohort." *Police Quarterly* 12: 192–213, p. 203.

MEDIA REPORTS

A fourth common source of data regarding police misconduct is media reports and stories. Studies have examined several types of police misconduct using this source, including police sexual violence and misconduct,[21] off-duty misconduct of officers,[22] criminal misconduct of female officers,[23] and domestic violence incidents in which officers were involved.[24] The major problem with the news media as a source of data is that the media are likely only to report the most serious and sensational stories. This may lead to an overall view of police misconduct that is biased toward the most serious types of incidents. The media also may not provide complete accounts of the incidents of interest.

From the analysis of media accounts, the following has been found:

- Between 2005 and 2011, there were 6,724 cases in which officers were arrested for crimes. These cases involved approximately 792 officers per year. The most common crimes were simple assault, driving under the influence, and aggravated assault. About 40% of crimes were committed on duty and nearly 95% involved male officers. The arrest rate for officers translates into approximately 1.7 arrests per 100,000 people. In comparison, in 2012 the arrest rate for the general population was 3,888 arrests per 100,000 people.[25]
- Between January 1, 2005, and December 31, 2007, 1,126 officers in the United States were arrested for off-duty conduct and 993 for on-duty conduct. Of the cases where a charge was specified in the media account (about half of the time), the most common off-duty offenses were driving under the influence and simple assault. The most common on-duty offenses were aggravated assault and simple

assault. The most common sanction for off-duty and on-duty offenses was suspension.[26]

- Between January 1, 2005, and December 31, 2007, 324 incidents occurred in the United States in which police were arrested for domestic violence. These cases involved 281 officers employed by 226 agencies. Nearly all of the officers who were arrested were male and held the position of police officer. The officers were between the ages of twenty-six and forty-three years and had zero to five years of service. In nearly all cases, they were off duty at the time of the incident.[27]

- Between 2005 and 2007, criminal conduct by policewomen was more often profit-driven compared to policemen and less likely to involve violent crime. Five percent of all off-duty arrests involved female officers.[28]

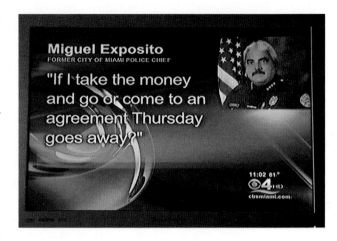

Photo 12.4

The media are often interested in stories of police corruption and misconduct. However, those stories may not always be reported fully or accurately. @Jeff Greenberg/Alamy Stock Photo

DECERTIFICATION STATISTICS

Decertification statistics are another indicator of police misconduct. When a police officer is decertified, it means the officer's training credentials are revoked, which makes him or her unemployable as a law enforcement officer. There are forty-five states that have the authority to decertify officers,[29] and these decisions are typically made by a POST (Peace Officer Standards and Training) board. POST boards receive reports of officer misconduct from police agencies. From the reports, decertification decisions are made. One study showed that of all reports of misconduct received by the Arizona POST board from 2000 to 2011, only 22% led to decertification.[30]

These statistics have limitations. The reporting of misconduct by police agencies to POST boards is largely voluntary and likely represents only serious cases of misconduct. Decertifications—cases where officers have been fired—represent the most serious of the serious cases. It is also difficult to interpret decertification as a measure of misconduct. More decertifications (or more reports of misconduct for that matter) do not necessarily mean more misconduct; it may mean less tolerance for misconduct. Finally, the rules vary across states as to the reasons why an officer can be decertified and vary across departments as to whether that reason needs to be reported to the state law enforcement licensing agency.[31]

A study that examined decertification statistics in 2011 found that, in most cases, the reason for decertification was the officer had received a felony conviction (43%). Other common reasons included being convicted of a misdemeanor (22%) or other misconduct (29%). In 2011, forty-four of the forty-five states with decertification authority reported certifications had been revoked from a total of 717 officers.[32] Georgia reported the greatest number of decertifications (178 officers decertified out of 25,332 officers employed), followed by Florida (72 officers out of 44,613). Several states had no officers decertified (Alaska, Kentucky, Louisiana, North Carolina, Virginia, Wisconsin). The most important takeaway here is that decertification is a potentially useful method of preventing officers dismissed from one law enforcement agency from being hired by other, but these statistics are not a good measure of police misconduct. Given the limitations of these data, one should be cautious about drawing conclusions from them.

decertification: The revoking of an officer's training credentials.

CAUSES OF POLICE MISCONDUCT AND CORRUPTION

There are two broad perspectives regarding the causes of and contributors to police misconduct. These perspectives are not necessarily mutually exclusive; they may interact and reinforce each other. The first is the belief something is wrong with the officers who engage in misconduct. They have personal shortcomings or character flaws, low morals, personal deficiencies, or other characteristics that lead to problematic behaviors. With this perspective, misconduct is viewed as being limited to officers who have these characteristics, and the solution is to identify these officers as soon as possible and provide appropriate interventions, such as retraining, or appropriate sanctions, such as employment termination. The prevention of the misconduct in the first place depends on a more thorough selection process. Additionally, it is important to identify and deal with problem officers early because they can influence other officers, creating a bigger problem. This phenomenon is sometimes referred to as the bad apple theory.

The other perspective suggests police corruption and misconduct are more a function of the nature of the job and the culture of the organization rather than strictly a problem with officers. Critical to this view is the belief that the environment in which officers work creates opportunities for misconduct. Officers may start their careers with integrity and the best of intentions, but the job changes them. Only when there are measures put into place to counter these opportunities and temptations can misconduct be prevented and organizational integrity maximized.

JOB AND ORGANIZATIONAL CHARACTERISTICS

There are several characteristics of the police job that create opportunities for officer corruption and misconduct. As you may recall, many of these issues were also discussed in Chapter 8.

POWER, AUTHORITY, AND DISCRETION Police officers have extraordinary discretion; they have the duty and authority to make decisions based on their own judgment. Whenever the police need to make a decision, they may make that decision based on their own self-interest or act on their own personal biases or desires. Some officers confront many temptations in the course of their work. These temptations are a direct path to misconduct. Some officers create such opportunities through outright abuse of their power—through coercion. Such is the case when officers demand bribes or other favors from citizens to avoid arrest. This is unfortunate, but it is a fact of police work. As explained by one officer,

> most cops are good guys and don't commit crimes or make a habit of breaking the rules, but this job sure does provide the opportunities and temptations. Let's face it, cops have a lot of power and if they decide to abuse that power, it's not that hard for them to get away with it.[33]

LOW-VISIBILITY WORK ENVIRONMENT Not only do the police confront temptations on a regular basis, they do so in low-visibility situations. In other words, no one actually sees what a police officer does, besides perhaps another officer or citizens involved in a particular situation. There are seldom credible witnesses to police misconduct, unless a citizen has cell

bad apple theory: Theory holding a small number of problem officers can influence other officers, thus creating a bigger problem.

phone video of the incident. In addition, rarely do (or can) police supervisors directly or continuously monitor the actions of officers. Even with body-worn cameras, supervisors are unlikely to review all of an officer's conduct over the course of an entire shift, and of course, if the camera is not on there is no record of the officer's actions. This low-visibility environment creates opportunities for some to act on their biases and desires. Think of your own work situation and how it is when a supervisor is not at work or not watching your work. There may be much more "goofing off" among some employees, at the very least. The police are no different in this regard, except the potential for serious misconduct is greater in the police occupation than in many others.

THE CODE OF SILENCE AND THE POLICE CULTURE Even if police officers and supervisors are aware of fellow officers' misconduct, it may not make much of a difference in stopping the behavior. This is due to what has been referred to as the code of silence that exists among police officers. As one officer explains,

> many cops will go a long way to help protect their own. I am not saying that cops cover for cops in all cases . . . but if it is something minor, like having sex on duty, it isn't likely cops will "rat out" other cops for something that doesn't hurt anyone.[34]

The code of silence is an important dimension of the police culture, and any attempt to control police misconduct must deal directly with it. As discussed in Chapter 8, a culture consists of a collection of unwritten rules, expectations, values, and attitudes shared and understood by members of a group. Violation of a rule typically results in informal sanction(s) being applied to the person who violated the rule.

code of silence: An unwritten rule of the police culture that holds officers should not report the misconduct of fellow officers. It is a human phenomenon, not just a police phenomenon.

culture: A collection of unwritten rules, expectations, values, and attitudes shared and understood by members of a group.

Photo 12.5

Many decisions that officers make are only visible to the officer and to the citizen(s) affected by the decision. This can create opportunities for misconduct and make the control of misconduct more difficult.

© iStockphoto.com/400tmax

EXHIBIT 12.3

Extraordinary Consequences of Breaking the Code of Silence

Officer Donna Watts of the Florida Highway Patrol pursued and eventually stopped a marked police vehicle that was traveling over 120 mph. She believed the police vehicle might have been stolen, but in actuality, the driver was an off-duty uniformed Miami police officer who was late for an off-duty job. The Miami officer asked for a professional courtesy to avoid arrest, but Officer Watts arrested the officer for reckless driving anyway. As a result of this action, she experienced all sorts of retaliation: prank phone calls, pizzas delivered that she did not order, unidentified vehicles parked outside her home, social media threats, and the access of her personal DMV records by officers from twenty-five different agencies.

According to Watts's supervisors, she could no longer be assigned to patrol duties because of fear that other officers would not respond to her calls for backup if necessary. As they had essentially ruined her career, Watts filed a lawsuit against the officers and agencies involved in the retaliation. Although this is clearly an extraordinary example, it speaks directly to the implications of reporting the misconduct of fellow officers.

Source: Geoffrey P. Alpert, Jeffrey J. Nobel, and Jeff Rojek. 2015. Solidarity and the Code of Silence. In Roger G. Dunham and Geoffrey P. Alpert, *Critical Issues in Policing: Contemporary Readings.* Long Grove, IL: Waveland.

There are certain things all police officers understand because of their shared experiences as police officers, but there are also understandings and rules unique to particular police organizations. Thus, there is a police *occupational* subculture and a police *organizational* subculture. In many instances the two overlap. Certain values are strongly emphasized in the police occupational subculture, including solidarity among officers. Police work involves danger, hostility from citizens, and scrutiny from outsiders. Most people do not understand the difficulties of police work—how emotionally challenging and frustrating it is and how easy it is to make mistakes. In response, the police tend to become suspicious of others outside their occupation. This is a police survival skill. Officers have to know their fellow officers have their back; their life and well-being may depend on it. From this solidarity comes a rule: Do not report on the misconduct of fellow officers. Trust requires loyalty, and people who are loyal to each other do not report on each other's misdeeds. A trustworthy friend does not reveal the secrets of another. This is the code of silence.

The violation of the code of silence can carry significant consequences from other members of the subculture group. At the very least, an officer who breaks it may find herself or himself viewed as an outsider within the department. Avoiding the consequences associated with breaking the code is a powerful inducement not to report misconduct (Exhibit 12.3).

THE CONTROL OF POLICE MISCONDUCT AND CORRUPTION

As Professor Robert Worden and his colleagues explain, "For all of the ink that has been spilled by social scientists on the subject of police misconduct, there is much that we do not know,"[35] including how to effectively control it. Police misconduct is a perfect example of a deeply entrenched and persistent problem in the occupation. It has existed since the first officers patrolled the streets back in the 1800s, and though numerous attempts have been made throughout history to address it, the issue is still present today.

Why is police misconduct such a difficult problem to confront? The short answer is for the reasons discussed above: (a) because power, authority, and discretion are undeniable aspects of the police job; (b) because police work is performed in a low-visibility environment; and (c) because the code of silence among officers is very difficult to crack. However, effectively controlling police misconduct should not be considered a lost cause; renewed and continuing efforts on initiatives already in place may help. Effectively controlling police misconduct depends heavily on managing the causes of the problem. The following remedies are discussed in the remaining pages of this chapter: (a) controlling police discretion and authority, (b) cracking the code of silence, (c) decertification of officers, and (e) early intervention systems.

CONTROLLING POLICE DISCRETION AND AUTHORITY

As discussed earlier in this book, it is not possible to eliminate police discretion or to revoke the authority of the police. The police are the police *because* of their authority; it is one of the fundamental reasons we have the police. It is not possible to *eliminate* discretion because police tasks are too complex, the police operate with limited resources, and we expect the police to treat people with some degree of flexibility. But it is possible to *control* police discretion. This can be done in indirect ways, such as through the selection of qualified and capable officers who demonstrate a history of integrity. It can be done by creating a virtuous culture within the police organization so that *everything* done in the organization (e.g., selection and training of officers, policy development, rule enforcement, and so forth) is done with the value of integrity in mind.

Police work occurs and will continue to occur in low-visibility situations. However, as discussed in previous chapters, police body-worn cameras can be helpful in this regard. With body-worn cameras, low-visibility environments are no longer quite so low. Although there are many issues, limitations, and controversies associated with the deployment of body-worn cameras, their impact on the integrity of police work may prove to be significant.

CRACKING THE CODE OF SILENCE

When it comes to combatting police corruption and misconduct, there is no greater enemy than the code of silence. It is an enemy that needs to be attacked from many different directions in order to be overcome. If anyone is aware of officers engaging in misconduct, it is their fellow officers. And if officers believe they can act improperly without being reported, there will be little control over misconduct. To fight the code of silence, officers must understand their responsibilities and be aware that consequences exist for remaining silent if they witness misconduct. It is probably unrealistic to think that the code of silence can be eliminated, but it *can* be minimized.

A Question to Consider 12.1

Have You Ever Reported the Misconduct of Another Student?

Throughout your educational career, have you ever been aware of a fellow student engaging in academic misconduct, such as cheating on an exam or plagiarizing a paper? If so, did you report the conduct of the student to the teacher or professor?[36] Why or why not? Do you think you would have been more or less likely to report the dishonest student if you were friends and/or coworkers with him or her? Do you think you would have been more or less likely to report the student if you feared serious retaliation from other students in the class for reporting the misconduct?

Part of the reason it is so difficult to break the code of silence is that the code is not just a police phenomenon, it is a human phenomenon. In this sense, trying to break the code of silence among police actually involves changing human nature. From an early age, children are often discouraged from "tattling" on others. And we all know "loose lips sink ships." Silence is a human value and is evident in many situations. Citizens who have information about criminals are sometimes reluctant to give that information to the police (in some cases because "snitches end up in ditches"). Workers in other occupations are also often reluctant to report on the misdeeds of coworkers. Even students may be hesitant to report on the dishonesty of fellow students (see the Question to Consider feature). The code of silence is not unique to the police, but what makes it more significant for the police than for many other occupations is police misconduct can result in an extraordinary amount of harm to citizens.

In spite of the difficulties of cracking the code of silence, several tools can be used to try to do exactly that.[37] These tools are discussed next.

RULES AND POLICIES Police departments have rules that govern much of the conduct of officers. They also need rules that define and prohibit police misconduct, but it is practically impossible to specify every police action that is forbidden. Policies regarding mandatory reporting of serious misconduct are necessary. Essentially, it must be made clear to officers that not reporting misconduct is in itself a form of misconduct. According to an article in *Critical Issues in Policing*,

> The goal of mandatory reporting is to remove any level of discretion from the officers by creating an affirmative duty to report serious misconduct. Officers who fail to report will be disciplined, thus removing the stigma that was attached to voluntary reporting.[38]

GOOD POLICING
The Importance of Police Honesty

Honesty is the best policy. Always tell the truth. Only losers lie. Although we have all heard these expressions, often we find ourselves in situations that are not so black or white. For instance, should you tell the truth when a spouse or friend asks, "Do these pants make you look fat?" Should you tell the truth when a murderer asks you the location of his intended victim? The issue of truthfulness gets even more complicated in police work,[39] where deception is not only sometimes accepted, it is expected! Police frequently lie to suspects on the street and in the interrogation room, and they are deceptive when conducting undercover investigations. They even sometimes lie to the community, such as when they post street signs indicating a "speed enforcement area" when in fact this is not true. In general, these deceptions are legal. However, it is *not* legal for police officers to lie in reports, in responses to other officers or supervisors regarding official matters, or in court when providing testimony. These lies represent police misconduct even if provided for "good" reasons. Deception in these circumstances can be serious.

The 1963 U.S. Supreme Court case *Brady v. Maryland* and related subsequent cases have held that officer credibility as a witness may be considered exculpatory evidence and given to the defense during discovery.[40] In other words, if there is evidence in an officer's personnel file that he or she was untruthful regarding any police-related matter, that evidence could be used in a defendant's defense. Ultimately, the judge in the case determines whether the evidence is admissible in court, but the point is that it could be and could cause major difficulties for the prosecution. Due to this, if an officer's deception is deemed serious enough to document and discipline, there is justification for that officer to be fired. Think back to the case presented earlier in which the officers who went sledding were fired. A major consideration in the decision to fire the officers involved the deceptive statements they made on top of the misconduct. Those officers would no longer have credibility as witnesses in any cases they might be involved in, which was not a risk the department could take. Their careers as police officers were over. By reviewing cases such as this, it is easy to see honesty is critical for police officers.

The objective is not to encourage mistrust among officers but to make clear that the reporting of misconduct is required. If this is understood, it stands to reason officers will be less likely to engage in misconduct. Policies that protect officers from retaliation due to the reporting of misconduct are also needed to reduce the influence of the code of silence. Guidelines regarding unacceptable deceitful conduct are also necessary. Deception is a complicated issue in police work that can have severe consequences for officers if improperly used (see Good Policing feature). The consequences for inappropriate deception need to be enforced. By using all these methods to address misconduct, a culture of integrity can overtake the code of silence in a police department.

A PROPER CITIZEN COMPLAINT PROCESS Besides rules and policies, another tool used to address the code of silence is an adequate citizen complaint process. Some police departments make it much more difficult for citizens to file a complaint against officers than others. Some departments require complainants to sign forms that specify a penalty for perjury, which means if the complainant is determined to have lied in the complaint, he or she can be criminally prosecuted. Some require complaints to be filed in person and/or at police headquarters. Departments may refuse to accept complaints from lawyers, other third parties, or juveniles. Each of these requirements represents a barrier to the citizen reporting of possible police misconduct. If officers believe complaints from citizens are very unlikely, some of them may act without fear of consequence.

It is important to understand the number of citizen complaints filed against officers in a department is not meaningful without an understanding of the complaint process itself. If complaints are easy for citizens to file in a particular department, it is likely that department will receive more complaints than one that makes it more difficult to file. Thus, the number of complaints filed does not necessarily speak to the quality of policing in a particular jurisdiction.

PROPER INVESTIGATIONS OF MISCONDUCT Another important aspect of cracking the code of silence is proper investigations of alleged misconduct. It is difficult, if not impossible, to specify what steps should be taken when investigating an allegation of police misconduct; it depends much on the nature of the allegation. As such, it is only possible here to emphasize the value of thorough investigations into misconduct and of the immediate administration of meaningful sanctions to officers who engage in misconduct.[41] The bottom line is that

> police officers who engage in misconduct must be held accountable for their actions. It is only through effective disciplinary actions that an employer sends an unequivocal message to the offending officer and others that serious misconduct will not be tolerated. Officers who protect other officers from allegations of criminal conduct or civil rights abuses by engaging in a code of silence should be separated from the organization.[42]

Accountability will not occur on its own. Effective supervision and leadership are also critical ingredients in the investigation of alleged misconduct.

ETHICS TRAINING A final tool to address the issues associated with the code of silence is ethics training. Ethics training should be ongoing throughout officers' careers and emphasize the situations officers may find themselves in and what actions should be taken. The training should include critical thinking, problem-solving, and decision-making strategies to use in difficult real-life work situations. However, as discussed in Chapter 8, ethics training must be built on a culture of integrity. All organizational processes must incorporate ethics and emphasize proper conduct of members in order for this training to have a chance of actually making a difference.

LOS ANGELES POLICE DEPARTMENT
COMPLAINT OF EMPLOYEE MISCONDUCT

This form should be used exclusively to report employee misconduct. Complaints regarding Los Angeles Police Department policies and procedures, or police response time to a location, should be discussed with the watch commander at your local police station. Upon completion of this form, you may either return it in person to the nearest police station, or mail the top copy to LOS ANGELES POLICE DEPARTMENT, Internal Affairs Group, P.O. Box 30158, Los Angeles, CA 90099-4896. A preaddressed business reply envelope has been provided for your convenience. Keep the second copy for your records.

Name _____   Phone _____   ☐ Day
                                                             ☐ Evening

Address _____   Language Spoken _____

Date of Occurrence _____   Time of Occurrence _____

Location of Occurrence _____

Names, Badge Numbers or Serial Numbers   Names, addresses, and telephone numbers of witnesses present at the
of Employees Involved (If known).   time of occurrence (If known).

_____   _____

_____   _____

_____   _____

(LIST ADDITIONAL EMPLOYEES AND/OR WITNESSES UNDER THE "DETAILS" SECTION.)

Details - (Please state your complaint, including names, times, locations, witnesses, and any other information that would help in investigating your complaint. If employee names are unknown, explain what each employee looked like.)

_____
_____
_____
_____
_____
_____
_____
_____
_____
_____
_____
_____
_____
_____

Date _____   Signature _____

DEPARTMENT USE ONLY

To be completed by the supervisor receiving this form.

Supervisor's name _____   Serial Number _____

Date and time received _____   Division _____

Final disposition _____
(i.e. forwarded to IAG; 01.28.00 initiated; sent correspondence to complainant, etc.)

(Attach additional sheets, if needed.) | CF NO. _____ | DIV. NO. _____ |

01.81.06 (12/98)

Photo 12.6

A well-managed citizen complaint process is essential to address the code of silence and reduce instances of police misconduct. Los Angeles Police Department

DECERTIFICATION OF OFFICERS AND THE NATIONAL DECERTIFICATION INDEX

When officers who are fired from one police department for serious misconduct (or resign prior to being fired) get hired by another agency, misconduct simply gets moved around rather than eliminated. Decertification has the potential to stop this from happening. As noted earlier, most states have the authority to revoke an officer's certification (decertification), essentially making it impossible for that person to be employed again as a police officer. An issue that arose, however, was how to share information about decertified officers across agencies. Enter the National Decertification Index (NDI), a list of decertified officers developed by the International Association of Directors of Law Enforcement Standards and Training. Hiring agencies can check the NDI to see if an applicant has been decertified and, if so, can request additional information about the applicant from the decertifying state. Ideally, the NDI can prevent the rehiring of problem officers and, as a result, enhance the integrity of law enforcement agencies.[43]

Although the NDI is useful, it has some limitations. First, very few officers are fired for misconduct; when this *does* happen, the misconduct was usually very severe. Thus, the NDI will likely only prevent the most serious violators from being rehired. Additionally, not all states have the authority to decertify officers, states decertify for different reasons, and only thirty states contribute to the NDI. Further, not all agencies report the dismissal of an officer to the state licensing agency for inclusion into the NDI, so not all officers who are fired for misconduct will be included in the index. And finally, not all agencies check the NDI when conducting background investigations on applicants. The full potential of the NDI will not be realized until (a) state law requires police departments to report officer dismissals for misconduct to their state licensing agencies and (b) federal legislation requires states to submit information to the NDI.

EARLY INTERVENTION SYSTEMS

An early intervention system (EIS or EI system) or early warning system, as it is sometimes known, is an information management system designed to monitor the performance of officers and identify those officers who show certain at-risk behaviors. Information on various aspects of the performances of a department's officers is continuously entered into the system. When officers are identified who might pose a problem, they receive some type of nondisciplinary remedial intervention, typically in the form of counseling or retraining. This can allow performance issues to be corrected before they get too serious.[44]

During the past twenty years, EISs have often been highly recommended as a way of ensuring increased police integrity and accountability,[45] and they are common in large police departments today.[46] Indeed, these systems hold much promise. Some, however, see them as a work in progress. According to Professor Worden,

> For all of the stock that has been placed in EI systems, social science has not provided much or very strong evidence on their effectiveness or on their unintended effects.... Very few evaluative studies have been conducted, and their research designs have not been strong. Moreover, scarcely any evidence exists on the hypothesized but unintended effect of EI systems.[47]

There are three critical components to the operation of an EIS: (1) the identification of the at-risk performance indicators, (2) the threshold at which these indicators are deemed problematic, and (3) the nature of the intervention in which the identified at-risk officers participate.[48] With regard to each component, there are questions that need to be resolved in order to maximize the effectiveness of future systems.

The first component is the performance indicators that are collected and entered into the system. These indicators consist of the factors that presumably indicate an officer is at risk—the predictors of future serious misconduct. Various indictors may be used in an EIS, including the following:[49]

National Decertification Index (NDI): A list of decertified officers in the United States.

- Number of force incidents involving an officer (*may* indicate over-zealousness)
- Citizen complaints received about an officer (*may* indicate lack of sympathy or over-zealousness)
- Civil suits (*may* reflect inappropriate or illegal behavior)
- Traffic stops and arrests made (*may* indicate bias)
- Traffic accidents involved in (*may* indicate lack of attention, carelessness)
- Use of sick leave (*may* indicate job dissatisfaction, chronic health concerns)

On its face, it seems reasonable to collect and monitor such information. However, the problem is that "none of the indicators used in or prescribed for EI systems is a valid and reliable indicator of misconduct."[50] For example, many citizen complaints are without merit and allege conduct that was neither in violation of law or policy. Use of force is sometimes a necessary aspect of the job and it is usually justified. Therefore, commonly used at-risk performance indicators are not necessarily good predictors of misconduct.

The second component of an EIS is the threshold at which officers are identified as exhibiting at-risk behaviors. For example, do three incidents of use of force in a year constitute at-risk behavior, or should it be four a year? Or five? The point is the cut-offs at which officers are identified as being at risk are usually arbitrary. There may be no connection between a certain number of at-risk behaviors and whether an officer is on a path to serious misconduct. Further, certain officers, due to their assignment (e.g., to a high-crime area) and/or overall activity level (e.g., make a lot of arrests), may be more likely than others to be involved in at-risk behaviors and thus more routinely identified as exhibiting at-risk performance. The use of force and citizen complaints may be a by-product of productive policing, not an indicator of bad conduct.

The third component of an EIS is the intervention or what is done with officers who are flagged as exhibiting the supposed at-risk behaviors. Usually the intervention consists of retraining or counseling. However, it is unclear what the content of the retraining or counseling should be in order for it to make a difference in the performance of officers going forward.

A recent study that examined the effects of an EIS had some surprising findings.[51] The study monitored the performance of two groups of officers: (1) at-risk officers who were flagged by the department's EIS and who participated in retraining (Officer–Civilian Interaction School) and (2) other officers who were not identified as exhibiting at-risk behaviors and who did not participate in retraining. Of particular interest was the number of complaints received against the two groups of officers and the number of arrests made by them. The authors found that after the introduction of the EIS, the number of complaints received against the at-risk officers and the number of arrests they made decreased, *but so did the number of complaints against and arrests made by the other group of officers.* Apparently, both groups of officers knew making arrests increased the likelihood of use of force and citizen complaints—two factors that could lead to enrollment in the Officer–Civilian Interaction School. Perhaps the "threat" of retraining carried with it a negative stigma, so all officers did what needed to be done to avoid that stigma: They made fewer arrests. The takeaway here appears to be the EIS had a chilling effect on arrests. If there are crime control benefits to arrests, this is clearly an unwelcomed consequence.

With these understandings in mind, Professor Worden and his colleagues conclude,

No one would dispute that police managers should have at their disposal information about their officers' patterns of performance, including citizen complaints, documented uses of force, and so forth—that is, the kinds of information that many EI systems capture. Indeed, it seems logical that this information—and more—is needed to effectively assess and manage officers' risk of misconduct. But how that risk can be effectively managed and whether EI systems have realized their promise remain largely unanswered questions.[52]

automatic vehicle locator (AVL) system: A GPS system that tracks the whereabouts and activities of officers when in their squad cars.

GPS, Tracking Police Vehicles, and Preventing Misconduct

Global positioning system (GPS) technology has many uses, including tracking the whereabouts and activities of officers when in their squad cars through an **automated vehicle locator (AVL) system**.[53] Knowing the location of squad cars can be useful when determining which squad car is closest to a crime in progress, identifying the location of squad cars for safety purposes, and identifying where patrol is being conducted when officers are not responding to calls for service. Some systems also record the real-time speed of police vehicles, deployment of air bags, use of seat belts, and use of lights and siren. GPS technology can also be used to monitor the whereabouts of officers for purposes of preventing misconduct.

While police departments that deploy GPS tracking of officers emphasize the benefits of the technology, some officers see it as surveillance.[54] Departments have reported the GPS tracking devices are sometimes rendered inoperable, presumably by the officers the technology is intended to track. This should not be too surprising as workers in most occupations tend to resist closer supervision. The technology by itself will not eliminate or even limit police misconduct; its effectiveness as an accountability device depends on a system of discipline and sanctions for verified misconduct.

How Does the Police Occupation Rank on Perceptions of Honesty and Ethics?

In 2017, a telephone survey was conducted by the Gallup organization. Callers asked a random sample of 1,049 adults ages eighteen and older to rate the honesty and ethical standards of the police as well as of people in other occupations. Specifically, the researchers said, "Please tell me how you would rate the honesty and ethical standards of people in these different fields." Answer options were "very high,"

"high," "average," "low," or "very low." For police officers, 56% of the respondents stated "very high" or "high" (see Table 12.1). Note police are on the more positive side of the continuum when judged in relation to other occupations. Nurses, military officers, and grade school teachers ranked the highest; members of Congress, car salespeople, and lobbyists were on the negative side of the continuum.

TABLE 12.1

Gallup's 2017 Honesty and Ethics of Professions Ratings

Please tell me how you would rate the honesty and ethical standards of people in these different fields—very high, high, average, low, or very low?			
Sorted by very high/high.			
	Very high / high %	Average %	Low / very low %
Nurses	82	16	2
Military officers	71	24	3
Grade school teachers	66	27	5
Medical doctors	65	31	4
Pharmacists	62	32	6

(Continued)

TABLE 12.1 (Continued)

	Very high / high %	Average %	Low / very low %
Police officers	56	32	12
Day care providers	46	43	7
Judges	43	41	15
Clergy	42	41	13
Auto mechanics	32	53	14
Nursing home operators	26	48	22
Bankers	25	54	21
Newspaper reporters	25	39	35
Local officeholders	24	53	20
TV reporters	23	39	37
State officeholders	19	47	33
Lawyers	18	53	28
Business executives	16	54	28
Advertising practioners	12	49	34
Members of Congress	11	29	60
Car salespeople	10	48	39
Lobbyists	8	31	58

Source: Gallop, December 26, 2017, "Nurses Keep Healthy Lead as Most Honest, Ethical Profession."

FIGURE 12.2

Overall Confidence in the Police 1993–2017

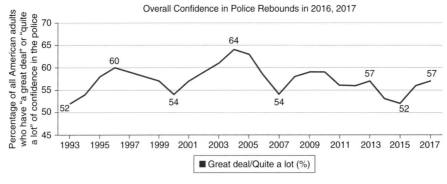

Source: https://news.gallup.com/poll/213869/confidence-police-back-historical-average.aspx.

It is also important to note that Gallup Polls conducted every year since 1993 show that confidence in the police rebounded in 2016 and 2017 from 2015 lows (Figure 12.2). The years preceding 2015 were especially turbulent for the police and citizens beginning with the 2013 acquittal of George Zimmerman for the shooting death of unarmed black teenager Trayvon Martin. This incident led to the creation of the Black Lives Matter movement. Even though overall ratings of the police have bounced back in 2016 and 2017, it is important to note that there have been and continue to be large differences in ratings of the police by race; blacks consistently rate the police less favorably with whites. In 2015 through 2017 for instance, 61% of white adults expressed "a great deal" or "quite a lot" of confidence in the police while only 30% of blacks had this same sentiment.[55]

Main Points

- Police misconduct can have very negative and far-reaching consequences for the police and the communities they serve. As a result, it is important to understand the complexities of this issue.
- Police corruption refers to a relatively narrow set of police behaviors: (a) illegal behaviors (b) committed by police officers who use their authority as police officers to commit crimes (c) that are motivated by personal gain.
- There is nothing noble about noble cause corruption.
- Police misconduct may include on-duty as well as off-duty conduct. It refers to corruption but also to other inappropriate or illegal behaviors of police officers.
- Police integrity can be defined as the inclination among offices to resist temptations to abuse the rights and privileges of their occupation. Police integrity is the opposite of misconduct.
- Five sources of data have been used to try to provide insight into the nature and extent of police misconduct: self-report surveys, citizen complaints, lawsuits against the police, media reports, and decertification statistics. Each has its limitations.
- Several factors can contribute to an environment in which police corruption can occur, including the extraordinary power and discretion of the police, the fact that police work in low-visibility situations, and the strong influence of the code of silence. These factors must be counterbalanced with other strategies.
- To combat the code of silence, officers must understand there are serious consequences for remaining silent about other officers' misconduct. Policies that protect officers from retaliation due to reporting of misconduct are also necessary.
- Ethics training is another important tool to address the issues associated with the code of silence. However, ethics training must be built on a culture of integrity.
- Most states have the authority to revoke an officer's certification (decertification), essentially making it impossible for a person to be employed again as a police officer. The National Decertification Index (NDI) is a list of decertified officers. The NDI can be useful but has limitations.
- An early intervention system (EIS) is an information management system designed to monitor the performance of officers and identify those officers who show certain at-risk behaviors. While these systems hold promise, there are critical questions that need to be resolved in order to maximize their effectiveness.

Important Terms

Review key terms with eFlashcards at **edge.sagepub.com/brandl2e.**

Questions for Discussion and Review

Take a practice quiz at **edge.sagepub.com/brandl2e.**

1. How does Denver's Citizen/Police Complaint Mediation Program operate? What do you see as the major strengths and limitations of such a program?
2. What is the difference between police corruption, police misconduct, police deviance, and police integrity?
3. True or False: The number of citizen complaints received by a police department is a good indicator of the quality of services provided by that agency. Explain.
4. What is the code of silence? Is it any more significant in police agencies than in other organizations? Why or why not?

5. How can the code of silence be cracked? Why is it so important to address this code when trying to create police department integrity?
6. Why is the issue of honesty and deception such a difficult one in police work? How does the U.S. Supreme Court decision of *Brady v. Maryland* (1963) relate to this issue?
7. How can the effective management of citizen complaints (their receipt, investigation, and disposition) help control police misconduct and corruption?
8. What is decertification? What are the strengths and limitations of decertification as a way of promoting police integrity?
9. What is an early intervention system? What are its intended and unintended effects?
10. How might GPS technology prevent police misconduct?

Fact or Fiction Answers

1. Fiction
2. Fiction
3. Fact
4. Fiction
5. Fact

6. Fact
7. Fact
8. Fact
9. Fact
10. Fiction

$SAGE edge™

Digital Resources

Get the tools you need to sharpen your study skills. SAGE Edge offers a robust online environment featuring an impressive array of free tools and resources.

Access practice quizzes, eFlashcards, video, and multimedia at **edge.sagepub.com/brandl2e**.

Media Library

View these videos and more in the interactive eBook version of this text!

Career Video
12.1: Common Public Misconceptions About Police Misconduct

Criminal Justice in Practice
12.1: Ethics—Gratuities

SAGE News Clip
12.1: US NYPD Corruption

Part IV
POLICE STRATEGIES AND THE FUTURE OF THE POLICE IN AMERICA

13
COMMUNITY AND PROBLEM-ORIENTED POLICING

Police Spotlight: Problem-Oriented Policing in Chula Vista[1]

The city of Chula Vista, California, had a problem: Crime was rampant at the city's more than two dozen motels. Because of this, visitors to the city were hesitant to stay at these motels, thus hurting the local economy. The city has an Olympic training center, but even the visiting athletes stayed farther away, in San Diego, for safety reasons. Business leaders and policymakers were very concerned. They brought the issue to the Chula Vista Police Department (CVPD) to see if it could do something to address the situation.[2]

The CVPD engaged in a multiyear effort to address the crime problem at the motels. Initial analysis of the problem focused on calls for service (CFS) in relation to motel rooms. An analysis of CFS revealed that five motels accounted for 24% of the Chula Vista motel rooms but 55% of the approximately 1,200 CFS.

The project staff also interviewed motel guests at problem properties and learned that 75% of those questioned were residents of San Diego County. Many were homeless, on probation, or on parole; very few were tourists. Next, the CVPD partnered with researchers at California State University, San Bernardino, to develop and administer a survey to the local motel managers to gain a better understanding of the crime problems from their perspective. The results revealed that motels that rented primarily to local customers and long-term guests produced more CFS. Based on this analysis, project staff believed educating motel managers about their crime problems and what could be done about them would reduce CFS. The project team met with local motel managers and provided technical assistance to improve the properties and increase safety through measures such as dead bolts on exterior doors. These efforts led to only a 7% decrease in CFS. The city then implemented an ordinance that prohibited hourly room rentals and required motel guests to present photo IDs when they checked in, but those measures had no impact on crime, disturbances, drugs, or assaults.

Then, working with the city's planning and building division (including code enforcement), community development, finance department, fire department, the city attorney's office, and community/business

(Continued)

Objectives

After reading this chapter you will be able to:

13.1 Identify the factors that gave rise to community and problem-oriented policing and the strategies tried by police departments prior to these efforts

13.2 Define community policing and discuss what is known about the overall effectiveness of it

13.3 Explain why it may be difficult for the police to change citizens' attitudes toward them

13.4 Define the concept of procedural justice and discuss why it may be difficult for the police to actually create it

13.5 Discuss the SARA model of problem solving and discuss examples of it being used in police departments to address crime problems

13.6 Evaluate what is known about the overall effectiveness of problem-oriented policing

Fact or Fiction

To assess your knowledge of community and problem-oriented policing prior to reading this chapter, identify each of the following statements as fact or fiction. (See page 300 at the end of this chapter for answers.)

1. Between 1960 and 1970, the crime rate in the United States was quite stable, due in large part to preventive patrol and the work of detectives.

2. Nearly all studies on the issue show people of color—especially African Americans—have more negative attitudes toward the police than whites.

3. Community relations bureaus in police departments and team policing were effective in reducing crime but not in improving the relationship between police and citizens.

4. Community policing and problem-oriented policing are different terms for the same concept.

5. The research is clear: Disorder causes crime.

(Continued)

(Continued)

6. Fear of crime is bad and should be eliminated.

7. It is relatively easy for the police to improve citizens' attitudes toward them.

8. If citizens have positive attitudes toward the police, then they will engage in behaviors that help the police.

9. Evaluations of community policing are mixed: Some studies show it works; some show it does not.

10. Problem-oriented policing has been identified as a strategy that works.

(Continued)

Photo 13.1
The Chula Vista Police Department used a problem-solving approach to address crime at the city's motels.
ZUMA Press. Inc./Alamy Stock Photo

groups, Chula Vista revisited the ordinance idea. This time the focus was on safety performance standards. To address this, the city council developed and passed an ordinance that required motels to obtain an annual permit to operate. Failure to have a valid permit could result in fines of up to $1,000 and/or six months in jail. Permits could be denied based on CFS levels, unsanitary rooms, or lack of basic crime prevention devices, such as window locks or dead bolts. As a result of these performance standards, motel CFS dropped 49%.

A follow-up survey of motel managers was then conducted. The survey found local clientele and long-term guests—both of which were responsible for many of the CFS—decreased substantially at the motels. The permit requirement and code inspections also led to the implementation of safety features (dead bolts, door peepholes, door chains) in every motel room, increasing the overall safety of the properties.

This effort was judged so exemplary it won the Herman Goldstein Award for Excellence in Problem-Oriented Policing, which is awarded by the Office of Community Oriented Policing Services of the U.S. Department of Justice.

SINCE the 1970s, most police and other law enforcement agencies have become more citizen and problem oriented. This chapter discusses these orientations and the value and effectiveness of them.

THE RISE OF COMMUNITY POLICING

If there is one universal truth in policing (and in life), it is that change is constant. As discussed earlier in this book, policing has undergone several major changes since the first formal police departments were created in the United States in the mid-1800s. Each change was prompted by a crisis. In the 1960s, the crisis took the form of a crime wave. From 1960 to 1970, the violent crime rate more than doubled and appeared to be out of control (Figure 13.1). The 1960s were a difficult time for the police for other extraordinary reasons: President John F. Kennedy; the president's brother, Sen. Robert Kennedy; and Martin Luther King Jr. were all assassinated, and citizens were protesting the Vietnam War and demonstrating and rioting in the name of the civil rights movement.

FIGURE 13.1

Violent Crime Rate, 1960–1970

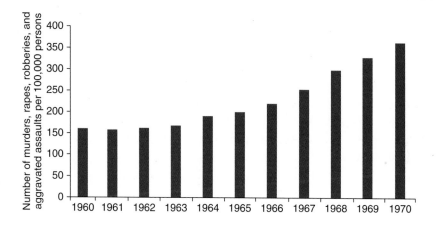

Source: Federal Bureau of Investigation, Uniform Crime Reporting Statistics, http://www.ucrdatatool.gov/Search/Crime/State/Run CrimeStatebyState.cfm.

A Question to Consider 13.1

Problem-Solving Efforts in Chula Vista

What do you think was the most important action taken by the Chula Vista Police Department to reduce crime at the city's motels? Why do you think that action made such a difference in CFS to the police department? Are there any other reasons why CFS from the motels may have declined?

To anyone who was paying attention, it appeared policing was not working. Prior to the 1960s, the police were operating under the belief they had what was necessary to control crime. Random preventive patrol was supposed to deter criminals, and for those criminals *not* deterred, rapid response to crime calls and detective-based criminal investigations were supposed to lead to their apprehensions. But it was clear in most large cities this was not happening, and research studies provided confirmation. Preventive patrol did not prevent crime,[3] rapid police response to calls for service seldom led to on-scene arrests,[4] and the actions of detectives rarely led to crimes being solved.[5] Undoubtedly, some of the crime problems of the 1960s were simply the result of population demographics—much of the "baby boomer" generation being in their crime-prone years—which had nothing to do with the operations of the police.

The police realized they needed to do something different. The immediate objective was somehow to fix the relationship between the police and the minority community. Secondly, the police needed to do something about the crime problem. The police and citizens had to somehow work together for there even to be a chance for order maintenance and crime prevention. But how was this to be done? Since this question was first asked in the early 1970s, several potential answers have been put forth.

IMPROVE THE RACIAL COMPOSITION OF POLICE DEPARTMENTS

One of the underlying reasons for the poor relationship between the police and racial minorities is that up until and through the 1960s, police departments were overwhelmingly represented by white officers—white, male officers to be precise. In the era of civil rights, this was especially problematic. For example, in Detroit in the mid-1960s, approximately 40% of the population was African American, but 95% of officers were white. So when the riots occurred in Detroit and other cities—riots often sparked by white officers shooting black citizens—it was predominantly white officers deployed to stop them. The white police who worked in the urban ghettos were seen by the minority community as "alien intruders" or an "occupying army."[6]

As discussed in Chapter 5, equal employment opportunity (EEO) and affirmative action laws put into place in the 1970s helped change the racial (and gender) composition of police forces. Although EEO was not a police-led effort, it impacted strongly on the occupation: Police departments have seen a significant rise in representation of minorities and women. For example, by the mid-1980s, almost 30% of Detroit police officers were African American,[7] and by 2017, the figure had risen to 67%.[8] Efforts at improving the racial diversity of police departments were intended to lower the resentment many in the African American community felt toward the police. It was believed more favorable attitudes would lead to other positive outcomes, such as cooperation with the police.

COMMUNITY RELATIONS BUREAUS

Another effort designed to improve the relationship between the citizens and the police—again, particularly between people of color and the police—was the creation of **community relations bureaus** in police departments. This simply involved a revision to the formal organizational chart of a department to include a new bureau and assigning a limited number of officers to that

community relations bureaus: Bureaus created to improve community–police relations through public relations activities.

Photo 13.2
During the urban riots of the 1960s, predominantly white police forces were seen as an "occupying army" in minority neighborhoods. The diversification of police departments was an effort to reduce police–minority conflict. Everett Collection Historical/ Alamy Stock Photo

bureau. These officers were supposed to engage in public relations activities designed to improve the race-relations problem. This approach had limited success as there was little substance to it, and community relations bureaus quickly disappeared or changed into crime prevention bureaus.

TEAM POLICING

A third effort designed to improve the relationship between citizens and the police was **team policing**.[9] This strategy became popular in several police departments in the 1970s. Team policing had many variations, but at its core it involved assigning small teams of patrol officers, investigators, and supervisors to neighborhoods on a long-term basis in order to increase the exchange of information and improve cooperation. Team policing did not last long because it was very difficult to implement and, as a result, its impact was limited. As team policing disappeared, yet another strategy arose to takes its place: community policing.

COMMUNITY POLICING AND PROBLEM-ORIENTED POLICING

According to the Office of Community Oriented Policing Services,

> Community policing is a philosophy that promotes organizational strategies that support the systematic use of partnerships and problem-solving techniques to proactively address the immediate conditions that give rise to public safety issues such as crime, social disorder, and fear of crime.[10]

The idea of community policing was born in the 1980s and reached the pinnacle of its popularity in the 1990s. Most research on this strategy was also conducted in the 1990s and much of it showed community policing to have few crime control effects. However, many police leaders still believe in the ideas of community policing today. In fact, community policing is still mentioned in the mission statements of most police departments, especially the largest ones.[11] The most recent research related to community policing has shifted from a direct focus on the strategy to other, separate but related issues, such as procedural justice.

Community policing first took the form of foot patrol, but over time other activities, programs, and tactics have been developed, including neighborhood watches, citizen police academies, bike patrol, park and walk, storefront police stations/offices, citizen surveys, citizen

team policing: Policing strategy that involved assigning small teams of patrol officers, investigators, and supervisors to neighborhoods on a long-term basis in order to increase the exchange of information and improve cooperation.

community policing: An orientation to policing that focuses on building relationships with citizens in order to achieve crime control and other outcomes such as citizen satisfaction.

EXHIBIT 13.1
Examples of Community Policing

The Newport News (Virginia) Police Department operates a citizen police academy. Through a series of lectures, field trips, and simulated activities, citizens are provided training similar to that of police officers.[13] Other police departments, such as the Carmel (Indiana) Police Department, operate an academy specifically for teenagers.[14]

Grand Rapids (Michigan) Police Department uses a community-policing model to serve the community. This means that the police department uses the same team of officers in neighborhoods, thus allowing the police to create better relationships with the communities they interact with every day.[15]

Among other community initiatives, the Atlanta (Georgia) Police Department offers subsidies for some officers to live in certain areas of the city. Called the "Secure Neighborhoods" program, officers are given an extra monthly stipend for several years, a marked vehicle to park

(Continued)

(Continued)

at home, and a down payment on a home. Officers attend neighborhood meetings and volunteer in mentorship programs and local events. The program is funded by an independent foundation.[16]

The Dayton (Ohio) Police Department operates a community–police council that brings together representatives of various community groups to provide input into police strategies and operations.[17]

Many police departments offer "coffee with a cop" where police officers and community members meet informally—over coffee—to discuss issues and get to know each other.[18]

The Houston (Texas) Police Department offers the "Teen and Police Service (TAPS) Academy" that provides police mentors and curriculum for at-risk teens. Topics include bullying, staying out of gangs, sexting, dating violence, safe driving, and more.[19]

The Illinois State Police is one of many agencies that collects survey data from citizens in order to provide feedback to the agency on citizens' evaluations of the police.[20]

Nashville (Tennessee) Police Department conducts foot and bicycle patrols in a public housing complex in the city several times a week. The emphasis is on engagement and interactions with citizens, not law enforcement. Numerous other departments offer foot patrols tailored to fit the needs of citizens and the police department.[21]

Jeffrey Greenberg/UIG/Getty Images

Photo 13.3
Bicycle patrol is often presented as a community policing activity.

The Dallas (Texas) Police Department has a Neighborhood Policing Team. According to the agency's website, officers who volunteer to serve on this team attend crime watch meetings, teach classes for Volunteers in Patrol, coordinate engraving of valuables, and participate in other related programs.[22]

The Lansing (Michigan) Police Department operates a Community Services Unit. Officers assigned to this unit organize neighborhood watch groups, provide senior citizens with safety tips, operate the Citizen Police Academy, offer crime prevention seminars to businesses and residents, and conduct home security surveys, among other activities. The department also designates certain officers as "community officers" who work in particular areas of the city and serve as a point of contact for residents in those areas.[23]

advisory committees, police–community meetings, and crime prevention education seminars (see Exhibit 13.1). Ideally and as suggested by the definition provided above, community policing goes beyond programs and reflects an overall orientation to policing or a philosophy of policing.[12] At the center of this philosophy is the idea that citizens matter to the police and citizen satisfaction is an important goal of policing.

Community policing in its true form has at least three important features.[24] The first is community engagement. A belief exists in community-policing departments that citizens play an important role in crime control. The concept of coproduction of crime prevention is at the

A Question to Consider 13.2

How Should Police Departments Use Social Networking Sites Most Effectively?

- From your perspective, what advantages and disadvantages do Facebook and Twitter have compared to police use of traditional media?
- How could police departments use Facebook and Twitter most effectively? Specifically, what types of information should police departments post on Facebook or tweet on Twitter. What types of information should they *not* post or tweet?
- What might be the unintended negative consequences of a police department's sharing information or certain types of information on Facebook or Twitter?

Facebook, Twitter, and the Internet

Increased interaction and improved communication between police and citizens are important aspects of community policing. Increased communication between citizens and the police may lead to citizens becoming more participatory in crime prevention and help police become aware of citizens' concerns and priorities. Police–community meetings and citizen advisory groups are designed to address communication issues. Also common among police departments today is the use of social networking sites, particularly Facebook and Twitter. One of the potential major advantages of such sites for the police is that the police, not the media, control the content of the information released. However, of course, the police have little control over other crime and police-related information on these sites.

A 2015 survey of police agencies asked about their use of social media. The survey found that 94% used Facebook, 71% used Twitter, and 40% used YouTube.[25] A 2016 survey found that the most common reasons that police agencies used social media was to notify the public of public safety concerns, community outreach and engagement, public relations, notifying the public of a non-crime issue (e.g., traffic), and soliciting tips on crime from citizens.[26] Another study showed that with regard to Facebook specifically, most "likes" were received for stories about officers injured in the line of duty, and the most comments were received about other miscellaneous stories. "Likes" and comments were relatively uncommon with posts about specific crimes and public relations, two of the most common types of posts.[27] Seldom do the police use social media to respond to citizens, thereby limiting interaction.[28]

Fort Worth Police @fortworthpd · Oct 14
Thank you to all citizens who helped us search the area near Sandy Ln Park. We do not need anymore volunteers today. This search ending soon

Fort Worth Police @fortworthpd · Oct 14
9080 Camp Bowie W Blvd. Luckily and thankfully no injuries.

Photo 13.4
Improved communication between citizens and the police is an important component of community policing. Social media can be used by the police for this purpose. Fort Worth Police/Twitter

Facebook and other social networking sites have a great deal of potential for increasing communication between police and citizens. That being said, social media is a tool police departments are still learning to use most effectively,[29] and its use raises several important questions (see A Question to Consider 13.2).

heart of community policing. That is, the prevention of crime requires that police and citizens work together; they share the responsibility for crime prevention. Second, community policing also involves modification of the traditionally conceived role of the police. With this type of policing, the police role goes beyond law enforcement and includes direct attempts to enhance citizen satisfaction with the police and thereby reduce actual crime. A third important feature of community-policing programs is decentralization. In many of these programs, officers are assigned to neighborhoods and encouraged to communicate with citizens, get to know neighborhood residents, and promote community-oriented programs. Officers also have the ability and authority to act on residents' concerns and priorities.

In the 1980s, a strategy closely related to community policing, **problem-oriented policing**, also began to be put into use. Problem-oriented policing is often considered part of community policing; the two approaches are congruent with each other but not necessarily synonymous. In most of its forms, problem-oriented policing is more focused than community policing in that problem-oriented policing involves the police trying to solve specific problems that affect the quality of life of community residents. Similar to

> **problem-oriented policing:** A type of policing that attempts to address specific problems that affect the quality of life of residents in a community.

community policing, problem-oriented policing requires building relationships among the police, citizens, and other community agencies; however, with problem-oriented policing, police–citizen cooperation is more likely to be focused on a particular neighborhood or community problems. While both community and problem-oriented policing seek to achieve outcomes besides the reduction of crime, such as increasing citizens' quality of life in the community, problem-oriented policing attempts to achieve these outcomes by addressing identified problems rather than through other programs or activities. The effectiveness of problem-oriented policing is typically measured by the extent to which the identified problem has been solved. As such, this type of policing is typically evaluated on a case-by-case or problem-by-problem basis,[30] unlike with community policing, where its success depends on the impact of programs on any number of outcomes. Both strategies are meant to be less controversial than many other crime control strategies (e.g., stop and frisk, traffic stops). Finally, the effectiveness of community policing as a crime control strategy is not well established compared to problem-oriented policing (see Table 13.1). The remainder of this chapter provides a more in-depth discussion of community and problem-oriented policing.

TABLE 13.1

Community Policing Versus Problem-Oriented Policing

Community Policing	Problem-Oriented Policing
Based on premise that citizens are important in the police role	Based on premise that citizens are important in the police role
Focused on community engagement and outreach	Specifically focused on building relationships with citizens and creating partnerships with community agencies to address particular neighborhood/community crime problems
May include a problem-solving component	Always includes a problem-solving component
Seeks to improve citizens' attitudes toward the police and provides an indirect pathway to crime control	Seeks to reduce crime and improve citizens' quality of life by addressing specific crime and disorder problems
Success based on impact of program on relevant outcomes	Success based on ability to solve the identified problem
Meant to be a non-controversial policing strategy	Meant to be a non-controversial policing strategy
Effectiveness at crime control is not well supported	Effectiveness at crime control/problem solving is well supported, depending on specific program

COMMUNITY POLICING: THE DETAILS

Many scholars trace the origins of community policing to the creation of the National Neighborhood Foot Patrol Center at Michigan State University (MSU) in 1982. In the early 1980s, the School of Criminal Justice at MSU evaluated a new type of foot patrol in Flint, Michigan, and the National Neighborhood Foot Patrol Center was built around that evaluation. Similar foot patrol programs were also implemented in other cities.

The federal government has had an important role in the creation and diffusion of community policing as a policing strategy. In 1994, the U.S. Department of Justice Office of **Community Oriented Policing Services (COPS)** was created as a result of the passage of

Community Oriented Policing Services (COPS): An office created by the Violent Crime Control and Law Enforcement Act of 1994 to distribute funds to departments for the development and fostering of community-policing efforts.

the Violent Crime Control and Law Enforcement Act. This law provided $9 billion to police departments to fund the hiring of new officers to foster problem solving and police–community interaction. COPS was given the responsibility for distributing these funds. COPS provides money to police departments to hire community-policing officers, develop and test innovative policing strategies, and provide training and technical assistance. Since 1994, the COPS Office has granted more than $14 billion to advance community-policing type initiatives.[31]

With support from the federal government and interest from universities and key police leaders, community policing clearly seemed to be the right move for police departments. As mentioned above, community policing reached the summit of its popularity among police departments in the 1990s—in fact, the 1990s are referred to as "the era of community policing."[32] Surveys of police departments during that decade reveal that the overwhelming majority either already had implemented or were implementing community-policing methods.[33] Community policing remains popular among many police departments; nearly 97% of police training academies provide training to recruits on community-policing topics, including the history of community policing, identifying community problems, and problem-solving methods.[34]

A THEORY OF COMMUNITY POLICING: BROKEN WINDOWS

In 1982, James Q. Wilson and George L. Kelling published an article titled "Broken Windows: Police and Neighborhood Safety."[35] Although they did not mention community policing in the article, the ideas of Wilson and Kelling became the foundation for some iterations of this type of policing, particularly foot patrol. The hypotheses outlined in the article are now simply known as the broken windows theory. As discussed in Chapter 2, according to broken windows theory, criminal behavior is the result of (a) disorder (e.g., broken windows), (b) anonymity among residents, and (c) anonymity between the police and residents. Further, disorder leads to more disorder: "If a window in a building is broken *and is left unrepaired*, all of the rest of the windows will soon be broken"[36] (italics in original). This condition signifies that no one cares about crime prevention. Residents who live in areas of disorder will feel fearful of crime. These feelings reinforce the anonymity among residents, and they tend to avoid one another. And because people don't get involved, criminals may believe their chances of apprehension are minimal. In the words of Wilson and Kelling, "Serious street crime flourishes in areas in which disorderly behavior goes unchecked."[37] The authors suggest that although the actions of citizens are important in maintaining or restoring order in their neighborhoods, "the police are plainly the key to order maintenance."[38] Officers should "enforce rules about smoking, drinking, disorderly conduct, and the like."[39] In doing so, police actions will reduce disorder, which will, in turn, reduce both fear of crime and actual crime.

Although the community-policing strategy is meant to be noncontroversial, broken windows theory has been the subject of criticism. It has often been interpreted by the police (correctly or not) as encouraging a zero tolerance approach to disorder offenses.[40] However, this approach has been denounced for sometimes inspiring an overly harsh response from the police to minor infractions of the law. It has also led to claims of overpolicing in which the poor and racial minorities are disproportionately targeted. As an extraordinary example, during attempts to reduce disorder on the streets of New York City in 2014, Eric Garner died of a chokehold while being arrested for illegally selling cigarettes. Steve Zeidman, director of the Criminal Defense Clinic at City University School of Law, said, "While broken windows doesn't lead inexorably to a homicide like Eric Garner's, the more you turn loose 35,000 officers with the mandate to restore order, the more you increase the chances for something to go horribly wrong."[41]

In addition to the possible implications of a zero tolerance approach to disorder offenses, broken windows theory has also been criticized based on its assumptions regarding the relationships between police activities, crime, disorder, and fear of crime. These and other assumptions of community policing are discussed below.

THE RELATIONSHIP BETWEEN DISORDER, CRIME, AND THE POLICE

Little research substantiates a direct link between disorder and actual crime, especially serious crime. The best evidence is that crime and disorder, while often occurring together in places, actually come from the same source: concentrated economic disadvantage (i.e., poverty).[42] However, while some areas of concentrated poverty have a great deal of crime and disorder, other such areas have very little. The reason for this is not police enforcement actions but the existence of something called **collective efficacy**. Collective efficacy can be defined as "cohesion among neighborhood residents combined with shared expectations for informal social control of public space."[43] In other words, collective efficacy refers to the degree to which neighbors are willing to help each other and to take action if they see crime or disorder. In neighborhoods where collective efficacy is strong, there are usually low levels of crime and disorder, even when there are high levels of economic disadvantage (the only serious crime that appears even with high levels of collective efficacy is robbery).[44] Accordingly, "policies intended to reduce crime by eradicating disorder solely through tough law enforcement tactics are misdirected."[45]

A recent review of thirty studies that examined the effects of police efforts to reduce disorder supports this conclusion: "Aggressive order maintenance strategies that target individual disorderly behaviors do not generate significant crime reductions."[46] However, what was effective were "community and problem-solving interventions designed to change social and physical disorder conditions at particular places."[47] Research suggests that in order to have the best chance of reducing crime, police efforts should focus most directly on creating and supporting a sense of cohesion, trust, and cooperation among neighborhood residents and working to solve particular disorder-related problems in neighborhoods.

collective efficacy:
Cohesion among neighborhood residents combined with shared expectations for informal social control of public space.

THE RELATIONSHIP BETWEEN CRIME, THE FEAR OF CRIME, AND THE POLICE

One concern in many community-policing programs is fear of crime. The common thinking is that fear of crime is bad and contributes to people avoiding certain areas and limiting contact with other people, including neighbors. It thus stands to reason that whatever the police can do to reduce citizens' fear of crime is good. But caution should rule here.[48] Fear of crime may be entirely justified in some places at some times. Fear serves as a self-protection measure. It keeps people from leaving their doors unlocked at night, from going for walks alone at midnight, and from leaving their cars running and unattended. To unjustifiably lower citizens' fear of crime could lead to an increase in actual crime, therefore the police must tread carefully in this area. Interestingly, though, reducing fear of crime is often one of the things community-policing programs do best.[49]

THE RELATIONSHIP BETWEEN CITIZENS' ATTITUDES TOWARD THE POLICE AND OTHER OUTCOMES

Another objective of many community-policing initiatives is to improve citizens' attitudes toward the police, the reasoning being that favorable attitudes will lead to other desirable outcomes. Although this reasoning may seem logical, it is actually based on many assumptions, including the following:

- If the police increase the quality of their interactions with citizens, then citizens will evaluate the police more favorably in those contacts.
- If citizens rate the police favorably in specific contacts with the police, these evaluations will lead to more positive *overall* attitudes toward the police.
- If citizens have overall positive attitudes toward the police, it will lead to other desirable outcomes, namely citizen cooperation with the police and lawful citizen behavior.

If these assumptions are true, improving citizens' attitudes could cure many ills; however, as we will now discuss, none of them are strongly supported by research.

CAN THE POLICE AFFECT CITIZENS' ATTITUDES TOWARD THE POLICE? Many studies past and present have measured the attitudes of citizens toward the police. A variety of facets have been examined, including overall satisfaction with the police, satisfaction with the police in particular encounters, confidence in the police, and judgments about police department performance. An attitude that has been the subject of much recent research involves **procedural justice**. Procedural justice refers to citizens' perceptions of police fairness in contacts with officers.[50] A related concept is **police legitimacy**. Police legitimacy is what procedural justice is supposed to result in; it is the "belief that the police ought to be allowed to exercise their authority to maintain social order, manage conflicts and solve problems in their communities."[51] Regardless of the attitude measured, the most consistent finding of this research is white citizens have more favorable attitudes toward the police than minority citizens, especially African American citizens (Figure 13.2).

One of the goals of community policing is to improve citizens' attitudes toward the police, especially the negative attitudes often expressed by African Americans. The assumption is that if the police have positive contacts with citizens, citizens will favorably evaluate the conduct of the police during those encounters. Then, if citizens evaluate the police favorably in particular encounters, citizens will adopt a more favorable attitude in general toward the police. In this way, officers may be able to change citizens' overall attitudes toward the police.

> **procedural justice:** Citizens' perceptions of police fairness in contacts with officers.
>
> **police legitimacy:** The belief the police ought to be allowed to exercise their authority to maintain social order, manage conflicts, and solve problems in their communities.

FIGURE 13.2
Confidence in the Police by Race

Surveys conducted by the Pew Research Center in 2014[52] and 2016[53] corroborate what has been found in many other studies of the issue: Significant differences in views of local police are found among whites, blacks, and Hispanics, with whites the most likely to have positive views of the police and blacks the least likely.

. . . to do a good job of enforcing the law?

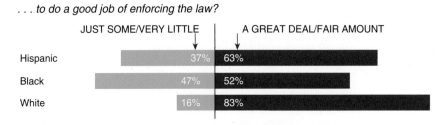

. . . to not use excessive force on suspects?

. . . to treat Hispanics and whites equally?

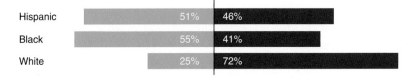

. . . to treat blacks and whites equally?

Source: "Latino confidence in local police lower than among whites," Pew Research Center, Washington, DC (August, 2014) http://www.pewresearch.org/fact-tank/2014/08/28/latino-confidence-in-local-police-lower-than-among-whites/

A Question of Ethics

How Involved Should the Police Be in Citizens' Lives?

Effective policing requires the police to take an active role in organizing citizens to work toward a common good and encouraging collective efficacy. But should this be the job of the police in our society? Is more involvement of the police in the private lives of citizens a good thing? And because the police will inevitably be more involved with citizens in some neighborhoods than in others, does that open the police to more criticisms about overpolicing some places and people and underpolicing others? Explain.

Unfortunately, only a few studies have adequately examined these assumptions. In one study[54] survey data were collected from respondents at two different points in time, six months apart. During the first survey (time #1), citizens were asked about their overall satisfaction with the police, among other questions. In the second survey (time #2), the same respondents were asked again about their overall satisfaction with the police and whether they had had contact with the police within the last six months (e.g., to request information or assistance, as the result of a stop, or because they became a victim of a crime). If contact had occurred, respondents were asked to evaluate that contact.

Two sets of analyses were conducted to test two different relationships. First, the researchers examined the factors that influenced citizens' general satisfaction with the police. They found that if citizens were generally satisfied (or dissatisfied) with the police at time #1, they were also likely to be generally satisfied (or dissatisfied) with the police at time #2, regardless if they had experienced a contact with the police between time #1 and time #2, and regardless of their evaluation of the police during that contact if one had occurred. Evaluations of the police during specific contacts seldom caused citizens to change their overall satisfaction with the police.

Second, the researchers examined the extent to which citizens' general satisfaction with the police (expressed during the time #1 survey) influenced specific evaluations of police performance (evaluations based on contacts that occurred between time #1 and time #2 and were asked about in the time #2 survey). The researchers found that preexisting general satisfaction with the police (time #1 satisfaction) had a substantial effect on subsequent evaluations of the police during contacts with them (reported at time #2).

These findings suggest general satisfaction with the police appears to influence specific evaluations of the police more than specific evaluations of the police influence general satisfaction. In other words, (a) citizens who hold generally favorable views of the police are more likely to evaluate contacts with the police more favorably, and (b) citizens who hold generally unfavorable views of the police are more likely to evaluate their contacts with the police unfavorably. The bottom line is that evaluations of the police in specific contacts tends to confirm how citizens already feel more generally about the police.

Another study took a similar approach and examined citizens' attitudes before and after encounters with the police in Chicago.[55] The authors found contacts with the police did not change citizens' overall attitudes toward the police. Finally, a study that examined officer behavior in interactions with citizens and the subsequent evaluations of the officers by those citizens concluded that "[an] improvement in officer behavior did not translate into improvements in citizens' judgments."[56] This conclusion "should temper our expectations for how police managers can enhance public trust."[57] The essence of this research is that, contrary to assumptions, improving the quality of contacts with citizens does not necessarily lead to more positive overall attitudes toward the police.

DO ATTITUDES TOWARD THE POLICE AFFECT COPRODUCTION? In the 1960s, especially in the urban ghettos of the time, it seemed more people were working against the police than working with them. One of the aims of community policing was to do something about this situation. It was believed if citizens' attitudes toward the police could be improved, then citizens' level of cooperation with the police would also improve. Sociologist Daniel Bell articulated this principle in 1979 when he wrote, "The manner in which police

Photo 13.6
Research has shown that how citizens generally feel about the police affects their evaluations of the police during specific contacts with them. ©iStockphoto.com/Anna Bryukhanova

What Can Be Done to Improve Perceptions of Procedural Justice Among Citizens Who File Complaints Against the Police?

For those citizens who file complaints against the police, what can be done to improve their satisfaction with the complaint process and their views about the fairness of the process? Professor Robert Worden and his colleagues addressed this question.[58] The researchers surveyed citizens about complaints they filed against the police in an American northeastern city. They asked respondents a series of questions about their views of the complaint process, including how easy it was to file the complaint, the timeliness of the settling the complaint, how they were treated throughout the process, the degree of perceived bias in the outcome of the compliant, and their satisfaction with the outcome of the complaint and with how the complaint was handled, among other related questions. In the analysis, the researchers also included whether the citizen's complaint was sustained (determined to be valid) or not. Only 16% of the citizens who were surveyed had their complaint sustained.

Overall, the researchers found that most citizens were dissatisfied with the outcome of their complaint and the process by which the outcome was reached. If complaints were not sustained, judgments of the process suffered; the outcome of the complaint affected citizens' evaluations of the whole process. As a result, the authors explain that besides sustaining more complaints, there may be little the police can do to improve complainants' views of the process, and sustaining more complaints may not be warranted. According to the researchers, it may be especially difficult to improve complainants' judgments of the police as this group of people—people who file complaints against the police—already likely do not have favorable views of the police. Improving citizens' perceptions of procedural fairness of the complaint process is a worthwhile goal, but it may be exceedingly difficult to achieve.

are perceived by citizens, to a great extent determines the quality and quantity of the cooperation police receive from the citizens."[59] Ultimately, the hope was this improved cooperation would allow the police to more effectively maintain order and control crime. These ideas became known as coproduction of crime prevention.

However, the link between citizens' attitudes toward the police and their cooperation with the police is not a strong one. Some studies show citizens who rate the police favorably are more likely to cooperate with officers,[60] but other studies show citizens who rate the police favorably are *less* likely to cooperate with officers.[61] All of the studies on this issue have used different measures of citizen attitudes and cooperation—some asked about coproduction-type behaviors the citizens engaged in (e.g., if they called the police); others asked about hypothetical actions (e.g., if they *would* call the police). Additionally, the studies were conducted in different cities. These variations could help explain the conflicting findings. It is also important to note that if results *do* show people who hold favorable attitudes toward the police cooperate with the police, it does not necessarily mean those attitudes are the cause of the cooperation or participation.

What factors other than citizens' attitudes may influence coproduction? A study in Chicago found citizens who attended the police beat meetings were more concerned about problems in their neighborhood than those who did not attend the meetings.[62] Another study in Detroit found citizens' perceptions of neighborhood conditions were especially important in understanding coproduction behaviors.[63] Specifically, when neighborhood conditions (e.g., kids hanging out, prostitution, people selling drugs, people drinking) were seen as a problem, people were more likely to contact the police and/or get involved with neighborhood organizations. The relationship between perceptions of neighborhood conditions and coproduction has been supported in other research studies.[64] The takeaway from this research is that favorable attitudes toward the police may or may not cause community members to engage in behaviors that support or assist the police.[65]

DO ATTITUDES TOWARD THE POLICE AFFECT LAW-ABIDING BEHAVIORS? It has also been argued that if citizens perceive officers as using just procedures when making decisions, the police will be viewed as legitimate by citizens, and, consequently, those citizens will be more likely to obey the law. If this is indeed the case, clearly it would help the police attain their goal of crime control.

Tom Tyler has done pioneering work on the topic of police legitimacy, the quality of services provided by the police, and the difference that legitimacy and quality of services make.[66] In a study he conducted in Chicago, legitimacy was measured by asking respondents to agree or disagree with various statements about their perceived obligation to obey the law (e.g., "People should obey the law even if it goes against what they think is right") and their support legal authorities (e.g., "I have a great deal of respect for the Chicago police"). Procedural justice was measured by asking respondents questions about the performance of the police and if they believed the police treated people fairly. Finally, compliance with the law was measured by asking respondents how frequently they had violated the law in the last year (e.g., drove over the speed limit, parked a car illegally, made noise to disturb the neighbors, littered, drove a car intoxicated, shoplifted). Overall, 22% of the respondents—most of whom were older women—said they had not broken any of these laws in the past year.

Tyler found that respondents who believed the police were fair also believed in the legitimacy of the law and the police. These respondents were also more likely to comply with the law. Tyler explained,

> People obey the law because they believe that it is proper to do so, they react to their experiences by evaluating their justice or injustice, and in evaluating the justice of their experiences they consider factors unrelated to the outcome, such as whether they have had a chance to state their case and be treated with dignity and respect.[67]

However, as with citizens' attitudes and coproduction, causality is also an issue when examining the relationship between legitimacy and compliance with the law. If a person believes the police to be legitimate and also obeys the law, it does not necessarily mean the belief in police legitimacy is what is *causing* the person to obey the law. If procedural justice and a belief in police legitimacy caused people to obey the law, the solution to crime would be simple: The police should merely treat people with respect.

Given the research already discussed, it seems likely people who hold generally favorable views of the police are in a sense predisposed to judge the police as acting fairly during specific contacts with the police. Conversely, when people who hold generally *unfavorable* views of the police have contact with the police, they will tend to judge the police as acting unfairly. General attitudes color specific evaluations, including evaluations about fairness. As a result, procedural justice (and police legitimacy) may not be separately or objectively determined by citizens; rather, their beliefs in these concepts may simply be a reflection of their preexisting overall views of the police.

A Question to Consider 13.3

Why Obey the Law?

Let's make a safe assumption that you are a law-abiding citizen. Why do you obey the law? Is a belief that the police are fair one of the reasons? How about a belief that the police are legitimate? Conversely, do you think murderers and rapists break the law because they think the police are unfair? Explain why or why not.

GOOD POLICING
Verbal Judo and Procedural Justice

Although research is yet to convincingly establish a strong causal relationship between procedural justice and (a) cooperation with the police and (b) law-abiding behavior,[68] it seems obvious that at least *trying* to create a sense of procedural justice can only be a good thing. To do this, police officers must communicate effectively and treat citizens with respect. Indeed, as stated by the President's Task Force on 21st Century Policing, "Because offensive or harsh language can escalate a minor situation, law enforcement agencies should underscore the importance of language used and adopt policies directing officers to speak to individuals with respect."[69] In their book *Verbal Judo: The Gentle Art of Persuasion,* George Thompson and Jerry Jenkins offer a method police officers can use to learn how to do this. (Thompson, a former police officer and English literature professor, held black belts in judo and karate.) The authors present five "universal truths" of human interaction:

1. All people want to be treated with dignity and respect.
2. All people want to be asked rather than told to do something.
3. All people want to be informed as to why they are being asked or ordered to do something.
4. All people want to be given options rather than threats.
5. All people want to be given a second chance when they make a mistake.[70]

The authors also provide the following example, in which a police officer is trying to calm a scared, potentially violent subject:

> In the classic macho approach, the cop would challenge the guy: "Put that knife down or I'll take you out! You haven't got a chance, I'll blow your head off," things like that. That virtually forces the man to attack, to defend his manhood, to save face. But what if that cop gently empathizes and says, "Hey friend, let's do each other a favor. You don't want to spend the night downtown with us, eating our food, sleeping on our steel cot, and missing your woman. And I don't want to sit at a typewriter for a couple of hours doing paperwork on this. If we can work this out, you can have dinner at your own table, be with your woman, and wake up in your own bed tomorrow morning. And I can go back to my own business."[71]

Treating citizens with respect is a very basic concept. It is not fancy, tricky, or flashy. It costs nothing. It is simply the right thing to do.

Another issue of concern is that most previous discussions of the relationship among procedural justice, police legitimacy, and compliance with the law focus on relatively minor law violations. Recall that Tyler asked citizens if they had committed such infractions as driving over the speed limit, littering, or shoplifting. While it would be good if a sense of procedural justice led to citizens avoiding behaviors such as these, it would be of even greater benefit if it led to the prevention of murders, rapes, and robberies. However, no claims have been made that instilling a sense of procedural justice in individuals would prevent these serious types of crimes.

Overall, given the complexities of this issue, additional research is necessary in order to draw confident conclusions about the relationship among citizens' perceptions of procedural justice, police legitimacy, and compliance with the law. Based on the research conducted thus far, more positive attitudes toward the police do not necessarily cause law-abiding behaviors.

OVERALL EFFECTIVENESS OF COMMUNITY POLICING

According to Professor Nick Tilley, "There are all sorts of problems in evaluating community policing as a whole. . . . It is too fuzzy a concept and too variable in its interpretation and implementation."[72] Asking about the effectiveness of community policing is in some ways akin to asking if food tastes good. This policing method comes in too many forms to make a simple judgment about its overall effectiveness. In addition, the quality of evaluations on community-policing programs has generally been poor. In the few studies that *have* been done well, the results were mixed. Overall, research shows most community-policing programs have minimal effects on crime but greater effects on citizens' perceptions of the police.[73] Research also shows these programs appear to work best in places that need them the least—low-crime, low-disorder areas.[74] Despite these findings, community-policing programs remain popular in many police departments today.

PROBLEM-ORIENTED POLICING: THE DETAILS

Problem-oriented policing is loosely related to community policing but is sometimes considered a separate form of policing. The idea behind problem-oriented policing is that, as its name suggests, the police more directly concern themselves with solving crime and disorder problems experienced by citizens in the community. As with community policing, police–citizen cooperation is important because the identification of and solutions to problems may require citizen input and assistance. An example of problem solving in action can be seen in how the Chula Vista Police Department addressed crime at city motels, as discussed in the Police Spotlight feature at the introduction to this chapter. Other examples are provided in Exhibit 13.3.

Photo 13.7

The relationship among citizens' perceptions of the police, views about police legitimacy, and criminal behavior is complicated. Positive attitudes toward the police may not necessarily cause people to abide by the law. Jonathan Wiggs/The Boston Globe/Getty Images

The problem-oriented strategy was first introduced in 1979 by Herman Goldstein in an article titled "Improving Policing: A Problem-Oriented Approach."[75] In this article, the author describes the means over ends syndrome, which we first discussed in Chapter 2. In this syndrome, the police essentially become preoccupied with the activities of policing and lose sight of the goals they are supposed to achieve. As an example provided by Goldstein, a great deal of time and effort have been devoted to making sure police respond quickly to calls for service; however, less attention has been given to the effects of those fast police responses. He explains, "The situation is somewhat like that of private industry that studies the speed of the assembly line, the productivity of its employees, and the nature of its public relations program, but does not examine the quality of its product."[76] Goldstein also asserts the police are traditionally incident driven, handling only one crime at a time. He argues they should become more problem oriented: They should seek to identify the underlying causes of crime problems and then focus directly on the ends of policing.

According to Goldstein, being problem oriented requires a new process. First, problems must be defined with specificity. For example, police efforts to "reduce crime" will be a failure simply because the concept is so inclusive and broad. Much variation can be found even among particular categories of crime, such as robbery. This category can include street robberies of college students near campuses, purse snatchings at shopping malls, robberies of convenience stories, and bank robberies. Robberies may occur in different areas, at disparate times, and involve victims and offenders with diverse characteristics. Each type of robbery may represent a distinct problem and require its own unique solution.

The second step of Goldstein's problem-solving process is to research the problem and collect information on its magnitude and forms. Other officers and citizens are a good source of information on crime-related problems, as is research conducted by criminologists.

Finally, new responses to problems need to be developed. It must be understood that "more law enforcement" is not necessarily the solution to every problem or the answer to every question. Other possible tactics include physical and technical changes (e.g., locks on doors, direct deposit of paychecks to thwart robberies); changes in the provision of governmental services (e.g., fixing playground equipment to keep children from playing in the streets); developing new skills among police officers (e.g., de-escalation training); introducing new community resources (e.g., crisis intervention teams); and increasing regulation (e.g., requiring security personnel and/or cameras at taverns). Such an approach requires the commitment of not just police administrators but also other agencies that may play an important role in solving problems in a community. However, because the problems being addressed are typically police-related concerns, commitment from the police and the capacity of the police to collect and analyze data and to evaluate the effectiveness of the policies and strategies are critical.

THE SARA MODEL OF PROBLEM SOLVING

Other discussions of problem-oriented policing have provided refinements to the problem-solving process. It has been proposed that problem solving follow the **SARA (scanning, analysis, response, and assessment) model**, as shown in Figure 13.3, which consists of scanning (identifying and describing the problem to be addressed), analysis (researching what is known about the causes of the problem), response (developing solutions to the problem), and assessment (evaluating the impact of the solutions).

Although the problem-oriented approach to policing makes logical sense, there are at least two noteworthy considerations to take into account. First, the extent to which this approach is new is debatable. One could reasonably argue the police have always tried to some degree to identify and solve problems. However, what differentiates the problem-oriented policing approach from previous efforts is that it is supposed to be more systematic in how its goals are accomplished, and individual officers are supposed to have the authority to undertake problem-solving tasks. Second, it is important to realize the most significant social problems, such as poverty, are beyond the control of the police. The police may be able to affect some problems marginally, but to actually solve them is unrealistic. According to Goldstein,

> Many of the problems coming to the attention of the police become their responsibility because no other means has been found to solve them. . . . It follows then that expecting the police to solve or eliminate them is expecting too much. It is more realistic to aim at reducing their volume, preventing repetition, alleviating suffering, and minimizing the other adverse effects they produce.[77]

SARA (scanning, analysis, response, and assessment) model: A method of problem solving that involves identifying a problem, researching what is known about the problem, developing solutions to the problem, and then evaluating the success of the solution.

FIGURE 13.3

The SARA Model

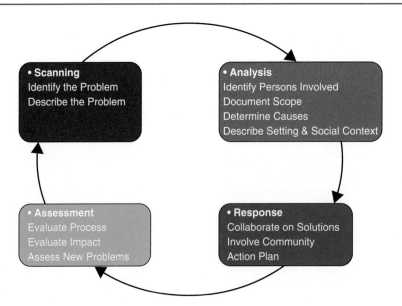

Source: Sarah Lawrence and Bobby McCarthy. 2013. *What Works in Community Policing.* University of California Berkeley: The Chief Justice Earl Warren Institute on Law and Social Policy.

EXHIBIT 13.2

Some Solutions to Common Community Problems

The Center for Problem-Oriented Policing has published many guides to aide police officers who are confronted with crime-related problems in need of solutions. Listed below are some things the center recommends officers consider when dealing with particular types of problems.

To address the safety of public parks, the center suggests the following:

- Control access to the parks through signs, gates, locks, use of natural boundaries like water-ways, etc.
- Ensure that park users can both see and be seen by means of lighting, landscaping, roads and paths, site orientation, equipment placement, etc.
- Clearly establish and promote legitimate park uses and prohibit and discourage illegitimate park uses through signs, landscaping, equipment, organized activities, enforcement, etc.
- Attract natural guardians to your parks, such as parents to safeguard their children, coaches to safeguard their players, and licensed park users to protect their park-use privilege.[78]

With regard to alcohol distribution and consumption, the center states,

> Alcohol abuse contributes perhaps more than any other factor to crime and disorder. It contributes strongly to noise complaints, disorderly conduct, public urination, litter, property damage, assaults, sexual assaults, domestic violence, drunken driving, and homicide. Strong policies governing alcohol distribution and consumption can have wide crime and disorder-control benefits.[79]

To control the distribution and consumption of alcohol, the center recommends the following:

- Ensure there is meaningful enforcement of alcohol regulations.
- Set a tone that promotes responsible alcohol distribution and consumption in your community.
- Publicly acknowledge both the legitimate interest that licensed establishments have in making a profit, as well as their responsibility to serve alcohol in ways that do not generate crime and disorder problems.
- Encourage and compel responsible licensed-establishment management. Responsible management is the most important factor in determining whether a licensed establishment is safe or unsafe.
- Ensure that sufficient alcohol detoxification and treatment services are available.[80]

The center has also addressed the problem of speeding in residential neighborhoods, stating that

> regardless of your jurisdiction's size, you are sure to hear complaints about speeders. Whether on a freeway, a county highway, a major arterial, or a residential street, excessive speed is dangerous and anxiety-provoking, particularly in residential areas and around schools. The most important principle in speed control is that motorists tend to drive at the speed at which they feel safe and comfortable, given the road conditions. Therefore, the key to reducing speed is to alter road conditions such that motorists feel uncomfortable speeding.[81]

The following specific measures are recommended for dealing with issues of speed control:

- Identify the most problematic areas based on complaints and crashes, and focus enforcement resources accordingly. Enforcing speed laws merely to generate revenue tends to alienate the driving public and is not particularly effective anyway, at least not for very long.
- Where permitted by law and warranted by complaint and crash data, use photo radar enforcement, varying the camera locations and operation hours. Bear in mind that photo radar enforcement is unpopular in some communities because it is viewed as unfair enforcement, too intrusive, or an unfair revenue generator.
- Install traffic-calming devices like roundabouts, traffic circles, and speed humps or tables. Be cautious of speed bumps, however, as they can be dangerous to drivers and are problematic for emergency vehicles. Pay attention to design details; they can mean all the difference in whether citizens support them.
- Have your traffic engineer evaluate parking patterns, traffic flow, and street widths. Narrower streets—or streets that appear to be narrow—slow drivers down.
- Encourage citizens to report speeding to police or conduct a publicity campaign to persuade motorists to slow down. Chronic neighborhood speeders—including the teenager with a new license, the commuter rushing to work, or the parent dropping children off at school—may respond to peer pressure from their neighbors.[82]

Source: Joel B. Plant and Michael S. Scott. 2009. *Effective Policing and Crime Prevention: A Problem-Oriented Guide for Mayors, City Managers, and County Executives.* Washington, DC: Center for Problem-Oriented Policing.

EXHIBIT 13.3

Examples of Problem-Oriented Policing

The Eureka (California) Police Department had a serious homeless population problem. Homeless persons were believed responsible for many thefts, robberies, arsons, and assaults, and prompted many calls about drug overdoses. The Eureka PD and its partners worked to develop housing solutions, homeless persons were informed about housing options, and much of the area in which many homeless camped was cleared. As a result of these efforts, the homeless population is still a concern in Eureka but is under control. [83]

The Arlington (Texas) Police Department and its community partners developed PROJECT RAISE (risk, assessment, inversion, safety, and engagement) to deal with the problem of repeat domestic violence calls to the police. Monthly home visits were conducted with the top five repeat domestic violence locations. The visits were conducted by a victim services counselor and a police officer in order to educate the victim (and perpetrator when present) about domestic violence escalation and provide information about resources and support. The police department reported that the number of repeat domestic violence and disturbance calls declined substantially as a result of the program.[84]

The Boston (Massachusetts) Police Department sought to improve homicide clearance rates, which were well below national averages. To do so, the PD increased the number of homicide investigators, provided additional training on investigative and forensic techniques, developed standard investigative procedures, and instituted monthly meetings of investigators to discuss all current homicide investigations. The department's evaluation of these efforts showed a substantial increase in homicide clearance rates.[85]

The Cincinnati (Ohio) Police Department recognized a problem with securing witness cooperation during investigations and during court proceedings. With community and other criminal justice agency partners, the CCROW Project was developed (Cincinnati Citizens Respect Our Witnesses). An important part of the program was to provide services to witnesses so that they would more often cooperate in the court process. For example, funds for emergency and temporary housing for witnesses were provided when necessary, conveniences such as parking money was provided, and panic alarm key-fobs with connection to police dispatch were provided. Community outreach organizations provided referral services for ongoing witness needs. The CPD reported success in securing the assistance of witnesses helped by the program.[86]

The Providence (Rhode Island) Police Department saw a need to address the problem of street-level prostitution in one of its neighborhoods. Partnering with a community outreach organization staffed by women who were former prostitutes, the police provided ride-a-longs with the outreach workers who would speak to prostitutes about services and support available to them to end their prostitution activities. The police department also enlisted the help of other city agencies and a property owner to make the area in which solicitation occurred more visible to the police. The police department reported that according to citizens, business owners, and their own observations, these efforts resulted in a noticeable drop in prostitution activity in the area.[87]

OVERALL EFFECTIVENESS OF PROBLEM-ORIENTED POLICING

Overall, research that has examined the effectiveness of the problem-solving approach indicates that "this approach can result in solutions to problems (though not in every instance) and that problem-oriented policing, on average, results in more prevented crime and disorder than do non-problem-oriented approaches with which it has been compared."[88] Even though difficulties may exist in successfully implementing problem-oriented policing,[89] it has been identified as a strategy that works.[90]

Main Points

- As a result of the crime-ridden and riotous 1960s, various police strategies were used to try to reduce crime and improve police–citizen relationships, especially with racial minorities. Community policing and problem-oriented policing are the most recent attempts to achieve these objectives.
- Community policing in its purest form has several important features, including community engagement, modification of the police role to include citizen satisfaction with the police, fear and disorder reduction, crime control, and decentralization of police efforts to the neighborhood level.
- Typically, problem-oriented policing is more focused than community policing. In particular, problem-oriented policing involves the police trying to solve specific problems that affect the quality of life of community residents. Similar to community policing, problem-oriented policing requires relationship building among the police, citizens, and other community agencies.
- Broken windows theory serves as a basis for some community-policing initiatives. According to this theory, criminal behavior is the result of (a) disorder, (b) anonymity among residents, and (c) anonymity between the police and residents. Research support for broken windows theory is limited.
- Research suggests that in order to have the best chance of reducing crime, police efforts should focus most directly on creating and supporting a sense of cohesion, trust, and cooperation among neighborhood residents and on working to solve particular disorder-related problems in neighborhoods.
- Contrary to conventional wisdom, it is not safe to assume improved police performance will lead to more favorable citizen evaluations of the police or more favorable attitudes toward the police.
- Favorable attitudes toward the police may or may not cause community members to engage in behaviors that support or assist the police.
- Positive attitudes toward the police may not necessarily translate into law-abiding behaviors.
- Although research has yet to convincingly establish a strong causal relationship between procedural justice and (a) cooperation with the police and (b) law-abiding behavior, it seems obvious that at least *trying* to create a sense of procedural justice can only be a good thing. To achieve this, police officers must communicate effectively and treat citizens with respect.
- Community policing comes in too many forms to make a simple judgment about its overall effectiveness. The existing research shows these programs have more effect on perceptions of the police than on actual crime.
- In problem-oriented policing, the police (a) concern themselves more directly with solving crime and disorder problems experienced by citizens in the community and (b) are more concerned with the ends of policing than with the means.
- The SARA model can be used in problem-solving situations: The model consists of scanning, analysis, response, and assessment.
- Even though successfully implementing problem-oriented policing can be difficult, and the success of problem solving depends much on the particular issue being addressed, problem-oriented policing has been identified as a strategy that works.

Important Terms

Review key terms with eFlashcards at **edge.sagepub.com/brandl2e.**

collective efficacy   288
Community Oriented Policing Services (COPS)   286
community policing   283
community relations bureaus   282
police legitimacy   289

problem-oriented policing   285
procedural justice   289
SARA (scanning, analysis, response, and assessment) model   296
team policing   283

Questions for Discussion and Review

Take a practice quiz at **edge.sagepub.com/brandl2e.**

1. What factors led to the "rethinking" of policing in the 1970s and ultimately to the creation of community and problem-oriented policing?
2. Why might people of color—especially African Americans—tend to have more negative attitudes toward the police than whites?
3. What are the important features of community policing?
4. To what extent are community and problem-oriented policing the same concept? How are they different?
5. What is broken windows theory? What are the criticisms of it?
6. Why should the police be careful when it comes to efforts intended to reduce fear of crime?
7. What is the relationship between citizens' evaluations of police officers in particular encounters and citizens' overall attitudes toward the police?
8. What is the assumed relationship between citizens' attitudes and coproduction? What does research say about this assumption?
9. Does a sense of procedural justice cause people to obey the law? Why or why not?
10. What is the SARA model of problem solving?

Fact or Fiction Answers

1. Fiction
2. Fact
3. Fiction
4. Fiction
5. Fiction
6. Fiction
7. Fiction
8. Fiction
9. Fact
10. Fact

$SAGE edge™

Digital Resources

Get the tools you need to sharpen your study skills. SAGE Edge offers a robust online environment featuring an impressive array of free tools and resources.

Access practice quizzes, eFlashcards, video, and multimedia at **edge.sagepub.com/brandl2e.**

Media Library

View these videos and more in the interactive eBook version of this text!

Career Video
13.1: Fostering Community Relationships in Multicultural Cities
13.2: Community Policing

SAGE News Clip
13.1: Chicago Gangs

14

EVIDENCE-BASED AND INTELLIGENCE-LED POLICING

Police Spotlight: Smart Policing in the Chicago Police Department

In October 2018, leaders of the Chicago Police Department (CPD) announced that the department would be expanding its "smart policing strategy" to include nearly all of Chicago's 22 police districts. According to police officials, this strategy as earlier implemented in 13 districts "contributed to nearly two years of consecutive declines in violent crime, including 18% fewer shooting victims citywide and a 20% decline in murders citywide over 2017." In addition, "car-jackings declined 13% and carjacking arrests increased 39% compared to 2017."

The smart policing strategy in Chicago consists of the following:

- Two hundred forty patrol vehicles equipped with license plate reader (LPR) technology, cameras that continuously scan vehicle license plates and identify vehicles reported as stolen so that officers can stop the vehicle and investigate further.
- Six districts will each operate a "Strategic Decision Support Center" that will allow analysts to access data sources such as criminal histories and crime data in order to help district command staff plan patrol deployments.
- The installation of more police observation devices (PODs; street cameras) to the 30,000 government operated cameras already in the city.[1]
- Officers who work in the districts with a strategic decision support center will be issued mobile phones that allow officers to access "district intelligence information" to help determine deployment strategies. "This mobile technology will allow for smarter, data driven patrols and significantly decreased response times to potential calls for service."

According to CPD Superintendent Eddie Johnson, "Our smart policing strategy leveraging the power of data, technology and precision community-centric policing is allowing us to make considerable progress to reduce gun violence across our city. I am pleased that we will be able to expand this vital model in our city."[2]

(Continued)

Objectives

After reading this chapter you will be able to:

14.1 Define smart policing, data-driven and evidence-based policing, CompStat, predictive policing, and intelligence-led policing

14.2 Identify the strengths and limitations of smart policing, data-driven and evidence-based policing, CompStat, predictive policing, and intelligence-led policing

14.3 Compare the similarities and differences in smart policing, data-driven and evidence-based policing, CompStat, predictive policing, and intelligence-led policing

14.4 Discuss the relative importance of research findings when making police policy decisions

14.5 Explain how geospatial crime analysis may lead to place-based crime prevention and how intelligence-led policing may lead to person-based crime prevention

14.6 Discuss how individual privacy can be threatened by intelligence-led policing and geospatial crime analysis

Fact or Fiction

To assess your knowledge of the newest approaches to policing prior to reading this chapter, identify each of the following statements as fact or fiction. (See page 318 at the end of this chapter for answers.)

1. Smart policing usually refers to the incorporation of crime-fighting technology into the operations of police departments.

2. Smart policing was invented in 2019; there was no reference to the concept before that year.

3. Data-driven policing focuses on the value of DNA in conducting criminal investigations.

4. Research findings are always the most important factor police executives should consider when making policy-related decisions.

5. CompStat is an approach to policing that uses data and accountability in an attempt to improve policing.

(Continued)

(Continued)

6. CompStat has significantly reduced crime wherever it has been used.

7. Geospatial crime analysis is based on the fact that most crime does not occur randomly but is concentrated in certain places.

8. Crime maps are a primary tool of geospatial crime analysis.

9. Intelligence-led policing refers to the principle that police executives must incorporate the latest technology into the operations of police agencies.

10. Intelligence-led policing and geospatial crime analysis both involve the prediction of crime.

(Continued)

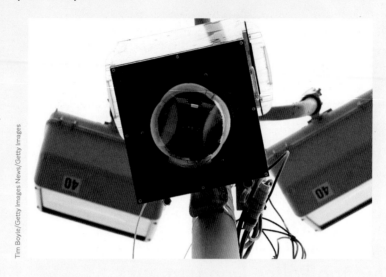

Photo 14.1
The smart policing strategy involves the incorporation of crime-fighting technology in an attempt to reduce the amount of crime.

AS departments continue to seek ways of achieving the crime control mandate, numerous new approaches to policing have emerged in recent years, including smart policing, evidence-based policing, CompStat, predictive policing, and intelligence-led policing. These approaches are separate but related; the common denominator among them is an increased reliance on various forms of information in attempts to improve policing. This chapter will introduce and discuss these strategies.

A Question to Consider 14.1

Smart Policing in Chicago

What is your assessment of smart policing in the Chicago Police Department? Would you expect the initiative to further reduce crime? Explain. What might be some of the other benefits and drawbacks to the strategy?

SMART POLICING

The idea of **smart policing** has evolved: It was first coined as a term in 2009 by the U.S. Department of Justice, Bureau of Justice Assistance, when it created the "Smart Policing Initiative" (SPI). The SPI made funding available to police agencies to create partnerships with college- or university-affiliated researchers. The research findings resulting from these partnerships could then be used when making decisions about police policy and strategies.[3] Funded projects addressed problems such as prescription drug abuse, homicide, violent crime, gangs, gun violence, drugs, domestic violence, property crime, repeat offending, and neighborhood disorder, among others.[4] However, since 2009 the concept of "smart policing" has also been used by police agencies to describe various initiatives, many having little or nothing to do with

smart policing: Any initiative on the part of a police department to incorporate technology into police operations with the goal of developing information that can be used to reduce crime.

police–researcher partnerships, such as the smart policing initiative of the Chicago Police Department described in the Police Spotlight feature. Smart policing is now simply known as any initiative on the part of a police department to incorporate technology into police operations with the goal developing information that can be used to reduce crime. Having realized how the smart policing term was being used by police departments, in 2015 the "Smart Policing Initiative" changed its name to "Strategies for Police Innovation," still using the SPI acronym.[5]

It is worthwhile to note that part of the reason for creating an initiative to support police–practitioner research partnerships in 2009 was the realization that much of the research being conducted on the effectiveness and efficiency of police strategies was of questionable quality. For example, when researchers sought to assess the overall effectiveness of problem-oriented policing based on the previous research conducted on it, only ten of 5,500 available articles were deemed to be of sufficient quality to be included in the assessment.[6] The goal of Strategies for Police Innovation remains to support and encourage police–researcher partnerships in order to improve the quality of research conducted on crime issues; however, it relates more to advancing problem-oriented policing than it does current conceptions of smart policing.

DATA-DRIVEN AND EVIDENCE-BASED POLICING

Over the last decade, there has been increased discussion of data-driven and evidence-based policing. Studies have shown that there is confusion about these terms even among police officers and police leaders,[7] so it is important to clarify here. A police department is data driven when its leaders collect and analyze data *from their department* to make informed decisions about how that agency should best operate, particularly with regard to policing policies and strategies. Evidence-based policing occurs when police leaders consider any research findings to help inform policy decisions, but these findings are not necessarily generated from the department using them. Data-driven policing and evidence-based policing share the belief that "police practices should be based on scientific evidence about what works best."[8] In some respects, the data-driven approach is more difficult because, in order for there to be data on which to base decisions, agencies must have some capability of collecting and analyzing such data and then turning it into usable information. This capability may come through partnerships with researchers or it may be internal, such as a department's own crime analysis center or research office. With evidence-based policing, it is not necessary for agencies to collect or analyze data, only to consider existing research findings.

Although many police departments have used data and research findings to inform decision making in the past, especially since the 1970s, with evidence-based and data-driven policing, there is increased emphasis on the practice. Some of the first police policy questions answered with research data related to the effectiveness of random patrol, rapid response, and reactive investigations. Since then and as discussed throughout this book, literally thousands of policy-related questions have been informed by research data. For example:

- How effective is hot spot patrol in reducing different types of crime? What activities of officers make a difference in this regard?
- How effective are crackdowns? For which offenses do they work best? Under what conditions are they most effective?
- What is the impact of DNA on the investigation of various types of crimes?
- What effects do community-policing programs have on crime? What effect do they have on citizens' attitudes toward the police?
- How do police actions affect citizens' perceptions of procedural justice and police legitimacy? What is the significance of these perceptions?
- How effective is problem-oriented policing in reducing crime and disorder?
- What types of training are necessary for police officers to perform their work competently? How is that training best delivered?

evidence-based policing: Policing that involves the use of research findings to inform policies and strategies.

data-driven policing: Policing that occurs when leaders collect and analyze departmental data to make informed decisions about how their departments should best operate, particularly with regard to policies and strategies.

- What measures can be taken to reduce the physical and psychological risks of police work?
- Under what circumstances is force most likely to be used by officers? What types of force lessen the likelihood of citizen and police injuries?
- How should patrol officers most effectively spend their uncommitted time?
- What effect do body-worn cameras have on police use of force and citizen complaints?

OTHER BASES FOR POLICY DECISIONS

While research findings based on data can be used to inform policy decisions, as made clear with data-driven and evidence-based policing, there are other factors that may explain why things are done the way they are in police departments. For example, decisions can be based on assumptions or what some people believe to be true. For decades, people believed that fast police responses to crime scenes significantly improved the chances of the police making on-scene arrests of offenders. Not until the 1980s was this assumption tested and found to be untrue. Decisions may be based on authority. For instance, things might be done the way they are in a department simply because the chief wants them done that way. Decisions can also be based on tradition. Some things are done a certain way today simply because that is how they have been done in the past. In contrast to assumptions, tradition, and authority, research findings provide a more analytic basis on which to make policy-level decisions.

SHOULD POLICY DECISIONS BE BASED ON RESEARCH FINDINGS?

It is reasonable to ask if decisions *should* be based on research evidence. Interestingly, there are some reasons to believe data and research evidence are *not* the most important—and certainly not the only—factors to consider when making decisions. First, it should be clear that using research evidence to make decisions is most relevant for police leaders making policy-related decisions, not for patrol officers making street-level discretionary decisions. There are few opportunities for patrol officers to consider research findings when making on-the-spot decisions, although the policies and training that guide those decisions may be research based.

There is, however, value in patrol officers being familiar with the research process.[9] When research is conducted in police departments, it is usually officers who are either the subjects of the research or the ones expected to carry out the research procedures. If officers have a familiarity with the research process and understand the value of it, they may be more likely to comply with it. There are many stories of police officers failing to comply with the procedures of a research study and the quality of the study suffering as a result. For example, in the Minneapolis Domestic Violence Experiment,[10] officers were instructed by the researchers to take one of three actions (arrest, warn, or separate) when responding to an eligible domestic violence incident. The action to be taken was predetermined through a procedure involving color-coded report forms placed in random order (e.g., a red form meant arrest, a yellow form meant warn, etc.). However, some officers subverted the randomization procedure by not following the color-coded reports. When they were supposed to warn the subject based on the color of the report, they made arrests. Or they warned when they were supposed to separate. This resulted in bias being introduced into the study.[11] Perhaps if officers had been made more aware of the importance of following procedure in producing accurate results, they would have been more likely to comply with the research protocol of the study.

Second, if police leaders are to base policy decisions on research findings, ideally that research should be conducted in that agency. Because of different community and police department characteristics, what is true in one setting may not be true in another. Again, as an example, the Minneapolis Domestic Violence Experiment was replicated in five other cities. Each study had at least slightly different results. Based on victim interviews, three of the studies found arrest deterred repeat domestic violence; three found it did not. In fact, in three of the cities, arrest was shown to *increase* the likelihood of repeat domestic violence. This has been referred to as the *different communities dilemma*.[12]

Third, even if a study is of high quality and is conducted in the community or police department where its findings are to be applied, one must be reasonably cautious about making policy decisions based on a single study. A policymaker should be more confident in making decisions based on research if there are consistent findings from many studies. However, this is not a luxury often afforded to police policymakers. If multiple studies do exist, there is a good chance they have produced different findings, as noted in the domestic violence studies.

Finally, research evidence may not be the most important consideration in policy development; many other issues and interests may need to be considered. For example, police executives must always consider the liability implications of policy decisions. Even if research indicates a particular policy is effective, setting policy strictly on this evidence could subject the police department to lawsuits. For a case in point, we again turn to the domestic violence studies. Even though some of the studies clearly showed arrest did not work best and may have actually increased the likelihood of repeat domestic violence in some instances, many departments proceeded to implement mandatory arrest policies primarily to protect themselves from liability claims.[13]

Resource constraints are another major factor to consider. Can the agency afford to implement the research-recommended solution? Political factors also may play a role. How might other constituent groups be affected by a proposed solution? Will citizens accept the solution? Will politicians support the solution? Will police officers accept it? It would be naïve to think research findings should be the sole or even primary consideration when making policy-level decisions. The data-driven and evidence-based approach to policing offers a scientific way of improving policies, but it is important to put its potential contribution in perspective and understand its limitations.

COMPSTAT

CompStat (usually identified as a short form of *computer statistics*) was first introduced in 1994 by Police Commissioner William Bratton in the New York City Police Department (NYPD). CompStat can be thought of as a data-driven approach to policing, the goal of which is to reduce crime and sometimes to solve other departmental problems.[14] It has been described as a data-driven management model, a management process, a management device, a tool of management accountability, a performance management system, and a management model. Clearly, CompStat has something to do with management!

There are various ways in which CompStat operates, but at its core, the system consists of two critical elements: data and accountability. Through the analysis of data, particularly crime data, serious crime problems are identified and then assigned to particular individuals for resolution. These individuals are usually high-level police managers, such as precinct commanders. Once a problem is assigned, it is "owned" by the assignee until it is resolved. The commander who is given the problem works with his or her subordinates, including sergeants and patrol officers, to identify potential solutions. After the initial assignment of the problem is made, follow-up is conducted to see what actions were taken and what results were obtained. In this way, particular people are held accountable for particular problems.

Shortly after the introduction of CompStat in the NYPD, crime in the city began to decline. Police leaders attributed this to CompStat.[15] In the face of this apparent success, many police

> **CompStat:**
> A management process that involves the analysis of data to identify problems and the assignment of responsibility to police personnel for the resolution of those problems.

Photo 14.2
While crime was reduced in New York City after the introduction of CompStat, it has limitations as a crime control strategy. ©iStockphoto.com/ Eloi_Omella

Photo 14.3
CompStat involves assigning responsibility for crime problems to particular officers.
Methods to address the problems are then identified and follow-up is conducted.
CompStat provides accountability for the reduction of crime. Jeff Greenberg/UIG via Getty Images

departments across the country also implemented CompStat or some version of it. However, the system's success in other cities has been mixed.[16] While accountability can certainly lead to positive outcomes, some people have argued a relentless focus on responsibility for addressing difficult crime problems may create incentive for the police to manipulate crime data to make it appear crime has been reduced when it has not.[17]

Aside from this possibility, the biggest issue with CompStat is that although using effective tactics to solve crime problems is central to the success of CompStat, those tactics are not specified by the system. They have to be discovered. In a study examining the use of CompStat in seven police departments, the results revealed sergeants and patrol officers were largely responsible for solving CompStat-identified crime problems, and they were expected to do so without any clear guidance. One of the sergeants interviewed in the study spoke about how his commander gave him little direction in how to solve a problem involving theft from autos:

> He is saying, O.K. guys, you have one hundred years of experience between the three of ya'. Let's fix it. That's what he is doing; it is not a direct, "this is what we are going to do." He throws this thing out there. Put your heads together and figure out how we are going to resolve this issue.[18]

Another sergeant explained how the solutions to identified crime problems were usually not very creative:

> Typically, I mean when you are talking about bar problems, construction area thefts, they don't demand a tremendous amount of creativity in how you approach them. . . . You either want to be seen or you don't want to be seen. If you want to be seen, you want to patrol to increase visibility, do traffic enforcements, do bar checks.[19]

GOOD POLICING
CompStat Versus Problem-Oriented Policing

One of the primary limitations of CompStat is that it is heavy on problem identification and accountability but light on the identification of solutions for the identified problems. The Lowell (Massachusetts) Police Department sought to test if a problem-solving approach to CompStat would provide a more effective approach to crime control. As an alternative to the traditional CompStat process, "problem-solving meetings" were used to generate ideas about how to address identified crime problems. Compared to CompStat meetings, the problem-solving meetings were conducted in a more informal manner with fewer police personnel in attendance. Also, in contrast to the CompStat process, at these meetings, free exchange of ideas and brainstorming were encouraged. Researchers found that the problem-solving meetings generated nearly seven times as many crime prevention measures as CompStat meetings did, even though many of the same people attended each of the meetings. These measures also produced more crime control gains compared to those derived through the CompStat process. The researchers suggest that, to be more effective, CompStat should include a strong problem-solving component.[20]

? A RESEARCH QUESTION

Are Some Retail Businesses Associated With Crime?

A study conducted in South Los Angeles (California) examined the amount of crime that occurred in close proximity to three types of retail establishments: tobacco shops (e.g., "smoke shops"), off-sale alcohol outlets (e.g., liquor stores), and medical marijuana dispensaries. The researchers found that property and violent crime rates near tobacco shops and alcohol outlets exceeded the crime rates around grocery/ convenience stores, but crime rates around medical marijuana dispensaries did not. This pattern was found even when controlling for important neighborhood factors, such as poverty and residential mobility. The authors explain that these findings may help inform zoning and licensing decisions, which may be part of an overall problem-oriented policing strategy, and help explain the formation of crime hot spots.[21]

In summary, CompStat is an accountability tool that involves the identification of particular crime problems and the assignment of those problems to specific police department personnel. Although the system has been successful in some departments, it is limited in the same way that data-driven policing and evidence-based policing are limited: It is not a method or strategy by which to control crime; it is merely an approach to policing.

PREDICTIVE POLICING

Predictive policing includes some aspects of data-driven policing, CompStat, and intelligence-led policing (see below) but with the specific purpose of trying to predict (a) where crimes will occur, (b) when they will occur, (c) against whom they will occur, and (d) who will commit them. Simply stated, predictive policing involves using data to try to predict and prevent crime.

No matter how sophisticated the analyses, predictive policing is not in the realm of science fiction, where the police make arrests of citizens for crimes before those crimes even occur (as in the film *The Minority Report*). Predictive policing also does not mean the police become psychic or use a crystal ball. Predictive policing is based on the simple fact that the majority of crime is not random. In particular, predatory street crime (e.g., murder, rape, robbery, shootings) is not distributed evenly across time, place, or people. Certain *times* are more likely to experience crimes than others (e.g., there is more crime at night than during the day); some *places* are more likely to experience crime than others (e.g., taverns have more crime than libraries); some *people* are more likely to become victims than others (e.g., drug dealers versus monks); and some people are more likely to commit certain crimes than others (e.g., men versus women). Because many crimes are not random, if patterns can be identified, interventions may be introduced to prevent these crimes. One of the problems with predictive policing, however, is the more specific the prediction, the more likely it is to be wrong. If the police could determine the precise location and time of a crime and the identities of the victim and offender crime prevention would be easy. Unfortunately, this is not possible.

CRIME ANALYSIS

Crime analysis is the primary method of predictive policing. Broadly defined, "crime analysis involves the collection and analysis of data pertaining to a criminal incident, offender, and target."[22] When the data being analyzed relate to *who* is involved in crimes (victims and offenders), it is often known as crime intelligence analysis.[23] When the analyses are focused on *where* crimes occur, it is often referred to as geospatial crime analysis.

> **predictive policing:** Policing strategy that uses data to try to predict and prevent crime.
>
> **geospatial crime analysis:** Analysis of crime data that focuses on where crimes occur.

GEOSPATIAL CRIME ANALYTICS Methods of geographic-based crime analysis vary considerably in their sophistication, but in all forms, **crime maps** are a commonly used tool. The most basic maps use color-coded pins to illustrate visually where various crimes have occurred in a particular jurisdiction over a particular period of time. Through the quick inspection of such a map, a basic understanding of the distribution of crime across space can be developed. Police departments have been using these types of maps as a basis for predicting crimes for decades. More elaborate crime analysis methodologies used today include the use of geographical information systems (GIS) or other predictive software that allow for the automated recording and plotting of criminal incidents on detailed computerized maps.[24] Some police departments provide interactive on-demand crime maps on their websites (see https://www.phillypolice.com/crime-maps-stats/ and Figure 14.1).[25] Some crime mapping technologies also allow for the inclusion of other data, such as the following:

- Demographic characteristics of a population
- Location of various businesses (e.g., taverns and other alcohol outlets)
- Residences of parolees (e.g., sex offenders)
- Gang territories
- Known and suspected drug houses
- Traffic stops and field interviews
- Calls for service[26]

Many police departments operate a computer-aided dispatch (CAD) system (see Chapter 6). CAD systems include data on calls for service, traffic stops, field interviews, and other patrol officer activities. Many police departments also use a records management system (RMS) to

FIGURE 14.1

An Example of a Crime Map From the Milwaukee Police Department

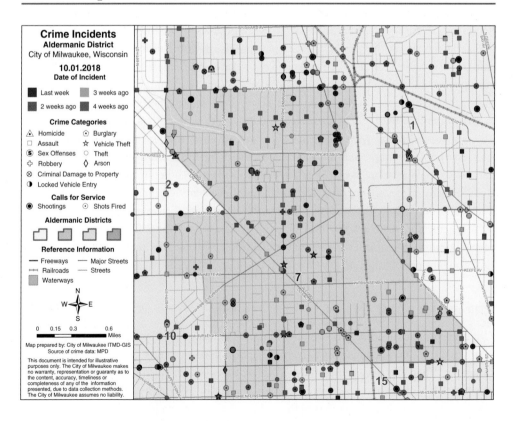

electronically store other information, such as crime reports, arrest reports, traffic citations, and/or other police-citizen contacts. When CAD data and RMS data are combined with GIS capabilities and software, crime analysts have a wealth of data to analyze and interpret. This can include everything from a city-wide analysis of crime over time to calls for service and police actions at particular addresses. The data may also be analyzed to reveal patterns of crime or certain types of crime across time (e.g., time of day, year) and space (e.g., in and around schools, at public transportation stops). Geospatial crime analysis can offer general predictions about when and where crime will occur and provide the necessary information to make informed patrol allocation decisions, thus providing a basis for place-based crime prevention.

Photo 14.4

The hiring of crime analysts is increasing among law enforcement agencies. As its title suggests, the job involves analyzing crime in an effort to predict and prevent it. tom carter/Alamy Stock Photo

The police and other law enforcement agencies are increasingly hiring crime analysts in their departments (Exhibit 14.1). In the past, if crime analysis was performed in departments, it was usually by officers with an interest in the issue. Now, given the complexity and technical knowledge required to collect, analyze, and interpret crime data, specially trained civilians are often hired for the positions. Several universities include crime analytics as a curricular option in criminal justice undergraduate and graduate degrees.

A Question of Ethics

Balancing the Right to Know With the Right to Privacy

As discussed earlier in this book, community and problem-solving policing involve developing a trusting and cooperative relationship with citizens in order to prevent crime. When this approach is combined with predictive policing, it often means the police share crime information, such as crime maps and statistics, with citizens. Indeed, citizens often appreciate being kept informed by the police about crimes in their neighborhoods and communities. Furthermore, citizens may take appropriate precautions if they are aware of a crime problem. If an agency fails to notify citizens about crime risks, it may be doing the community a major disservice. However, while citizens have a right to know about crime in their communities and there can be benefits to sharing that information, victims also have a right to privacy. For example, if the location of a crime is provided, then the identity of the victim may be determined. If the crime is a sexual assault or some other sensitive crime, clearly the violation of the victim's privacy is a major issue.

So how should police departments balance the right to know with the right to privacy? Specifically, should information about crimes be released to the community? If yes, what types of information should be released and what types should not? If a department releases only limited information, do you think that information will still be useful to citizens and potential victims? Explain.

THE IMPACT OF GEOSPATIAL CRIME ANALYSIS Geospatial crime analysis has the potential to impact crime, but it must be understood that it is a method of developing and analyzing information, not a crime reduction strategy by itself. This is analogous to how magnetic resonance imaging (MRI) produces information to diagnose an illness, but it does not cure the illness.[27] Indeed, geospatial crime analysis is not a cure for crime. Crime analysis can only lead to a crime reduction when combined with effective police strategies.

EXHIBIT 14.1

Crime and Intelligence Analyst Job Description, Arlington (Texas) Police Department[28]

This position will close December 26, 2018.
SALARY: 53,030.00–66,287.00 Salary USD

JOB SUMMARY:

Under general direction, reviews and analyzes crime reports; interprets pattern and trends; disseminates information and makes recommendations related to possible suspects, potential crimes, location of future crimes and methods of operation.

ESSENTIAL JOB FUNCTIONS:

- Ability to monitor, analyze and evaluate patterns and trends of criminal activity using statistical analysis and other analytical methodologies to assist the division(s) in meeting objectives.
- Ability to prepare comprehensive analytical products and reports including crime bulletins and summaries; recommend investigative direction, provide oral briefings and presentations; facilitate intelligence and crime analysis meetings.
- Ability to research, retrieve and evaluate confidential information related to complex criminal investigations; create automated files, databases and spreadsheets for information management functions.
- Ability to interpret information regarding criminal activity in internal and external reports, registers, records and databases.

OTHER JOB FUNCTIONS:

- Ability to establish and maintain professional contacts with experts, analytical counterparts, educators and researchers; liaison with other agencies and the public.
- Ability to respond to requests for crime information from the news media, civic organizations, general public and outside agencies.
- Ability to participate in training various criminal investigative techniques.

MINIMUM QUALIFICATIONS:

Knowledge, Skills and Abilities Required:

Knowledge of offense elements as specified by Texas Penal Code; Knowledge of TCIC/NCIC computerized database and proper procedures for use; Knowledge of computers including, relevant spreadsheet, graphic, mapping, mainframe and network applications; word processing, the Internet and other computer software; Knowledge of research techniques; Knowledge of interviewing techniques and methods; Knowledge of modern police record systems; Knowledge of Business English, spelling and punctuation; Knowledge of modern office practices and procedures; Ability to interpret and understand criminal law, Code of Criminal Procedure, city ordinances, policies and procedures; Ability to communicate effectively orally and in writing and prepare comprehensive reports; Ability to operate a variety of office equipment including, but not limited to PC, fax machine, typewriter, calculator and copier; Ability to perform a variety of physical skills including, but not limited to seeing, sorting, typing and writing; Ability to receive detailed information through oral communication and make fine discriminations in sound.

Qualifying Education and Experience:

A Bachelor's Degree in Criminal Justice, Statistics, Computer Science or related field. One year experience in crime research, crime analysis, law enforcement, criminal records maintenance or related area.

Preferred:

Advanced skill in MS Access, Excel and ArcGIS; Crime Analyst Certifications (ex: CLEA)

Research suggests much of the potential of crime analysis and predictive policing more generally has yet to be realized in the nation's police departments.[29] While nearly three-fourths of agencies report having personnel who conduct crime analysis, overall, crime analysis results have not been optimally utilized by police departments.[30] This is generally because the crime information is not useful, it is unclear how the information can be used, or a combination of the two.[31] In any case, in order for crime analysis to be helpful, the analysts must work closely with those who will be using the information. Ideally, analysts should be able to provide answers to questions about crime patterns and trends posed by officers and commanders. And if officers and commanders do not know what questions to ask, the analysts should be able to educate them about the potential of crime analysis information in improving the police response to crime. Crime analysis is also not without controversy. Some people argue that it further encourages the police to target black and Latino communities, leading to familiar criticisms of overpolicing.[32]

INTELLIGENCE-LED POLICING

Intelligence-led policing has been described as a management philosophy that uses data and criminal intelligence to focus enforcement activities.[33] Using this definition, it is apparent intelligence-led policing is closely related to what has been discussed in this chapter as data-driven policing. However, what separates the two is a focus on criminal intelligence. Criminal intelligence refers to information about criminal offenders, particularly who they are and what methods they use. A central aspect of intelligence-led policing is intelligence-based crime analysis, which is person-focused analysis. This is in contrast to geospatial crime analysis, which is place focused. For intelligence-led policing to be a management philosophy, intelligence-based crime analysis needs to be well integrated into the criminal investigation operations of an organization. Intelligence-led policing also requires priority be placed on information development and sharing.[34]

Person-based crime analysis makes fundamental sense because it is people who commit crimes. With this approach, the focus is directly on identifying and targeting offenders and potential offenders. In contrast, with a place-based approach, the focus is on places; places that include offenders *and* nonoffenders. Arguably, then, the intelligence-based approach is more focused than the place-based approach when it comes to addressing crime. In addition, a small proportion of people in a population account for a relatively large proportion of all crimes in that setting. If the high-rate offenders could be identified and deterred/incapacitated from committing additional crimes, crime rates could be substantially affected. While intelligence-based crime analysis is often oriented toward gang and drug enforcement (Exhibit 14.2) and terrorism, it can be used to target other crimes as well.[35]

> **intelligence-led policing:** An approach to policing that involves the collection and analysis of information on likely offenders and their methods of operation.
>
> **person-focused analysis:** In this type of analysis, the focus is on identifying and targeting offenders and potential offenders.

EXHIBIT 14.2

The Drug Enforcement Administration as an Intelligence-Led Agency

The Drug Enforcement Administration (DEA) is a good example of an intelligence-led agency. As explained on the agency's website,

> Since its establishment in 1973, the DEA, in coordination with other federal, state, local, and foreign law enforcement organizations, has been responsible for the collection, analysis, and dissemination of drug-related intelligence. The role of intelligence in drug law enforcement is critical. The DEA Intelligence Program helps initiate new investigations of major drug organizations, strengthens ongoing ones and subsequent prosecutions, develops information that leads to seizures and arrests, and provides policy makers with drug trend information upon which programmatic decisions can be based. The specific functions of the DEA's intelligence mission are:
>
> - Collect and produce intelligence in support of the Administrator and other federal, state, and local agencies;
> - Establish and maintain close working relationships with all agencies that produce or use narcotics intelligence;
> - Increase the efficiency in the reporting, analysis, storage, retrieval, and exchange of such information and undertake a continuing review of the narcotics intelligence effort to identify and correct deficiencies.
>
> The DEA's Intelligence Program has grown significantly since its inception. From only a handful of Intelligence Analysts (I/A) in the domestic offices and Headquarters in 1973, the total number of I/As worldwide is now over 680. DEA's Intelligence Program consists of several entities that are staffed by both I/As and Special Agents: Intelligence Groups/Functions in the domestic field divisions, district, resident and foreign offices, the El Paso Intelligence Center, and the Intelligence Division at DEA Headquarters. Program responsibility for the DEA's intelligence mission rests with the DEA Assistant Administrator for Intelligence.[36]

Source: Drug Enforcement Administration website, https://www.dea.gov/ops/intel.shtml.

Intelligence-led policing also incorporates an element of prediction. The prediction is not about which *places* are likely to experience crime in the future, as is the case with geospatial crime analysis, but about which *people* are likely to offend in the future. This approach is oriented toward identifying offenders before they commit crimes. The horrific terrorist attacks of September 11, 2001, clearly highlighted the potential value of identifying offenders before they act. Other events, including the Boston Marathon bombing in 2013 and the shootings in San Bernardino in 2015, Orlando in 2016, Las Vegas in 2017, and Pittsburgh in 2018, also reinforce the value of early identification of offenders. In all these cases, law enforcement authorities were unable to identify the offenders prior to their crimes and were criticized as a result.

Due to technological advances, it is possible for law enforcement agencies to collect increasing amounts of information on people in an effort to prevent crimes and identify perpetrators. In addition to national networks of crime-related information, such as the National Crime Information Center (see Technology on the Job feature), most police departments collect information on crimes and criminals. For example, urban police departments typically collect a lot of information relating to street gangs that operate in their jurisdiction. This is appropriate and necessary given the high involvement of gang members in drug sales and other violent crimes[37] (see Figure 14.2).

In this age of information, many other sources of intelligence are available to law enforcement. For example, social networking sites (e.g., Facebook, Twitter) are a potentially rich source of information on offenders and their criminal plans. Confidential informants can play a critical role in the identification of criminals and provide important information about criminal activity. There are also a multitude of databases used by law enforcement agencies for criminal intelligence purposes. Most large police departments maintain databases that include such things as pawnshop records, modus operandi (MO) files, and the street names or aliases used by individuals. Many other databases maintained by federal agencies are also available to law enforcement agencies (Exhibit 14.3).

TECHNOLOGY ON THE JOB
National Crime Information Center

There is probably no greater source of intelligence for the police than the National Crime Information Center (NCIC). The NCIC, which is maintained by the FBI, is the largest and most well-known crime information network system in the United States. The NCIC began operations in 1967, and by 1971 police agencies in all fifty states were linked to the system. Today, more than 80,000 law enforcement and criminal justice agencies have access to the NCIC database. At the end of 2015, the system contained more than twelve million records. Nearly thirteen million queries ("transactions') are made of the system daily.[38]

The NCIC consists of a centralized database and a network of connecting computers. Representatives of agencies may enter information into the database and make queries of the database. The originating agency also has the responsibility for removing records once the information is no longer valid. The system contains a multitude of files, including records on stolen guns, vehicles, boats, and license plates and missing and wanted persons, among other information. [39]

The NCIC has the capability of quickly putting vast and critical information in the hands of police and investigators. As such, the center can be a powerful investigative tool and can have an important role in intelligence-led policing efforts.

Source: Federal Bureau of Investigation, National Crime Information Center website, https://www.fbi.gov/services/cjis/ncic.

National Crime Information Center (NCIC): The largest, most well-known crime information network system in the country.

LIMITATIONS OF INTELLIGENCE-LED POLICING

As with the other approaches discussed in this chapter, intelligence-led policing has the potential to improve the operations of law enforcement agencies, particularly the criminal detection and crime prevention functions of these agencies. While using information to identify and apprehend offenders is certainly nothing new, intelligence-led policing as a management

FIGURE 14.2
Street Gang Involvement in Criminal Activity

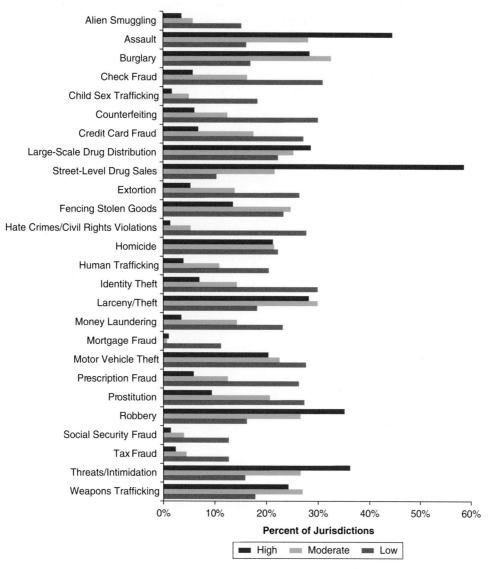

Source: National Gang Intelligence Center. "National Gang Report: 2015 Gangs," 2016. https://www.fbi.gov/stats-services/publications/national-gang-report-2015.pdf, p. 14.

philosophy is a relatively new idea, and "next to nothing is known about this area and best practice does not exist."[40] Much has yet to be figured out in intelligence-led policing before its full potential can be realized.

Although it is possible for law enforcement agencies to collect much information on people, it is still exceedingly difficult to make random information actionable.[41] It has to be usable and used. A great deal of media attention has focused on the inability of law enforcement to "connect the dots" of information in order to prevent mass shootings and other attacks. For example, Tamerlan Tsarnaev, the mastermind of the 2013 Boston Marathon bombing, was on the national terrorism database watch list, yet this information was of no use in the prevention of the bombings or in his identification. Clearly, computers and databases cannot do everything, but they—and those who operate them—play a critical role in directing and focusing investigative activities. This, by definition, is the meaning of *intelligence-led.*

EXHIBIT 14.3

National Databases for Law Enforcement Intelligence Purposes

Besides the NCIC, which was discussed earlier, many other intelligence databases are also available to law enforcement. These databases include the following:

National Law Enforcement Telecommunications System (NTETS): NLETS is a network that links law enforcement agencies, other criminal justice agencies in the United States, and motor vehicle and licensing departments. The information available through the system includes vehicle registrations by license or vehicle identification number, driver's license and driving record information, other vehicle registration information, parole/probation and corrections information, and sex offender registration information.

INTERPOL Case Tracking System (ICTS): The ICTS contains information about persons, property, and organizations involved in international criminal activity.

Terrorist Watch List: The Terrorist Screening Center maintains the U.S. government's consolidated Terrorist Watch List, which is a single database of identifying information about those known or reasonably suspected of being involved in terrorist activity.

Central Index System (CIS) and related databases: The CIS is operated and maintained by the Bureau of U.S. Citizenship and Immigration Services (USCIS). It contains information on legal immigrants, naturalized citizens, and aliens who have been formally deported or excluded from the United States. The Nonimmigrant Information System of the USCIS contains information on the entry and departure of nonimmigrants (aliens) in the United States for a temporary

stay. The Law Enforcement Support Center is also operated by the USCIS and provides information to local, state, and federal law enforcement agencies about aliens who have been arrested. The National Alien Information Lookout System consists of a USCIS index of names of individuals who may be excludable from the United States. The Consular Lookout and Support System is a related database operated by the U.S. Department of State. It contains information on several million individuals who have been determined to be ineligible for visas, those who need additional investigation prior to issuance of a visa, and those who would be ineligible for visas should they apply for one.

El Paso Intelligence Center (EPIC): EPIC is designed to collect, process, and disseminate information concerning drug trafficking, alien smuggling, weapons trafficking, and related criminal activity on the southwest border as well as in the entire Western Hemisphere.

Sentry: Sentry is operated by the Federal Bureau of Prisons and contains information on all federal prisoners incarcerated since 1980. The available information includes the inmate's physical description, location, release information, custody classification, and sentencing information, among other items.

Equifax and TransUnion: Equifax and TransUnion are companies that provide credit information on individuals. Databases within their operation may be used to collect various information on people, including recent addresses and demographic information.

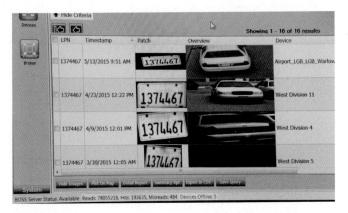

Photo 14.5

Intelligence-led policing depends not only on the collection of information but also the use of this information in order to prevent crimes and apprehend offenders. Sometimes determining the value of information is more difficult than collecting it. AP Photo/Damian Dovarganes

Although a person-based approach to crime reduction has the potential to make law enforcement organizations more focused on serious and repeat offenders (and therefore more efficient), significant resources are necessary to operate an effective intelligence-gathering and production process. Personnel are required to collect, evaluate, and produce useful intelligence. Personnel cost money. Technology to store and process information costs money. Unfortunately, the ability and willingness of agencies to pay for technological advancements is lagging behind the pace of those advancements.

Another issue is determining what legal actions can be taken to incapacitate a person after that person is identified as being a possible threat but prior to criminal acts having occurred. A final concern with intelligence-gathering activities in law enforcement agencies is their possible effect on citizens' privacy. This important issue is discussed in detail in the next chapter.

Main Points

- Smart policing, evidence-based policing, data-driven policing, CompStat, predictive policing, and intelligence-led policing are separate but related approaches to policing. A common denominator among them is an increased reliance on various forms of information in an attempt to improve policing.

- Smart policing, data-driven policing, CompStat, predictive policing, and intelligence-led policing share a common limitation: They are not methods or strategies of crime control, they are merely approaches to policing.

- At its inception, smart policing referred to police–researcher partnerships. Now smart policing refers to any initiative on the part of a police department to incorporate technology into police operations with the goal of developing information that can be used to reduce crime.

- Data-driven and evidence-based policing occur when police leaders use data or research findings to make informed decisions about how their departments should best operate.

- There are several bases for police policy making; the data-driven and evidence-based approaches provide an analytic basis on which to make policy-level decisions.

- CompStat is a data-driven approach to policing that involves identifying crime problems and assigning them to particular commanders for resolution. By assigning responsibility for problem solutions to particular police personnel, CompStat provides for accountability.

- Predictive policing involves using data to try to predict and prevent crime. Crime analysis is the primary method of predictive policing. Crime analysis involves the collection and analysis of data relating to crime, offenders, and targets. When the analysis relates to the people involved in crimes (victims and offenders), it is often known as crime intelligence analysis. When the analysis relates to where crimes occur, it is often referred to as geospatial (geographically based) crime analysis.

- Geospatial crime analysis can offer general predictions about when and where crime will occur and provide the necessary information to make informed patrol allocation decisions.

- Intelligence-led policing uses data and criminal intelligence to focus enforcement activities. Criminal intelligence refers to information about criminal offenders, particularly who they are and what methods they use. An important aspect of intelligence-led policing is intelligence-based crime analysis.

Important Terms

Review key terms with eFlashcards at **edge.sagepub.com/brandl2e.**

Questions for Discussion and Review

Take a practice quiz at **edge.sagepub.com/brandl2e.**

1. Compare data-driven policing with evidence-based policing.
2. What is smart policing and how does it differ from data-driven and evidence-based policing?
3. Should police policy decisions be based on research findings? Explain.
4. What is CompStat? Is it effective? What is its major limitation?
5. Can crime be predicted? Why or why not?
6. What is geospatial crime analysis? How can it be used to prevent crime? How might it be controversial?
7. How does intelligence-led policing differ from evidence-based policing?
8. Explain how intelligence-led policing is a form of predictive policing.
9. Identify and discuss two limitations/concerns with intelligence-led policing.
10. What is the National Crime Information Center (NCIC)? Identify and discuss other sources of criminal intelligence.

Fact or Fiction Answers

1. Fact
2. Fiction
3. Fiction
4. Fiction
5. Fact

6. Fiction
7. Fact
8. Fact
9. Fiction
10. Fact

$SAGE edge™

Digital Resources

Get the tools you need to sharpen your study skills. SAGE Edge offers a robust online environment featuring an impressive array of free tools and resources.

Access practice quizzes, eFlashcards, video, and multimedia at **edge.sagepub.com/brandl2e.**

Media Library

View these videos and more in the interactive eBook version of this text!

Career Video
14.1: Technology Students Should Learn
14.2: Police Division Chief Discusses Technology

SAGE News Clip
14.1: Light Based Intervention System (LBIS)

US-CRIME-SHOOTING-RELIGION/BRENDAN SMIALOWSKI/AFP/Getty Images

15
TERRORISM, TECHNOLOGY, ACCOUNTABILITY, AND THE FUTURE OF AMERICAN POLICING

Fact or Fiction

To assess your knowledge of these issues prior to reading the chapter, identify each of the following statements as fact or fiction. (See page 335 at the end of this chapter for answers.)

1. Terrorism may be the most significant crime demand on the U.S. police forces in the future.

2. Police organizations in the United States are quasi-military in structure and function; however, the parallels are becoming stronger and will likely continue to do so in the future.

3. The militarization of the police is clearly a positive development in the evolution of the police and has few if any implications for police–community relations.

4. The incorporation of accountability oriented technology (such as body-worn cameras and early intervention systems) has been proven to enhance the accountability of officers and police departments to the communities they serve.

5. Most legal challenges of the future that relate to technology will likely center on privacy issues.

Police Spotlight: The Significance of September 11 for Law Enforcement

As discussed in detail in Chapter 2, the police in America have evolved over time as significant new crime demands caused departments to change their priorities and operational strategies. It would be naïve to think the police are now done evolving—that no new significant demands will be confronted by law enforcement. To the contrary, the events of September 11, 2001, and other terrorist acts since then have made terrorism in all of its various forms a major concern to law enforcement and may spark the creation of a new way of policing. Given the significant impact of the September 11 attacks on law enforcement practice and policy, any informed discussion of the future of American policing would be incomplete without an examination of these events.

On the morning of September 11, 2001, nineteen men armed with box-cutter knives and logistical, financial, and training support from the al-Qaeda terrorist organization changed America and the world. At 7:59 a.m., American Airlines Flight 11 departed Boston Logan Airport on its way to Los Angeles with ninety-two people aboard. It was hijacked by the terrorists and flown into the North Tower of the World Trade Center at 8:45 a.m. At 8:14 a.m., United Airlines Flight 175 left Boston Logan en route to Los Angeles with sixty-five people aboard. It was hijacked and flown into the South Tower of the World Trade Center at 9:03 a.m. At 8:10 a.m., American Flight 77 left Dulles International Airport in Washington, D.C., for Los Angeles with sixty-four people aboard. It was hijacked and flown into the Pentagon at 9:39 a.m. United Flight 93 left New Jersey's Newark International Airport for San Francisco at 8:01 a.m. with forty-four people aboard. It crashed in a cornfield in Pennsylvania at 10:03 a.m. It is believed the aircraft was headed for the White House or the Capitol when passengers attempted to take control of it from the hijackers. In total, more than 3,000 people, including sixty police officers, were killed that day.

The attacks led to a massive law enforcement response and investigation that has evolved into what we know now as the war on terror. They also prompted

(Continued)

(Continued)

the creation of the Department of Homeland Security and an unprecedented reorganization of federal law enforcement agencies. Investigations were carried out in numerous foreign countries, and intelligence that was uncovered led to war in Afghanistan and Iraq.

Over the years, the United States has been successful in killing key members of the al-Qaeda terrorist organization, including Osama bin Laden, the group's leader at the time of the attacks. However, this may not have reduced the threat posed by terrorists. Still an important terrorism concern today is the Islamic State in Iraq and Syria (ISIS). ISIS and people who pledge their support of it is responsible for the murder of thousands of people in the Middle East and a multitude of terrorist attacks across the world and the United States, including the 2016 Orlando, Florida, night club mass shooting that killed 49 people and injured 53 (the perpetrator was killed, ISIS claimed responsibility for attack).

Since 2001, several other terrorist incidents have occurred in the United States, including the following:

- In 2009, a U.S. Army major shot and killed thirteen people and wounded more than thirty others at Fort Hood in Texas. (Note: Officially, the U.S. government has not defined this as an act of terrorism.)
- In 2009, a Nigerian man on a flight from Amsterdam to Detroit attempted to set off an explosion with materials concealed in his underwear.
- In 2010, a bomb was ignited in a car parked in New York's Times Square, but it failed to detonate.
- In 2013, two perpetrators with radical ties placed bombs at the Boston Marathon that killed three people and injured nearly 200 others.
- In 2013, a gunman targeting Transportation Security Administration officers opened fire at Los Angeles International Airport.
- In 2015, a Pakistani national couple who spoke of jihad killed fourteen people in San Bernardino, California.
- In 2017, eight people were killed when a subject used a rented pickup truck to drive down a busy bicycle path in New York City. The attack was believed to have been made in the name of ISIS.

Photo 15.1

The terrorist attacks of September 11, 2001, changed the United States and the world forever. Stacy Walsh Rosenstock/Alamy Stock Photo

The rest of the world has seen hundreds of successful terrorist acts since 2001, many of which have led to the deaths of Americans, such as the 2012 attack on the U.S. consulate in Benghazi, Libya. The U.S. ambassador and three others were killed in this incident. In 2016, ISIS claimed responsibility for suicide bombings in Belgium that killed thirty-two people. Also in 2016, an individual used a truck to run over people celebrating in Nice, France. Over eighty-five people were killed in this attack; ISIS claimed responsibility. A suicide bomber killed twenty-two people at a 2017 Ariana Grande concert in England; ISIS claimed responsibility. In 2018, a series of bombings at churches in Indonesia resulted in the death of 28 people. ISIS also claimed responsibility for this attack.

Terrorism is undoubtedly alive and well in the world, as is the war against it.

CHAPTER 15 provides a discussion about what policing in the future may look like and the important role of technology in how policing will be performed.

HISTORY AS A GUIDE TO THE FUTURE

One reason for studying history is that it may serve as a guide to the future. In fact, some people argue that history tends to repeat itself, so to know the past is to know the future. And if we know the future, we can prepare for it.

CRISIS AND CHANGE IN POLICE HISTORY

As we have discussed, American police departments have progressed through three eras: the political era, the reform era, and the community problem-solving era.[1] During the political era, from the mid-1800s to the early 1900s, politicians controlled virtually every aspect of policing. The police were corrupt and generally inefficient. The political era came to an end in the early 1900s due to rising concern about serious crime and the inability of the police to deal with it. People realized a new way of policing was required.

This new way of policing became known as the reform era. The reform era was marked by police professionalism, the police as experts, and distance maintained between the police and citizens. The use of patrol cars allowed the police to institute preventive patrol and to offer fast responses to crime scenes in order to make more arrests. State and federal law enforcement agencies were created to assist local police departments. This style of policing worked well until the 1960s. With the dramatic rise in crime, the urban riots, and research that showed the ineffectiveness of preventive patrol during this decade, the police found themselves in the midst of another crisis. And yet again, crisis led to change.

This time, the new style of policing took the form of the community problem-solving era, which many scholars say remains the predominant style of policing today. This style is characterized by police–citizen cooperation and the belief that citizens share the responsibility for crime prevention with the police. However, it is unreasonable to expect this era of policing will last forever. If history repeats itself, it will come to an end as the result of a crisis, a crisis caused by new and increased demands being placed on the police. What might these demands be? What will be the new style of policing?

A Question of Ethics

Information at What Cost?

A critical component in the investigation and prevention of terrorism is information, and there may be no better source of information than terrorists themselves. Unfortunately, apprehended terrorists may not be willing to reveal what they know. Enter waterboarding. The practice of waterboarding emerged in the aftermath of September 11 as a technique used by Central Intelligence Agency (CIA) and military officials to obtain information from suspected non-U.S. citizen terrorists.[2] The procedure involves restraining a subject, covering the subject's face with cloth, and then pouring water over the subject's mouth and nose. Waterboarding creates the sensation of drowning, and, because the subject is unable to breathe, it has the potential to cause injuries and brain damage. Opinions differ as to whether waterboarding is legal, moral, necessary, or effective. Especially troubling is the notion it might be used on individuals who do not actually possess the sought-after information. From your perspective, under what circumstances, if any, would the use of torture be an ethically acceptable way of obtaining critical information from criminal suspects? Explain. Be specific.

NEW DEMANDS ON THE POLICE

As we move deeper into the twenty-first century, technology is creating new criminal threats. Terrorism has become a serious concern. An additional major concern for the police is renewed demands from citizens for increased police accountability. These three demands—terrorism, technology, and accountability—may prompt changes to policing in the future. Each is discussed below.

TERRORISM: DEFINITIONS AND VARIATIONS

A perpetrator's motivation for committing a crime determines whether that crime is terrorism. The Code of Federal Regulations (28 CFR 0.85) defines **terrorism** as "the unlawful use of force or violence against persons or property to intimidate or coerce a government, the civilian population, or any segment thereof, in furtherance of political or social objectives." The Global Terrorism Database defines terrorism as "the threatened or actual use of illegal force and violence by a non-state actor to attain a political, economic, religious or social goal through fear, coercion or intimidation."[3] Hundreds of other definitions are available from various sources. Not surprisingly, the number and characteristics of terrorist incidents believed to have occurred depends on the definition used. For example, the Global Terrorism Database, with its rather broad definition, includes nearly 2,836 terrorism incidents in the United States from 1970 to 2017.[4] It is important to note the newest form of terrorism, cyberterrorism, is not included in these definitions or statistics. Cyberterrorism involves attacks on computer systems in order to steal information or to disrupt infrastructure operations, such as utility or transportation systems.

Terrorism can be classified as internationally based (committed by foreigners or foreign nationals; such as the acts on September 11) or domestic (committed by American citizens in the United States). Terrorism perpetrated by foreigners is typically motivated by extreme religious views and hatred toward the United States. There is much political debate about the labels used to describe these individuals. Some people are reluctant to use the terms *radical Islam* or *Islamic extremists*, as they wish to avoid equating the Islam religion and Muslims (persons who follows Islam) with terrorists. Instead they use terms such as *violent extremists* or **jihadists**. Jihadists are people who believe in the duty of Muslims to spread their religion through militant means.

Domestic terrorist events, such as the 2015 San Bernardino shootings by Pakistani nationals, have led to increasing concern about American citizens who have become radicalized. Although there is no widely accepted definition of the term, to be *radicalized* means to learn, accept, and believe in extreme ideas that promote the legitimacy of violent actions to accomplish political

---

terrorism: According to the Code of Federal Regulations, terrorism is the unlawful use of force or violence against persons or property to intimidate or coerce a government, the civilian population, or any segment thereof, in furtherance of political or social objectives.

jihadists: People who believe in the duty of Muslims to spread their religion through militant means.

EXHIBIT 15.1

Variations in Terrorism

International terrorism: Terrorism committed by foreigners or foreign nationals. It is typically motivated by extreme religious views and hatred toward the United States.

Domestic terrorism: Usually motivated by sentiment against government, abortion, or race/gender equality, or by hatred of some group or cause; occurs in the United States.

State-sponsored terrorists: Terrorists who receive funding, support, training, and/or protection from a government.

Lone wolf (or lone actor) terrorists: Terrorists who act on their own without help from a terrorist group or a government but may still have ideological attachment to a larger group.

ends. A person who has become radicalized has learned to believe in the value of violence to achieve the goals of his or her group. One of the ways to identify individuals with terrorist motivations—and thereby perhaps prevent terrorist acts—is to monitor the various hallmarks of radicalization, such as association with other known or suspected terrorists; travel to terrorism hot spots (e.g., Syria); Internet activity (e.g., YouTube, Twitter); and prison radicalization experiences.

In contrast to international terrorism, domestic terrorism is usually motivated by sentiment against government, abortion, or race/gender equality, or by hatred of some group or cause. Some domestic terrorists are described as **right-wing extremists**. Sometimes referred to as the alt-right, these individuals act on their extreme beliefs, which can include racism, homophobia, anti-Semitism, and neo-Nazism. For example, Timothy McVeigh, who bombed the Alfred

Photo 15.2
When people think of terrorism, many think of internationally based terrorists, such as Islamic extremists. However, most law enforcement agencies identify domestic terrorism, especially right-wing extremists, as a top threat. Barcroft Media/Getty Images

P. Murrah federal building in Oklahoma City in 1995, killing 168 people and injuring 700, was described as a right-wing extremist motivated by antigovernment views. A survey of American law enforcement agencies conducted by the *New York Times* and the Police Executive Research Forum in 2014 found 74% of agencies identified right-wing extremists as one of the top terrorist threats in their jurisdiction. In contrast, only 39% identified internationally based terrorist organizations as a top threat. Since the time of this survey in 2014, the threat and concern posed by right-wing extremists has not diminished. In fact, right-wing extremists were responsible for more murders in 2018 (at least 50) in the United States than in any year since 1995.[5] Among the murders included here are the eleven people killed in the Tree of Life synagogue shooting in Pittsburgh in 2018. Indeed, while many headlines are devoted to Islamic extremism, a potentially greater threat is posed by American extremist groups.[6] Besides the Ku Klux Klan and Aryan Nations, most domestic terrorist groups are not familiar to the American public.

Another distinction that can be made with regard to terrorists is that they can be either state-sponsored or **lone wolf (or lone-actor) terrorists**. If terrorists are state sponsored, they receive funding, support, training, or protection from a government. *State sponsored* is sometimes a fuzzy concept: For example, ISIS claims to be a state, but it is not recognized as such. The al-Qaeda terrorist group coexisted with the Taliban in Afghanistan, but al-Qaeda is not considered a government. Individuals acting on their own without help from a terrorist group or a government are often referred to as lone wolves or lone-actors, although inspiration for the attack may be rooted in ideological attachment to a larger group. With the exception of the September 11, 2001, attacks and cyber-security related incidents, it is believed that all recent terrorist attacks in the United States have been perpetrated by lone-actors.

It is important to make clear that the prevention of terrorism is not an easy task, to say the least. Lone wolf terrorists act alone, and there may be few if any opportunities for law enforcement to discover their plans. Terrorist groups are usually structured into cells, which are small groups of individuals. Each cell may have a particular purpose in the group's overall mission, and the members of one cell may not be aware of the individuals in other cells. Cells may be located in different areas of a country or in different countries altogether. No one person in the group knows everything, so that if one cell or person is discovered by law enforcement, the overall plan may still go forward. Due to this group structure and the limited information often available to authorities, preattack activities may go undetected.

right-wing extremists: Domestic terrorists who act on their extreme beliefs, which can include racism, homophobia, anti-Semitism, and neo-Nazism.

lone wolf terrorists: Terrorists who act on their own without help from a terrorist group or a government. Also known as lone-actors.

What Are the Characteristics of Far-Right Extremism?

Based on an analysis of the United States Extremist Crime Database (ECDB), researchers from the National Consortium for the Study of Terrorism and Responses to Terrorism (START) report the following:

- Between 1990 and 2018 there were 217 ideologically motivated homicide incidents committed by far-right extremists in the United States.
- Most of these incidents (76%) were committed by white supremacists against persons in racial or ethnic minority groups (e.g., African Americans), persons of certain religious affiliations (e.g., Jewish, Islamic), or sexual orientations (e.g., homosexual).
- The remaining 26% of incidents were committed by other far-right extremists who targeted government officials, other ideological enemies, and abortion providers.
- Terrorist attacks attributed to formal organizations were relatively rare. Most attacks were typically carried out by individuals who were only loosely linked to a specific organization or ideological movement.[7]
- Law enforcement officers, corrections officers, private security guards, and a judge have been killed during these incidents.

- Considering the time period 1990 to 2017, only five states had more than ten far-right ideologically motivated homicides: Oregon ($n = 11$), Florida ($n = 14$), Pennsylvania ($n = 15$), Texas ($n = 21$), and California ($n = 33$).
- Between 1990 and 2014, there were nearly 100 violent plots against Jewish individuals and/or targets that were planned by more than 25 far-right individuals or groups. Nearly 85% of those plots were discovered and stopped by law enforcement before the act was carried out.
- Between 1990 to 2014, there were 15 ideologically motivated plots against Muslim individuals and/or Islamic targets by far-right extremists or groups of far-right extremists. Law enforcement foiled more than 85% of these anti-Islamic plots.

Source: National Consortium for the Study of Terrorism and Responses to Terrorism. 2018. *Far-Right Fatal Ideological Violence Against Religious Institutions and Individuals in the United States: 1990-2018*. https://www.start.umd.edu/pubs/START_ECDB_FarRightFatalIdeologicalViolenceAgainstReligiousTargets1990-2018_Oct2018.pdf.

American power, influence, and foreign policy, which are the source of much of the hatred directed against the country, are not likely to change or vanish, and the political stability of Middle Eastern countries is not likely to improve. Unfortunately, antigovernment sentiment and hatred of some groups by others within the United States is also not likely to disappear. As a result, the threat of terrorism is likely to be with us for a while and law enforcement may have to adapt to more effectively counter this threat.

TECHNOLOGY

Another trend that may have implications for law enforcement operations is the rapid advancement and deployment of technology. Just as electricity, the telephone, the automobile, and the airplane did in decades past, the technology of today dramatically affects the nature of our lives. As seen throughout history, technology can place incredible demands on the police. However, technology also has the capability to *reduce* demands, as might be the case with the self-driving (and presumably safer) cars of the future.

Arguably, the most significant technology of today and the foreseeable future is the Internet. The Internet is a global computerized network that allows for the exchange and dissemination of information. The Internet has the potential to change *everything*; it affects nearly all facets of life, including the nature of work and where people live, methods of governing, medical care, the content and delivery of education, how people communicate with each other, and media and entertainment. But the Internet has a dark side: It has created a multitude of criminal

EXHIBIT 15.2

Common Types of Cyberattacks and Crimes

- A computer virus is a program that is unwittingly downloaded to a computer, causing that computer to shut down. Most often viruses are transmitted through email.
- Spyware is software that collects and transmits information about computer usage to another person without the computer user's consent.
- Ransomware is unknowingly downloaded software that takes control of a computer or renders it inoperable until payment is made to the person responsible for the attack.
- Hacking involves the infiltration of a computer or computer system to gain control over that system in

order to steal user information or other sensitive information.
- Phishing (pronounced *fishing*) involves obtaining usernames, passwords, and other personal information through trickery.
- Spoofing involves a person misrepresenting his or her identity through an email in order to obtain user information.
- Cyberbullying involves using the Internet to intimidate or threaten another person.
- Sextortion refers to the use of the Internet or other electronic means to coerce an individual into a sexual relationship or behaviors.

opportunities. The Internet can be used to disseminate information of hate through websites and email. It can be used to facilitate crimes (see Exhibit 15.2). It can be used as an educational tool of terrorists (e.g., publishing bomb making instructions); to perform cyberterrorism; to facilitate communication among criminals; and to recruit members of criminal/terrorist groups (e.g., inspirational videos and propaganda). The Internet can be used to gain illegal access to and steal protected governmental and corporate information or to facilitate fraud and identity theft. And the Internet is still in its infancy. One of the critical issues facing law enforcement agencies with regard to Internet-based crimes is determining their proper role in preventing and investigating such crimes. With cybercrimes, the perpetrator and victim may not be in the same jurisdiction or even the same country. Given the likely magnitude of the threat, these issues will need to be addressed. There is no question the Internet poses a tremendous challenge to law enforcement today and will continue to do so.

ACCOUNTABILITY

Two critical deadly force incidents sparked recent demands for increased police accountability. First, Eric Garner was killed by New York City Police Department officers on July 17, 2014. Garner, who was unarmed, resisted arrest for selling individual cigarettes and was forcibly taken to the ground by several officers. In an attempt to gain control of Garner, one of the officers placed his arm around Garner's neck, essentially executing a chokehold. Because chokeholds have been demonstrated to be lethal, they are seldom allowed as a force technique. While on the ground Garner repeatedly stated, "I can't breathe." The incident was partially caught on cell phone video. Garner was pronounced dead at the hospital around one hour later. The officers involved in the incident were eventually cleared of any criminal wrongdoing.

The second incident occurred approximately one month later, on August 9, 2014. Officer Wilson of the Ferguson (Missouri) Police Department shot and killed an unarmed eighteen-year-old named Michael Brown. Brown and a friend had just stolen several packs of cigarillos from a store and were walking in the street when Officer Wilson confronted them. A struggle erupted and Officer Wilson ended up shooting Brown at the squad car and several more times about 150 feet from the squad car. Officer Wilson was cleared of criminal wrongdoing by a St. Louis County grand jury and in a U.S. Department of Justice investigation. In response, intense rioting took place in Ferguson and was the focus of national media attention for months.

Photo 15.3

Conflict between police and citizens may be reduced through increased accountability of the police to citizens. Jochen Tack/Alamy Stock Photo

Since these two events, a steady stream of media reports has focused on incidents of police officers killing civilians. Beginning in 2015, the *Washington Post* has reported incidents where police have killed citizens; these reports have raised awareness of these incidents among citizens and have led to demands for increased police accountability. The significant demands the police are facing—terrorism (domestic and international), new forms of crime related technology, and calls for increased accountability—may represent a crisis of enough magnitude to lead to changes in how policing is conducted in the future. Increased intelligence collection and management, adoption of technology, and enhanced accountability may become top priorities of law enforcement agencies in the coming years.

A Question to Consider 15.1

Police and Progress

In their book *The New World of Accountability*, authors Samuel Walker and Carol Archbold wrote,

> A reasonable person . . . may conclude that there has been little progress in American policing since the strife-torn decade of urban riots in the 1960s . . . that racial and ethnic discrimination, excessive force, and unjustified shootings are as prevalent as they were 50 years ago . . . that the many police reforms of the past half century have accomplished nothing.[8]

Do you think that the police have made progress in the last fifty years? Explain why or why not.

THE NEW POLICE

In the face of these increased demands, what might be the next adaptation for the police? Based on what has already been discussed, six predictions can be offered.

First, at a minimum it is likely policing will be more complex and global than in the past. When crimes are simple and local, detection strategies may be simple and local, but when crimes are complex and global, it stands to reason detection strategies should also be complex and global.[9] To support the increased scope of police operations to global crimes, it is probable more resources and legal authority will be provided to law enforcement agencies, particularly federal agencies.[10] This change of priorities has implications for local police agencies and the amount of federal funds available to them for other needs.

Second, to support this new orientation, more powers will be given to law enforcement agencies. The first example of this was the **USA PATRIOT Act of 2001**, which stands for Uniting and Strengthening America by Providing Appropriate Tools Required to Intercept and Obstruct Terrorism. This law was signed approximately one month after September 11, 2001. It gave law enforcement agencies additional authority to gather electronic evidence, amended money laundering laws, changed immigration laws, and expanded the authority of law enforcement officers to seize individuals' records held by a third party. In 2015, and after some modifications, the USA PATRIOT Act became the **USA Freedom Act.**

A third prediction is that in the future citizens will remain powerful players in the police enterprise, especially with regard to fighting crime and providing information. Reward and tip lines will be increasingly important as the police continue to rely on citizens for information about crimes and the people who committed them. This includes the investigation of terrorism.

USA PATRIOT Act of 2001: Act signed after the September 11 attacks; it gave law enforcement agencies additional authority to gather electronic evidence, amended money laundering laws, changed immigration laws, and expanded the authority of law enforcement officers to seize individuals' records held by a third party.

USA Freedom Act: A modified version of the USA PATRIOT Act enacted in 2015.

The Department of Homeland Security's media campaign, "If You See Something, Say Something," speaks directly to the issue.[11]

Fourth, as discussed in Chapter 14, in an effort to identify criminals and prevent their crimes, intelligence-led policing is likely to become better defined and developed. It is safe to say that policing of the future will be massively dependent on data and criminal intelligence.[12] In addition and relatedly, agencies will continue to adopt technology at a fast rate in an attempt to deal with the increasingly sophisticated criminal threat, a strategy earlier described as smart policing.

A fifth prediction about policing in the future is that the police will continue to increase their transparency so officers and agencies are held responsible for their conduct.[13] Much of this effort will be advanced through technology, in particular, body-worn cameras. The goal of this is to make the police less controversial in carrying out their responsibilities. However, this is an enormously difficult, if not unrealistic, task. When Professor Egon Bittner explained that the police are inherently controversial, the word *inherently* was included for a reason. Even if the police of the future make additional strides to become more accountable to the citizens they police and protect, it is questionable if they will become any less controversial.

MILITARIZATION

A final prediction is that the lines separating the police and the military may become increasingly blurry. In particular, the police may become more military-like and the military may become more police-like. With regard to the latter, since September 11 there have been numerous instances of the military becoming involved in domestic law enforcement activities. For example, an immediate response to September 11 was the placement of fully equipped National Guard personnel in airports to provide security. Efforts in recent years to tighten security at the U.S.-Mexico border have included the deployment of the military. The National Guard has been deployed to assist the police in controlling riots triggered by deadly force incidents, including the deployment of thousands of National Guard troops in Ferguson, Missouri.

Along with the military becoming more police-like, the police are also becoming more military-like. This trend has

Photo 15.4
On some occasions the police appear to be indistinguishable from the military. Jim West/Alamy Stock Photo

been referred to as the militarization of the police.[14] This is particularly obvious with police tactical units and their routine deployments. Relatedly, police departments in the twenty-first century are increasing their adoption of military tactics and technologies, at least in part due to the availability of "free" military equipment (e.g., military vehicles, weapons, other equipment) through the Department of Defense's Excess Property Program.[15] While the trend of the police becoming more military-like is likely to continue, law enforcement agencies may still be sometimes more community oriented than military-like. The police will have a "velvet glove" and an "iron fist."[16]

THE NEW TECHNOLOGY OF CRIME DETECTION AND ACCOUNTABILITY

To combat criminal activity and to provide accountability, police of the future will increasingly use technology. Accountability-oriented technologies include video recording, GPS, and early intervention systems. Crime fighting technology will focus on identification (e.g., **biometrics**), "seeing" and scanning, computer/Internet applications, and information management and access. Each of these technologies is discussed below.

biometrics:
Technologies capable of identifying a person by measuring a feature of a person's unique physical characteristics.

THE TECHNOLOGY OF ACCOUNTABILITY

Since the high-profile deadly force incidents of 2014, many calls have been made for the rapid incorporation of police body-worn cameras into the daily work of officers. The reasoning is twofold: (1) If the conduct of officers is being recorded and they are aware of this, they will be less likely to engage in unlawful conduct; and (2) if officers engage in unlawful conduct, video of it will serve as powerful evidence of their guilt. While this reasoning seems logical, there is mixed evidence as to its truth.[17] Even in cases where video shows an officer using force, the video obviously does not show the officer's state of mind, understanding of the situation, or intent. Nevertheless, the police are likely to listen to the demands for body-worn cameras as this shows a commitment to transparency and accountability. Transparency and accountability are not easily achieved, however. For example, in some recent cases, the police have been criticized by citizens for not releasing video footage of the use of force in question quickly enough (see Good Policing feature); accusations have been made that the delayed release was due to police altering the video to protect unjust actions.

As discussed earlier in this book, the use of GPS as an accountability tool may rise in the future. GPS can be attached to police vehicles to show their real-time location or installed on police equipment, such as badges or radios, to show the real-time location of officers themselves. This technology could prevent officers from engaging in inappropriate or illegal behaviors (recall the story from Chapter 12 about the officers who went sledding on duty). It could also serve a valuable safety function for officers, particularly if an officer becomes incapacitated.

Early intervention systems will likely become more common in police departments. As discussed, an EIS stores and monitors data on police officer performance and can be used to identify potentially problematic performance. Interventions such as counselling or retraining can be explored with an officer who displays patterns of worrisome behavior. Although the use of an EIS has been repeatedly identified as a best practice in policing,[18] serious research on the proper operation, functioning, performance indicators, thresholds on which to flag officers, and appropriate and effective interventions associated with these systems is still lacking.[19] As the concept holds much promise, however, and as accountability of officers takes on even greater importance in the coming years, more frequent and effective use of early intervention systems in police departments will probably occur.

GOOD POLICING

The Impact of Legislation

Technology needs to be supported by good policy in order to have a chance at improving police operations. Case in point is body-worn cameras and Bill 748. Bill 748 was signed into law in California in 2018. It requires all police departments in California that deploy body-worn cameras to release body-worn camera video and audio recordings of officer shootings and other serious use of force incidents within 45 days unless doing so would interfere with an ongoing investigation.[20] This requirement is intended to increase the transparency of police department actions in these critical incidents.

Photo 15.5
Timely release of body camera footage to the public is important even if it shows questionable police conduct. AP Photo/Uncredited

THE CRIME-FIGHTING TECHNOLOGY OF IDENTIFICATION

Biometrics refers to technologies capable of identifying a person by measuring a feature of a person's unique physical characteristics.[21] Although the concept of biometrics has appeared only relatively recently, attempts at identifying people based on their characteristics have been around for a long time. Of course, most significant today is the science of DNA. Since DNA was first used for criminal identification purposes in the late 1980s, it has become a powerful tool of justice. However, the value of DNA for purposes of criminal identification still largely depends on having a suspect for comparison. With the development of CODIS, a nationwide network that includes DNA samples from convicted offenders and crime scenes, DNA has become a more useful tool for identifying suspects, but it still has major limitations. In the future, to enhance the value of DNA in criminal identification, law enforcement agencies may collect and store DNA samples from all *arrested* offenders (not just convicted offenders). This was common practice in England until recently when millions of DNA profiles that were collected from people who were arrested but not convicted were removed from police databases due to ethical and legal concerns.[22] On the more distant horizon may be the construction and operation of a massive databank that would include DNA from every individual who was born or resides in the United States. The technology to store this amount of data is available, but serious legal and ethical issues have inhibited the development of such a system.[23] Presumably, with such a database, DNA would become a much more effective tool for identifying perpetrators.

Progress continues to be made in the science of collecting and analyzing DNA. New technologies are being developed that will allow DNA to be collected from difficult objects and environments and be analyzed more efficiently and at lower cost. The miniaturization of collection and testing devices will allow DNA evidence to be collected and analyzed at crime scenes and allow for faster solving of crimes. Technology is being developed and tested whereby DNA can be analyzed at a cost of $20 per test, compared to the current cost of $600 to $1,600.[24] As technology evolves, on-the-spot field testing (versus laboratory testing) will be common practice. DNA analysis is also evolving to more reliably identify instances of secondary DNA transfers.[25]

Familial DNA typing, whereby offenders are identified through analysis of their DNA left at crime scenes and their relatives DNA stored in genealogy databases, is an emerging but controversial practice (see Chapter 7).[26] DNA technology is advancing whereby physical characteristics of individuals can be determined on the basis of a DNA profile. This field of scientific inquiry is referred to as **DNA intelligence or phenotyping**. To date, this has been limited to identifying sex, hair color, eye color, and race of the subject, but future applications may involve other externally visible characteristics.[27]

The technology that supports fingerprints as a method of identification is also likely to continue to evolve. In many police departments today, the old system of using ink and paper for fingerprinting has been replaced with optical sensing or other electronic methods. Future verification and identification of suspects will more frequently be performed through portable fingerprint collection technologies. Fingerprint collection at crime scenes will also likely become more efficient and less labor intensive. Advances are being made that will allow fingerprints to be recovered from difficult surfaces, such as skin and burned items, and from items exposed to water, weather, and sunshine.[28] Fingerprint databases are also being improved to provide faster and more accurate search results.

Facial recognition systems are also being used as a method of identification and will continue to evolve. Sometimes discussed in terms of artificial intelligence, these systems go beyond passive camera surveillance and include a capability whereby a facial image can be captured on camera and compared to digital images stored in a reference database. When a match is made, the operator of the system is alerted. In the United States, facial recognition systems have been deployed in casinos for several years to assist security personnel in identifying and apprehending known gambling cheats and scam artists. Such systems have also been implemented on a

DNA intelligence or phenotyping: Determining the physical characteristics of an individual based on a DNA profile.

facial recognition systems: Systems whereby a facial image can be captured on camera and compared to digital images stored in a reference database.

limited basis in airports, on public streets, and in sports stadiums. In the future, facial recognition systems may be used instead of public video cameras to allow for quick identification of known offenders. Facial recognition technology paired with police body-worn cameras may allow the police to easily identify individuals with whom they are interacting.[29] Relatedly, cameras that do not require human monitoring are being developed that can assess and identify suspicious or criminal behavior.[30] As with other technologies, facial recognition systems raise a multitude of issues regarding privacy. In addition, research has demonstrated numerous reliability problems where false identifications have been made.[31] However, given the potential value and usefulness of the technology, it is likely that efforts to refine it will continue.

Future biometric technologies may include heartbeat and pulse recognition, voice biometrics, and handwritten signature recognition. Of these, voice biometrics probably has the widest application and greatest potential. Common today are telephone voice response systems in which navigation can be made via speech prompts. Much more sophisticated, however, are systems that can recognize *individual* voices for purposes of identification. While such applications have been deployed, additional work remains to be done to improve the reliability of the technology.

A Question to Consider 15.2

What's So Great About Privacy?

One of the major impediments to the widespread deployment of crime detection and criminal identification technology is concern about individual privacy. This is also an issue with intelligence-led policing more generally. Is the threat to privacy really something to be concerned about? After all, what is so great about privacy? Should we not take whatever steps are necessary to prevent crime and keep our society safe? Explain.

THE CRIME FIGHTING TECHNOLOGY OF "SEEING"

Advances are also being made in the technology of "seeing." Perhaps most significant is the use of low-level X-rays to facilitate the detection of weapons, explosives, drugs, and other contraband. For example, machines now allow security personnel to easily see through clothing and identify items that may be hidden underneath. These systems are deployed at airports, and future installations may include shopping malls, schools, theaters, and other public buildings. Similar technology allows personnel to see through luggage to detect explosives and through cargo containers. Future applications may also include portable or hand-held devices that allow police officers to detect weapons in the possession of a subject from a distance.

Many urban police departments presently deploy gunshot detection systems. These units are mounted on buildings or poles and have the capability to detect gunshots in a defined geographic area. Ideally, these systems shorten police response times to shooting incidents and thus allow for more on-scene apprehensions of offenders. Work is being done to increase the sophistication of these systems including their ability to identify the number of different firearms present and the type and caliber of the weapons used.[32] These systems cost in the range of $40,000 per square mile.[33] It is probable in coming years this cost will decline and deployment of these systems will increase.

Other sensing technologies, such as **thermal imaging**, which can detect heat from closed structures, thus allowing one to "see" through walls and barriers, already exist but have encountered several legal obstacles to their use. Currently this technology is most often used to detect illegal indoor marijuana farms, to assist in search and rescue operations, and to detect fleeing suspects and find missing persons.[34] Successful navigation of legal issues will allow the police to use sensing capabilities in many other situations.[35] As a result, it could become a useful tool of crime detection and evidence collection.

thermal imaging:
Technology that detects heat from within closed structures.

It is certainly no stretch to predict the use of **closed-circuit television (CCTV)** surveillance will increase dramatically in the future of the nation. The United States may follow the lead of the United Kingdom in this regard. The United Kingdom has an estimated 4.2 million cameras in place—nearly 500,000 in London alone.[36] It has been suggested that the typical Briton is recorded by about 300 cameras every day. No city in the United States even comes close to this level of camera deployment. Most common now is the installation of cameras at intersections and roadways as a way to remotely enforce traffic laws, particularly red-light violations and speed limits. Compared to many other technologies, CCTVs are relatively inexpensive to purchase, operate, and maintain.

Like CCTV, **unmanned aerial vehicles (UAVs)** with camera technology may also see widespread use in future law enforcement efforts. UAVs, or drones, could be used for routine police patrol and to obtain quick visuals at crime scenes. UAVs could be equipped with facial recognition or gunshot detection systems, further enhancing their usefulness. UAVs will continue to shrink and probably reach the size of a small bird. This technology will take video surveillance to a new level, literally.

Photo 15.6

UAVs (drones) are likely to have many uses in police work, from routine patrol to search operations. picture alliance/Getty Images

In addition to "seeing" technology, law enforcement also utilizes "sniffing" technology. Electronic "sniffer" microchips are designed to detect microscopic amounts of substances, such as chemicals, explosives, radiation, and drugs. In the future, it is probable that this technology will replace canines. These sniffers could also be mounted on buildings or have mobile applications.

Experiments are being conducted with so-called smart cameras that can detect increased body temperature, high pulse and blood pressure, and heavy respiration, all of which are indicators of stress and deception.[37] This application may be particularly useful in detecting drug smugglers and people with terrorist intentions in airports. With regard to lie detection, the emerging use of brain scans could replace other more unreliable methods of deception detection.[38]

THE CRIME FIGHTING TECHNOLOGY OF COMPUTER AND INTERNET APPLICATIONS

Other investigative tools are being deployed to deal more effectively with Internet and computer-related crimes. For example, technology has been developed to search email traffic for specific senders, recipients, and keywords. Law enforcement can direct a computer virus to a particular individual via email to record key strokes made by that user, which is especially useful when investigating terrorists and hackers. Technology to identify cell phones in a particular area and to track the location of specific phones is also continuing to progress.

TECHNOLOGIES FOR INFORMATION MANAGEMENT AND ACCESS

One of the primary functions of technology in the workplace is to enhance productivity by making work tasks less time consuming. Common today is the use of laptop computers in police cars for communication and deployment purposes.[39] In the future, police personnel may have immediate access to a multitude of information through portable or wearable computers—information such as photographs of suspects or stolen property, fingerprint databases, and

closed-circuit television (CCTV): A system of cameras that transmit their signals to a restricted set of monitors.

unmanned aerial vehicles (UAVs): Also known as drones, UAVs are remotely controlled flying machines that can be used for surveillance.

DNA profiles. Officers and investigators will be able to construct and conduct photographic lineups using computers in their vehicles. Computer and smartphone applications will continue to be developed for law enforcement purposes. For example, numerous agencies in Minnesota are using an app that alerts officers when they are in proximity to a person who has a condition such as autism, bipolar disorder, or schizophrenia and is wearing a beacon to alert police of their presence.[40] Voice recognition, still in its infancy, will likely control these technologies.

THE IMPLICATIONS OF TECHNOLOGY

There is no question that the widespread adaptation of most of the technologies discussed here could produce a substantial reaction, mostly with regard to the issue of privacy. Indeed, most legal challenges of the future that relate to technology will probably center on privacy issues.[41] For example, the law relating to the constitutionality of strip and body cavity searches is well developed; however, do body examinations through the use of low-level X-rays alter notions of privacy and expectations of it? Do through-the-wall surveillance devices alter the meaning of search and seizure? What are the parameters of privacy in cell phone and email correspondence? The courts have begun to address these issues, but clearly technology has opened a new set of questions relating to the reasonable expectation of privacy.

Another factor that may inhibit the incorporation of the newest technology into policing is its effectiveness. Research studies designed to assess the impact and effects of emerging technologies are necessary but, in many instances, lacking.[42] If a particular technology is not effective for purposes of crime control, then it just leads to less privacy for citizens. Finally, the cost of technology cannot be dismissed. Police departments continue to operate on lean budgets. If money is spent on technology, it cannot be spent on other things (e.g., personnel, training). Concerns about privacy, narrow resources, and the limitations of technology may have implications for the effectiveness of the police.

Main Points

- One reason for studying history is that it may serve as a guide to the future. The purpose of studying the future is to prepare for it.
- American police departments have progressed through three eras: the political era, the reform era, and the community problem-solving era. The political era and the reform era ended as the result of a crisis. The community problem-solving era may end in a similar fashion.
- The most significant new demands on the police of the future may relate to terrorism, technology, and an increased demand for accountability.
- Different terrorist groups have different goals. Terrorism can be defined as "the unlawful use of force or violence against persons or property to intimidate or coerce a government, the civilian population, or any segment thereof, in furtherance of political or social objectives."
- Terrorism can be classified in several ways: international or domestic, state-sponsored or lone actor.

- It seems reasonable to predict that in the near future the lines separating the police and the military will become increasingly blurry.
- Technology not only makes certain types of crime possible, it also provides the police with new tools for fighting these, and other, crimes. It may also provide opportunities to enhance the accountability of the police.
- Much of the new crime-fighting technology will focus on identification (e.g., biometrics); "seeing"; computer/Internet applications; and information management and access.
- Most legal challenges of the future that relate to technology will probably center on privacy issues.
- In addition to privacy concerns, the costs associated with developing and adopting technologies may inhibit their incorporation into law enforcement agencies.

Important Terms

Review key terms with eFlashcards at **edge.sagepub.com/brandl2e.**

biometrics   329

closed circuit television (CCTV)   333

DNA intelligence or phenotyping   331

facial recognition systems   331

jihadists   324

lone wolf terrorists   325

right-wing extremists   325

terrorism   324

thermal imaging   332

unmanned aerial vehicles (UAVs)   333

USA Freedom Act   328

USA PATRIOT Act of 2001   328

Questions for Discussion and Review

Take a practice quiz at **edge.sagepub.com/brandl2e.**

1. What is terrorism? What forms does it take?
2. What is right-wing extremism? According to law enforcement leaders, what is the significance of this threat?
3. How might technology represent a new demand on the police?
4. How and why might accountability be a significant new demand on the police? How might the police address this demand?
5. What are the most important ways in which police of the future may differ from the police of today?
6. What is meant by the militarization of the police? Is militarization good or bad? Explain.
7. How could DNA become even more useful as a tool in identifying criminals?
8. What is biometrics? How might such technology make criminal investigations more effective?
9. Why is it important to determine the effects and effectiveness of police technology?
10. What is likely to be the most significant legal challenge of the future? Why?

Fact or Fiction Answers

1. Fact
2. Fact
3. Fiction
4. Fiction
5. Fact

$SAGE edge™

Digital Resources

Get the tools you need to sharpen your study skills. SAGE Edge offers a robust online environment featuring an impressive array of free tools and resources.

Access practice quizzes, eFlashcards, video, and multimedia at **edge.sagepub.com/brandl2e.**

Media Library

View these videos and more in the interactive eBook version of this text!

Career Video

15.1: How Local Police Departments Work With Federal Agencies

Criminal Justice in Practice

15.1: Terrorism

SAGE News Clip

15.1: Bob Marley's Granddaughter Wants Accountability After NY Airbnb Police Stop

APPENDIX

The Bill of Rights, United States Constitution

IN the United States, some of the rights of citizens are defined by the Constitution—in particular, the first ten amendments to the Constitution. It is important to realize many of these laws have undergone modification since the time of the ratification of the Constitution. For example, according to the Fourth Amendment, searches must be based on probable cause and supported with a warrant. However, for reasons discussed in Chapter 9, most legal searches conducted by the police today are made without either of these things. When studying the police, the Fourth, Fifth, and Sixth Amendments are most relevant.

Amendment 1: Congress shall make no law respecting an establishment of religion, or prohibiting the free exercise thereof; or abridging the freedom of speech, or of the press; or the right of the people peaceably to assemble, and to petition the government for a redress of grievances.

Amendment 2: A well regulated militia, being necessary to the security of a free state, the right of the people to keep and bear arms, shall not be infringed.

Amendment 3: No soldier shall, in time of peace be quartered in any house, without the consent of the owner, nor in time of war, but in a manner to be prescribed by law.

Amendment 4: The right of the people to be secure in their persons, houses, papers, and effects, against unreasonable searches and seizures, shall not be violated, and no warrants shall issue, but upon probable cause, supported by oath or affirmation, and particularly describing the place to be searched, and the persons or things to be seized.

Amendment 5: No person shall be held to answer for a capital, or otherwise infamous crime, unless on a present-ment or indictment of a grand jury, except in cases arising in the land or naval forces, or in the militia, when in actual service in time of war or public danger; nor shall any person be subject for the same offense to be twice put in jeopardy of life or limb; nor shall be compelled in any criminal case to be a witness against himself, nor be deprived of life, liberty, or property, without due process of law; nor shall private property be taken for public use, without just compensation.

Amendment 6: In all criminal prosecutions, the accused shall enjoy the right to a speedy and public trial, by an impartial jury of the state and district wherein the crime shall have been committed, which district shall have been previously ascertained by law, and to be informed of the nature and cause of the accusation; to be confronted with the witnesses against him; to have compulsory process for obtaining witnesses in his favor, and to have the assistance of counsel for his defense.

Amendment 7: In suits at common law, where the value in controversy shall exceed twenty dollars, the right of trial by jury shall be preserved, and no fact tried by a jury, shall be otherwise reexamined in any court of the United States, than according to the rules of the common law.

Amendment 8: Excessive bail shall not be required, nor excessive fines imposed, nor cruel and unusual punishments inflicted.

Amendment 9: The enumeration in the Constitution, of certain rights, shall not be construed to deny or disparage others retained by the people.

Amendment 10: The powers not delegated to the United States by the Constitution, nor prohibited by it to the states, are reserved to the states respectively, or to the people.

GLOSSARY

affirmative action: Action intended to (a) proactively recruit, hire, and promote women, minorities, disabled individuals, and other protected groups of people and (b) correct past discrimination in order to eliminate its present-day effects.

AMBER (America's Missing: Broadcast Emergency Response) Alert: This alert is activated when a child abduction has occurred or is suspected; it includes a description and photo of the missing child and information about the suspected perpetrator, the suspected vehicle, a tip line phone number, and any other information that may assist in locating the child.

arrest: An arrest occurs when the police take a person into custody for the purposes of criminal prosecution and interrogation.

arrest warrant: A document issued by a magistrate that authorizes the arrest of an individual.

assault: A physical attack.

assessment center: A center that allows officer applicants or officers seeking promotion to be judged on their performance in role-playing exercises.

automatic vehicle locator (AVL) system: A GPS system that tracks the whereabouts and activities of officers when in their squad cars.

bad apple theory: Theory holding a small number of problem officers can influence other officers, thus creating a bigger problem.

Bertillonage: A system wherein physical measurements were used to identify and differentiate suspects.

Bill of Rights: The first ten amendments to the U.S. Constitution; these amendments articulate several basic freedoms of the nation's citizens.

biological evidence: Physical evidence that contains DNA, such as blood, semen, or saliva.

biometrics: Technologies capable of identifying a person by measuring a feature of a person's unique physical characteristics.

black codes: Codes designed to limit the rights of freed slaves in the post–Civil War South.

bodily force: Force that involves physical restraining maneuvers, such as vertical stuns, takedowns, and punches or kicks.

body armor: Protective vest worn by officers to help prevent injury or death, particularly from gunshot wounds.

bona fide occupational qualifications (BFOQs): Qualifications such as skills, traits, or abilities that are required to correctly perform a certain job.

bribery: The acceptance of money, services, or goods in exchange for some consideration, such as not issuing a speeding ticket.

broken windows theory: A theory that posits minor incidents of disorder lead to a fear among the populace that keeps citizens from interacting with each other, which in turn leads to a reduction in the fear of apprehension among criminals.

brutality: Use of force that is consciously and purposefully cruel and harsh.

bureaucracy: An organization characterized by many departments and divisions operating through a complicated structure system of rules and regulations.

burnout: A prolonged response to chronic and interpersonal stressors on the job.

call priority: A system in which a faster response and more resources are given to more critical calls for service.

chain of command: This principle holds that every person in the organization has a supervisor, and supervisors have more authority than their subordinates.

chain of custody: The record of individuals who maintained control (custody) over evidence from the time it was obtained by the police to when it was introduced in court.

circadian rhythm: A natural biological rhythm that regulates sleeping patterns, alertness, and brain wave activity, among other things.

civil service protections: Protections that ensure government employees are recruited, hired, and promoted based on qualifications and merit and demoted or dismissed only for cause.

closed circuit television (CCTV): A system of cameras that transmit their signals to a restricted set of monitors.

code of silence: An unwritten rule of the police culture that holds officers should not report the misconduct of fellow officers. It is a human phenomenon, not just a police phenomenon.

cold case investigations: Investigations of past serious, unsolved crimes that have been reopened, typically due to the availability of new evidence or witnesses.

collective efficacy: Cohesion among neighborhood residents combined with shared expectations for informal social control of public space.

Combined DNA Index System (CODIS): An electronic database operated by the FBI that allows federal, state, and local crime laboratories to share DNA profiles electronically.

Community Oriented Policing Services (COPS): An office created by the Violent Crime Control and Law Enforcement Act of 1994 to distribute funds to departments for the development and fostering of community policing efforts.

community policing: An orientation to policing that focuses on building relationships with citizens in order to achieve crime control and other outcomes such as citizen satisfaction.

community problem-solving era: An era of policing that emphasizes the assistance and support of the community in fighting crime.

community relations bureaus: Bureaus created to improve community-police relations through public relations activities.

community service officer (CSO) programs: Programs wherein college students are hired as nonsworn police personnel. CSOs may assist in nonemergency situations and perform administrative duties.

CompStat: A management process that involves the analysis of data to identify problems and the assignment of responsibility to police personnel for the resolution of those problems.

computer-aided dispatch (CAD): A computer system and database that tracks calls for service as they are received, monitors the status of patrol units, and provides various reports relating to calls for service, the activities of officers, and the calls to which officers respond.

confidential informants: Members of the public who assist law enforcement in an active and ongoing capacity; they are typically associated with the criminal underworld.

consent: The granting of permission for police to take requested action, such as to enter a home or conduct a search.

consent decree: A decree that provides specific goals or timetables to an agency in order to increase the agency's employment representation of protected groups.

constable: The first appointed law enforcement officers in colonial America. They often organized and supervised the watch.

continuum of force: The principle that police can and should only use as much force as necessary to overcome the resistance offered by the subject.

control talk: A police officer's verbalization to induce compliance from a subject.

coproduction: A concept in which the police and the community work together to prevent crime.

crackdown: The allocation of additional police resources to the enforcement of laws with the intent of deterring illegal conduct.

crime analysis: The collection and analysis of data pertaining to crimes, perpetrators, and targets of crimes.

crime hot spots: Small geographic areas with a high concentration of crime.

crime maps: Maps that display where various crimes have occurred in a particular jurisdiction over a particular period of time.

crime scene profile: Information about an offender that includes such details as race, age (or age range), employment status, type of employment, marital status, level of education, and location of residence.

criminal evidence: Knowledge or information that relates to a particular crime or perpetrator.

culture: A collection of unwritten rules, expectations, values, and attitudes shared and understood by members of a group.

curbside justice: The use of force by police rather than arrest to deal with law breakers.

custodial interrogation: An interrogation that takes place when a subject is in the custody of the police.

cyberattack: Crimes in which a computer system is the target. Examples include computer viruses, hacks, and denial of service attacks.

cyberbullying: Bullying that occurs through social media sites, email, text messages, and other digital media.

data-driven policing: Policing that occurs when leaders collect and analyze departmental data to make informed decisions about how their departments should best operate, particularly with regard to policies and strategies.

deadly force: Forms of force used by the police that have a high likelihood of resulting in the death of a subject.

decertification: The revoking of an officer's training credentials.

decision: When a person chooses a particular option based on consideration of the available information.

decoy operation: An operation in which an undercover police officer presents the opportunity to commit a crime to an individual. Once the crime has been attempted, officers who are standing by can arrest the would-be perpetrator.

de-escalation training: Training that is designed to teach officers how to resolve potentially violent situations without the use of force.

demeanor: The behavior of a person.

de-policing: A phenomenon where the police withdraw from active police work and avoid interactions with citizens.

deterrence: Making someone decide not to do something.

differential police response (DPR): A response to a call for service other than immediately dispatching an officer to the scene.

digital evidence: Information or data relating to a crime that is located on an electronic device.

direct evidence: Crime-related information that immediately demonstrates the existence of a fact in question.

dirty means: Unethical or illegal means used by police officers.

discovery crimes: Crimes discovered after their completion.

discretion: A police officer's personal judgment of how best to handle a situation.

DNA analysis: The analysis of DNA in order to include or exclude a particular person as the source of that DNA.

DNA intelligence or phenotyping: Determining the physical characteristics of an individual based on a DNA profile.

dragnet: An historical reference to the process wherein when a crime occurred, the police would bring in for questioning all the suspects usually associated with that type of crime.

drive-stun mode: Taser mode in which the weapon itself is placed in contact with the subject's body, delivering the electrical current directly.

early intervention system (EIS): Data management system designed to record and monitor certain behaviors of officers with the objective of identifying potentially problematic patterns of behaviors while they can still be corrected.

entrapment: This occurs when the police induce a person to commit a crime that he or she would not have committed otherwise.

equal employment opportunity (EEO): Provision for employment regardless of employees' characteristics. Laws related to equal employment opportunity are intended to stop arbitrary discrimination.

ethical standards of conduct: Doing what is right based on moral standards.

ethics: Rules of behavior that are influenced by a person's perception of what is morally good or bad.

evidence-based policing: Policing that involves the use of research findings to inform policies and strategies.

excited delirium: A potentially deadly medical condition involving psychotic behavior, an elevated temperature, and an extreme fight-or-flight response by the nervous system.

exclusionary rule: If a search is determined to be unreasonable, the evidence obtained must be excluded from trial.

exculpatory evidence: Evidence that tends to exclude or eliminate someone from consideration as a suspect.

exigent circumstances exception: Emergency situations that allow the police to conduct a search without a warrant.

extortion: The use of threats to obtain something, usually money.

facial recognition systems: Systems whereby a facial image can be captured on camera and compared to digital images stored in a reference database.

Ferguson effect: The unsupported notion that the police enforce the law less often and, as a consequence, crime has increased and there is more violence directed against the police.

field training: This training involves the probationary officer working alongside a field training officer (FTO) to learn and perform the tasks of policing while actually on the street.

forensic science: The field of science that addresses legal questions.

free society: A society in which the government recognizes that human beings have certain basic human rights.

gate keepers: The role the police play in determining who comes into the criminal justice system and who stays out.

genderized occupation: An occupation that tends to be represented by one gender over another.

geospatial crime analysis: Analysis of crime data that focuses on where crimes occur.

good ends: The desired goals of policing.

gratuities: Small gifts provided for services.

hot-pursuit exception: If the police are in pursuit of a subject and have probable cause to believe that subject committed a crime and is in a home, then the police may enter the home to make an arrest and conduct a search.

hot spot policing: Concentrating police in areas where crime is most likely to occur.

identity theft: A crime that occurs when one person steals the personal information from another and uses it without permission.

impact weapons: Weapons used to induce compliance; they include batons, flashlights, chemical sprays, and Tasers.

implicit bias: Thoughts and feelings about social groups that can influence people's perceptions, decisions and actions without awareness.

impossible mandate: This term reflects the idea that the police have been assigned the task of crime control, but because they cannot control the factors that *cause* crime, this task is difficult—if not impossible—to accomplish.

incapacitation: Making it impossible for people to commit crimes.

inculpatory evidence: Evidence that tends to include or incriminate a person as the perpetrator.

indirect evidence: Crime-related information in which inferences and probabilities are needed to draw an associated conclusion.

Industrial Revolution: A period during the eighteenth and nineteenth centuries marked by new manufacturing processes and a transition from rural means of production to urban ones.

initial deterrence decay: Decay that occurs when an initial deterrent effect dissipates or disappears.

in-service training: Training that occurs throughout an officer's career.

Integrated Automated Fingerprint Identification System (IAFIS): National database of fingerprints and criminal histories maintained by the FBI.

intelligence-led policing: An approach to policing that involves the collection and analysis of information on likely offenders and their methods of operation.

internal affairs unit: A unit responsible for investigating citizen complaints against officers and internally generated complaints among police department members.

interrogation: Any questioning or other action that is intended to elicit incriminating information from a suspect when this information is intended to be used in a criminal prosecution.

investigative interview: Any questioning that is intended to produce information regarding a particular crime or a person believed to be responsible for a crime.

involvement crimes: Crimes witnessed as they are occurring.

jihadists: People who believe in the duty of Muslims to spread their religion through militant means.

Jim Crow laws: Laws that mandated racial segregation in public facilities.

knock and talk search: A search in which the police talk with the occupant of a home in an attempt to get consent to conduct a search.

lateral entry: When officers move from one department to another and retain their position and seniority.

law enforcement: Tasks performed by the police that involve conducting investigations and making arrests.

legalistic style of policing: Policing style in which enforcing the law is the top priority.

less lethal weapons: Weapons less likely to cause death to a subject. They commonly include OC spray and Tasers.

lone wolf terrorists: Terrorists who act on their own without help from a terrorist group or a government. Also known as lone-actors.

low-visibility situations: Situations in which not many people can see the decisions of the police that are being made.

means over ends syndrome: When police are more concerned with how things are done than with the goals they are supposed to achieve.

Miranda warnings: A list of rights that must be provided to an individual when in custody and prior to questioning.

mobile data computers (MDCs): Computers mounted in police vehicles that are connected wirelessly to a department's computer network and computer-aided dispatch system.

monopoly: A company or agency that does not have competitors.

morality: A person's internal beliefs about what is right or wrong conduct.

National Crime Information Center (NCIC): The largest, most well-known crime information network system in the country.

National Decertification Index (NDI): A list of decertified officers in the United States.

noble cause corruption: A belief the police may take illegal action to achieve a greater good.

non-stress academy training: Police academy training that is more relaxed and oriented toward academic achievement.

offender-focused strategies: Strategies in which the police depend on criminal intelligence to identify high-rate offenders on whom the police then focus enforcement efforts.

oleoresin capsicum (OC) spray: Spray containing an inflammatory agent that causes an inflammation of the respiratory tract and a burning and swelling of the eyes; also known as pepper spray.

order maintenance: Activities of the police that involve keeping the peace without resorting to citations or arrests.

organizational culture: The practices, rules, and principles of conduct that govern how people in an organization behave.

other places exception: A warrant exception that allows police to search certain places and things not afforded Fourth Amendment protection.

overpolicing: The perception of too much police presence and action in a neighborhood.

person-focused analysis: In this type of analysis, the focus is on identifying and targeting offenders and potential offenders.

physical evidence: Tangible items that can be held or seen that are produced as a direct result of a crime having been committed.

physiological response: A reaction that triggers a bodily response to a stimulus.

plain view exception: If the police conduct a search with a warrant or are legally present at a particular place and evidence is observed in plain sight, that evidence may be seized.

police corruption: Illegal behaviors committed by police officers who use their authority as police officers to commit crimes motivated by personal gain.

Police Explorer programs: Voluntary programs for young people aged fourteen to twenty designed to educate them about policing activities and to provide them training and hands-on experience with matters related to a law enforcement career.

police integrity: The inclination among police to resist temptations to abuse the rights and privileges of their occupation.

police legitimacy: The belief the police ought to be allowed to exercise their authority to maintain social order, manage conflicts, and solve problems in their communities.

police matrons: Female police department employees whose duties usually involved only female prisoners.

police misconduct: Inappropriate or illegal behaviors of police officers; also sometimes referred to as police deviance.

political era: The period from the mid-1800s to the early 1900s during which policing was heavily influenced by politics.

positional asphyxia: A dangerous condition that occurs when a person's body position prevents normal and adequate breathing.

post-traumatic stress disorder (PTSD): Mental health condition triggered by witnessing or experiencing an extremely upsetting event. Symptoms may include flashbacks, sleep problems, and anxiety.

predictive policing: Policing strategy that uses data to try to predict and prevent crime.

pretext traffic stop: A traffic stop made for any traffic offense that may then allow for other law enforcement action.

probable cause: A standard of proof that is generally required in order for police to justify a search or arrest.

probe mode: Taser mode in which two probes are fired from the Taser into the body of a subject; an electrical current then flows from the Taser through the wires attached to the probes and into the subject.

problem-oriented policing: A type of policing that attempts to address specific problems that affect the quality of life of residents in a community.

procedural justice: Citizens' perceptions of police fairness in contacts with officers.

proof: Something, such as evidence, that proves something else is true.

reasonable force: The minimum amount of force necessary to overcome the resistance offered by the subject.

reasonable suspicion: Level of proof required for police to legally stop and frisk a person. It relates to the likelihood of a person's involvement in a criminal act. It is less of a hurdle than probable cause.

reform era: An era in policing that centered on removing the police from the control of politicians and making departments more professional and efficient.

residual deterrence decay: Decay that occurs when a residual deterrent effect dissipates and the normal level of crime resumes.

response time: The amount of time that elapses between when a crime occurs and when officers arrive at the scene. It includes both citizen reporting time and police response time.

reverse discrimination: Discrimination that occurs when women or minorities are hired (or promoted) not because of their qualifications but because of their sex or race.

right-wing extremists: Domestic terrorists who act on their extreme beliefs, which can include racism, homophobia, anti-Semitism, and neo-Nazism.

rogues gallery: A collection of photographs of known criminals.

SARA (scanning, analysis, response, and assessment) model: A method of problem solving that involves identifying a problem, researching what is known about the problem, developing solutions to the problem, and then evaluating the success of the solution.

search: A governmental infringement into a person's reasonable expectation of privacy for the purpose of discovering things that could be used as evidence in a criminal prosecution.

search incident to arrest exception: This exception allows a warrantless search of an individual as a result of that person's arrest.

search warrant: A document that authorizes the search and seizure of an individual's property; it specifies the person, place, or vehicle to be searched and the types of items to be seized by the police.

seizure: When the police take control of a person or thing because of a violation of the law.

service: Other duties performed by police, such as assisting stranded motorists and attending to other people in need.

service-oriented style of policing: Policing style that maintains citizen satisfaction is the top priority.

sexual harassment: Unwelcome sexual advances, requests for sexual favors, verbal or physical harassment of a sexual nature, and/or offensive remarks about a person's gender.

sheriff: In early American policing, the police figure who typically worked in a less populated area. The primary responsibilities of the sheriff were to apprehend criminals, assist the justice of the peace, collect taxes, and supervise elections.

slave patrols: Patrols tasked with looking for runaway slaves, policing the whereabouts of slaves, and making sure slaves were not in possession of weapons or property they were not allowed to have.

smart policing: Any initiative on the part of a police department to incorporate technology into police operations with the goal of developing information that can be used to reduce crime.

specialization: A focus on certain responsibilities in order to develop expertise and efficiency in those tasks.

stakeout: Watching a location and monitoring activities at that location.

standard operating procedures (SOPs): The official policies and procedures of an organization.

sting: A police operation that involves an investigator posing as someone who wishes to buy or sell illicit goods, such as drugs or sex, or to execute some other sort of illicit transaction.

stop and frisk exception: If there is reasonable suspicion criminal activity is afoot and the subject may be armed, the police may conduct a search of the outer clothing of a person without a warrant even though an arrest of that person may not be justified (see also Terry stops).

stopping, questioning, and frisking (SQF): SQFs occur when an officer has reasonable suspicion that a subject is involved in criminal behavior so the officer conducts a stop of that person and searches that person for weapons.

street-level bureaucracy: A public agency that serves primarily low-income clients and whose workers have substantial discretion in processing those clients.

stress-based academy training: Police academy training that reflects military training; often includes drills and intense physical demands.

subject resistance: Subject resistance to police; can involve (a) passive resistance, (b) verbal resistance, (c) physically defensive resistance, (d) physically active or aggressive resistance, and/or (e) deadly resistance.

suicide by cop (SBC): Intentional subject behavior that seeks to coerce police officers to respond with lethal force.

surveillance: Watching a person to monitor their activities.

symbolic assailant: People who an officer believes may pose a danger.

tactical enforcement unit (TEU): A police unit that handles high-risk criminal situations, such as hostage situations and the execution of certain arrests and search warrants.

Taser: An electronic stun device that resembles a gun. It delivers an electrical current to incapacitate a subject.

team policing: Policing strategy that involved assigning small teams of patrol officers, investigators, and supervisors to neighborhoods on a long-term basis in order to increase the exchange of information and improve cooperation.

terrorism: According to the Code of Federal Regulations, terrorism is the unlawful use of force or violence against persons or property to intimidate or coerce a government, the civilian population, or any segment thereof, in furtherance of political or social objectives.

Terry stops: Brief detentions by the police if there is suspicion of criminal activity (see also stock and frisk exception).

theft: The taking of property without permission.

thermal imaging: Technology that detects heat from within closed structures.

third degree: An historical reference to the physically brutal interrogation of suspects by police.

threat assessment: An officer's assessment of the degree of danger posed by a subject; factors involved in the assessment can be physical or situational.

turnover: The rate at which people leave a job and are replaced by others.

twenty-one-foot rule: A guideline used by some departments stating a subject can pose a significant threat to a police officer when that person is within a twenty-one-foot boundary of the officer.

undercover fencing operation: In this type of operation, the police put out word that someone is willing to buy stolen goods and then arrest those who come in to sell.

underpolicing: The perception of too little police presence and action in a neighborhood.

unmanned aerial vehicles (UAVs): Also known as drones, UAVs are remotely controlled flying machines that can be used for surveillance.

unnecessary force: Force that violates the principles of the continuum of force due to training errors or good faith mistakes.

USA Freedom Act: A modified version of the USA PATRIOT Act enacted in 2015.

USA PATRIOT Act of 2001: Act signed after the September 11 attacks; it gave law enforcement agencies additional authority to gather electronic evidence, amended money laundering laws, changed immigration laws, and expanded the authority of law enforcement officers to seize individuals' records held by a third party.

value statement: A set of values important for an agency's employees to consider when performing their duties.

vehicle exception: A warrant exception that allows police to search a vehicle if there is probable cause to believe the vehicle contains evidence or contraband.

watch: Men in larger villages in colonial America who were tasked with guarding the town, especially at night.

watchman style of policing: Policing style that focuses on order maintenance.

ENDNOTES

CHAPTER 1

1. David C. Couper, "Great Expectations," *Improving Police*, November 18, 2014, https://improvingpolice.wordpress.com/2014/11/18/great-expectations/.

2. Arthur Neiderhoffer, *Behind the Shield: The Police in Urban Society* (New York, NY: Doubleday and Company, 1967).

3. Ray Surette, *Media, Crime and Criminal Justice: Images, Realities, and Policies* (Stamford, CT: Cengage, 2015). Dennis P. Rosenbaum et al., "Attitudes Toward the Police: The Effects of Direct and Vicarious Experience," *Police Quarterly* 8 (2005): 343–365. With specific regard to the effect of media on perceptions of the police, see: Kathleen M. Donovan and Charles F. Klahm IV, "The Role of Entertainment Media in Perceptions of Police Use of Force," *Criminal Justice and Behavior* 42 (2015), 1261–1281. Jonathan Intravia, Kevin T. Wolff, and Alex R. Piquero, "Investigating the Effects of Media Consumption on Attitudes Toward Police Legitimacy," *Deviant Behavior* 39 (2018): 963–980.

4. Bureau of Justice Statistics, "Contacts Between Police and the Public, 2015," https://www.bjs.gov/content/pub/pdf/cpp15_sum.pdf.

5. Steven G. Brandl et al., "Global and Specific Attitudes Toward the Police: Disentangling the Relationship," *Justice Quarterly* 11 (1994): 119–134.

6. Amnesty International, http://www.amnestyusa.org/research/reports/state-of-the-world-201.

7. Amnesty International, https://www.amnesty.org/en/countries/middle-east-and-north-africa/saudi-arabia/.

8. Amnesty International, "6 of President Vladimir Putin's Most Oppressive Laws," January 15, 2014, http://blog.amnestyusa.org/europe/6-of-president-vladimir-putins-most-oppressive-laws-2/.

9. Amnesty International, https://www.amnesty.org/en/countries/asia-and-the-pacific/north-korea/report-korea-democratic-peoples-republic-of/.

10. Shira A. Scheindlin and Peter K. Manning, "Will the Widespread Use of Police Body Cameras Improve Police Accountability?" *Americas Quarterly* 9 (2015): 24.

11. Stephan G. Grimmelikhuijsen and Albert J. Meijer, "Does Twitter Increase Perceived Police Legitimacy?" *Public Administration Review* 75 (2015): 598–607.

12. As one of many examples, see *The Force Report* by NJ Advance Media, 2018 https://force.nj.com/.

13. Shelley S. Hyland, *Body-Worn Cameras in Law Enforcement Agencies, 2016* (Washington, DC: Bureau of Justice Statistics, 2018), 2.

14. Bureau of Justice Assistance, *Body-Worn Camera Toolkit, FAQ* (Washington DC: U.S. Department of Justice, 2015).

15. Shelley S. Hyland, *Body-Worn Cameras in Law Enforcement Agencies, 2016* (Washington, DC: Bureau of Justice Statistics, 2018).

16. Ibid.

17. Ibid.

18. Rashawn Ray, Kris Marsh, and Connor Powelson. "Can Cameras Stop the Killings? Racial Differences in Perceptions of the Effectiveness of Body-Worn Cameras in Police Encounters," *Sociological Forum* 32 (2017): 1032–1050.

19. Cynthia Lum, Megan Stoltz, Christopher S. Koper, and J. Amber Scherer. "Research on Body-Worn Cameras: What We Know, What We Need to Know," *Criminology & Public Policy* 18 (2019): 93–118.

20. Peter K. Manning, "The Police: Mandate, Strategies, and Appearances," in *Policing: A View from the Street*, ed. Peter K. Manning and John Van Maanen (Santa Monica, CA: Goodyear, 1978), 7–31; Peter K. Manning, "Role and Function of the Police," in *The Encyclopedia of Criminology and Criminal Justice*, ed. Gerben Bruinsma and David Weisburd (New York, NY: Springer, 2014), 4510–4529.

21. *FBI Uniform Crime Report*, 2017. https://ucr.fbi.gov/crime-in-the.u.s/2017/crime-in-the.u.s.-2017/topic-pages/clearances.

22. Egon Bittner, *The Function of the Police in Modern Society* (Chevy Chase, MD: National Institute of Mental Health, 1967).

23. Ibid.

24. Surette, *Media, Crime and Criminal Justice*.

25. Matthew B. Robinson, *Media Coverage of Crime and Criminal Justice* (Durham, NC: Carolina Academic Press, 2011).

26. Surette, *Media, Crime and Criminal Justice*.

27. Intravia, Wolff, and Piquero, "Investigating the Effects of Media Consumption," 963–980.

28. For a similar finding see: Valerie J. Callanan and Jared S. Rosenberger, "Media and Public Perceptions of the Police: Examining the Impact of Race and Personal Experience," *Policing & Society* 21 (2011): 167–189.

29. Brian D. Fitch, *Law Enforcement Ethics* (Thousand Oaks, CA: Sage, 2014).

30. Peter Parker's uncle in the movie *Spiderman* was not the first person to come up with this saying; it is believed that the axiom was first stated by French philosopher Voltaire in the early 1800s.

31. Carl B. Klockers, "The Dirty Harry Problem," *American Academy of Political and Social Science* 452 (1985): 33–47.

32. Associated Press, "Facebook Unfriends the DEA," Politico, October 17, 2014, http://www.politico.com/story/2014/10/facebook-dea-fake-profiles-112003.

CHAPTER 2

1. Milwaukee Police Department, *The First One Hundred Years* (Milwaukee, WI: City of Milwaukee, 1955), 3.

2. Thomas S. Weaver, *Historical Sketch of the Police Service of Hartford: From 1636 to 1801* (Hartford, CT: The Hartford Police Mutual Aid Association, 1901).

3. National Law Enforcement Museum Insider, "The Early Days of American Law Enforcement: The Watch," April 2012, http://www.nleomf.org/museum/news/newsletters/online-insider/2012/April-2012/early-days-american-law-enforcement-april-2012.html.

4. Henry Louis Gates Jr., "How Many Slaves Landed in the U.S.?," The Root, January 6, 2014, https://www.theroot.com/how-many-slaves-landed-in-the-us-1790873989.

5. Stephen D. Behrendt, "Transatlantic Slave Trade," in *Africana: The Encyclopedia of the African and African American Experience* (Cambridge, MA: W. E. B. Du Bois Institute for African and African-American Research, Harvard University, 1999); Michael Tadman, "The Demographic Cost of Sugar: Debates on Slave Societies and Natural Increase in the Americas," *The American Historical Review* 105 (2005): 1534–1575.

6. Sally E. Hadden, "Slave Patrols," *New Georgia Encyclopedia*, January 10, 2014, http://www.georgiaencyclopedia.org/articles/history-archaeology/slave-patrols.

7. Ibid.

8. Samuel Walker, *Popular Justice: A History of American Criminal Justice* (New York, NY: Oxford University Press, 1999).

9. Hadden, "Slave Patrols."

10. Ibid.

11. Larry D. Ball, *The United States Marshals of New Mexico and Arizona Territories, 1846–1912* (Albuquerque: University of New Mexico Press, 1978).

12. Jerome H. Skolnick and James J. Fyfe, *Above the Law: Police and the Excessive Use of Force*, 2nd ed. (New York, NY: The Free Press), 28.

13. Ibid.

14. Williams and Murphy, "The Evolving Strategy of Police."

15. Hubert Williams and Patrick V. Murphy, "The Evolving Strategy of Police: A Minority View," *Perspectives on Policing* 13 (January 1990), https://www.ncjrs.gov/pdffiles1/nij/121019.pdf.

16. George L. Kelling and Mark H. Moore, "The Evolving Strategy of Policing," *Perspectives on Policing* 4 (November 1988), https://www.innovations.harvard.edu/sites/default/files/114213.pdf.

17. Weaver, *Historical Sketch of the Police Service of Hartford*.

18. Williams and Murphy, "The Evolving Strategy of Police."

19. Ibid.

20. "A History of Women in Law Enforcement," *Community Policing Dispatch, A Newsletter of the COPS Office* 10, no. 3 (2017), https://cops.usdoj.gov/html/dispatch/03-2017/history_of_women_in_LE.asp.

21. Kelling and Moore, "The Evolving Strategy of Policing."

22. Ibid.

23. Roger Lane, *Policing the City: Boston 1822–1885* (Cambridge, MA: Harvard University Press, 1967).

24. Mark Haller, "Historical Roots of Police Behavior: Chicago 1890–1925," *Law and Society Review* 10 (1976): 303–323.

25. Steven G. Brandl, *An Analysis of 2018 Use of Force Incidents in the Milwaukee Police Department* (Milwaukee, WI: City of Milwaukee Fire and Police Commission, 2019).

26. Lane, *Policing the City*.

27. Ibid.

28. Jack Kuykendall, "The Municipal Police Detective: An Historical Analysis," *Criminology* 24 (1986): 175–200.

29. Donald C. Dilworth, *Identification Wanted: Development of the American Criminal Identification System, 1893–1943* (Gaithersburg, MD: International Association of Chiefs of Police, 1977).

30. Gallus Muller, trans., *Alphonse Bertillon's Instructions for Taking Descriptions for the Identification of Criminals and Others* (Chicago, IL: American Bertillon Prison Bureau, 1889).

31. Emanuel Lavine, *The Third Degree: A Detailed and Appalling Exposé of Police Brutality* (New York, NY: Garden City Publishing, 1930).

32. Haller, "Historical Roots of Police Behavior," 303–323.

33. Kuykendall, "The Municipal Police Detective," 175–200.

34. Lavine, *The Third Degree*.

35. Kuykendall, "The Municipal Police Detective," 175–200.

36. Kelling and Moore, "The Evolving Strategy of Policing."

37. Skolnick and Fyfe, *Above the Law*.

38. Philip M. Conti, *The Pennsylvania State Police: A History of Service to the Commonwealth, 1905 to Present* (Harrisburg, PA: Stackpole, 1977).

39. Smithsonian.com, "The First Criminal Trail That Used Fingerprints as Evidence," December 5, 2018, https://www.smithsonianmag.com/history/first-case-where-fingerprints-were-used-evidence-180970883/.

40. Advisory Commission on Civil Disorders, *Report of the Advisory Commission on Civil Disorders* (Washington, DC: National Criminal Justice Reference Service, 1968).

41. George L. Kelling et al., *The Kansas City Preventive Patrol Experiment: A Summary Report* (Washington, DC: The Police Foundation, 1974).

42. Jan M. Chaiken, Peter W. Greenwood, and Joan Petersilia, "The Criminal Investigation Process: A Summary Report," *Policy Analysis* 3 (1977): 187–217.

43. Kelling and Moore, "The Evolving Strategy of Policing."

44. Lawrence W. Sherman, Catherine H. Milton, and Thomas V. Kelly, *Team Policing: Seven Case Studies* (Washington, DC: The Police Foundation, 1973).

45. Herman Goldstein, "Improving Policing: A Problem-Oriented Approach," *Crime and Delinquency* 25 (1979): 236–258.

46. National Academy of Sciences, *Fairness and Effectiveness in Policing: The Evidence* (Washington, DC: National Academy Press, 2004), 10.

47. Peter Martin and Lorraine Mazerolle, "Police Leadership in Fostering Evidence-Based Agency Reform," *Policing: A Journal of Policy and Practice* 10 (2015): 34–43; National Academies of Sciences, Engineering, and Medicine. *Proactive Policing: Effects on Crime and Communities* (Washington, DC: The National Academies Press, 2018); Cody W. Telep and David Weisburd, "What Is Known About the Effectiveness of Police Practices in Reducing Crime and Disorder?" *Police Quarterly* 15 (2012): 331–357; National Research Council, *Strengthening the National Institute of Justice* (Washington, DC: National Academies Press, 2010).

48. Federal Bureau of Investigation, *Uniform Crime Report* (Washington, DC: Department of Justice, 2017), https://ucr.fbi.gov/crime-in-the-u.s/2017/crime-in-the-u.s.-2017/topic-pages/tables/table-74.

49. Tony Platt et al., *The Iron Fist and the Velvet Glove: An Analysis of the U.S. Police* (San Francisco, CA: Crime and Social Justice Associates, 1982).

CHAPTER 3

1. Jim Seida, "Police Pay Gap: Many of America's Finest Struggle on Poverty Wages," *NBC News*, October 26, 2014, https://www.nbcnews.com/feature/in-plain-sight/police-pay-gap-many-americas-finest-struggle-poverty-wages-n232701.

2. Shelly Bradbury, "At Local Police Departments, Inequality Abounds," *Pittsburgh Post-Gazette*, December 18, 2018, https://newsinteractive.post-gazette.com/blog/allegheny-county-police-departments-inequality-budgets/.

3. Norb Franz, "Warren Touts Diversity and Experience in Hiring Dozen Police Officers," *Macomb Daily*, July 25, 2018, https://www.macombdaily.com/news/local/warren-touts-diversity-experience-in-hiring-dozen-police-officers/article_19a13ab4-8ec4-11e8-af51-1bddbcc8b7a1.html.

4. H. H. Gerth and C. W. Mills, *From Max Weber: Essays in Sociology* (New York, NY: Oxford University Press, 1946).

5. Michael Allen, "California Highway Patrol Calls Itself 'Paramilitary' Organization," Opposing Views, December 18, 2014, http://www.opposingviews.com/i/society/california-highway-patrol-calls-itself-paramilitary-organization; Peter B. Kraska, "Militarization and Policing—Its Relevance to 21st Century Police," *Policing* 1 (2007): 501–513; Peter B. Kraska, *Militarizing the American Criminal Justice System: The Changing Roles of the Armed Forces and the Police* (Boston, MA: Northeastern University Press, 2001); Charles J. Dunlap, "The Thick Green Line: The Growing Involvement of Military Forces in Domestic Law Enforcement," in Kraska, *Militarizing the American Criminal Justice System*. See also Garth den Heyer, "Mayberry Revisited: A Review of the Influence of Police Paramilitary Units on Policing," *Policing and Society* 24 (2014): 346–361.

6. U.S. Department of Justice, Office of Justice Programs, "Strengthening Law Enforcement-Community Relations," September 18, 2015, https://content.govdelivery.com/accounts/USDOJOJP/bulletins/1142a62.

7. Michael Lipsky, *Street-Level Bureaucracy: Dilemmas of the Individual in Public Services* (New York, NY: Russell Sage Foundation, 2010).

8. Ibid., 24.

9. James J. Fyfe, "Good Policing," in *The Socioeconomics of Crime and Justice*, ed. Brian Forst (Armonk, NY: M. E. Sharpe, Inc.), 269–290; E. Reuss-Ianni, *Two Cultures of Policing: Street Cops and Management Cops* (New Brunswick, NJ: Transaction Books, 1983).

10. Lipsky, *Street-Level Bureaucracy*.

11. Ibid.

12. Lorraine Mazerolle et al., "Managing Citizen Calls to the Police: The Impact of Baltimore's 3-1-1 Call System," *Criminology and Public Policy* 2 (2002): 97–124.

13. Herman Goldstein, "Improving Policing: A Problem-Oriented Approach," *Crime and Delinquency* 25 (1979): 236–258.

14. Brian A. Reaves, *Local Police Departments, 2013.* (Washington DC: Bureau of Justice Statistics, 2015).

15. Amie M. Schuck, "Female Representation in Law Enforcement: The Influence of Screening, Unions, Incentives, Community Policing, CALEA, and Size," *Police Quarterly* 17 (2014): 54–78.

16. Watertown, Massachusetts, Police Department website: http://watertownpd.org/.

17. Houston, Texas, Police Department website: http://www.houstonpolice.org/go/doc/2133/289249/.

18. Reaves, *Local Police Departments, 2013.*

19. Ibid.

20. Ibid.

21. Ibid.

22. Ibid.

23. Shelly Hyland, *Body-Worn Cameras in Law Enforcement Agencies, 2016* (Washington, DC: Bureau of Justice Statistics, 2018).

24. Shelly Hyland, *Full-Time Employees in Law Enforcement Agencies, 1997-2016* (Washington DC: Bureau of Justice Statistics, 2018).

25. Hyland, *Body-Worn Cameras.*

26. Ibid.

27. Hyland, *Full-Time Employees.*

28. Hyland, *Body-Worn Cameras.*

29. Andrea M. Burch, *Sheriffs' Office Personnel, 1993–2013* (Washington, DC: Bureau of Justice Statistics, 2016).

30. Ibid.

31. Brian A. Reaves, *Census of State and Local Law Enforcement Agencies, 2008* (Washington, DC: Bureau of Justice Statistics, 2011).

32. *Uniform Crime Report, 2017* (Washington, DC: Federal Bureau of Investigation, 2018).

33. Ibid.

34. Ibid.

35. Brian A. Reaves, *Campus Law Enforcement, 2011–12* (Washington, DC: Bureau of Justice Statistics, 2015).

36. Ibid.

37. Ibid.

38. Ibid.

39. Ibid.

40. Ibid.

41. Ibid.

42. *Uniform Crime Report, 2017*, Table 81.

43. Dara Lind, "The Trump Administration's Separation of Families at the Border, Explained," Vox, August 14, 2018, https://www.vox.com/2018/6/11/17443198/children-immigrant-families-separated-parents.

44. Homeland Security Special Agent Brochure, https://www.ice.gov/doclib/careers/pdf/investigator-brochure.pdf.

45. U.S. Customs and Border Protection, https://www.cbp.gov/about.

46. Immigration and Customs Enforcement, http://www.ice.gov/about.

47. U.S. Secret Service, https://www.secretservice.gov/join/careers/.

48. Transportation and Security Administration, https://www.tsa.gov/about/strategy.

49. U.S. Department of Justice, https://www.justice.gov/doj/page/file/1033761/download#mission.

50. Federal Bureau of Investigation, http://www.fbi.gov/about/mission.

51. Drug Enforcement Administration, https://www.dea.gov/special-agent-careers.

52. U.S. Marshals Service, https://www.usmarshals.gov/duties/factsheets/facts.pdf.

53. Bureau of Alcohol, Tobacco, Firearms, and Explosives, https://www.justice.gov/sites/default/files/jmd/pages/attachments/2015/02/01/26._atf_exhibits.pdf.

CHAPTER 4

1. Elizabeth Van Brocklin, "Where Cop Cars Double as Ambulances," *The Trace*, November 14, 2018, https://www.thetrace.org/features/philadelphia-police-scoop-and-run-shooting-victims/.

2. "Weird and Strange Laws in Every State," KCRA.com, August 8, 2014, http://www.kcra.com/news/weird-and-strange-laws-in-every-state/27358590.

3. Stephen D. Mastrofski and R. Richard Ritti, "You Can Lead a Horse to Water. . . : A Case Study of a Police Department's Response to Stricter Drunk Driving Laws," *Justice Quarterly* 9 (1992): 465–491; Stephen D. Mastrofski and R. Richard Ritti, "Police Training and the Effects of Organization on Drunk Driving Enforcement," *Justice Quarterly* 13 (1996): 291–320.

4. Jason Pohl, "Montgomery Body-Cam Policy 'Undermines' Trust in Police, Phoenix Leaders Say," *The Republic*, AZCentral.com, May 25, 2018, https://www.azcentral.com/story/news/local/phoenix-breaking/2018/05/25/phoenix-city-leaders-oppose-bill-maricopa-county-attorney-bill-montgomery-public-records-rules/646368002/.

5. Egon Bittner, *The Function of the Police in Modern Society* (Chevy Chase, MD: National Institute of Mental Health, 1967), 94.

6. Eric J. Scott, *Calls for Service: Citizen Demand and Initial Police Response* (Washington, DC: National Institute of Justice, 1981).

7. Katie Sheperd, "Portlanders Call 911 to Report 'Unwanted' People More Than Any Other Reason. We Listened In," *Willamette Week*, February 6, 2019, https://www.wweek.com/news/2019/02/06/portlanders-call-911-to-report-unwanted-people-more-than-any-other-reason-we-listened-in/.

8. James J. Fyfe, "Good Policing," in *The Socioeconomics of Crime and Justice*, ed. Brian Frost (New York, NY: Routledge, 1993), 269–285, 274.

9. Roger B. Parks et al., "How Officers Spend Their Time With the Community," *Justice Quarterly* 16 (1999): 483–518.

10. Christine N. Famega, James Frank, and Lorraine Mazerolle, "Managing Police Patrol Time: The Role of Supervisor Directives," *Justice Quarterly* 22 (2005): 540–559.

11. James Frank, Steven Brandl, and Cory Watkins, "The Content of Community Policing: A Comparison of the Early Activities of Community and 'Beat' Officers," *Policing: An International Journal of Police Strategies and Management* 20 (1997): 716–728.

12. Jack R. Greene and Carl B. Klockars, "What Police Do," in *Thinking About Police: Contemporary Readings*, ed. Carl B. Klockars and Stephen D. Mastrofski (New York, NY: McGraw-Hill, 1991), 273–284.

13. Steven G. Brandl, *An Analysis of Use of Force Incidents in the Milwaukee Police Department* (Milwaukee, WI: City of Milwaukee Fire and Police Commission, 2019).

14. Nancy La Vigne et al., eds., *Key Issues in the Police Use of Pedestrian Stops and Searches: Discussion From an Urban Institute Roundtable* (Washington, DC: The Urban Institute, 2012), iv.

15. David Weisburd et al., "Do Stop, Question, and Frisk Practices Deter Crime?" *Criminology and Public Policy* 15 (2016): 31–56.

16. Joseph Ferrandino, "The Efficiency of Frisks in the NYPD, 2004–2010," *Criminal Justice Review* 38 (2013): 149–168.

17. Andrew Gelman, Jeffrey Fagan, and Alex Kiss, "An Analysis of the New York City Police Department's 'Stop-and-Frisk' Policy in the Context of Claims of Racial Bias," *Journal of the American Statistical Association* 102 (2012): 813–823; La Vigne et al., *Key Issues in the Police Use of Pedestrian Stops and Searches*.

18. Weston J. Morrow, Michael D. White, and Henry F. Fradella, "After the Stop: Exploring the Racial/Ethnic Disparities in Police Use of Force During Terry Stops," *Police Quarterly* 20 (2017): 367–396.

19. La Vigne et al., *Key Issues*.

20. Peter K. Manning, "The Police: Mandate, Strategies, and Appearances," in *Policing: A View From the Street*, ed. Peter K. Manning and John Van Maanen (Santa Monica, CA: Goodyear, 1978), 7–31; Peter K. Manning, "Role and Function of the Police," in *The Encyclopedia of Criminology and Criminal Justice*, ed. Gerben Bruinsma and David Weisburd (New York, NY: Springer, 2014), 4510–4529.

21. Fyfe, "Good Policing," 269–285.

22. Robert C. Davis, Christopher W. Ortiz, Samantha Euler, and Lorrianne Kuykendall, "Revisiting 'Measuring What Matters' Developing a Suite of Standardized Performance Measures for Policing." *Police Quarterly* 18 (2015): 469–495.

23. Bittner, *The Function of the Police in Modern Society.*

24. Jennifer D. Wood, Amy C. Watson, and Anjali J. Fulambarker, "The 'Gray Zone' of Police Work During Mental Health Encounters: Findings From an Observational Study in Chicago," *Police Quarterly* 20 (2017): 81–105.

25. Ibid.

26. Egon Bittner, "Police Discretion in Emergency Apprehension of Mentally Ill Persons," *Social Problems* 14 (1967): 278–292.

27. Amy C. Watson, Michael T. Compton, and Jeffrey N. Draine, "The Crisis Intervention Team (CIT) Model: An Evidence-Based Policing Practice?" *Behavioral Sciences & The Law* 35 (2017): 431–441.

28. Carl B. Klockars, *The Idea of Police* (Newbury Park, CA: Sage, 1985).

29. Fyfe, "Good Policing," 269–285.

30. Michelle Perin, "Law Enforcement's Role in the Opioid Epidemic," *Officer.com*, April 20, 2018, https://www.officer.com/investigations/drug-alcohol-enforcement/article/20996096/law-enforcements-role-in-the-opioid-epidemic.

31. Jessica Rando, Derek Broering, James E. Olson, Catherine Marco, and Stephen B. Evans. "Intranasal Naloxone Administration by Police First Responders Is Associated With Decreased Opioid Overdose Deaths," *The American Journal of Emergency Medicine* 33 (2015): 1201–1204.

32. Mackenzie Bean, "U.S. States Ranked by Number of Police Departments Carrying Naloxone," *Becker's Hospital Review*, July 18, 2016, https://www.beckershospitalreview.com/population-health/us-states-ranked-by-number-of-police-departments-carrying-naloxone.html.

33. Kevin Litten, "N.O. City Council Endorses Civilian Accident Investigations," Nola.com, December 6, 2018, https://www.nola.com/politics/2018/12/no-city-council-clears-way-for-civilian-accident-investigations.html.

34. James Q. Wilson, *Varieties of Police Behavior: The Management of Law and Order in Eight Communities* (Cambridge, MA: Harvard University Press, 1978).

35. Robert E. Worden, "Police Officers' Belief Systems: A Framework for Analysis," *American Journal of Police* 14 (1995): 49–81.

CHAPTER 5

1. David Garrick, "San Diego Police Using Social Media to Attract Millennials," The San Diego Union Tribune, October 11, 2018, http://www.sandiegouniontribune.com/news/politics/sd-me-police-recruit-20181011-story.html.

2. Ronald Weitzer and Steven A. Tuch, *Race and Policing in America* (New York, NY: Cambridge University Press, 2006).

3. Center for State and Local Government Excellence, *State and Local Government Workforce: 2017 Trends* (Washington DC: Center for State and Local Government Excellence, 2017).

4. Daniel Denvir, "Who Wants to Be a Police Officer?," CityLab, April 21, 2015, http://www.citylab.com/crime/2015/04/who-wants-to-be-a-police-officer/391017/.

5. See Generational Differences Chart at http://www.wmfc.org/uploads/GenerationalDifferencesChart.pdf.

6. Anthony J. Raganella and Michael D. White, "Race, Gender, and Motivation for Becoming a Police Officer: Implications for Building a Representative Police Department," *Journal of Criminal Justice* 32 (2004): 501–513.

7. Michael D. White et al., "Motivations for Becoming a Police Officer: Re-assessing Officer Attitudes and Job Satisfaction After Six Years on the Street," *Journal of Criminal Justice* 38 (2010): 520–530.

8. Elizabeth Linos, "More Than Public Service: A Field Experiment on Job Advertisements and Diversity in the Police," *Journal of Public Administration Research and Theory* 28 (2017): 67–85.

9. Brian A. Reaves, "Hiring and Retention of State and Local Law Enforcement Officers, 2008—Statistical Tables," Bureau of Justice Statistics, U.S. Department of Justice, October 2012, http://www.bjs.gov/content/pub/pdf/hrslle008st.pdf.

10. Ibid.

11. Mike Maciag, "With Fewer Applicants, Police Departments Engage in Bidding Wars," Governing.com, April 24, 2018, http://www.governing.com/topics/public-justice-safety/gov-hiring-police-officers.html.

    Patrick Jonsson, "Desperate for Officers, a Georgia Police Chief Hits the Road," *Christian Science Monitor*, August 14, 2018, https://www.csmonitor.com/USA/Justice/2018/0814/Desperate-for-officers-a-Georgia-police-chief-hits-the-road.

12. The Behavioral Insights Team, "Behavioral Insights for Building the Police Force of Tomorrow," The Behavioral Insights Team, January 23, 2019, https://www.bi.team/publications/behavioral-insights-for-building-the-police-force-of-tomorrow/.

13. Reaves, "Hiring and Retention."

14. Ibid.

15. Bureau of Labor Statistics, Police and Sheriff's Patrol Officers, May 2018, https://www.bls.gov/oes/2017/may/oes333051.htm.

16. Brian Reaves, *Local Police Departments, 2013: Personnel, Policies, and Practices* (Washington, DC: Bureau of Justice Statistics, U.S. Department of Justice, 2015).

17. Bob Bauder, "How Do Pittsburgh Police Salaries Stack Up to Suburban Departments?" *Pittsburgh Tribune-Review*, February 16, 2018, https://triblive.com/local/allegheny/13313811-74/how-do-pittsburgh-police-salaries-stack-up-to-suburban-departments.

18. Eli Hager and Gerald Rich, "In Blue, but Not Blue-Collar," The Marshall Project, April 6, 2015, https://www.themarshallproject.org/2015/04/06/in-blue-but-not-blue-collar#.AchlM3RgU.

19. Mike Maciag, "With Fewer Police Applicants, Departments Engage in Bidding Wars," Governing, April, 2018, http://www.governing.com/topics/public-justice-safety/gov-hiring-police-officers.html.

20. Ibid.

21. Lisa Kashinsky, "Methuen Begins Layoffs of 50 Police Officers," Boston Herald, January 25, 2019, https://www.bostonherald.com/2019/01/24/methuen-begins-layoffs-of-50-police-officers/.

22. Eli Hager and Gerald Rich, "In Blue, but Not Blue-Collar," *The Marshall Project*, April 6, 2015, https://www.themarshallproject.org/2015/04/06/in-blue-but-not-blue-collar#.AchlM3RgU.

23. James Barragan, "Bills to Help Cops Pay Back Student Loans Approved in Texas House, Senate," Dallas News, April 17, 2019, https://www.dallasnews.com/news/politics/2019/04/16/bills-help-cops-pay-back-student-loans-approved-texas-house-senate.

24. Jennifer Wareham, Brad W. Smith, and Eric G. Lambert, "Rates and Patterns of Law Enforcement Turnover: A Research Note," *Criminal Justice Policy Review* 26 (2015): 345–370.

25. U.S. Equal Employment Opportunity Commission, "Milestones: 1972," http://www.eeoc.gov/eeoc/history/35th/milestones/1972.html.

26. Modified from U.S. Equal Employment Opportunity Commission, "Laws Enforced by the EEOC," http://www.eeoc.gov/laws/statutes/.

27. Richard T. Schaefer, *Reverse Discrimination. Encyclopedia of Race, Ethnicity, and Society* (Berkeley, CA: Sage, 2008), 1160–1162.

28. Jason Rydberg and William Terrill, "The Effect of Higher Education on Police Behavior," *Police Quarterly* 13 (2010): 92–120.

29. Reaves, "Hiring and Retention."

30. Norma M. Riccucci and Margaret Riccardelli, "The Use of Written Exams in Police and Fire Departments: Implications for Social Diversity," *Review of Public Personnel Administration* 35 (2015): 352–366.

31. For reasoning through spelling questions, see http://www.csmonitor.com/USA/Justice/2015/0302/Can-you-pass-the-written-police-officer-exam/license-plate. For reading comprehension questions, see http://www.tests.com/practice/police-test.

32. Jonathan Lough and Kathryn Von Treuer, "A Critical Review of Psychological Instruments Used in Police Officer Selection," *Policing: An International Journal of Police Strategies and Management* 36 (2013): 737–751.

33. Riccucci and Riccardelli, "The Use of Written Exams."

34. Federal Bureau of Investigation, "Uniform Crime Reports: Crime in the United States," 2017, https://ucr.fbi.gov/crime-in-the-u.s/2017/crime-in-the-u.s.-2017/tables/table-74

35. Ibid.

36. Gary Cordner and AnnMarie Cordner, "Stuck on a Plateau? Obstacles to Recruitment, Selection, and Retention of Women Police," *Police Quarterly* 14 (2011): 207–226. Quotation appears on page 213.

37. Ibid., 215.

38. Josie Francesca Cambareri and Joseph B. Kuhns. "Perceptions and Perceived Challenges Associated With a Hypothetical Career in Law Enforcement: Differences Among Male and Female College Students," *Police Quarterly* 21 (2018): 335–357.

39. Ibid.

40. U.S. Census Bureau, "Quick Facts: United States," http://quickfacts.census.gov/qfd/states/00000.html.

41. See James Frank et al., "Reassessing the Impact of Race on Citizens' Attitudes Toward the Police: A Research Note," *Justice Quarterly* 13 (1996): 321–334.

42. Candice Williams, "Report: 'Growing' Racial Problem in Detroit Police Department," *The Detroit News*, January 12, 2017, https://www.detroitnews.com/story/news/local/detroit-city/2017/01/12/report-says-growing-racial-problem-police-dept/96520140/.

43. U.S. Census Bureau, Quick Facts Detroit City Michigan, 2018, https://www.census.gov/quickfacts/fact/table/detroitcitymichigan,mi/PST045217#qf-headnote-a

44. Jihong Zhao and Nicholas Lovrich, "Determinants of Minority Employment in American Municipal Police Agencies: The Representation of African American Officers," *Journal of Criminal Justice* 26 (1998): 267–277.

45. Brinck Kerr and Kenneth R. Mladenka, "Does Politics Matter? A Time-Series Analysis of Minority Employment Patterns," *American Journal of Political Science* 38 (1994): 918–943.

46. For police and populations estimates in various cities see: http://www.governing.com/gov-data/safety-justice/police-department-officer-demographics-minority-representation.html.

47. Zhao and Lovrich, "Determinants of Minority Employment"; Joseph Gustafson, "Diversity in Municipal Police Agencies: A National Examination of Minority Hiring and Promotion," *Policing: An International Journal of Police Strategies and Management* 36 (2013): 719–736.

48. For a review see Anne Li Kringen and Jonathan Allen Kringen, "Identifying Barriers to Black Applicants in Police Department Screening," *Policing* 9 (2015): 15–25.

49. Ibid.

50. For a discussion see William T. Jordan et al., "Attracting Females and Racial/Ethnic Minorities to Law Enforcement," *Journal of Criminal Justice* 37 (2009): 333–341.

51. Denvir, "Who Wants to Be a Police Officer?"

52. Reaves, *Local Police Departments, 2013.*

53. Christine M. Galvin-White and Eryn Nicole O'Neal, "Lesbian Police Officers' Interpersonal Working Relationships and Sexuality Disclosure: A Qualitative Study," *Feminist Criminology* 11 (2016): 253–284.

54. Kimberly D. Hassell and Steven G. Brandl, "An Examination of the Workplace Experiences of Police Patrol Officers: The Role of Race, Sex, and Sexual Orientation," *Police Quarterly* 12 (2009): 408–430.

55. Brian Reaves, *State and Local Law Enforcement Training Academies, 2013* (Washington, DC: Bureau of Justice Statistics, U.S. Department of Justice, 2016).

56. Ibid.

57. Ibid.

58. Reaves, "Hiring and Retention"; Michael D. White, "Identifying Good Cops Early: Predicting Recruit Performance in the Academy," *Police Quarterly* 11 (2008): 27–49.

59. Allison Chappell, "Police Academy Training: Comparing across Curricula," *Policing: An International Journal of Police Strategies and Management* 31 (2008): 36–56.

60. Reaves, *State and Local Law Enforcement Training Academies, 2013.*

61. Corey Jones, "New TPD Recruit Training Focuses on Empathy and Cultural Competency as Well as Catching Bad Guys," *Tulsa World*, October 10, 2018, https://www.tulsaworld.com/news/local/new-tpd-recruit-training-focuses-on-empathy-and-cultural-competency/article_ed07c551-e84c-51dd-99da-dc0e386c48f6.html.

62. Courtesy of the Appleton (Wisconsin) Police Department.

63. Ibid.

64. Ibid.

65. Reaves, *State and Local Law Enforcement Training Academies, 2013.*

CHAPTER 6

1. B. M. Cowell and A. L. Kringen, "Engaging Communities One Step at a Time: Policing's Tradition of Foot Patrol as an Innovative Community Engagement Strategy." Washington, DC: Police Foundation, 2016.

2. Ibid.

3. "Park, Walk and Talk Is St. Petersburg Police Chief's Signature Program. Is It Working?" *Tampa Bay Times*, 2017, www.tampabay.com/news/publicsafety/Park-Walk-and-Talk-is-St-Petersburg-police-chief-s-signature-program-Is-it-working-_162286992, https://www.cocoafl.org/1362/Park-Walk-and-Talk.

4. "San Francisco Auto Break-Ins: UC Study Finds Police Foot Patrol Helps Decrease Larceny Crimes," *San Francisco Chronicle*, 2018, Police https://www.sfchronicle.com/crime/article/San-Francisco-auto-break-ins-UC-study-finds-13443374.php.

5. Jerry H. Ratcliffe et al., "The Philadelphia Foot Patrol Experiment: A Randomized Controlled Trial of Police Patrol Effectiveness in Violent Crime Hotspots," *Criminology* 49 (2011): 795–831.

6. Ibid.

7. Community Oriented Policing Services (COPS), "Glossary of Terms," U.S. Department of Justice, http://www.cops.usdoj.gov/default.asp?Item=632.

8. Brian A. Reaves, *Local Police Departments, 2013: Equipment and Technology* (Washington, DC: National Institute of Justice, 2015).

9. COPS, "Glossary of Terms."

10. Reaves, *Local Police Departments, 2013.*

11. Meghan Dwyer and Stephen Davis, "'It Has Been a Problem for Quite Some Time:' 911 Abuse Wasting Time and Resources," Fox 6 News, February 17, 2017, https://fox6now.com/2017/02/16/it-has-been-a-problem-for-quite-some-time-911-abuse-wasting-time-and-resources/; Rana Sampson, *Misuse and Abuse of 911, Guide Number 19* (Washington, DC: Center for Problem-Oriented Policing, 2002).

12. Lorraine Mazerolle et al., "Managing Citizen Calls to the Police: The Impact of Baltimore's 3-1-1 Call System," *Criminology & Public Policy* 2 (2002): 97–124.

13. Robert E. Worden, "Toward Equity and Efficiency in Law Enforcement: Differential Police Response," *American Journal of Police* 12 (1993): 1–32.

14. Leanne F. Alarid and Kenneth J. Novak, "Citizens' Views on Using Alternate Reporting Methods in Policing," *Criminal Justice Policy Review* 19 (2008): 25–39.

15. Rana Sampson, *False Burglar Alarms* (Washington, DC: Office of Community Oriented Policing Services, 2001).

16. Christine N. Famega, James Frank, and Lorraine Mazerolle, "Managing Police Patrol Time: The Role of Supervisor Directives," *Justice Quarterly* 22 (2005): 540–559.

17. William Spelman and Dale K. Brown, *Calling the Police: A Replication of the Citizen Reporting Component of the Kansas City Response Time Analysis* (Washington, DC: Police Executive Research Forum, 1981), iii.

18. Ibid.

19. Abdullah Cihan, Yan Zhang, and Larry Hoover, "Police Response Time to In-Progress Burglary: A Multilevel Analysis," *Police Quarterly* 15 (2012): 308–327.

20. George L. Kelling et al., *The Kansas City Preventive Patrol Experiment: A Summary Report* (Washington, DC: The Police Foundation, 1974).

21. Lawrence W. Sherman, Patrick R. Gartin, and Michael E. Buerger, "Hot Spots of Predatory Crime: Routine Activities and the Criminology of Place," *Criminology* 27 (1989): 27–56.

22. Steven G. Brandl, *The Effects of the Elimination of the Late Power Shift in District Two and District Five in the Milwaukee Police Department: Final Report* (Milwaukee, WI: City of Milwaukee Police Department, 2004).

23. Wesley Skogan and Kathleen Frydl, eds., *Fairness and Effectiveness in Policing: The Evidence* (Washington, DC: National Academies Press, 2004), 250.

24. David Weisburd and Cody W. Telep, "Hot Spots Policing: What We Know and What We Need to Know," *Journal of Contemporary Criminal Justice* 30 (2014): 200–220.

25. Evan T. Sorg, et al., "Foot Patrol in Violent Crime Hot Spots: The Longitudinal Impact of Deterrence and Posttreatment Effects of Displacement," *Criminology* 51 (2013): 65–101.

26. Elizabeth R. Groff et al., "Does What Police Do at Hot Spots Matter? The Philadelphia Policing Tactics Experiment," *Criminology* 53 (2015): 23–53.

27. Lawrence W. Sherman and Dennis P. Rogan, "Effects of Gun Seizures on Gun Violence: 'Hot Spots' Patrol in Kansas City," *Justice Quarterly* 12 (1995): 673–693.

28. Ibid., 689.

29. Anthony A. Braga, Andrew V. Papachristos, and David M. Hureau, "The Effects of Hot Spots Policing on Crime: An Updated Systematic Review and Meta-analysis," *Justice Quarterly* 31 (2014): 633–663; Weisburd and Telep, "Hot Spots Policing"; William Wells, Yan Zhang, and Jihong Zhao, "The Effects of Gun Possession Arrests Made by a Proactive Police Patrol Unit," *Policing: An International Journal of Police Strategies and Management* 35 (2012): 253–271; David Weisburd, "Hot Spots of Crime and Place-Based Prevention," *Criminology & Public Policy* 17 (2018): 5–25.

30. As reported in Tammy Riehart Kochel, "Constructing Hot Spots Policing: Unexamined Consequences for Disadvantaged Populations and for Police Legitimacy," *Criminal Justice Policy Review* 22 (2011): 350–374.

31. Weisburd and Telep, "Hot Spots Policing," 207.

32. Jerry H. Ratcliffe, Elizabeth R. Groff, Evan T. Sorg, and Cory P. Haberman. "Citizens' Reactions to Hot Spots Policing: Impacts on Perceptions of Crime, Disorder, Safety and Police," *Journal of Experimental Criminology* 11, no. 3 (2015): 393–417.

33. Kevin Rector, "Hopkins Study Finds Disbanded Baltimore Police Plainclothes Unit Provided Most Effective Violence Strategy, Despite Problems," *The Baltimore Sun*, January 11, 2018, https://www.baltimoresun.com/news/maryland/crime/bs-md-ci-hopkins-violence-study-20180111-story.html.

34. Weisburd and Telep, "Hot Spot Policing."

35. Eric L. Piza, Joel M. Caplan, Leslie W. Kennedy, and Andrew M. Gilchrist. "The Effects of Merging Proactive CCTV Monitoring With Directed Police Patrol: A Randomized Controlled Trial," *Journal of Experimental Criminology* 11, no. 1 (2015): 43–69.

36. Lawrence W. Sherman, "Police Crackdowns: Initial and Residual Deterrence," *Crime and Justice* 12 (1990): 1–48.

37. Steven G. Brandl, *An Analysis of 2018 Use of Force Incidents in the Milwaukee Police Department* (Milwaukee, WI: City of Milwaukee Fire and Police Commission, 2019).

38. Sharad Goel, Amy Shoemaker, Ravi Shroff, and Alex Chohlas-Wood, "An Analysis of the Metropolitan Nashville Police Department's Traffic Stop Practices," *Stanford Computational Policy Lab*, November 19, 2018.

39. Alexander Weiss and Edmund McGarrell, "Traffic Enforcement and Crime: Another Look," *Police Chief* 66 (1999): 25–29.

40. Goel, Shoemaker, Shroff, and Chohlas-Wood, "An Analysis of the Metropolitan Nashville Police."

41. Jacinta M. Gau, "Consent Searches as a Threat to Procedural Justice and Police Legitimacy: An Analysis of Consent Requests during Traffic Stops," *Criminal Justice Policy Review* 24 (2013): 759–777.

42. Chris L. Gibson et al., "The Impact of Traffic Stops on Calling Police for Help," *Criminal Justice Policy Review* 21 (2010): 139–159.

43. Jacinta M. Gau, "Consent Searches as a Threat to Procedural Justice and Police Legitimacy: An Analysis of Consent Requests during Traffic Stops," *Criminal Justice Policy Review* 24 (2013): 759–777.

44. Emma Pierson, Camelia Simoiu, Jan Overgoor, Sam Corbett-Davies, Vignesh Ramachandran, Cheryl Phillips, and Sharad Goel, "A Large-Scale Analysis of Racial Disparities in Police Stops Across the United States," Cornell University, 2017, *arXiv preprint arXiv:1706.05678*.

45. Goel et al., "An Analysis of the Metropolitan Nashville Police."

46. Joel Rubin and Ben Poston, "L.A. County Deputies Stopped Thousands of Innocent Latinos on the 5 Freeway in Hopes of Their Next Drug Bust," *The Los Angeles Times*, October 4, 2018, https://www.latimes.com/local/lanow/la-me-sheriff-latino-drug-stops-grapevine-20181004-htmlstory.html.

47. Goel et al., "An Analysis of the Metropolitan Nashville Police."

48. Lydia Chavez, "Two Words Edited Into Bias Policy Setback SFPD Reform," Mission Local, September 12, 2018, https://missionlocal.org/2018/09/two-words-edited-into-bias-policy-can-setback-sfpd-reform/.

49. Richard Rosenfeld and Robert Fornango, "The Impact of Police Stops on Precinct Robbery and Burglary Rates in New York City, 2003–2010," *Justice Quarterly* 31 (2014): 96–122.

50. Weston J. Morrow, Michael D. White, and Henry F. Fradella, "After the Stop: Exploring the Racial/Ethnic Disparities in Police Use of Force During Terry Stops," *Police Quarterly* 20 (2017): 367–396.

51. Frank Main, "Chicago Police and ACLU Agree to Major Changes in Stop-and-Frisk Policy," *Chicago Sun Times*, August 6, 2015, http://chicago.suntimes.com/news/7/71/464407/cpd-stopped-over-250000-people-mostly-black-without-arrests-last-summer-aclu.

52. Ibid.

53. Chuck Goudie, "CPD 'Stop and Frisks' Down 80 Percent in 2016," ABC 7 Eyewitness News, February 1, 2016, https://abc7chicago.com/news/cpd-stop-and-frisks-down-80-percent-in-2016/1182604/.

Jeremy Gorner, "While Pedestrian Stops by Chicago Police Plummeted, Officers' Traffic Stops Soarded, ACLU Says," *Chicago Tribune*, January 14, 2019, https://www.chicagotribune.com/news/local/breaking/ct-met-chicago-police-traffic-stops-20190111-story.html.

54. Steven G. Brandl, *An Analysis of 2018 Use of Force Incidents in the Milwaukee Police Department* (Milwaukee, WI: City of Milwaukee Fire and Police Commission, 2019); Steven G. Brandl, *An Analysis of 2015 Use of Force Incidents in the Milwaukee Police Department* (Milwaukee, WI: City of Milwaukee Fire and Police Commission, 2016).

55. Rosenfeld and Fornango, "The Impact of Police Stops."

56. Richard Rosenfeld and Robert Fornango. "The Relationship Between Crime and Stop, Question, and Frisk Rates in New York City Neighborhoods," *Justice Quarterly* 34 (2017): 931–951.

57. Juan Del Toro et al., "The Criminogenic and Psychological Effects of Police Stops on Adolescent Black and Latino Boys," *Proceedings of the National Academy of Sciences* 116 (2019): 8261–8268.

58. John E. Boydstun, Michael E. Sherry, and Nicholas P. Moelter, *Patrol Staffing in San Diego: One-or Two-Officer Units* (Washington, DC: Police Foundation, 1977).

59. Ibid., 5.

60. Brian A. Reaves, *Local Police Departments, 2007* (Washington, DC: National Institute of Justice, 2010).

61. Ibid.

62. George L. Kelling et al., *The Newark Foot Patrol Experiment* (Washington, DC: Police Foundation, 1981).

63. Ibid., 40.

64. Robert C. Trojanowicz, "An Evaluation of a Neighborhood Foot Patrol Program," *Journal of Police Science and Administration* 11 (1983): 410–419.

65. Eric L. Piza and Brian A. O'Hara, "Saturation Foot-Patrol in a High-Violence Area: A Quasi-Experimental Evaluation," *Justice Quarterly* 31 (2014): 693–718.

66. Ibid., 714.

67. Jerry H. Ratcliffe et al., "The Philadelphia Foot Patrol Experiment: A Randomized Controlled Trial of Police Patrol Effectiveness in Violent Crime Hotspots," *Criminology* 49 (2011): 795–831.

68. Allan F. Abrahamse et al., "An Experimental Evaluation of the Phoenix Repeat Offender Program," *Justice Quarterly* 8 (1991): 141–168; Susan E. Martin and Lawrence W. Sherman, "Selective Apprehension: A Police Strategy for Repeat Offenders," *Criminology* 24 (1986): 155–174.

69. Groff et al., "Does What Police Do at Hot Spots Matter?"

CHAPTER 7

1. "In the Footsteps of a Killer," *Los Angeles Magazine*, 2013, https://www.lamag.com/longform/in-the-footsteps-of-a-killer/.

2. "Accused Serial Killer's Sick Words as He Raped Women", News.com, 2018, https://www.news.com.au/lifestyle/relationships/marriage/accused-serial-killers-sick-words-as-he-raped-women/news-story/681701cb27217e28eb69a0928b4ad187.

3. "We Will Find You," *Science Magazine*, 2018, https://www.sciencemag.org/news/2018/10/we-will-find-you-dna-search-used-nab-golden-state-killer-can-home-about-60-white.

4. "Golden State Killer Suspect's DNA Taken From Car as He Shopped at Hobby Lobby," *San Francisco Chronicle*, June 1, 2018, https://www.sfchronicle.com/crime/article/Golden-State-Killer-suspect-s-DNA-taken-from-12961700.php.

5. "How an Unlikely Family History Website Transformed Cold Case Investigations," *New York Times*, October 15, 2018, https://www.nytimes.com/2018/10/15/science/gedmatch-genealogy-cold-cases.html.

6. Peter W. Greenwood, Jan M. Chaiken, and Joan Petersilia, *The Criminal Investigation Process* (Lexington, MA: D. C. Heath, 1977).

7. Dennis Jay Kenney, Michael D. White, and Marc A. Ruffinengo, "Expanding the Role of Patrol in Criminal Investigations: Houston's Investigative First Responder Project," *Police Quarterly* 13 (2010): 136–160.

8. David Hirschel and Donald Faggiani, "When an Arrest Is Not an Arrest: Exceptionally Clearing Cases of Intimate Partner Violence," *Police Quarterly* 15 (2012): 358–385.

9. Robert C. Davis, Carl Jensen, and Karin E. Kitchens, *Cold Case Investigations: An Analysis of Current Procedures and Factors Associated with Successful Outcomes* (Santa Monica, CA: Rand, 2011).

10. Ibid.

11. Robert D. Keppel and Joseph G. Weis, "Time and Distance as Solvability Factors in Murder Cases," *Journal of Forensic Sciences* 39 (1994): 386–400.

12. John E. Eck, *Solving Crimes: The Investigation of Burglary and Robbery* (Washington, DC: Police Executive Research Forum, 1983); Steven G. Brandl and James Frank, "The Relationship between Circumstances, Evidence, Investigative Effort, and the Disposition of Burglary and Robbery Investigations," *American Journal of Police* 13 (1994): 149–168.

13. Tom McEwen and Wendy Regoeczi. "Forensic Evidence in Homicide Investigations and Prosecutions," *Journal of Forensic Sciences* 60, (2015): 1188–1198.

    James M. Anderson, Carl F. Matthies, Sarah Michal Greathouse, and Amalavoyal V. Chari, "The Unrealized Promise of Forensic Science: An Empirical Study of its Production and Use" (working paper, RAND Corporation, 2018).

14. For a more detailed discussion of this issue, see Steven G. Brandl, *Criminal Investigation* (Thousand Oaks, CA: Sage, 2019).

15. National Research Council, *Strengthening Forensic Science in the United States: A Path Forward* (Washington, DC: The National Academy Press, 2009).

16. Cynthia M. Cale, Madison E. Earll, Krista E. Latham, and Gay L. Bush, "Could Secondary DNA Transfer Falsely Place Someone at the Scene of a Crime?" *Journal of Forensic Sciences* 61 (2016): 196–203.

17. Greg Hampikian, "The Dangers of DNA Testing," *The New York Time*, September 21, 2018, https://www.nytimes.com/2018/09/21/opinion/the-dangers-of-dna-testing.html.

18. National Institute of Justice, *DNA for the Defense Bar* (Washington, DC: U.S. Department of Justice, 2012).

19. David Schroeder and Michael White, "Exploring the Use of DNA Evidence in Homicide Investigations: Implications for Detective Work and Case Clearance," *Police Quarterly* 12 (2009): 319–342.

20. National Institute of Justice, "Forensic Evidence and Criminal Justice Outcomes in Sexual Assault Cases," (Washington DC: National Institute of Justice, 2018), https://nij.gov/topics/law-enforcement/investigations/sexual-assault/Pages/forensic-evidence-and-criminal-justice-outcomes-in-sexual-assault-cases.aspx; Tasha A. Menaker, Bradley A. Campbell, and William Wells. "The Use of Forensic Evidence in Sexual Assault Investigations: Perceptions of Sex Crimes Investigators," *Violence Against Women* 23, (2017): 399–425.

21. John K. Roman et al., "The DNA Field Experiment: A Randomized Trial of the Cost-Effectiveness of Using DNA to Solve Property Crimes," *Journal of Experimental Criminology* 5 (2009): 345–369.

22. Heather Murphy, "Coming Soon to a Police Station Near You: The DNA 'Magic Box,'" *The New York Times*, January 21, 2019, https://www.nytimes.com/2019/01/21/science/dna-crime-gene-technology.html.

23. Sara Debus-Sherrill and Michael B. Field, "Familial DNA Searching—An Emerging Forensic Investigative Tool," *Science & Justice* 59 (2019): 20–28.

24. Federal Bureau of Investigation, "Frequently Asked Questions on CODIS and NDIS," https://www.fbi.gov/about-us/lab/biometric-analysis/codis/codis-and-ndis-fact-sheet.

25. Federal Bureau of Investigation, *CODIS-NDIS Statistics*, 2018, https://www.fbi.gov/services/laboratory/biometric-analysis/codis/ndis-statistics.

26. Joseph Goldstein, "Jailing the Wrong Man: Mug Shot Searches Persist in New York, Despite Serious Risks," *The New York Times*, January 5, 2019, https://www.nytimes.com/2019/01/05/nyregion/nypd-mug-shots-false-identification.html.

27. Rachel Wilcock, Ray Bull, and Rebecca Milne, *Witness Identification in Criminal Cases: Psychology and Practice* (New York, NY: Oxford University Press, 2008).

28. Michael Ollove, "Police Are Changing Lineups to Avoid False IDs," Pew Charitable Trusts, July 13, 2018, https://www.pewtrusts.org/en/research-and-analysis/blogs/stateline/2018/07/13/police-are-changing-lineups-to-avoid-false-ids; Gary L. Wells et al., "Eyewitness Identification Procedures: Recommendations for Lineups and Photospreads," *Law and Human Behavior* 22 (1998): 603; Maureen McGough, "To Err Is Human: Using Science to Reduce Mistaken Eyewitness Identifications in Police Lineups," *National Institute of Justice Journal* 270 (2012): 30–34; *60 Minutes*, "Eyewitness Cotton," YouTube, April 14, 2015, https://www.youtube.com/watch?v=snopWGqTIF8.

29. Arye Rattner, "Convicted but Innocent: Wrongful Conviction and the Criminal Justice System," *Law and Human Behavior* 12 (1988): 283–293.

30. National Institute of Justice, *Eyewitness Evidence: A Guide for Law Enforcement* (Washington, DC: National Institute of Justice, U.S. Department of Justice, 1999); National Institute of Justice, *Eyewitness Evidence: A Guide for Law Enforcement* (Washington, DC: National Institute of Justice, U.S. Department of Justice, 2003).

31. Michael Ollove, "Police Are Changing Lineups"; Police Executive Research Forum, *A National Survey of Eyewitness Identification Procedures in Law Enforcement Agencies* (Washington, DC: Police Executive Research Forum, 2013).

32. Wells et al., *Eyewitness Identification Procedures*, 627.

33. Ibid., 629.

34. Ibid., 630.

35. Ibid., 635.

36. John R. Schafer and Joe Navarro, *Advanced Interviewing Techniques: Proven Strategies for Law Enforcement, Military and Security Personnel* (Springfield, IL: Charles C. Thomas, 2010).

37. Richard A. Leo, *Police Interrogation and American Justice* (Cambridge, MA: Harvard University Press, 2008).

38. Ibid., 6.

39. Schafer and Navarro, *Advanced Interviewing Techniques*.

40. Leo, *Police Interrogation and American Justice*.

41. Schafer and Navarro, *Advanced Interviewing Techniques*; Fred E. Inbau et al., *Criminal Interrogations and Confessions* (Burlington, MA: Jones and Bartlett Learning, 2013).

42. David Vessel, "Conducting Successful Interrogations," *FBI Law Enforcement Bulletin* 67 (1998): 1–6.

43. John Douglas et al., "Criminal Profiling From Crime Scene Analysis," *Behavioral Sciences and the Law* 4 (1986): 401–421, 413.

44. Ronald M. Holmes and Stephen T. Holmes, *Profiling Violent Crimes: An Investigative Tool* (Thousand Oaks, CA: Sage, 2009).

45. Kim D. Rossmo, "Geographic Profiling," in *Encyclopedia of Criminology and Criminal Justice*, ed. Gerben Bruinsma and Davis Weisburd (New York, NY: Springer, 2014), 1934–1942.

46. Timothy Griffin et al., "A Preliminary Examination of AMBER Alert's Effects," *Criminal Justice Policy Review* 18 (2007): 378–394.

47. J. Mitchell Miller, "Becoming an Informant," *Justice Quarterly* 28 (2011): 203–220.

48. Rachael Trost, "Woman Sought in Connection with Boyfriend's Killing, Allegedly Confesses in Facebook Posting, NBC News, January 12, 2016, http://www.nbcnews.com/dateline/woman-sought-connection-boyfriend-s-killing-allegedly-confesses-facebook-posting-n495116.

49. Robert Botsch, "Developing Street Sources: Tips for Patrol Officers," *FBI Law Enforcement Bulletin* 77 (2008): 24–27.

50. Philip Canter, "Using a Geographical Information System for Tactical Crime Analysis," in *Analyzing Crime Pattern: Frontiers of Practice*, ed. Victor Goldsmith et al. (Thousand Oaks, CA: Sage, 2000), 4.

CHAPTER 8

1. President's Task Force on 21st Century Policing, *Final Report of the President's Task Force on 21st Century Policing* (Washington, DC: Community Oriented Policing Services, 2015), 12.

2. Ibid.

3. Ben Bradford, "After Stephon Clark Shooting Sacramento Police Can No Longer Turn Off Body Cameras," Capital Public Radio, April 9, 2018, http://www.capradio.org/articles/2018/04/09/after-stephon-clark-shooting-sacramento-police-can-no-longer-turn-off-body-cameras/.

4. City of Chicago Police Department, "Body Worn Cameras," October 30, 2018, http://directives.chicagopolice.org/directives/data/a7a57b38-151f3872-56415-1f38-89ce6c22d026d090.html.

5. There are some exceptions to this overall conclusion. For example, white officers have been found to be more likely to make arrests: See Robert A. Brown and James Frank, "Police-Citizen Encounters: Do Encounter Characteristics Influence Ticketing?," *Policing: International Journal of Police Strategies and Management* 28 (2005): 435–454; and Robert A. Brown and James Frank, "Race and Officer Decision Making: Examining Differences in Arrest Outcomes Between Black and White Officers," *Justice Quarterly* 23 (2006): 96–126.

6. Steven G. Brandl and Meghan S. Stroshine, "The Role of Officer Attributes, Job Characteristics, and Arrest Activity in Explaining Police Use of Force," *Criminal Justice Policy Review* 24 (2013): 551–572.

7. Kenneth J. Novak, Robert A. Brown, and James Frank, "Women on Patrol: An Analysis of Differences in Officer Arrest Behavior," *Policing: An International Journal of Police Strategies and Management* 34 (2011): 506–587.

8. Jason Rydberg and William Terrill, "The Effect of Higher Education on Police Behavior," *Police Quarterly* 13 (2010): 92–120; Richard Rosenfeld, Thaddeus L. Johnson, and Richard

Wright, "Are College-Educated Police Officers Different? A Study of Stops, Searches, and Arrests," *Criminal Justice Policy Review* (2018): First published online 19 Dec. 2018.

9. Randy Garner, "Police Attitudes: The Impact of Experience After Training," *Applied Psychology in Criminal Justice* 1 (2005): 56–70.

10. Robert E. Worden, "The 'Causes' of Police Brutality: Theory and Evidence on Police Use of Force," *ER Maguire, & DE Duffee, Criminal Justice Theory: Explaining the Nature and Behavior of Criminal Justice* 2 (2015): 149–204.

   James Frank and Steven G. Brandl, "The Police-Attitude Relationship: Methodological and Conceptual Considerations," *American Journal of Police* 10 (1991): 83–103.

11. Robert E. Worden, "Police Officers Belief Systems: A Framework for Analysis," *American Journal of Police* 14 (1995): 49–81.

12. John Van Maanen, "The Asshole," in *Policing: A View from the Street*, ed. Peter K. Manning and John Van Maanen (Santa Monica, CA: Goodyear, 1978).

13. See Robert E. Worden and Robin L. Shepard, "Demeanor, Crime, and Police Behavior: A Reexamination of the Police Services Study Data," *Criminology* 34 (1996): 83–105; and Robin Shepard Engle, James J. Sobol, and Robert E. Worden, "Further Exploration of the Demeanor Hypothesis: The Interaction Effects of Suspects' Characteristics and Demeanor on Police Behavior," *Justice Quarterly* 17 (2000): 235–258.

14. Kenneth J. Novak and Robin S. Engel, "Disentangling the Influence of Suspects' Demeanor and Mental Disorder on Arrest," *Policing: An International Journal of Police Strategies and Management* 28 (2005): 493–512.

15. Robin S. Engel et al., "From the Officer's Perspective: A Multilevel Examination of Citizens' Demeanor During Traffic Stops," *Justice Quarterly* 29 (2012): 650–683; Robin S. Engel, "Explaining Suspects' Resistance and Disrespect toward Police," *Journal of Criminal Justice* 31 (2003): 475–492.

16. Michael T. Rossler and William Terrill, "Police Responsiveness to Service-Related Requests," *Police Quarterly* 15 (2012): 3–24.

17. Daniel J. Lytle, "The Effects of Suspect Characteristics on Arrest: A Meta-Analysis," *Journal of Criminal Justice* 42 (2014): 589–597, 595.

18. Tammy Rinehart Kochel, David B. Wilson, and Stephen D. Mastrofski, "Effects of Suspect Race on Officers' Arrest Decisions," *Criminology* 49 (2011): 473–513.

19. Emma Pierson, Camelia Simoiu, Jan Overgoor, Sam Corbett-Davies, Vignesh Ramachandran, Cheryl Phillips, and Sharad Goel. "A Large-Scale Analysis of Racial Disparities in Police Stops Across the United States," arXiv.org (2017), arXiv preprint arXiv:1706.05678.

20. Rob Voigt, Nicholas P. Camp, Vinodkumar Prabhakaran, William L. Hamilton, Rebecca C. Hetey, Camilla M. Griffiths, David Jurgens, Dan Jurafsky, and Jennifer L. Eberhardt. "Language From Police Body Camera Footage Shows Racial Disparities in Officer Respect," *Proceedings of the National Academy of Sciences* 114, no. 25 (2017): 6521–6526.

21. Principled Policing Training: Procedural Justice and Implicit Bias (white paper, Stanford SPARQ and California Department of Justice, 2015), https://oag.ca.gov/sites/all/files/agweb/pdfs/law_enforcement/principled-policing-white-paper.pdf.

22. Patricia G. Devine and Andrew J. Elliot, "Are Racial Stereotypes Really Fading? The Princeton Trilogy Revisited," *Personality and Social Psychology Bulletin* 21, no. 11 (1995): 1139–1150.

23. Justin Nix, Bradley A. Campbell, Edward H. Byers, and Geoffrey P. Alpert. "A Bird's Eye View of Civilians Killed by Police in 2015: Further Evidence of Implicit Bias," *Criminology & Public Policy* 16 (2017): 309–340. But also see David A. Klinger and Lee Ann Slocum, "Critical Assessment of an Analysis of a Journalistic Compendium of Citizens Killed by Police Gunfire," *Criminology & Public Policy* 16 (2017): 349–362.

24. Joshua Correll, Sean M. Hudson, Steffanie Guillermo, and Debbie S. Ma. "The Police Officer's Dilemma: A Decade of Research on Racial Bias in the Decision to Shoot," *Social and Personality Psychology Compass* 8 (2014): 201–213.

25. Katherine B. Spencer, Amanda K. Charbonneau, and Jack Glaser, "Implicit Bias and Policing," *Social and Personality Psychology Compass* 10 (2016): 50–63.

26. Steven G. Brandl, "The Impact of Case Characteristics on Detectives' Decision Making," *Justice Quarterly* 10 (1993): 395–415.

27. Donald Black, *The Manners and Customs of the Police* (New York, NY: Academic Press, 1980).

28. Thomas W. McCahill, Linda C. Meyer, and Arthur M. Fischman, *The Aftermath of Rape* (Lexington, MA: D. C. Heath, 1979); Vicki McNickle Rose and Susan Carol Randall, "The Impact of Investigator Perceptions of Victim Legitimacy on the Processing of Rape/Sexual Assault Cases," *Symbolic Interaction* 5 (1982): 23–36; Julie Horney and Cassia Spohn, "The Influence of Blame and Believability Factors on the Processing of Simple Versus Aggravated Rape Cases," *Criminology* 34 (1996): 135–162.

29. Virtually every study that has examined the influence of offense seriousness on arrest decisions has shown it to be influential. For example, see Kimberly M. Tatum and Rebecca Pence, "Factors That Affect the Arrest Decision in Domestic Violence Cases," *Policing: An International Journal of Police Strategies and Management* 38 (2015): 56–70. Also see Donald Black, *The Manners and Customs of the Police* (New York, NY: Academic Press, 1980).

30. Black, *The Manners and Customs of the Police*. Andrea Allen, "Campus Officers' Sanctioning of Alcohol-Involved Crime: Influences on Discretionary Decision-Making," *Police Practice and Research* 17, (2016): 249–262.

31. Joongyeup Lee, Yan Zhang, and Larry T. Hoover, "Police Response to Domestic Violence: Multilevel Factors of Arrest Decision," *Policing: An International Journal of Police Strategies and Management* 36 (2013): 157–174; Richard Rosenfeld and Robert Fornango, "The Impact of Police Stops on Precinct Robbery and Burglary Rates in New York City, 2003–2010," *Justice Quarterly* 31 (2014): 96–122; Jason R. Ingram, "The

Effects of Neighborhood Characteristics on Traffic Citation Practices of the Police," *Police Quarterly* 10 (2007): 371–393.

32. Scott W. Phillips and James J. Sobol, "Police Decision Making: An Examination of Conflicting Theories," *Policing: An International Journal of Police Strategies and Management* 35 (2012): 551–565.

33. See John P. Crank, *Understanding Police Culture* (Cincinnati, OH: Anderson, 2004); and William Terrill, Eugene E. A. Paoline, and Peter K. Manning, "Police Culture and Coercion," *Criminology* 41 (2003): 1003–1034. Jason Ingram, William Terrill, and Eugene A. Paoline III, "Police Culture and Officer Behavior: Application of a Multilevel Framework," *Criminology* 56, (2018): 780–811.

34. Robert W. Worden, "Police Officers Belief Systems: A Framework for Analysis," *American Journal of Police* 14 (1995): 49–81.

35. Peter K. Manning, "The Police Occupational Culture in Anglo-American Societies," in *The Encyclopedia of Police Science*, ed. L. Hoover and J. Dowling (New York, NY: Garland, 1989), 360.

36. For example, see Crank, *Understanding Police Culture*; Michael K. Brown, *Working the Street: Police Discretion and the Dilemmas of Reform* (New York, NY: Russell Sage Foundation, 1988); Jerome H. Skolnick, *Justice Without Trial: Law Enforcement in Democratic Society* (New York, NY: Wiley, 1994). Jason Ingram, William Terrill, and Eugene A. Paoline III, "Police Culture and Officer Behavior: Application of a Multilevel Framework," *Criminology* 56, (2018): 780–811.

37. Elizabeth Reuss-Ianni, *Two Cultures of Policing: Street Cops and Management Cops* (New Brunswick, NJ: Transaction Publishers, 1993), 1–6.

38. Skolnick, *Justice Without Trial*; Crank, *Understanding Police Culture*.

39. Skolnick, *Justice Without Trial*.

40. James J. Fyfe, "Administrative Interventions on Police Shooting Discretion: An Empirical Examination," *Journal of Criminal Justice* 7 (1979): 309–323.

41. Geoffrey P. Alpert, *Police Pursuit: Policies and Training* (Washington, DC: U.S. Department of Justice, 1997).

42. William Terrill and Eugene A. Paoline, "Police Use of Less Lethal Force: Does Administrative Policy Matter?" *Justice Quarterly* 34 (2017): 193–216.; Stephen A. Bishopp, David A. Klinger, and Robert G. Morris, "An Examination of the Effect of a Policy Change on Police Use of Tasers," *Criminal Justice Policy Review* 26 (2015): 727–746.

43. Richard R. Johnson, "Making Domestic Violence Arrests: A Test of Expectancy Theory," *Policing: An International Journal of Police Strategies and Management* 33 (2010): 531–547.

44. Jerry R. Sparger and David J. Giacopassi, "Memphis Revisited: A Reexamination of Police Shootings After the *Garner* Decision," *Justice Quarterly* 9 (1992): 211–225.

45. Jihong Zhao, Ling Ren, and Nicholas Lovrich, "Wilson's Theory of Local Political Culture Revisited in Today's Police Organizations: Findings From Longitudinal Panel Study," *Policing: An International Journal of Police Strategies and Management* 33 (2010): 287–304.

46. John A. Shjarback, David C. Pyrooz, Scott E. Wolfe, and Scott H. Decker, "De-policing and Crime in the Wake of Ferguson: Racialized Changes in the Quantity and Quality of Policing Among Missouri Police Departments," *Journal of Criminal Justice* 50 (2017): 42–52; David C. Pyrooz, Scott H. Decker, Scott E. Wolfe, and John A. Shjarback, "Was There a Ferguson Effect on Crime Rates in Large US Cities?" *Journal of Criminal Justice* 46 (2016): 1–8; Edward R. Maguire, Justin Nix, and Bradley A. Campbell, "A War on Cops? The Effects of Ferguson on the Number of US Police Officers Murdered in the Line of Duty," *Justice Quarterly* 34 (2017): 739–758.

47. Willard M. Oliver, "Depolicing: Rhetoric or Reality?" *Criminal Justice Policy Review* 28, (2017): 437–461.

48. Joscha Legewie, "Racial Profiling and Use of Force in Police Stops: How Local Events Trigger Periods of Increased Discrimination," *American Journal of Sociology* 122 (2016): 379–424.

49. Danielle Wallace, Michael D. White, Janne E. Gaub, and Natalie Todak, "Body-Worn Cameras as a Potential Source of Depolicing: Testing for Camera-Induced Passivity," *Criminology* 56 (2018): 481–509.

50. Cynthia Lum, Megan Stoltz, Christopher S. Koper, and J. Amber Scherer, "Research on Body-Worn Cameras: What We Know, What We Need to Know," *Criminology & Public Policy* 18 (2019): 93–118.

51. American Bar Association, *Standards Relating to the Urban Function* (Boston, MA: Little, Brown, 1980), Part IV, Standard 1–4.3.

52. Peter K. Manning, *Police Work: The Social Organization of Policing* (Cambridge, MA: MIT Press, 1977), 165.

53. Seattle Police Department, "Vehicle/Eluding Pursuits," January 1, 2018, https://www.seattle.gov/police-manual/title-13---vehicle-operations/13031---vehicle-eluding/pursuits.

54. Police Executive Research Forum, *Promoting Excellence in First-Line Supervision: New Approaches to Selection, Training, and Leadership Development*, (Washington DC: Police Executive Research Forum, 2018).

55. Seattle Police Department, "Title 8: Use of Force," http://www.seattle.gov/police-manual/title-8.

56. Ibid.

57. The Leadership Conference, *Police Body-Worn Cameras: A Policy Scorecard*, https://www.bwcscorecard.org/.

58. International Association of Chiefs of Police, *IACP In-Service Training Manual* (Alexandria, VA: IACP, 1996), 12.

59. Warren Christopher et al., *Report of the Independent Commission on the Los Angeles Police Department*, 1991, http://michellawyers.com/wp-content/uploads/2010/06/Report-of-the-Independent-Commission-on-the-LAPD-re-Rodney-King_Reduced.pdf.

60. Seattle Police Department, *Police Policy Manual*, https://www.seattle.gov/police-manual/general-policy-information.

61. Robert Salonga, "San Jose Police Department Makes Accountability Push With Slew of Training, Data Initiatives," May 8, 2016, http://www.mercurynews.com/2016/05/08/san-jose-police-department-makes-accountability-push-with-slew-of-training-data-initiatives/.

62. City of Milwaukee, Fire and Police Commission, *Reports on MPD Practices and Policies*, February 25, 2019, http://city.milwaukee.gov/fpc/Reports/ReportsonMPD.htm#.Vwf8z6QrLIU.

63. Citizens Police Data Project, https://cpdp.co/.

64. New York City Police Department, Stop, Question, and Frisk Data, https://www1.nyc.gov/site/nypd/stats/reports-analysis/stopfrisk.page.

65. International Association of Chiefs of Police, http://www.iacp.org/-Ethics-Training-in-Law-Enforcement.

66. Los Angeles Police Department, "Law Enforcement Code of Ethics," http://www.lapdonline.org/lapd_manual/code_of_ethics.htm.

67. International Association of Chiefs of Police, "Police Ethics: Problems and Solutions, Part II," https://www.theiacp.org/resources/training-key/476-police-ethics-problems-and-solutions-part-ii.

CHAPTER 9

1. Cornell University Law School, Legal Information Institute, "*Arizona v. Gant*," https://www.law.cornell.edu/supct/cert/07-542; LeeAnn Freeman, "Vehicle Searches in the Wake of *Arizona v. Gant*," *Police Chief* 78 (2011): 12–13.

2. Cornell University Law School, "*Arizona v. Gant*."

3. Rolando V. del Carmen, *Criminal Procedure: Law and Practice* (Boston, MA: Cengage, 2013).

4. Michael Pettry, "The Emergency Aid Exception to the Fourth Amendment's Warrant Requirement," *FBI Law Enforcement Bulletin* 80 (2011): 26–32; Edward M. Hendrie, "Warrantless Entries to Arrest: Constitutional Considerations," *FBI Law Enforcement Bulletin* 67 (1998): 25–32.

5. Richard Schott, "The Supreme Court Reexamines Search Incident to Lawful Arrest," *FBI Law Enforcement Bulletin* 78 (2009): 22–31; Kenneth A. Myers, "Searches of Motor Vehicles Incident to Arrest in a Post-*Gant* World," *FBI Law Enforcement Bulletin* 80 (2011): 24–32.

6. Interestingly, as noted, an anonymous 911 call can provide reasonable suspicion to justify a traffic stop (*Navarette v. California*, 2014).

7. Carl Benoit, "Questioning 'Authority': Fourth Amendment Consent Searches," *FBI Law Enforcement Bulletin* 77 (2008): 23–32.

8. Jayme Holcomb, "Knock and Talks," *FBI Law Enforcement Bulletin* 75 (2006): 22–32.

9. Ibid.

10. Jayme Walker-Holcomb, "Consent Searches," *FBI Law Enforcement Bulletin* 73 (2004): 22–32.

11. Del Carmen, *Criminal Procedure*.

12. Paul Sutton, "The Fourth Amendment in Action: An Empirical View of the Search Warrant Process," *Criminal Law Bulletin* 22 (1986): 405–429, 405.

13. Ibid.

14. Christine Martin, *Illinois Municipal Officers' Perceptions of Police Ethics* (Chicago, IL: Criminal Justice Information Authority, 1994).

15. Sutton, *The Fourth Amendment in Action*.

16. Craig D. Uchida and Timothy S. Bynum, "Search Warrants, Motions to Suppress, and 'Lost Cases': The Effects of the Exclusionary Rule in Seven Jurisdictions," *Journal of Criminal Law and Criminology* 81 (1991): 1034–1066.

17. Sutton, *The Fourth Amendment in Action*.

18. Martin, *Illinois Municipal Officers' Perceptions of Police Ethics*.

19. Ibid.; Larry Cunningham, "Taking on Testifying: The Prosecutor's Response to In-Court Police Deception," *Criminal Justice Ethics* 18 (1999): 26–40.

20. General Accounting Office, *Report of the Comptroller General of the United States, Impact of the Exclusionary Rule on Federal Prosecutions* (Washington, DC: U.S. Government Printing Office, 1979); Peter F. Nardulli, "The Societal Costs of the Exclusionary Rule: An Empirical Assessment," *American Bar Foundation Research Journal* 8 (1983): 585–609; Uchida and Bynum, "Search Warrants, Motions to Suppress, and 'Lost Cases.'"

21. Carl Benoit, "The 'Public Safety' Exception to Miranda," *FBI Law Enforcement Bulletin* 80 (2011): 25–32.

22. Nardulli, *The Societal Costs of the Exclusionary Rule*.

23. Richard A. Leo, "The Impact of *Miranda* Revisited," in *The Miranda Debate: Law, Justice, and Policing*, ed. George C. Leo and George C. Thomas (Boston, MA: Northeastern University Press, 1998); Paul G. Cassell and Bret S. Hayman, "Police Interrogation in the 1990s: An Empirical Study of the Effects of *Miranda*," in Leo and Thomas, *The Miranda Debate*; George C. Thomas, "Miranda: The Crime, the Man, and the Law of Confessions," in Leo and Thomas, *The Miranda Debate*.

24. Saul M. Kassin and Rebecca J. Norwick, "Why People Waive Their Miranda Rights: The Power of Innocence," *Law and Human Behavior* 28 (2004): 211–221.

25. Ibid., 218.

CHAPTER 10

1. Mary Huber, "Austin Police Reviewing Mental Health Services Amid Mounting Stress," Statesman, August 3, 2018, https://www.statesman.com/NEWS/20180806/Austin-police-reviewing-mental-health-services-amid-mounting-stress.

2. Julia Hill et al., *Making Officer Safety and Wellness Priority One: A Guide to Educational Campaigns*, Community Oriented Policing Services, 2014, https://ric-zai-inc.com/Publications/cops-p300- pub.pdf.

3. Kimberly D. Hassell and Steven G. Brandl, "An Examination of the Workplace Experiences of Police Patrol Officers: The Role of Race, Sex, and Sexual Orientation," *Police Quarterly* 12 (2009): 408–430.

4. Michael S. Christopher, Richard J. Goerling, Brant S. Rogers, Matthew Hunsinger, Greg Baron, Aaron L. Bergman, and David T. Zava, "A Pilot Study Evaluating the Effectiveness

of a Mindfulness-Based Intervention on Cortisol Awakening Response and Health Outcomes Among Law Enforcement Officers," *Journal of Police and Criminal Psychology* 31 (2016): 15–28.

5. Terry A. Beehr, Leanor B. Johnson, and Ronie Nieva, "Occupational Stress: Coping of Police and Their Spouses," *Journal of Organizational Behavior* 16 (1995): 3–25.

6. Ibid.

7. Jennifer H. Webster, "Police Officer Perceptions of Occupational Stress: The State of the Art," *Policing: An International Journal of Police Strategies and Management* 36 (2013): 636–652.

8. Ibid., 644.

9. Cheryl Regehr et al., "Acute Stress and Performance in Police Recruits," *Stress and Health* 24 (2008): 295–303.

10. John J. Violanti et al., "Post-Traumatic Stress Symptoms and Cortisol Patterns Among Police Officers," *Policing: An International Journal of Police Strategies and Management* 30 (2007): 189–202.

11. Gregory S. Anderson, Robin Litzenberger, and Darryl Plecas, "Physical Evidence of Police Officer Stress," *Policing: An International Journal of Police Strategies and Management* 25 (2002): 399–420; Mathew J. Hickman et al., "Mapping Police Stress," *Police Quarterly* 14 (2011): 227–250.

12. John M. Violanti, Desta Fekedulegn, Tara A. Hartley, Luenda E. Charles, Michael E. Andrew, Claudia C. Ma, and Cecil M. Burchfiel. "Highly Rated and Most Frequent Stressors Among Police Officers: Gender Differences," *American Journal of Criminal Justice* 41 (2016): 645–662. Also see John Violanti and Fred Aron, "Police Stressors: Variations in Perception Among Police Personnel," *Journal of Criminal Justice* 23 (1995): 287–294.

13. Hickman et al., "Mapping Police Stress"; Hassell and Brandl, "An Examination of the Workplace Experiences of Police Patrol Officers"; Merry Morash, Robin Haarr, and Dae-Hoon Kwak, "Multilevel Influences on Police Stress," *Journal of Contemporary Criminal Justice* 32 (2006): 631–641; Stephen A. Bishopp, Nicole Leeper Piquero, John L. Worrall, and Alex R. Piquero, "Negative Affective Responses to Stress Among Urban Police Officers: A General Strain Theory Approach," *Deviant Behavior* (2018): 1–20.

14. John P. Crank and Michael Caldero, "The Production of Occupational Stress in Medium-Sized Police Agencies: A Survey of Line Officers in Eight Municipal Departments," *Journal of Criminal Justice* 19 (1991): 339–349.

15. Hassell and Brandl, "An Examination of the Workplace Experiences of Police Patrol Officers." Also see John M. Violanti et al., "Highly Rated and Most Frequent Stressors," 645–662.

16. U.S. Equal Opportunity Employment Commission, "Sexual Harrasment," http://www.eeoc.gov/laws/types/sexual_harassment.cfm.

17. Kimberly A. Lonsway, Rebecca Paynich, and Jennifer N. Hall, "Sexual Harassment in Law Enforcement: Incidence, Impact, and Perception," *Police Quarterly* 16 (2013): 177–210.

18. Ibid.

19. Ibid.

20. Robin N. Haarr and Merry Morash, "The Effect of Rank on Police Women Coping With Discrimination and Harassment," *Police Quarterly* 16 (2013): 395–419.

21. Leonard B. Bell, Thomas B. Virden, Deborah J. Lewis, and Barry A. Cassidy, "Effects of 13-hour 20-Minute Work Shifts on Law Enforcement Officers' Sleep, Cognitive Abilities, Health, Quality of Life, and Work Performance: The Phoenix Study," *Police Quarterly* 18, (2015): 293–337.

22. Mike Maciag, "The Alarming Consequences of Police Working Overtime," Governing, October, 2017, http://www.governing.com/topics/public-justice-safety/gov-police-officers-overworked-cops.html.

23. Forensic Science Institute, "Force Science News #172: Anti-Fatigue Measures Could Cut Cop Deaths 15%, Researcher Claims," February 25, 2011, http://www.forcescience.org/fsnews/172.html.

24. Ibid.

25. Ibid.

26. Ibid.

27. R. Morgan Griffin, "The Health Risks of Shift Work," WebMD, 2010, http://www.webmd.com/sleep-disorders/excessive-sleepiness-10/shift-work?page=2.

28. Lois James, "The Stability of Implicit Racial Bias in Police Officers," *Police Quarterly* 21, (2018): 30–52.

29. Bell et al., "Effects of 13-hour 20-Minute Work Shifts," 293–337.

30. John M. Violanti, Sherry L. Owens, Desta Fekedulegn, Claudia C. Ma, Luenda E. Charles, and Michael E. Andrew, "An Exploration of Shift Work, Fatigue, and Gender Among Police Officers: The BCOPS Study," *Workplace Health & Safety* (2018).

31. Mora L. Feidler, *Officer Safety and Wellness: An Overview of the Issues* (Washington, DC: Community Oriented Policing Services, 2011).

32. Shira Maguen et al., "Routine Work Environment Stress and PTSD Symptoms in Police Officers," *The Journal of Nervous and Mental Disease* 197 (2009): 754.

33. Tahera Darensburg et al., "Gender and Age Differences in Posttraumatic Stress Disorder and Depression Among Buffalo Police Officers," *Traumatology* 12 (2006): 220–228; Tara A. Hartley et al., "Health Disparities in Police Officers: Comparisons to the U.S. General Populations," *International Journal of Emergency Mental Health* 13 (2011): 211–220.

34. Wallace Mandell et al., "Alcoholism and Occupations: A Review and Analysis of 104 Occupations," *Alcoholism: Clinical and Experimental Research* 16 (1992): 734–746; Frederick S. Stinson, Samar Farha DeBakey, and Rebecca A. Steffens, "Prevalence of DSM-III-R Alcohol Abuse and/or Dependence Among Selected Occupations," *Alcohol Health and Research World* 16 (1992): 165–172; A. R. Meyers and M. W. Perrine, "Drinking by Police Officers, General Drivers and Late-Night Drivers," *Journal of Studies on Alcohol* 57 (1996): 187–192; Vicki Lindsay, "Police Officers and Their Alcohol Consumption: Should We Be Concerned?" *Police Quarterly* 11 (2008): 74–87.

35. Steven G. Brandl and Brad W. Smith, "An Empirical Examination of Retired Police Officers' Length of Retirement and Age at Death: A Research Note," *Police Quarterly* 16 (2013): 113–123; John M. Violanti, John E. Vena, and Sandra Petrolia, "Mortality of a Police Cohort: 1950–1990," *American Journal of Industrial Medicine* 33 (1998): 366–373; Richard A. Raub, "Death of Police Officers After Retirement," *American Journal of Police* 6 (1988): 91–102.

36. Sari D. Holmes et al., "Mental Stress and Coronary Artery Disease: A Multidisciplinary Guide," *Progress in Cardiovascular Diseases* 49 (2006): 106–122.

37. Elizabeth A. Mumford, Bruce G. Taylor, and Bruce Kubu, "Law Enforcement Officer Safety and Wellness," *Police Quarterly* 18 (2015): 111–133.

38. Mandell et al., "Alcoholism and Occupations," 734–746; Meyers and Perrine, "Drinking by Police Officers," 187–192; Lindsay, "Police Officers and Their Alcohol Consumption," 74–87.

39. Brandl and Smith, "An Empirical Examination," 113–123; Violanti et al., "Mortality of a Police Cohort," 366–373.

40. Raub, "Death of Police Officers,"91–102.

41. Hartley, "Health Disparities in Police Officers."

42. Mumford et al., "Law Enforcement Officer Safety and Wellness," 111–133.

43. See L. M. Rouse et al., "Law Enforcement Suicide: Discerning Etiology Through Psychological Autopsy," *Police Quarterly* 18 (2015): 79–108. Robert Loo, "A Meta-Analysis of Police Suicide Rates: Findings and Issues," *Suicide and Life Threatening Behavior* 33 (2003): 313–325.

44. Rouse et al., "Law Enforcement Suicide."

45. Ibid.

46. Ibid.

47. Ibid.

48. Ibid.

49. William P. McCarty and Wesley G. Skogan, "Job-Related Burnout Among Civilian and Sworn Personnel," *Police Quarterly* 16 (2013): 66–84, 66.

50. Ibid.

51. Ibid.

52. Ibid.

53. William P. McCarty, Hani Aldirawi, Stacy Dewald, and Mariana Palacios. "Burnout in Blue: An Analysis of the Extent and Primary Predictors of Burnout Among Law Enforcement Officers in the United States," *Police Quarterly* (2019): on-line first.

54. McCarty and Skogan, "Job-Related Burnout," 66–84.

55. American Psychiatric Association, *Diagnostic and Statistical Manual of Mental Disorders* (Arlington, VA: American Psychiatric Association, 2013).

56. Erin McCanlies et al., "Posttraumatic Stress Disorder Symptoms, Psychobiology, and Coexisting Disorders in Police Officers," in *Dying for the Job: Police Work Exposure and Health*, ed. John M. Violanti (Springfield, IL: Charles C. Thomas, 2014), 155–165, 156.

57. John M. Violanti, Claudia C. Ma, Anna Mnatsakanova, Desta Fekedulegn, Tara A. Hartley, Ja Kook Gu, and Michael E. Andrew, "Associations Between Police Work Stressors and Posttraumatic Stress Disorder Symptoms: Examining the Moderating Effects of Coping," *Journal of Police and Criminal Psychology* 33, (2018): 271–282.

58. Mumford et al., "Law Enforcement Officer Safety and Wellness," 111–133.

59. Ibid.

60. Madhulika A. Gupta, "Review of Somatic Symptoms in Post-Traumatic Stress Disorder," *International Review of Psychiatry* 25 (2013): 86–99.

61. Jitender Sareen et al., "Physical and Mental Comorbidity, Disability, and Suicidal Behavior Associated with Posttraumatic Stress Disorder in a Large Community Sample," *Psychosomatic Medicine* 69 (2007): 242–248.

62. Deborah B. Maia et al., "Post-Traumatic Stress Symptoms in an Elite Unit of Brazilian Police Officers: Prevalence and Impact on Psychosocial Functioning and on Physical and Mental Health," *Journal of Affective Disorders* 97 (2007): 241–245.

63. Erin McCanlies et al., "Treating Trauma in Law Enforcement," in Violanto, *Dying for the Job*, 176.

64. Kerry M. Karaffa and Julie M. Koch, "Stigma, Pluralistic Ignorance, and Attitudes Toward Seeking Mental Health Services Among Police Officers," *Criminal Justice and Behavior* 43, (2016): 759–777.

65. Community Oriented Policing Services, Officers' Physical and Mental Health and Safety (Washington DC: Community Oriented Policing Services, 2018).

66. Brandl and Smith, "An Empirical Examination," 113–123; Violanti et al., "Mortality of a Police Cohort," 366–373; Anthony J. Sardinas, Julia Wang Miller, and Holger Hansen, "Ischemic Heart Disease Mortality of Firemen and Policemen," *American Journal of Public Health* 76 (1986): 1140–1141.

67. Raub, "Death of Police Officers after Retirement," 91–102.

68. Webster, "Police Officer Perceptions of Occupational Stress," 645.

69. Mark H. Chae and Douglas J. Boyle, "Police Suicide: Prevalence, Risk, and Protective Factors," *Policing: An International Journal of Police Strategies and Management* 36 (2013): 91–118.

70. Federal Bureau of Investigation, "2017 Law Enforcement Officers Killed and Assaulted," 2017, https://ucr.fbi.gov/leoka/2017/resource-pages/methodology_assaults_-2017.

71. Ibid.

72. Hope M. Tiesman, Melody Gwilliam, Srinivas Konda, Jeff Rojek, and Suzanne Marsh. "Nonfatal Injuries to Law Enforcement Officers: A Rise in Assaults," *American Journal of Preventive Medicine* 54 (2018): 503–509.

73. Steven G. Brandl, "In the Line of Duty: A Descriptive Analysis of Police Assaults and Accidents," *Journal of Criminal Justice* 24 (1996): 255–264; Steven G. Brandl and Meghan S. Stroshine, "The Physical Hazards of Police Work Revisited," *Police Quarterly* 15 (2012): 262–282. Also see Hope M.

Tiesman, Melody Gwilliam, Srinivas Konda, Jeff Rojek, and Suzanne Marsh, "Nonfatal Injuries to Law Enforcement Officers: A Rise in Assaults," *American Journal of Preventive Medicine* 54 (2018): 503–509.

74. Brandl, "In the Line of Duty," 255–264.

75. Ibid.

76. Brandl and Stroshine, "The Physical Hazards of Police Work Revisited," 262–282.

77. National Institute of Justice, "Body Armor" (2018), https://www.nij.gov/topics/technology/body-armor/Pages/welcome.aspx.

78. Federal Bureau of Investigation, "2018 Law Enforcement Officers Killed and Assaulted," https://ucr.fbi.gov/leoka/2018/tables/table-36.xls.

79. Federal Bureau of Investigation, "2017 Law Enforcement Officers Killed and Assaulted."

80. Taylor et al., *A Practitioner's Guide to the 2011 National Body Armor Survey of Law Enforcement Officers* (2012), https://www.ncjrs.gov/pdffiles1/nij/grants/240225.pdf.

81. Brandl and Stroshine, "The Physical Hazards of Police Work Revisited," 262–282.

82. Ibid.

83. Steven G. Brandl, *An Analysis of 2018 Use of Force Incidents in the Milwaukee Police Department* (Milwaukee, WI: City of Milwaukee Fire and Police Commission, 2019).

84. Meghan S. Stroshine and Steven G. Brandl, "The Use, Effectiveness, and Hazards Associated With Police Use of Force: The Unique Case of Weaponless Physical Force," *Police Practice and Research* (2019; on-line first): 1–18.

85. Ibid.

86. Ibid.

87. Brandl and Stroshine, "The Physical Hazards of Police Work Revisited," 262–282.

88. Robert J. Kaminski et al., "Correlates of Foot Pursuit Injuries in the Los Angeles County Sheriff's Department," *Police Quarterly* 15 (2012): 177–196.

89. Tanya Eiserer, "Dallas Police Tightening Foot Chase Policy to Save Lives," *The Dallas Morning News*, December 25, 2012, http://www.dallasnews.com/news/news/2012/12/25/dallas-police-tightening-foot-chase-policy-to-save-lives.

90. Brandl and Stroshine, "The Physical Hazards of Police Work Revisited," 262–282.

91. Tom LaTourrette, "Risk Factors for Injury in Law Enforcement Officer Vehicle Crashes," *Policing: An International Journal of Police Strategies and Management* 38 (2015): 478–504.

92. Brandl and Stroshine, "The Physical Hazards of Police Work Revisited," 262–282.

93. Federal Bureau of Investigation (FBI), "2017 Law Enforcement Officers Killed and Assaulted"; LaTourrette, "Risk Factors for Injury," 478–504.

94. Geoffrey P. Alpert and Roger G. Dunham, "Policing Hot Pursuits: The Discovery of Aleatory Elements," *The Journal of Criminal Law and Criminology* 80 (1989): 521–539.

95. Frederick P. Rivara and Chris D. Mack, "Motor Vehicle Crash Deaths Related to Police Pursuits in the United States," *Injury Prevention* 10 (2004): 93–95.

96. Scott E. Wolfe et al., "Characteristics of Officer-Involved Collisions in California," *Policing: An International Journal of Police Strategies and Management* 38 (2015): 458–477.

97. Ibid.

98. LaTourrette, "Risk Factors for Injury," 478–504.

99. Wolfe et al., "Characteristics of Officer-Involved Collisions in California," 458–477; Thomas M. Rice, Lara Troszak, and Bryon G. Gustafson, "Epidemiology of Law Enforcement Vehicle Collisions in the U.S. and California," *Policing: An International Journal of Police Strategies and Management* 38 (2015): 425–435.

100. LaTourrette, "Risk Factors for Injury," 478–504.

101. Ibid.

102. LaTourrette, "Risk Factors for Injury," 478–504; Wolfe et al., "Characteristics of Officer-Involved Collisions in California," 458–477; Rice et al., "Epidemiology of Law Enforcement Vehicle Collisions."

103. LaTourrette, "Risk Factors for Injury," 478–504.; Rice et al., "Epidemiology of Law Enforcement Vehicle Collisions"; Wolfe et al., "Characteristics of Officer-Involved Collisions in California," 458–477.

104. LaTourrette, "Risk Factors for Injury," 478–504.

105. Ibid.

106. Stephen M. James, "Distracted Driving Impairs Police Patrol Officer Driving Performance," *Policing: An International Journal of Police Strategies and Management* 38 (2015): 505–516.

107. Commission on POST, "California Law Enforcement: Vehicle Pursuit Guidelines," 2006, http://lib.post.ca.gov/Publications/vp_guidelines.pdf.

108. Ibid.

109. Ibid.

110. Ibid.

111. Geoffrey P. Alpert and Cynthia Lum, *Police Pursuit Driving: Policy and Research* (New York, NY: Springer, 2014).

CHAPTER 11

1. Egon Bittner, *The Functions of Police in Modern Society* (Washington, DC: U.S. Government Printing Office, 1970).

2. Timothy Williams, "Long Taught to Use Force, Police Warily Learn to De-escalate," *New York Times*, June 27, 2015, https://www.nytimes.com/2015/06/28/us/long-taught-to-use-force-police-warily-learn-to-de-escalate.html.

3. Ibid.

4. *Graham v. Connor*, 490 U.S. 386, 1989.

5. William Terrill and Eugene A. Paoline, "Examining Less Lethal Force Policy and the Force Continuum," *Police Quarterly* 16 (2013): 38–65.

6. Ibid.

7. Ibid.

8. Ibid.; Geoffrey P. Alpert et al., *Police Use of Force, Tasers, and Other Less-Lethal Weapons* (Washington, DC: U.S. Department of Justice, 2011).

9. William Terrill and Eugene A. Paoline III. "Police Use of Less Lethal Force: Does Administrative Policy Matter?" *Justice Quarterly* 34 (2017): 193–216.

10. Matthew S. Crow and Brittany Adrion, "Focal Concerns and Police Use of Force: Examining Factors Associated With Taser Use," *Police Quarterly* 14 (2011): 366–387; E. V. Morabito and W. G. Doerner, "Police Use of Less-Than-Lethal Force: Oleoresin Capsicum (OC) Spray," *Policing: An International Journal of Police Strategies and Management* 20 (1997): 680–697.

11. Terrill and Paoline, "Examining Less Lethal Force Policy," 38–65.

12. Wisconsin Department of Justice, *Defensive and Arrest Tactics: A Training Guide for Law Enforcement Officers* (Madison: State of Wisconsin Department of Justice, 2007), 28.

13. Ibid.

14. Megan Cassidy, "Phoenix Police Rethinking Traditional Foot Pursuits," *The Republic*, October 11, 2015, https://www.azcentral.com/story/news/local/phoenix/2015/10/11/phoenix-police-rethinking-traditional-foot-pursuits/73585196/.

15. Ron Martinelli, "The 21 Foot Rule: Forensic Fact or Police Myth?" *Law Officer*, February 12, 2016, http://lawofficer.com/2016/02/21footrule/; Beth Schwartzapfel, "Will the '21 Foot' Defense Work for he Chicago Cop Who Shot Laquan McDonald? Revisiting a 30-Year Concept That Is Used to Justify Deadly Force," The Marshall Project, November 25, 2015, https://www.themarshallproject.org/2015/11/25/will-the-21-foot-defense-work-for-the-chicago-cop-who-shot-laquan-mcdonald.

16. New York City Police Department (NYPD), *Annual Use of Force Report, 2017*. https://www1.nyc.gov/assets/nypd/downloads/pdf/use-of-force/use-of-force-2017.pdf.

17. "Justice Department Trails System to Count Killings by US Law Enforcement," *The Guardian*, October 5, 2015, https://www.theguardian.com/us-news/2015/oct/05/justice-department-trials-system-count-killings-us-law-enforcement-the-counted.

18. "FBI Announces the Official Launch of the National Use-Of-Force Data Collection," November 20, 2018, FBI National Press Office, Washington, DC, https://www.fbi.gov/news/pressrel/press-releases/fbi-announces-the-official-launch-of-the-national-use-of-force-data-collection.

19. Jerome H. Skolnick and James J. Fyfe, *Above the Law: Police and the Excessive Use of Force* (New York, NY: The Free Press, 1993).

20. Ibid., 19.

21. Ibid., 20.

22. Ibid.

23. Meghan S. Stroshine and Steven G. Brandl, "The Use, Effectiveness, and Hazards Associated With Police Use of Force: The Unique Case of Weaponless Physical Force," *Police Practice and Research* (2019): 1–18, on-line first.

Steven G. Brandl, *An Analysis of 2018 Use of Force Incidents in the Milwaukee Police Department* (Milwaukee, WI: City of Milwaukee Fire and Police Commission, 2019).

24. Stroshine and Brandl. "The Use, Effectiveness, and Hazards Associated," 1–18; Geoffrey P. Alpert and Roger Dunham, "The Force Factor: Measuring and Assessing Police Use of Force and Subject Resistance," in *Use of Force by the Police: Overview of National and Local Data*, ed. K. Adams (Washington, DC: National Institute of Justice, 1999), 45–60.

25. Victoria Police, "Positional/Restraint Asphyxia," 2012, http://www.police.vic.gov.au/content.asp?Document_ID=37779.

26. Skolnick and Fyfe, *Above the Law*, 41.

27. New York City Police Department, *2015 Annual Firearm Discharges Report*, New York: City of New York Police Department, 2016, 4.

28. Brandl, *An Analysis of 2018 Use of Force Incidents.*

29. NYPD, *Annual Use of Force Report, 2017.*

30. Alpert and Dunham, "The Force Factor," 45–60.

31. "Fatal Force," *The Washington Post*, 2018, https://www.washingtonpost.com/graphics/2018/national/police-shootings-2018/?utm_term=.9543ea235170.

32. "Fatal Force," *The Washington Post*, 2019, https://www.washingtonpost.com/graphics/2019/national/police-shootings-2019/?utm_term=.3176b12044be.

33. Christina L. Patton and William J. Fremouw, "Examining 'Suicide by Cop': A Critical Review of the Literature," *Aggression and Violent Behavior* 27 (2016): 107–120.

34. "Note in Suspected 'Suicide by Cop' Case: 'I Used You,'" Crimesider, January 26, 2015, http://www.cbsnews.com/news/note-in-suspected-suicide-by-cop-case-i-used-you/.

35. Vivian B. Lord, "Factors Influencing Subjects' Observed Level of Suicide by Cop Intent," *Criminal Justice and Behavior* 39 (2012): 1633–1646.

36. Kris Mohandie, J. Reid Meloy, and Peter I. Collins, "Suicide by Cop Among Officer-Involved Shooting Cases," *Journal of Forensic Sciences* 54 (2009): 456–462.

37. Ibid.

38. Lauren Dewey et al., "Suicide by Cop: Clinical Risks and Subtypes," *Archives of Suicide Research* 17 (2013): 448–461.

39. Patton and Fremouw, "Examining 'Suicide by Cop,'" 107–120.

40. Brian A. Reaves, *Local Police Departments, 2013: Equipment and Technology* (Washington, DC: National Institute of Justice, 2015).

41. Cynthia Lum and Paul C. Friday, "Impact of Pepper Spray Availability on Police Officer Use of Force Decisions," *Policing: An International Journal of Police Strategies and Management* 20 (1997): 173–185.

42. Brandl, *An Analysis of 2018 Use of Force Incidents.*

43. Mike Wood, "5 Tactical Considerations for Throwable Robot Deployment," PoliceOne.com, March 23, 2017, https://www.policeone.com/policing-in-the-video-age/articles/320406006-5-tactical-considerations-for-throwable-robot-deployment/.

44. Sidney Fussell, "Kentucky Is Turning to Drones to Fix Its Unsolved-Murder Crisis," *The Atlantic*, November 6, 2018, https://www.theatlantic.com/technology/archive/2018/11/police-drone-shotspotter-kentucky-gun-911-ai/574723/.

45. Alina Selyukh and Gabriel Rosenberg, "Bomb Robots: What Makes Killing in Dallas Different and What Happens Next?" NPR.org, July 8, 2016, http://www.npr.org/sections/thetwo-way/2016/07/08/485262777/for-the-first-time-police-used-a-bomb-robot-to-kill.

46. Ibid.

47. Lum and Friday, "Impact of Pepper Spray Availability."

48. Ibid.

49. Steven G. Brandl and Meghan S. Stroshine, "Oleoresin Capsicum Spray and TASERs: A Comparison of Factors Predicting Use and Effectiveness," *Criminal Justice Policy Review* 27 (2015), published online before print.

50. Johan Bertilsson, Ulf Petersson, Peter J. Fredriksson, Måns Magnusson, and Per-Anders Fransson, "Use of Pepper Spray in Policing: Retrospective Study of Situational Characteristics and Implications for Violent Situations," *Police Practice and Research* 18, (2017): 391–406.

51. Wisconsin Department of Justice, *Defensive and Arrest Tactics*, 28.

52. For a review see Brandl and Stroshine, "Oleoresin Capsicum Spray and TASERs."

53. Brandl, *An Analysis of 2018 Use of Force Incidents*; Brandl and Stroshine, "Oleoresin Capsicum Spray and TASERs."

54. Mark W. Kroll et al., "TASER Electronic Control Devices and Cardiac Arrests: Coincidental or Causal?" *Circulation* 129 (2014): 93–100; Charles S. Petty, *Deaths in Police Confrontations When Oleoresin Capsicum Is Used: Final Report* (Washington, DC: National Institute of Justice, 2004).

55. Reaves, *Local Police Departments, 2013.*

56. Alpert et al., *Police Use of Force*; Randy R. Means and Eric Edwards, "Electronic Control Weapons: Liability Issues," *The Police Chief* 72 (2005), http://www.policechiefmagazine.org/magazine/index.cfm?fuseaction=archivecontents&issue_id=22005.

57. Michael D. White and Justin Ready, "The Impact of the Taser on Suspect Resistance: Identifying Predictors of Effectiveness," *Crime and Delinquency* 56 (2010): 70–102.

58. Ibid.; Brandl and Stroshine, "Oleoresin Capsicum Spray and TASERs."

59. Means and Edwards, "Electronic Control Weapons."

60. Alpert et al., *Police Use of Force*; American Civil Liberties Union (ACLU) of Southern California, *Pepper Spray: A Magic Bullet under Scrutiny* (Los Angeles, CA: ACLU of Southern California, 1993); Amnesty International, *Amnesty International's Continued Concern about Taser Use* (New York, NY: Amnesty International, 2006); National Institute of Justice (NIJ), *Study of Deaths Following Electro Muscular Disruption* (Washington, DC: U.S. Department of Justice, National Institute of Justice, 2011).

61. ACLU of Southern California, *Pepper Spray*; Amnesty International, *Amnesty International's Continued Concern about Taser Use.*

62. William Terrill and Stephen D. Mastrofski, "Situational and Officer-Based Determinants of Police Coercion," *Justice Quarterly* 19 (2002): 215–248.

63. Amnesty International, *Amnesty International's Continued Concern about Taser Use.*

64. Brandl and Stroshine, "Oleoresin Capsicum Spray and TASERs."

65. Jacinta M. Gau, Clayton Mosher, and Travis C. Pratt, "An Inquiry Into the Impact of Suspect Race on Police Use of Tasers," *Police Quarterly* 13 (2013): 27–48.

66. Ibid.

67. Alpert et al., *Police Use of Force*; White and Ready, "The TASER as a Less Lethal Force Alternative," 70–102.

68. Cheryl W. Thompson and Mark Berman, "Improper Techniques, Increased Risks: Deaths Have Raised Questions About the Risk of Excessive or Improper Deployment of Tasers," *The Washington Post*, November 26, 2015, http://www.washingtonpost.com/sf/investigative/2015/11/26/improper-techniques-increased-risks/.

69. Ibid.; White and Ready, "The TASER as a Less Lethal Force Alternative," 70–102.

70. Brian Roach, Kelsey Echols, and Aaron Burnett, "Excited Delirium and the Dual Response: Preventing In-Custody Deaths," *FBI Law Enforcement Bulletin*, July 2014, https://leb.fbi.gov/2014/july/excited-delirium-and-the-dual-response-preventing-in-custody-deaths.

71. Thompson and Berman, "Improper Techniques, Increased Risks."

72. Brandl and Stroshine, "Oleoresin Capsicum Spray and TASERs."

73. NIJ, *Study of Deaths Following Electro Muscular Disruption.*

74. Kenneth Adams, "What We Know About Police Use of Force," in *Use of Force by Police: Overview of National and Local Data*, ed. Kenneth Adams et al. (Washington, DC: U.S. Department of Justice, National Institute of Justice, 1999), 1–14.

75. Steven G. Brandl and Meghan S. Stroshine, "The Role of Officer Attributes, Job Characteristics, and Arrest Activity in Explaining Police Use of Force," *Criminal Justice Policy Review* 24 (2012): 551–572.

76. Ibid.

77. Independent Commission on the Los Angeles Police Department, *Report of the Independent Commission on the Los Angeles Police Department*, 1991, http://michellawyers.com/wp-content/uploads/2010/06/Report-of-the-Independent-Commission-on-the-LAPD-re-Rodney-King_Reduced.pdf.

78. Brandl and Stroshine, "The Role of Officer Attributes," 551–572; Robert J. Kaminski, Clete DiGiovanni, and Raymond Downs, "The Use of Force Between the Police and Persons With Impaired Judgment," *Police Quarterly* 7 (2004): 311–338; John D. McCluskey and William Terrill, "Departmental and Citizen Complaints as Predictors of Police Coercion," *Policing: An International Journal of Police Strategies and Management* 28 (2005): 513–529; John D. McCluskey, William Terrill, and Eugene A. Paoline III, "Peer Group Aggressiveness and the Use of Coercion in Police-Suspect Encounters," *Police Practice and Research: An*

International Journal 6 (2005): 19–37; Eugene A. Paoline and William Terrill, "Police Education, Experience, and the Use of Force," *Criminal Justice and Behavior* 34 (2007): 179–196; Ivan Y. Sun and Brian K. Payne, "Racial Differences in Resolving Conflicts: A Comparison Between Black and White Police Officers," *Crime and Delinquency* 50 (2004): 516–541; Terrill and Mastrofski, "Situational and Officer-Based Determinants," 215–248.

79. Brandl and Stroshine, "The Role of Officer Attributes," 551–572; Brian A. Lawton, "Levels of Nonlethal Force: An Examination of Individual, Situational, and Contextual Factors," *Journal of Research in Crime and Delinquency* 44 (2007): 163–184; James P. McElvain and Augustine J. Kposowa, "Police Officer Characteristics and Internal Affairs Investigations for Use of Force Allegations," *Journal of Criminal Justice* 32 (2004): 265–279; McCluskey and Terrill, "Departmental and Citizen Complaints"; Paoline and Terrill, "Police Education, Experience, and the Use of Force," 179–196; Terrill and Mastrofski, "Situational and Officer-Based Determinants," 215–248; Robert E. Worden, "The 'Causes' of Police Brutality: Theory and Evidence on Police Use of Force," in *And Justice for All: Understanding and Controlling Police Abuse of Force*, ed. W. A. Geller and H. Toch (Washington, DC: Police Executive Research Forum, 1995), 31–60.

80. Brandl and Stroshine, "The Role of Officer Attributes," 551–572; Joel H. Garner, Christopher D. Maxwell, and Cedrick G. Heraux, "Characteristics Associated With the Prevalence and Severity of Force Used by and Against the Police," *Justice Quarterly* 19 (2002): 705–746; McElvain and Kposowa, "Police Officer Characteristics," 265–279.

81. D. H. Bayley and J. Garofalo, "Patrol Officers' Effectiveness in Managing Conflict During Police-Citizen Encounters," in *Report to the New York State Commission on Criminal Justice and the Use of Force*, Vol. III (Albany, NY: New York State Commission on Criminal Justice and the Use of Force, 1987), B1–88; McElvain and Kposowa, "Police Officer Characteristics," 265–279; Paoline and Terrill, "Police Education, Experience, and the Use of Force," 179–196.

82. Hans Toch, "The 'Violence-Prone' Police Officer," in Geller and Toch, *And Justice for All*, 99–112.

83. Brandl and Stroshine, "The Role of Officer Attribute," 551–572.

84. Adams, "What We Know about Police Use of Force," 1–14.

85. Brandl and Stroshine, "The Role of Officer Attributes," 551–572.

86. PERF, *Critical Issues in Policing Series*, 22.

87. Terrill and Paoline III. "Police Use of Less Lethal Force," 193–216; Samuel Walker and Lorie Fridell, "Forces of Change in Police Policy: The Impact of *Tennessee v. Garner* on Deadly Force Policy," *American Journal of Police* 11 (1993): 97–112.

88. Janet R. Oliva, Rhiannon Morgan, and Michael T. Compton, "A Practical Overview of De-escalation Skills in Law Enforcement: Helping Individuals in Crisis While Reducing Police Liability and Injury," *Journal of Police Crisis Negotiations* 10 (2010): 15–29.

89. Ibid.

90. FM Coalition for Homeless Persons web site, http://www.fmhomeless.org/sites/default/files/De-escalation%20Techniques.pdf.

91. Oliva et al., "A Practical Overview of De-escalation Skills," 15–29.

92. Ibid. Also see FM Coalition for Homeless Persons website, http://www.fmhomeless.org/sites/default/files/De-escalation%20Techniques.pdf.

93. For example: Owen Price, John Baker, Penny Bee, and Karina Lovell, "The Support-Control Continuum: An Investigation of Staff Perspectives on Factors Influencing the Success or Failure of De-escalation Techniques for the Management of Violence and Aggression in Mental Health Settings," *International Journal of Nursing Studies* 77 (2018): 197–206. Jill E. Spielfogel and J. Curtis McMillen. "Current Use of De-escalation Strategies: Similarities and Differences in De-escalation across Professions," *Social Work in Mental Health* 15 (2017): 232–248.

94. Natalie Todak and Lois James, "A Systematic Social Observation Study of Police De-escalation Tactics," *Police Quarterly* 21 (2018): 509–543.

95. Robert E. Worden, Moonsun Kim, Christopher J. Harris, Mary Anne Pratte, Shelagh E. Dorn, and Shelley S. Hyland, "Intervention With Problem Officers: An Outcome Evaluation of an EIS Intervention," *Criminal Justice and Behavior* 40 (2013): 409–437.

96. Eric C. Hedberg, Charles M. Katz, and David E. Choate, "Body-Worn Cameras and Citizen Interactions With Police Officers: Estimating Plausible Effects Given Varying Compliance Levels," *Justice Quarterly* 34 (2017): 627–651.

97. Barak Ariel, William A. Farrar, and Alex Sutherland, "The Effect of Police Body-Worn Cameras on Use of Force and Citizens' Complaints Against the Police: A Randomized Controlled Trial," *Journal of Quantitative Criminology* 31 (2015): 509–535. For similar findings see Anthony Braga, James R. Coldren, William Sousa, Denise Rodriguez, and Omer Alper, "The Benefits of Body-Worn Cameras: New Findings From a Randomized Controlled Trial at the Las Vegas Metropolitan Police Department" (Washington DC: U.S. Department of Justice, National Institute of Justice, December 2017).

98. Michael D. White, Janne E. Gaub, and Natalie Todak, "Exploring the Potential for Body-Worn Cameras to Reduce Violence in Police–Citizen Encounters," *Policing: A Journal of Policy and Practice* 12 (2017): 66–76.

99. Barak Ariel, Alex Sutherland, Darren Henstock, Josh Young, Paul Drover, Jayne Sykes, Simon Megicks, and Ryan Henderson. "Report: Increases in Police Use of Force in the Presence of Body-Worn Cameras Are Driven by Officer Discretion: A Protocol-Based Subgroup Analysis of Ten Randomized Experiments," *Journal of Experimental Criminology* 12 (2016): 453–463.

CHAPTER 12

1. Lonnie M. Schaible et al., "Denver's Citizen/Police Complaint Mediation Program: Officer and Complainant Satisfaction," *Criminal Justice Policy Review* 24 (2013): 626–650.

2. City of Denver, Office of the Independent Monitor web site, https://www.denvergov.org/content/denvergov/en/office-of-the-independent-monitor/mediation.html.

3. Schaible et al., "Denver's Citizen/Police Complaint Mediation Program," 633.

4. Schaible et al., "Denver's Citizen/Police Complaint Mediation Program," 626–650. Other studies have come to similar conclusions. For example, see Elizabeth C. Bartels and Eli B. Silverman, "An Exploratory Study of the New York City Civilian Complaint Board Mediation Program," *Policing: An International Journal of Police Strategies and Management* 28 (2005): 519–630.

5. Thomas J. Martinelli, "Unconstitutional Policing: The Ethical Challenges in Dealing With Noble Cause Corruption," *The Police Chief*, August 2016, http://www.policechiefmagazine.org/magazine/index.cfm?fuseaction=display&article_id=1025& issue_id=102006.

6. Carl B. Klockars, Sanja Kutnjak Ivković, and Maria R. Haberfeld, *Enhancing Police Integrity* (Dordrecht, Netherlands: Springer, 2006), 1.

7. Neal Trautman, "Police Code of Silence Facts Revealed," 2000 International Association of Chiefs of Police annual conference, http://www.aele.org/loscode2000.html.

8. Timothy M. Maher, "Police Sexual Misconduct: Officers' Perceptions of Its Extent and Causality," *Criminal Justice Review* 28 (2003): 355–381.

9. Sanja Kutnjak Ivković, Maria Haberfeld, and Robert Peacock, "Rainless West: The Integrity Survey's Role in Agency Accountability," *Police Quarterly* 16 (2012): 148–176.

10. Ibid.

11. Lynn Langton and Matthew Durose, *Police Behavior During Traffic and Street Stops, 2011* (Washington, DC: Bureau of Justice Statistics, 2013).

12. Christopher J. Harris, "Exploring the Relationship Between Experience and Problem Behaviors: A Longitudinal Analysis of Officers From a Large Cohort," *Police Quarterly* 12 (2009): 192–213, 209.

13. Kim M. Lersch and Tom Mieczkowski, "Who Are the Problem-Prone Officers? An Analysis of Citizen Complaints," *American Journal of Police* 15 (1996): 23–44; Steven G. Brandl, Meghan S. Stroshine, and James Frank, "Who Are the Complaint-Prone Officers? An Examination of the Relationship Between Police Officers' Attributes, Arrest Activity, Assignment, and Citizens' Complaints About Excessive Force," *Journal of Criminal Justice* 29 (2001): 521–529.

14. Harris, "Exploring the Relationship," 195.

15. For a review see Christopher J. Harris and Robert E. Worden, "The Effect of Sanctions on Police Misconduct," *Crime and Delinquency* 60 (2014): 1258–1288.

16. Ibid.

17. Brandl, Stroshine, and Frank, "Who Are the Complaint-Prone Officers?" 521–529; Kimberly D. Hassell and Carol A. Archbold, "Widening the Scope of Complaints of Police Misconduct," *Policing: An International Journal of Police Strategies and Management* 33 (2010): 473–489; Kim M. Lersch, "Are Citizen Complaints Just Another Measure of Officer Productivity? An Analysis of Citizen Complaints and Officer Activity Measures," *Police Practice and Research* 3 (2002): 135–147.

18. Harris and Worden, "The Effect of Sanctions," 1258–1288.

19. Jonah Newman, "Chicago Spent More Than $113 Million on Police Misconduct Lawsuits in 2018," *The Chicago Reporter*, March 17, 2019, https://www.chicagoreporter.com/chicago-spent-more-than-113-million-on-police-misconduct-lawsuits-in-2018/.

20. Jonah Newman, "Chicago Does Little to Control Police Misconduct—Or Its Costs," *The Chicago Reporter*, June 22, 2016, http://chicagoreporter.com/chicago-does-little-to-control-police-misconduct- or-its-costs/.

21. Cara E. Rabe-Hemp and Jeremy Braithwaite, "An Exploration of Recidivism and the Officer Shuffle in Police Sexual Violence," *Police Quarterly* 16 (2012): 127–147. Kristina M. Lopez, David R. Forde, and J. Mitchell Miller, "Media Coverage of Police Sexual Misconduct in Seven Cities: A Research Note," *American Journal of Criminal Justice* 42 (2017): 833–844.

22. Philip Matthew Stinson Sr., John Liederbach, and Tina L. Freiburger, "Off-Duty and Under Arrest: A Study of Crimes Perpetrated by Off-Duty Police," *Criminal Justice Policy Review* 23 (2012): 139–163.

23. Philip Matthew Stinson, Natalie Erin Todak, and Mary Dodge, "An Exploration of Crime by Policewomen," *Police Practice and Research* 16 (2015): 79–93.

24. Phillip Matthew Stinson Sr. and John Liederbach, "Fox in the Henhouse: A Study of Police Officers Arrested for Crimes Associated With Domestic and/or Family Violence," *Criminal Justice Policy Review* 24 (2012): 601–625.

25. Tom Jackman, "Study Finds Police Officers Arrested 1,100 Times per Year, or 3 per Day, Nationwide," *The Washington Post*, June 22, 2016, https://www.washingtonpost.com/news/true-crime/wp/2016/06/22/study-finds-1100-police-officers-per-year-or-3-per-day-are-arrested-nationwide/.

26. Stinson, Liederbach, and Freiburger, "Off-Duty and Under Arrest," 139–163.

27. Stinson and Liederbach, "Fox in the Henhouse," 601–625.

28. Philip Matthew Stinson, Natalie Erin Todak, and Mary Dodge, "An Exploration of Crime by Policewomen," *Police Practice and Research* 16 (2015): 79–93.

29. Loren T. Atherley and Matthew J. Hickman, "Officer Decertification and the National Decertification Index," *Police Quarterly* 16 (2013): 420–437.

30. Jessica Huff, Michael D. White, and Scott H. Decker, "Organizational Correlates of Police Deviance: A Statewide Analysis of Misconduct in Arizona, 2000–2011," *Policing: An International Journal* 41 (2018): 465–481.

31. Loren T. Atherley and Matthew J. Hickman, "Officer Decertification and the National Decertification Index," *Police Quarterly* 16 (2013): 420–437.

32. Ibid.

33. Jack McDevitt, Amy Farrell, and Russell Wolff, *Creating a Culture of Integrity* (Washington, DC: Community Oriented Policing Services, 2008).

34. Timothy M. Maher, "Police Sexual Misconduct: Officers' Perceptions of Its Extent and Causality," *Criminal Justice Review* 28 (2003): 355–381. Quote appears on page 372.

35. Robert E. Worden et al., "Intervention With Problem Officers: An Outcome Evaluation of an EIS Intervention," *Criminal Justice and Behavior* 40 (2013): 409–437, 432.

36. A similar question is posed in Jerome H. Skolnick and James J. Fyfe, *Above the Law: Police and the Excessive Use of Force* (New York, NY: The Free Press, 1993), 111.

37. This discussion draws heavily from Alpert, Nobel, and Rojek, "Solidarity and the Code of Silence."

38. Alpert et al., "Solidarity and the Code of Silence," 115.

39. For a more extensive discussion of these issues, see Geoffrey P. Alpert and Jeffrey J. Nobel, "Lies, True Lies, and Conscious Deception: Police Officers and the Truth," *Police Quarterly* 12 (2009): 237–254.

40. Jeff Noble, "Police Officer Truthfulness and the *Brady* Decision," *Police Chief* 70 (2003), http://www.policechiefmagazine.org/police-officer-truthfulness-and-the-brady-decision/.

41. Harris and Worden, "The Effect of Sanctions," 1258–1288.

42. Alpert et al., "Solidarity and the Code of Silence," 117.

43. Loren T. Atherley and Matthew J. Hickman, "Officer Decertification and the National Decertification Index," *Police Quarterly* 16 (2013): 420–437.

44. Samuel Walker, Geoffrey P. Alpert, and Dennis J. Kenney, "Early Warning Systems for the Police: Concept, History, and Issues," *Police Quarterly* 3 (2000): 132–152.

45. U.S. Department of Justice, *Principles for Promoting Police Integrity: Examples of Promising Police Practices and Policies* (Washington, DC: U.S. Department of Justice, 2001), 10–11; Walker et al., "Early Warning Systems," 132–152.

46. Worden et al., "Intervention With Problem Officers," 409–437.

47. Ibid., 410.

48. Ibid.

49. Karen L. Amendola and Robert C. Davis, *Best Practices in Early Intervention System Implementation and Use in Law Enforcement Agencies* (Washington DC: National Police Foundation, 2018).

50. Worden et al., "Intervention With Problem Officers," 430.

51. Ibid.

52. Ibid.

53. Neil Wain and Barak Ariel, "Tracking of Police Patrol," *An International Journal of Police Strategies and Management* 8 (2014): 274–283.

54. David Weisburd et al., *The Dallas AVL Experiment: Evaluating the Use of Automated Vehicle Locator Technologies in Policing* (Washington, DC: Police Foundation, 2012).

55. Jim Norman, "Confidence in Police Back at Historical Average," Gallup, July 10, 2017, https://news.gallup.com/poll/213869/confidence-police-back-historical-average.aspx.

CHAPTER 13

1. "Chula Vista Wins Herman Goldstein Excellence in Problem-Oriented Policing Award," Community Oriented Policing Services, U.S. Department of Justice, http://www.cops.usdoj.gov/html/dispatch/December_2009/print/goldstein_award_print.htm.

2. Federal Bureau of Investigation, Uniform Crime Reporting Statistics, http://www.ucrdatatool.gov/Search/Crime/State/RunCrimeStatebyState.cfm.

3. George L. Kelling et al., *The Kansas City Preventive Patrol Experiment: A Summary Report* (Washington, DC: The Police Foundation, 1974).

4. William Spelman and Dale K. Brown, *Calling the Police: Citizen Reporting of Serious Crime* (Washington, DC: Police Executive Research Forum, 1981).

5. Jan M. Chaiken, Peter W. Greenwood, and Joan Petersilia, "The Criminal Investigation Process: A Summary Report," *Policy Analysis* 3 (1977): 187–217.

6. Scott Decker and R. L. Smith, "Police Minority Recruitment: A Note on Its Effectiveness in Improving Black Evaluations of the Police," *Journal of Criminal Justice* 8 (1980): 387–393; James Q. Wilson, "The Police in the Ghetto," in *The Police and the Community*, ed. J. Rubin Garmire and J. Q. Wilson (Baltimore, MD: Johns Hopkins University Press, 1972), 51–90.

7. Eileen Ogintz, "Order to Rehire Cops Handcuffing Detroit," *Chicago Tribune*, August 5, 1984, http://archives.chicagotribune.com/1984/08/05/page/3/article/order-to-rehire-cops-handcuffing-detroit/index.html.

8. Candice Williams, "Report: 'Growing' Racial Problem in Detroit Police Department," *The Detroit News*, January 12, 2017, https://www.detroitnews.com/story/news/local/detroit-city/2017/01/12/report-says-growing-racial-problem-police-dept/96520140/.

9. Lawrence W. Sherman, Catherine H. Milton, and Thomas V. Kelly, *Team Policing: Seven Case Studies* (Washington, DC: Police Foundation, 1973).

10. Community Oriented Policing Services, *Community Policing Defined* (Washington, DC: U.S. Department of Justice, 2014), 1.

11. Bureau of Justice Statistics, *Law Enforcement Management and Administrative Statistics (LEMAS) Survey* (Washington DC: Bureau of Justice Statistics, 2013).

12. Herman Goldstein, "Toward Community-Oriented Policing: Potential, Basic Requirements, and Threshold Questions," *Crime and Delinquency* 33 (1987): 6–30.

13. Newport News, Virginia, website, https://www.nngov.com/346/Citizen-Police-Academy-CPA.

14. City of Carmel, Indiana, website, http://www.carmel.in.gov/department-services/police/community-resource-programs/teen-academy.

15. City of Grand Rapids, Michigan, website, https://www.grandrapidsmi.gov/Government/Programs-and-Initiatives/Community-Policing.

16. Cassie Spodak, "Atlanta Police Program Bringing More Cops Home," CNN, June 1, 2018, https://www.cnn.com/2018/06/01/politics/bridging-the-divide-atlanta-police/index.html.

17. City of Dayton, Ohio, website, http://daytonhrc.org/community-relations/community-police-relations-initiatives/community-police-council-cpc/.

18. Coffee with a Cop website, https://coffeewithacop.com/.

19. Houston, Texas, Police Officers' Union website, https://hpou.org/building-a-bridge-between-police-and-teens/.

20. Illinois State Police Citizen Survey website, http://www.isp.state.il.us/SurveySite/.

21. Natalie Allison, "Violent Crime Is Down in Cayce Homes a Year After Increased Foot Patrols," *The Tennessean*, September 7, 2018, https://www.tennessean.com/story/news/crime/2018/09/07/nashville-crime-cayce-homes-down-foot-patrol-program-continues/1079677002/.

22. Dallas, Texas, Police Department website, http://www.dallaspolice.net/division/central/neighborhood-police-team.

23. City of Lansing, Michigan, website, https://www.lansingmi.gov/464/Community-Services.

24. Wesley G. Skogan, *Community Policing: Can It Work?* (Belmont, CA: Wadsworth, 2004).

25. International Association of Chiefs of Police, *2015 Social Media Survey Results* (Alexandria, VA: International Association of Chiefs of Police, 2017).

26. K. Kim, A. Oglesby-Neal, and E. Mohr, *2016 Law Enforcement Use of Social Media Survey* (Washington DC: Urban Institute, 2017).

27. Joel D. Lieberman, Deborah Koetzle, and Mari Sakiyama, "Police Departments' Use of Facebook: Patterns and Policy Issues," *Police Quarterly* 16 (2013): 438–462.

28. Lori Brainard and Mariglynn Edlins, "Top 10 US Municipal Police Departments and Their Social Media Usage," *The American Review of Public Administration* 45 (2015): 728–745.

29. Emily Tiry, Ashlin Oglesby-Neal, and KiDeuk Kim, *Social Media Guidebook for Law Enforcement Agencies* (Washington DC: Urban Institute, 2019).

30. Edward R. Maguire, Craig D. Uchida, and Kimberly D. Hassell, "Problem-Oriented Policing in Colorado Springs: A Content Analysis of 753 Cases," *Crime and Delinquency* 61 (2015): 71–95.

31. U.S. Department of Justice website, https://www.justice.gov/opa/pr/attorney-general-sessions-announces-98-million-hire-community-policing-officers.

32. Skogan, *Community Policing*.

33. Edward R. Maguire, "Structural Change in Large Municipal Police Organizations During the Community Policing Era," *Justice Quarterly* 14 (1997): 547–576; Robert Trojanowicz, *Community Policing: A Survey of Police Departments in the United States* (Washington, D.C.: U.S. Department of Justice, 1994).

34. Brian A. Reaves, *State and Local Law Enforcement Training Academies, 2013* (Washington, DC: National Institute of Justice, 2016).

35. James Q. Wilson and George L. Kelling, "Broken Windows: Police and Neighborhood Safety," *Atlantic Monthly* 249 (1982): 29–38.

36. Ibid., 31.

37. Ibid., 34

38. Ibid., 36.

39. Ibid., 38.

40. Robert J. Sampson and Stephen W. Raudenbush, *Disorder in Urban Neighborhoods—Does It Lead to Crime?* (Washington, DC: National Institute of Justice, 2001).

41. Sam Roberts, "Author of 'Broken Windows' Policing Defends His Theory," *The New York Times*, August 10, 2014, http://www.nytimes.com/2014/08/11/nyregion/author-of-broken-windows-policing-defends-his-theory.html?_r=0.

42. Sampson and Raudenbush, *Disorder in Urban Neighborhoods*.

43. Ibid., 1.

44. Sampson and Raudenbush, *Disorder in Urban Neighborhoods*; John R. Hipp and Rebecca Wickes, "Violence in Urban Neighborhoods: A Longitudinal Study of Collective Efficacy and Violent Crime," *Journal of Quantitative Criminology* 2016 (2016): 1–26; John R. Hipp and James C. Wo, "Collective Efficacy and Crime," *International Encyclopedia of the Social and Behavioral Sciences* 4 (2015): 169–173.

45. Sampson and Raudenbush, *Disorder in Urban Neighborhoods*, 5.

46. Anthony A. Braga, Brandon C. Welsh, and Cory Schnell, "Can Policing Disorder Reduce Crime? A Systematic Review and Meta-analysis," *Journal of Research in Crime and Delinquency* 52 (2015): 567–588, 567.

47. Ibid., 567.

48. See Stephen D. Mastrofski, "Community Policing as Reform: A Cautionary Tale," in *Community Policing: Rhetoric or Reality?*, ed. Jack R. Greene and Stephen D. Mastrofski (New York, NY: Praeger, 1991), 47–67.

49. Gary W. Cordner, "Community Policing: Elements and Effects," in *Critical Issues in Policing: Contemporary Readings*, ed. Roger G. Dunham and Geoffrey P. Alpert (Prospect Heights, IL: Waveland, 2015), 481–498. Justin N. Crowl, "The Effect of Community Policing on Fear and Crime Reduction, Police Legitimacy and Job Satisfaction: An Empirical Review of the Evidence," *Police Practice and Research* 18 (2017): 449–462.

50. Tom Tyler, "What Are Legitimacy and Procedural Justice in Policing? And Why Are They Becoming Key Elements of Police Leadership?" in *Legitimacy and Procedural Justice: A New Element of Police Leadership*, ed. Craig Fisher (Washington, DC: Police Executive Research Forum, 2014). Tal Jonathan-Zamir, Stephen D. Mastrofski, and Shomron Moyal, "Measuring Procedural Justice in Police-Citizen Encounters," *Justice Quarterly* 32 (2015): 845–871.

51. Tyler, "What Are Legitimacy and Procedural Justice in Policing?" 9.

52. Pew Research Center, "Latino Confidence in Local Police Lower Than Among Whites," August 28, 2014, http://www.pewresearch.org/fact-tank/2014/08/28/latino-confidence-in-local-police-lower-than-among-whites/.

53. Pew Research Center, "The Racial Confidence Gap in Police Performance," September 29, 2016, https://www.pewsocialtrends.org/2016/09/29/the-racial-confidence-gap-in-police-performance/.

54. Steven G. Brandl et al., "Global and Specific Attitudes Toward the Police: Disentangling the Relationship," *Justice Quarterly* 11 (1994): 119–134.

55. Dennis P. Rosenbaum et al., "Attitudes Toward the Police: The Effects of Direct and Vicarious Experience," *Police Quarterly* 8 (2005): 343–365.

56. Robert E. Worden and Sarah J. McLean, "Measuring, Managing, and Enhancing Procedural Justice in Policing: Promise and Pitfalls," (2016): 36. http://www.jjay.cuny.edu/sites/default/files/contentgroups/p2ph/Worden%20%26%20McLean%20Measuring%20%26%20Managing%20Procedural%20Justice.pdf.

57. Ibid., 36.

58. Worden, Robert E., Heidi S. Bonner, and Sarah J. McLean, "Procedural Justice and Citizen Review of Complaints Against the Police: Structure, Outcomes, and Complainants' Subjective Experiences." *Police Quarterly* 21 (2018): 77–108.

59. Daniel J. Bell, "Police and Public Opinion," *Journal of Police Science and Administration* 7 (1979): 196–205, 196–197.

60. Wesley G. Skogan, "Representing the Community in Community Policing," in *Community Policing: Can It Work?*, ed. Wesley G. Skogan (Belmont, CA: Wadsworth, 2004), 57–75; Tom Tyler, "Policing in Black and White: Ethnic Group Differences in Trust and Confidence in the Police," *Police Quarterly* 8 (2005): 322–342; Jeffrey Fagan and T. Tyler, "Legitimacy and Cooperation: Why Do People Help the Police Fight Crime in Their Communities?" *Ohio State Journal of Criminal Law* 6 (2008): 231–275.

61. James Frank et al., "Citizen Involvement in the Coproduction of Police Outputs," *Journal of Crime and Justice* 19 (1996): 1–30. William Wells et al., "Neighborhood Residents' Production of Order: The Effects of Collective Efficacy on Responses to Neighborhood Problems," *Crime and Delinquency* 52 (2006): 523–550.

62. Skogan, "Representing the Community in Community Policing," 57–75.

63. Frank et al., "Citizen Involvement," 1–30.

64. Ling Ren et al., "Participation Community Crime Prevention: Who Volunteers for Police Work?," *Policing: An International Journal of Police Strategies and Management* 29 (2006): 464–481.

65. Frank et al., "Citizen Involvement," 24.

66. Tom R. Tyler, *Why People Obey the Law* (Princeton, NJ: Princeton University Press, 2006).

67. Ibid., 178.

68. Ibid.

69. President's Task Force on 21st Century Policing, *Final Report of the President's Task Force on 21st Century Policing* (Washington, DC: Office of Community Oriented Policing Services, 2015), 45.

70. George J. Thompson and Jerry B. Jenkins, *Verbal Judo: The Gentle Art of Persuasion* (New York, NY: William Morrow, 2013), ix.

71. Ibid., 12.

72. Nick Tilley, "Community Policing and Problem Solving," in Skogan, *Community Policing*, 174.

73. Crowl, "The Effect of Community Policing on Fear and Crime Reduction," 449–462.

74. Tilley, "Community Policing and Problem Solving."

75. Herman Goldstein, "Improving Policing: A Problem-Oriented Approach," *Crime and Delinquency* 25 (1979): 236–258.

76. Ibid., 243.

77. Goldstein, "Improving Policing," 243.

78. Joel B. Plant and Michael S. Scott, *Effective Policing and Crime Prevention: A Problem-Oriented Guide for Mayors, City Managers, and County Executives* (Washington, DC: Center for Problem-Oriented Policing, 2009), 41.

79. Ibid., 38.

80. Ibid.

81. Ibid., 46.

82. Ibid., 46–47.

83. Eureka Police Department "The Vacation of Devil's Playground," 2017 Herman Goldstein Award submission, https://popcenter.asu.edu/sites/default/files/17-06.pdf.

84. Arlington Police Department, "Project RAISE," 2017 Herman Goldstein Award submission, https://popcenter.asu.edu/sites/default/files/18-04_arlington_tx.pdf.

85. Boston Police Department, "Improving Homicide Clearance Rates," *2017 Strategies for Policing Innovation Spotlight Report*, http://www.strategiesforpolicinginnovation.com/sites/default/files/spotlights/Boston_SPI_Spotlight_%282017%29_FINAL.pdf.

86. Cincinnati Police Department, "The CCROW Project," https://popcenter.asu.edu/sites/default/files/17-08.pdf.

87. Providence Rhode Island Police Department, "Reducing Street-Level Prostitution," https://popcenter.asu.edu/sites/default/files/18-10_providence.pdf.

88. John E. Eck, "Why Don't Problems Get Solved?," in Skogan, *Community Policing*, 189. According to the influential National Research Council, problem-oriented policing shows "moderate evidence of effectiveness" (Frydl and Skogan, *Fairness and Effectiveness in Policing*).

89. Maguire et al., "Problem-Oriented Policing in Colorado Springs," 71–95.

90. Cody W. Telep and David Weisburd, "What Is Known About the Effectiveness of Police Practices in Reducing Crime and Disorder?" *Police Quarterly* 15 (2012): 331–357.

CHAPTER 14

1. Timothy Williams, "Can 30,000 Cameras Help Solve Chicago's Crime Problem?" *New York Times*, May 26, 2018, https://www.nytimes.com/2018/05/26/us/chicago-police-surveillance.html.

2. City of Chicago, Office of the Mayor website, "Mayor Emanuel Announces Expansion of Smart Policing Strategy Supporting Nearly Two Years of Consecutive Declines in Crime," October 10, 2018, https://www.cityofchicago.org/city/en/depts/mayor/press_room/press_releases/2018/october/101018_ExpansionSmartPolicingStrategy.html.

3. James R. Coldren Jr., Alissa Huntoon, and Michael Medaris, "Introducing Smart Policing: Foundations, Principles, and Practice," *Police Quarterly* 16 (2013): 275–286.

4. James R. Coldren Jr., Michael D. White, and Craig Uchida, "Smart Policing: What's in a Name?" *The Criminologist* 38 (2013): 54–55.

5. Strategies for Police Innovation website, http://www.strategiesforpolicinginnovation.com/.

6. Coldren et al., "Introducing Smart Policing,"275–286.

7. Cody W. Telep and Logan J. Somers, "Examining Police Officer Definitions of Evidence-Based Policing: Are We Speaking the Same Language?" *Policing and Society* (2017): 1–17.

8. Cody W. Telep and Cynthia Lum, "The Receptivity of Officers to Empirical Research and Evidence-Based Policing: An Examination of Survey Data From Three Agencies," *Police Quarterly* 17 (2014): 359–385, 360.

9. Telep and Lum, "Receptivity of Officers," 359–385.

10. Lawrence W. Sherman and Richard A. Berk, "The Specific Deterrent Effects of Arrest for Domestic Assault," *American Sociological Review* 49 (1984): 261–272.

11. Lawrence W. Sherman, "Experiments in Police Discretion: Scientific Boon or Dangerous Knowledge?," *Law and Contemporary Problems* 47 (1984): 61–80.

12. Lawrence W. Sherman, *Policing Domestic Violence* (New York, NY: The Free Press, 1992), 125.

13. Sherman, *Policing Domestic Violence*.

14. James J. Willis, "First-Line Supervision and Strategic Decision Making Under Compstat and Community Policing," *Criminal Justice Policy Review* 24 (2011): 235–256. Rebecca Neusteter, "CompStat 360," Vera Institute, 2019, https://www.vera.org/projects/compstat-360.

15. William J. Bratton, "Crime Is Down in New York City: Blame the Police," in *Policing: Key Issues*, ed. Tim Newburn (Portland, OR: Willan Publishing, 2005), 472–482.

16. Jacqui Karn, *Policing and Crime Reduction: The Evidence and Its Implications for Practice* (Washington, DC: Police Foundation, 2013); Hyunseok Jang, Larry T. Hoover, and Hee-Jong Joo, "An Evaluation of Compstat's Effect on Crime: The Fort Worth Experience," *Police Quarterly* 13 (2010): 387–412. Obed Pasha, Alexander Kroll, and Michael Ash, "Assessing Police Performance Systems: The Impact of CompStat on Crime," In *Academy of Management Proceedings*, vol. 2018, no. 1 (Briarcliff Manor, NY: Academy of Management, 2018), 15169.

17. Peter K. Manning, "The Police: Mandate, Strategies, and Appearances," in *Policing: A View from the Street*, ed. P. K. Manning and John Van Maanen (Santa Monica, CA: Goodyear, 1978), 7–31; Karn, *Policing and Crime Reduction*.

18. Willis, "First-Line Supervision," 242.

19. Ibid., 245.

20. Brenda J. Bond, and Anthony A. Braga, "Rethinking the Compstat Process to Enhance Problem-Solving Responses: Insights From a Randomized Field Experiment," *Police Practice and Research* 16, no. 1 (2015): 22–35.

21. Andrew M. Subica, Jason A. Douglas, Nancy J. Kepple, Sandra Villanueva, and Cheryl T. Grills, "The Geography of Crime and Violence Surrounding Tobacco Shops, Medical Marijuana Dispensaries, and Off-Sale Alcohol Outlets in a Large, Urban Low-Income Community of Color," *Preventive Medicine* 108 (2018): 8–16.

22. Philip Canter, "Using a Geographic Information System for Tactical Crime Analysis," in *Analyzing Crime Pattern: Frontiers of Practice*, ed. Victor Goldsmith (Thousand Oaks, CA: Sage, 2000), 4.

23. International Association of Crime Analysts, *Definition and Types of Crime Analysis*. (white paper 2014–02, Overland Park, KS: IACA, 2014), http://www.iaca.net/Publications/Whitepapers/iacawp_2014_02_definition_types_crime_analysis.pdf.

24. Jessica Fitterer, Trisalyn A. Nelson, and Farouk Nathoo, "Predictive Crime Mapping," *Police Practice and Research* 16 (2015): 121–135.

25. William V. Pelfrey Jr., "Geographic Information Systems: Applications for the Police," in *Critical Issues in Policing: Contemporary Readings*, ed. Roger G. Dunham and Geoffrey P. Alpert (Long Grove, IL: Waveland Press, 2015), 289–301.

26. Allison Rojek, "The Practice of Crime Analysis," in Dunham and Alpert, *Critical Issues in Policing: Contemporary Readings*, 302–325.

27. Rachel Boba Santos, "The Effectiveness of Crime Analysis for Crime Reduction: Cure or Diagnosis?" *Journal of Contemporary Criminal Justice* 30 (2014): 147–168.

28. City of Arlington, Texas, website, https://cityofarlington.csod.com/ats/careersite/JobDetails.aspx?site=1&id=322.

29. Albert Meijer and Martijn Wessels, "Predictive Policing: Review of Benefits and Drawbacks," *International Journal of Public Administration* (2019): 1–9.

30. Rachel Boba Santos and Bruce Taylor, "The Integration of Crime Analysis Into Police Patrol Work: Results From a National Survey of Law Enforcement Agencies," *Policing: An International Journal of Police Strategies and Management* 37 (2014): 501–520.

31. Carla Lewandowski, Jeremy G. Carter, and Walter L. Campbell, "The Role of People in Information-Sharing: Perceptions From an Analytic Unit of a Regional Fusion Center," *Police Practice and Research* 18 (2017): 174–193.

32. Mark Puente, "LAPD Ends Another Data-Driven Crime Program Touted to Target Violent Offenders," *Los Angles Times*, April 12, 2019, https://www.latimes.com/local/lanow/la-me-laser-lapd-crime-data-program-20190412-story.html.

33. Jerry Ratcliffe, *Intelligence-Led Policing* (Cullompton, Devon, UK: Willan, 2008).

34. Jerry H. Ratcliffe and Ray Guidetti, "State Police Investigative Structure and the Adoption of Intelligence-Led Policing," *Policing: An International Journal of Police Strategies and Management* 31 (2008): 109–128.

35. Ibid.

36. Drug Enforcement Administration, "Intelligence," https://www.dea.gov/ops/intel.shtml.

37. National Gang Intelligence Center, "National Gang Report: 2015 Gangs," 2016, https://www.fbi.gov/stats-services/publications/national-gang-report- 2015.pdf.

38. Federal Bureau of Investigation, National Crime Information Center, https://www.fbi.gov/services/cjis/ncic.

39. Ibid.

40. Ratcliffe and Guidetti, "State Police Investigative Structure," 125. Morgan Burcher and Chad Whelan, "Intelligence-Led Policing in Practice: Reflections From Intelligence Analysts," *Police Quarterly* (2018): 1–22.

41. Joel A. Capellan and Carla Lewandowski. "Can Threat Assessment Help Police Prevent Mass Public Shootings? Testing an Intelligence-Led Policing Tool," *Policing: An International Journal* 42 (2019): 16–30.

CHAPTER 15

1. George L. Kelling and Mark H. Moore, "The Evolving Strategy of Policing," *Perspectives on Policing* 4 (1988): 1–15.

2. Scott Shane, "Abu Zubaydah, Tortured Guantánamo Detainee, Makes Case for Release," *The New York Times*, August 23, 2016, http://www.nytimes.com/2016/08/24/us/abu-zubaydah-torture-guantanamo- bay.html?rref=collection%2Ftimestopic%2FWaterboarding&action=click&content Collection=timestopics&region=stream&module= stream_unit&version=latest&contentPlacement= 2&pgtype=collection.

3. Global Terrorism Database, "Data Collection Methodology," 2016, http://www.start.umd.edu/gtd/using-gtd/.

4. Global Terrorism Database, https://www.start.umd.edu/gtd/search/Results.aspx?country=217.

5. Homeland Security News Wire, "Right-wing Extremism Linked to Every 2018 Extremist Murder in the U.S.," January 25, 2019, http://www.homelandsecuritynewswire.com/dr20190125-rightwing-extremism-linked-to-every-2018-extremist-murder-in-the-u-s.

6. Charles Kurzman and David Schanzer, "The Growing Right-Wing Terror Threat," *The New York Times*, June 16, 2015, http://www.nytimes.com/2015/06/16/opinion/the-other-terror-threat.html?_r=0.

7. National Consortium for the Study of Terrorism and Responses to Terrorism, "Ideological Motivations of Terrorism in the United States 1970-2016," November 2017, https://www.start.umd.edu/pubs/START_IdeologicalMotivationsOfTerrorismInUS_Nov2017.pdf.

8. Samuel Walker and Carol A. Archbold, *The New World of Police Accountability* (Thousand Oaks, CA: Sage, 2014).

9. Gary Marx, *Undercover: Police Surveillance in America* (Berkeley: University of California Press, 1988).

10. David Bayley and David Weisburd, *Cops and Spooks: The Role of Police in Counterterrorism* (New York, NY: Springer, 2011).

11. U.S. Department of Homeland Security, "If You See Something, Say Something," 2016, https://www.dhs.gov/see-something-say-something.

12. Doug Irving, "How Will Technology Change Criminal Justice?," The RAND Corporation, 2016, http://www.rand.org/blog/rand-review/2016/01/how-will-technology-change-criminal-justice.html.

13. Walker and Archbold, *The New World*.

14. Peter B. Kraska and Victor E. Kappeler, "Militarizing American Police: The Rise and Normalization of Paramilitary Units," *Social Problems* 44 (1997): 1–18.

15. Aaron C. Davenport et al., *An Evaluation of the Department of Defense's Excess Property Program* (Santa Monica: RAND Corporation, 2018).

16. Tony Platt et al., *The Iron Fist and the Velvet Glove: An Analysis of the U.S. Police* (San Francisco, CA: Crime and Social Justice Associates, 1982).

17. Michael D. White, *Police Officer Body-Worn Cameras: Assessing the Evidence* (Washington, DC: Community Oriented Policing Services, 2014).

18. Robert E. Worden et al., "Intervention With Problem Officers: An Outcome Evaluation of an EIS Intervention," *Criminal Justice and Behavior* 40 (2013): 409–37.

19. Liam Dillon and Maya Lau, "'This Is Revolutionary': New California Laws Address Its Bad Rap for Police Secrecy," *The Guardian*, October 1, 2018, https://www.governing.com/topics/public-justice-safety/tns-california-police-secrecy-brown-misconduct.html.

20. U.S. Department of Justice, *Principles for Promoting Police Integrity* (Washington, DC: U.S. Department of Justice, 2001). Karen L. Amendola and Robert C. Davis, *Best Practices in Early Intervention System Implementation and Use in Law Enforcement Agencies* (Washington DC: National Police Foundation, 2018).

21. Nissan Moradoff, "Biometrics: Proliferation and Constraints to Emerging and New Technologies," *Security Journal* 23 (2010): 276–298.

22. Aaron Opoku Amankwaa and Carole McCartney, "The Effectiveness of the UK National DNA Database," *Forensic Science International: Synergy* (2019).

23. Marcus Smith, "Universal Forensic DNA Databases: Balancing the Costs and Benefits," *Alternative Law Journal* 43 (2018): 131–135.

24. John K. Roman et al., "The DNA Field Experiment: A Randomized Trial of the Cost-Effectiveness of Using DNA to Solve Property Crimes," *Journal of Experimental Criminology* 5 (2009): 345–369.

25. Christopher Rigano, "Using Artificial Intelligence to Address Criminal Justice Needs," *National Institute of Justice Journal* 280 (2019): 1–10.

26. Natalie Ram, Christi J. Guerrini, and Amy L. McGuire, "Genealogy Databases and the Future of Criminal Investigation," *Science* 360 (2018): 1078–1079.

27. Susan Walsh et al., "IrisPlex: A Sensitive DNA Tool for Accurate Prediction of Blue and Brown Eye Colour in the Absence of Ancestry Information," *Forensic Science International: Genetics* 5 (2011): 170–180; Susan Walsh et al., "The HIrisPlex System for Simultaneous Prediction of Hair and Eye Colour From DNA," *Forensic Science International: Genetics* 7 (2013): 98–115.

28. Ashley Halsey III, "Ready for a Spy in the Sky? Panel Says FAA Should Expand Commercial Drone Flights," *The Washington Post*, April 6, 2016, https://www.washingtonpost.com/local/trafficandcommuting/ready-for-a-spy-in-the-sky-panel-says-faa-should-expand-commercial-drone-flights/2016/04/06/cd649ef0-fc0a-11e5-9140-e61d062438bb_story.html; Hope Reese, "Police Are Now Using Drones to Apprehend Suspects and

Administer Non-Lethal Force: A Police Chief Weighs In," Tech Republic, November 25, 2015, http://www.techrepublic.com/article/police-are-now-using-drones-to-apprehend-suspects-and-administer-non-lethal-force-a-police-chief/.

29. Ian Wren and Scott Simon, "Body Camera Maker Weighs Adding Facial Recognition Technology," NPR, May 12, 2018, https://www.npr.org/2018/05/12/610632088/what-artificial-intelligence-can-do-for-local-cops.

30. Christopher Rigano, "Using Artificial Intelligence to Address Criminal Justice Needs," *National Institute of Justice Journal* 280 (2019): 1–10.

31. Jon Sharman, "Metropolitan Police's Facial Recognition Technology 98% Inaccurate, Figures Show," *The Independent*, May 13, 2018, https://www.independent.co.uk/news/uk/home-news/met-police-facial-recognition-success-south-wales-trial-home-office-false-positive-a8345036.html. Nissan Moradoff, "Biometrics: Proliferation and Constraints to Emerging and New Technologies," *Security Journal* 23 (2010): 276–298.

32. Christopher Rigano, "Using Artificial Intelligence to Address Criminal Justice Needs," *National Institute of Justice Journal* 280 (2019): 1–10.

33. Don Walker, "ShotSpotter Gets Milwaukee Cops to Crime Scenes Quickly," *Milwaukee Journal Sentinel*, June 10, 2013, http://archive.jsonline.com/news/crime/shotspotter-gets-milwaukee-cops-to-crime-scenes-quickly-b9928666z1–210940151.html; Don Walker, "Police Officials Detail High-Tech Shot Sensors in High-Crime Areas," *Milwaukee Journal Sentinel*, January 24, 2013, http://archive.jsonline.com/news/milwaukee/police-officials-detail-hightech-shot-sensors-in-high-crime-areas-bn8gp6i-188247881.html.

34. Lars Ericson et al., *Through-the-Wall Sensors (TTWS) for Law Enforcement: Use Case Scenarios* (Washington, DC: National Institute of Justice, 2014); National Institute of Justice, *Through-the-Wall Sensors for Law Enforcement* (Washington, DC: National Institute of Justice, 2012).

35. Sensor, Surveillance, and Biometric Technologies Center of Excellence, *Through-the-Wall Sensors for Law Enforcement: Best Practices* (Washington, DC: National Institute of Justice, 2014).

36. Moradoff, "Biometrics," 276–298.

37. Ralf Kolbel and Susanne Selter, "Hostile Intent—The Terrorist's Achilles Heel? Observations on Pre-Crime Surveillance by Means of Thought Recognition," *European Journal of Crime, Criminal Law, and Criminal Justice* 18 (2010): 237–259.

38. Adam Janos, "Forget Polygraphs. Brain Scan Lie Detectors are Coming," A&E TV, July 6, 2018, https://www.aetv.com/real-crime/forget-polygraphs-brain-scan-lie-detectors-are-coming.

39. Brian Reaves, *Local Police Departments, 2013: Equipment and Technology* (Washington, D.C.: Bureau of Justice Statistics, 2015).

40. Clairissa Baker, "Game Changing' App Helps Police Helps Police Interact with People with Disabilities," Saint Cloud Times, February 24, 2019, https://www.sctimes.com/story/news/local/2019/02/24/app-helps-police-interact-people-disabilities/2904837002/.

41. Moradoff, "Biometrics," 276–298.

42. Cynthia Lum, Christopher S. Koper, and James Willis. "Understanding the Limits of Technology's Impact on Police Effectiveness," *Police Quarterly* 20 (2017): 135–163.

INDEX